DNA Methylation: Current Research

DNA Methylation:
Current Research

Edited by Hope Miller

AMERICAN
MEDICAL PUBLISHERS
www.americanmedicalpublishers.com

American Medical Publishers,
41 Flatbush Avenue,
1st Floor, New York,
NY 11217, USA

Visit us on the World Wide Web at:
www.americanmedicalpublishers.com

ISBN: 978-1-63927-247-1

Cataloging-in-Publication Data

DNA methylation : current research / edited by Hope Miller.
    p. cm.
Includes bibliographical references and index.
ISBN 978-1-63927-247-1
1. DNA--Methylation. 2. Medical genetics. 3. Genes. 4. Genetics. I. Miller, Hope.
QR46 .D53 2022
611.018 166--dc23

Table of Contents

Preface

DNA is a molecule made up of two chains that coil around each other in order to form a double helix. It carries genetic instruction for the functioning, growth, development and reproduction of all organisms and viruses. The process by which methyl groups are added to the DNA molecule is known as DNA methylation. The process of methylation can change the activity of a DNA segment without impacting the sequence. In human beings, DNA methylation is a vital process for normal development. Various processes like genomic imprinting, X-chromosome inactivation, repression of transposable elements, etc. are also associated with DNA methylation. DNA has four bases out of which two cytosines and adenine can be methylated. This book unravels the recent studies in the field of DNA methylation. From theories to research to practical applications, case studies related to all contemporary topics of relevance to this field have been included herein. Those in search of information to further their knowledge will be greatly assisted by this book.

The information shared in this book is based on empirical researches made by veterans in this field of study. The elaborative information provided in this book will help the readers further their scope of knowledge leading to advancements in this field.

Finally, I would like to thank my fellow researchers who gave constructive feedback and my family members who supported me at every step of my research.

Editor

DNA methylation at modifier genes of lung disease severity is altered in cystic fibrosis

Milena Magalhães[1], Isabelle Rivals[2], Mireille Claustres[1,3], Jessica Varilh[1,3], Mélodie Thomasset[1], Anne Bergougnoux[1,3], Laurent Mely[4], Sylvie Leroy[5], Harriet Corvol[6,7,8], Loïc Guillot[6,7], Marlène Murris[9], Emmanuelle Beyne[1,3], Davide Caimmi[10], Isabelle Vachier[10], Raphaël Chiron[10] and Albertina De Sario[1]*

Abstract

Background: Lung disease progression is variable among cystic fibrosis (CF) patients and depends on DNA mutations in the *CFTR* gene, polymorphic variations in disease modifier genes, and environmental exposure. The contribution of genetic factors has been extensively investigated, whereas the mechanism whereby environmental factors modulate the lung disease is unknown. In this project, we hypothesized that (i) reiterative stress alters the epigenome in CF-affected tissues and (ii) DNA methylation variations at disease modifier genes modulate the lung function in CF patients.

Results: We profiled DNA methylation at *CFTR*, the disease-causing gene, and at 13 lung modifier genes in nasal epithelial cells and whole blood samples from 48 CF patients and 24 healthy controls. CF patients homozygous for the p.Phe508del mutation and ≥18-year-old were stratified according to the lung disease severity. DNA methylation was measured by bisulfite and next-generation sequencing. The DNA methylation profile allowed us to correctly classify 75% of the subjects, thus providing a CF-specific molecular signature. Moreover, in CF patients, DNA methylation at specific genes was highly correlated in the same tissue sample. We suggest that gene methylation in CF cells may be co-regulated by disease-specific *trans*-factors. Three genes were differentially methylated in CF patients compared with controls and/or in groups of pulmonary severity: *HMOX1* and *GSTM3* in nasal epithelial samples; *HMOX1* and *EDNRA* in blood samples. The association between pulmonary severity and DNA methylation at *EDNRA* was confirmed in blood samples from an independent set of CF patients. Also, lower DNA methylation levels at *GSTM3* were associated with the *GSTM3*B* allele, a polymorphic 3-bp deletion that has a protective effect in cystic fibrosis.

Conclusions: DNA methylation levels are altered in nasal epithelial and blood cell samples from CF patients. Analysis of *CFTR* and 13 lung disease modifier genes shows DNA methylation changes of small magnitude: some of them are a consequence of the disease; other changes may result in small expression variations that collectively modulate the lung disease severity.

Keywords: DNA methylation, Co-methylation, Nasal epithelial cells, Cystic fibrosis, Modifier gene, Pulmonary function, Polymorphism, Next-generation sequencing, Pyrosequencing

Background

Environmental factors (i.e., nutrition, maternal diet, pollution, exercise, and lifestyle) influence the phenotype of living organisms by shaping their epigenome and consequently by affecting gene expression [1]. Change in the epigenome could contribute to human diseases and might explain the incomplete penetrance

of some mutations as well as the age of appearance of symptoms [2].

Cystic fibrosis (CF) is a monogenic disease that results from mutations in the *cystic fibrosis transmembrane conductance regulator* (*CFTR*) gene that encodes a cAMP-regulated epithelial chloride channel. This life-threatening disease is characterized by recurrent pulmonary infections, chronic inflammation, pancreatic insufficiency, and male infertility. Although multiple organs are affected, morbidity and mortality are mainly due to the lung disease because chronic infections and abnormal inflammation lead to

* Correspondence: albertina.de-sario@inserm.fr
[1]Laboratoire de Génétique de Maladies Rares, EA7402 Montpellier University, Montpellier, France
Full list of author information is available at the end of the article

progressive airway destruction. Lung disease progression is variable among CF patients and depends on the combination of three factors: (i) DNA mutations in the *CFTR* gene, (ii) polymorphic variations in other genes, and (iii) environmental exposure.

The contribution of genetic factors to CF phenotype has been extensively investigated by previous studies [3]. DNA mutations have been classified in six groups, depending on the mechanism by which they alter CFTR synthesis, traffic, and function [4]. The p.Phe508del mutation (deletion of the phenylalanine residue at position 508) leads to protein misfolding and degradation. This mutation is very frequent in the Caucasian population (it is homozygous in 40% of CF patients) and is generally, but not always, associated with a severe phenotype. Genetic and transcriptomic studies have provided a rich compilation of genes that can modify the CF outcome and are responsible for the disease variability [5–7]. Genotype-phenotype correlations in CF twins showed that environmental factors also contribute to pulmonary function variation in CF patients [3, 8], but the precise mechanism whereby these factors modulate the lung disease is unknown. The respiratory system is exposed to environmental stimuli (e.g., chemicals, dust, bacteria, or viruses). Of note, CF airway tissues are exposed not only to these external pollutants but also to the high cellular stress generated by the inflammatory and immune responses. Oxidative products generated by the inflammatory response can alter DNA methylation in both directions. Oxidation of 5-methylcytosines and 8-guanosines hinders MBP and DNMT1 binding, favoring loss of DNA methylation [9]. Oxidative compounds produced by the neutrophilic response generate halogenated cytosines that, because they mimic CpG methylation, are recognized by the methyl-binding proteins (MBP) and by the DNMT1 and, hence, favor methylation gain [10, 11]. In CF airway tissues, the oxidative stress is high and the neutrophil response particularly strong. Therefore, we hypothesized that (i) reiterative stress alters the epigenome in CF-affected tissues and (ii) DNA methylation changes at CF modifier genes contribute to the lung function variations observed in CF patients. To test our hypotheses, we profiled DNA methylation in healthy controls and homozygous p.Phe508del CF patients stratified according to their pulmonary function. We analyzed *CFTR*, the disease-causing gene, and 13 lung modifier genes. Ten genes were identified by genetic association studies. They encode proteins involved in inflammatory and immune responses (*TLR2*, *TLR5*, *TGFβ2*, and *IFRD1*), oxidative stress (*HMOX1*, *GSTM1*, and *GSTM3*), bronchoconstriction (*EDNRA*), and mucus structure and hydration (*MUC5AC* and *ENaCγ*). Three genes (*ATF1*, *DUOX2*, and *YY1*) were differentially expressed in nasal epithelial cells collected from CF patients characterized by extreme disease phenotypes [5].

A major hurdle when addressing the epigenome effects on disease severity is to gather appropriate tissue samples from the patients. Here, we used nasal epithelial cells (NEC), which are an informative model to study DNA methylation in airway diseases [12], and blood cells because most of the analyzed genes encode proteins that are involved in the inflammatory and immune responses.

Results
DNA methylation analysis in NEC and blood samples
The study was carried out in NEC and blood samples from the METHYLCF cohort that includes 48 CF patients and 24 healthy controls (Table 1). Using bisulfite and next-generation sequencing (BS-NGS), we analyzed DNA methylation at CpG islands associated with *CFTR* and 13 CF lung modifier genes (Table 2). The analyzed regions ranged from 133 to 264 bp, included from 5 to 26 CpG dinucleotides, and were less than 1550 bp away from the transcriptional start site (TSS), except for the *MUC5AC* CpG island that was in the gene body (from exon 35 to intron 35–36). In each region, we measured the methylation at single CpG dinucleotides and the mean DNA methylation. DNA methylation was profiled in 72 blood samples (all patients and controls) and in 63 NEC samples (39 CF patients and all healthy controls). Six patient NEC samples did not provide enough genomic DNA and three samples were reserved for further analyses. To evaluate the repeatability of the BS-NGS methylation analyses, we duplicated the measurements of the 14 genes in four CF patients in both tissues. The estimated standard deviation of DNA methylation reached 0.44 in the logit unit used for variance homogenization. This is quite high because the standard deviation of the methylation around its mean value was of the order of 0.43 in blood and of 0.52 in NEC samples, for both the CF patients and the healthy controls. Nevertheless, the correlation between DNA methylation data in the two independent experiments was excellent (Spearman's $r = 0.97$, $p = 0$) (Additional file 1: Figure S1).

In the control samples, DNA methylation was very high at *MUC5AC* (median value 95% in blood and 83% in NEC samples), high at *TLR5* (median value 38% in blood and 26% in NEC samples), and low in the other genes (<20% for both sample types) (Additional file 2: Figure S2). A partial least square discriminant analysis of the mean DNA methylation (the descriptors were the percentage of DNA methylation at the 14 genes in blood and NEC samples) provided 75% of correct classification of CF patients versus controls (Fig. 1). The percentage of correct classification was slightly lower when we used DNA methylation data from NEC samples alone (72%) and even lower with data from blood samples (69%) (Fig. 1).

Table 1 Demographic and relevant clinical features of CF patients and controls

| | Discovery set (METHYLCF) | | | | | Replication set (FrGMC) | | |
	C ($n = 24$)	CF ($n = 48$)	Mild ($n = 23$)	Intermediary ($n = 13$)	Severe ($n = 12$)	CF ($n = 30$)	Mild ($n = 18$)	Severe ($n = 12$)
Age, year[a]	37	26	34	25.5	22	24.5	27.0	23.5
Sex, M:F	13:11	32:16	17:6	11:2	4:8	19:11	14:4	5:7
BMI (kg/m^2)[a]		20.9	22.1	20.5	19.8	19.9	21.4	18.0
Weight (kg)[a]		60.0	62.0	60.0	52.0	56.0	60.0	45.5
Height (cm)[a]		170	170	171	168	168.0	170.0	161.5
FEV$_1$%[a]		48.0	64.8	48.0	41.5	91.0	102.0	24.0
FVC %[a]		74.0	87.0	67.0	66.5	98.5	105.0	46.0
% PI		100	100	100	100			
% Diabetes		36.7	47.8	23.0	33.3			
% Atopy		28.3	31.8	23.1	27.3			
% P. Aeruginosa		93.9	91.3	100	90.9			
% MRSA		34.1	22.7	30.8	60.0			
% Aspergillus		18.2	26.1	14.3	20.0			

PI pancreatic insufficiency
[a]Median

Table 2 CFTR and CF modifier genes

| Gene symbol | Gene name | Genomic coordinates[a] | nb. CpG[b] | Amplicon size (bp) | Differentially methylated CpG sites[c] | |
					Blood	NEC
ATF1	Activating transcription factor1	chr12:50,764,850-50,765,098	12	249		
CFTR	Cystic fibrosis transmembrane conductance regulator	chr7:117,479,627-117,479,759	10	133	1(−)	
DUOX2	Dual oxidase 2	chr15:45,114,541-45,114,722	11	182		
EDNRA	Endothelin receptor type A	chr4:147,480,957-47,481,216	21	260	2(−) 3(−) 4(−) 8(−) 9(−) 16(−)	5(+) 10(+)
ENaCγ	Epithelial sodium channel	chr16:23,182,420-23,182,665	23	246	2(−) 9(−) 11(+)	2(−) 6(+) 16(−)
GSTM1	Glutathione S-transferase mu 1	chr1:109,687,687-109,687,897	13	211		
GSTM3	Glutathione S-transferase mu 3	chr1:109,740,573-109,740,793	9	221	1(−) 3(−) 4(−) 5(−) 6(−) 7(−) 8(−)	4(−) 9(−)
HMOX1	Heme oxygenase 1	chr22:35,381,269-35,381,436	5	168	2(−) 3(−) 4(+) 5(−)	2(+)
IFRD1	Interferon-related developmental regulator 1	chr7:112,450,883-112,451,040	12	158		10(−)
MUC5AC	Mucine 5 AC	chr11:1,194,622-1,194,807	13	186	1(+) 10(+) 12(+) 13(−)	1(+) 3(+) 4(+) 5(−) 8(+) 10(+) 11(+) 12(+) 13(+)
TGFβ	TGFβ1 Transforming growth factor	chr19:41,353,542-41,353,740	19	199		10(−)
TLR2	Toll-like receptor 2	chr4:153,684,576-153,684,704	12	129	8(−)	
TLR5	Toll-like receptor 5	chr1:223,142,813-223,142,967	8	155	8(−)	
YY1	Yin-Yang 1 transcription factor	chr14:100,240,497-100,240,751	26	255	8(−) 22(−)	

[a]Human Genome GRCh38/hg38 build
[b]nb of CpG in the analyzed region
[c]Position of the CpG in the analyzed sequence. Plus signs mean hypermethylated and minus signs mean hypomethylated CpG in CF patients

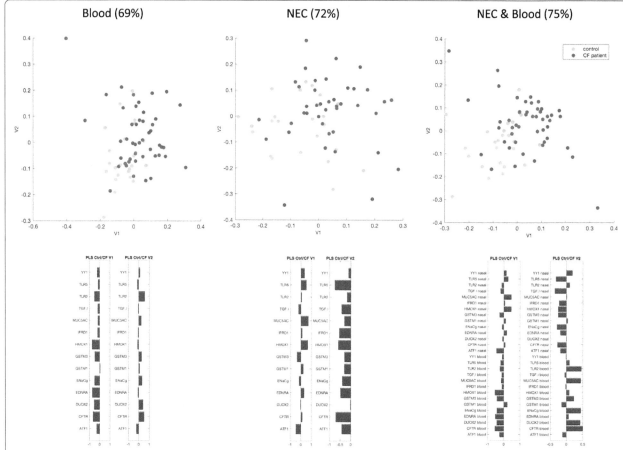

Fig. 1 (*Top*) Partial least square (PLS) discriminant analyses of CF patients and controls in blood, NEC, and both tissue samples and the percentage of correct classification of the subjects in each analysis. *v1,v2* are the scores on the first two PLS axes. (*Bottom*) The descriptors are mean DNA methylation of 14 genes

Besides the mean DNA methylation, we calculated the DNA methylation at individual CpG dinucleotides ($n = 194$ in the fourteen genes). Forty-two CpG sites (21%) in nine genes were differentially methylated between patients and controls in at least one tissue (Table 2). Specifically, 19 CpG sites were differentially methylated in NEC samples and 29 in blood samples. In NEC samples, most CpG sites were more methylated in CF patients than in controls (12 out of 19). Conversely, in blood samples, most of the differentially methylated CpG sites (24 out of 29) were less methylated in CF patients than in controls.

DNA methylation correlations in CF cells
Next, we looked for inter-tissue (DNA methylation of a gene in both cell types) and intra-tissue (DNA methylation of two genes in the same tissue) correlations. Data from CF patients and controls were analyzed separately using stringent criteria (Bonferroni-controlled family-

wise error rate (FWER) = 10%). Correlations were calculated using the mean DNA methylation of each gene region. Interestingly, DNA methylation at *GSTM3* was highly correlated in NEC and blood samples collected from the same individuals, both in controls and CF patients (Fig. 2). This finding suggests that methylation level at *GSTM3* is under genetic control.

Moreover, a few intra-tissue correlations were found in genomic DNA from CF patients (Fig. 2). Specifically, in NEC samples, we found two co-methylation modules: (i) the DNA methylation level of *TLR5* correlated with that of *MUC5AC*, *CFTR*, and *HMOX1* and (ii) the DNA methylation level of *HMOX1* correlated with that of *EDNRA* and *CFTR*. In blood samples, the DNA methylation level of *HMOX1* correlated with that of *CFTR* and with *TLR2*.

In control samples, no intra-tissue correlations were significant with a FWER of 10%. All genes were expressed in the tissue where their co-methylation was found, except

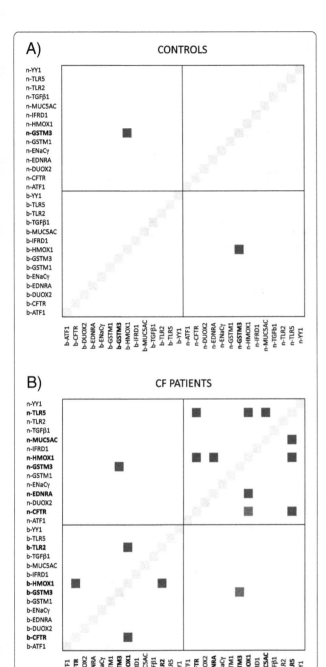

Fig. 2 The matrices show inter-tissue (mean DNA methylation of a gene in both cell types) and intra-tissue (mean DNA methylation of two genes in one tissue) correlations in controls (**a**) and CF patients (**b**). The prefix n- or b- in front of the gene name indicates that DNA methylation was measured in NEC or blood samples, respectively. Significant correlations (*black square*) were calculated using Spearman's test with a Bonferroni-controlled family-wise error rate (FWER) = 10%

for *CFTR* that was not expressed in blood samples. Thus, gene expression does not seem to be an essential pre-requisite for co-methylation. We assessed gene expression by RT-PCR using NEC and blood mRNAs from two healthy individuals (data not shown).

HMOX1 was differentially methylated in NEC and blood samples from CF patients

Next, we focused on genes that were differentially methylated in CF patients or groups of patients. *HMOX1* was previously identified as a CF modifier gene by genetic association studies [13]. We measured DNA methylation at the CpG island that overlaps exon 2 (Fig. 3). Using the mean DNA methylation in the region, we found that *HMOX1* was differentially methylated in NEC samples (Student $p = 0.018$) and blood cell samples (Wilcoxon $p = 0.009$) of CF patients compared with controls, but the direction of the methylation change was not the same in the two tissue models (Fig. 4a, b). Moreover, DNA methylation was associated with lung disease severity (ANOVA $p = 0.052$ in NEC samples; Kruskal-Wallis $p = 0.035$ in blood samples) (Fig. 4c, d). One CpG dinucleotide (CpG#2) was more methylated than the other four CpG in the region (approximately 30% compared with <10%) (Fig. 4e, f). CpG#2 was differentially methylated in CF patients compared with controls in both tissues (Bonferroni corrected $q = 0$ in blood and $q = 2.7 \ 10^{-3}$ in NEC) and was associated with pulmonary severity in blood samples (Kruskal-Wallis $p = 0.0019$).

DNA methylation at *HMOX1* was not associated with nearby polymorphisms

Previous studies showed that two polymorphic sequences in the 5′ untranslated region of *HMOX1* were associated with lung function in airway diseases. Specifically, the minor allele of the A(-413)T variant (rs2071746) was associated with CF lung disease severity in two independent cohorts [13] (Fig. 3). Next to this single-nucleotide polymorphism (SNP), the length of a $(GT)_n$ microsatellite correlated with pulmonary severity in airway (emphysema and COPD) and cardiovascular diseases [14, 15] (Fig. 3). Long microsatellites (>32 repeats) were associated with lower levels of transcription in vitro and with an adverse clinical phenotype in patients [16]. Because these polymorphisms were close (600 bp upstream) to the region analyzed in this study, we asked whether they affected DNA methylation. We assessed the A(-413)T SNP and the microsatellite length in CF patients and healthy volunteers of the METHYLCF cohort and found that DNA methylation levels (mean methylation of the amplicon and methylation at CpG#2) in NEC and blood samples did not correlate with any genotype (Spearman's correlation test) (Additional file 3: Table S2).

DNA methylation at *HMOX1* was not associated with a significant change of gene expression

Next, we asked whether DNA methylation at *HMOX1* affected gene expression. In blood cells, DNA methylation

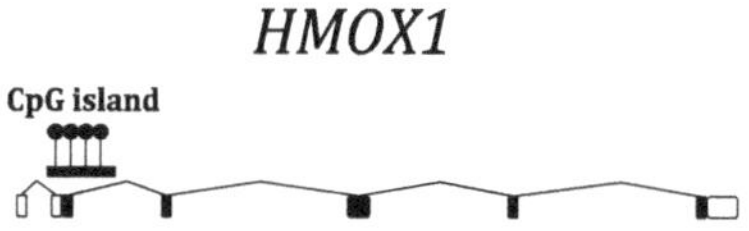

HMOX1

CpG island

```
TATGACAGGTGTCCCAGTCCAGGCGGATACCAGGTGCTGCCAGAGTGTGGAGGAGGC
AGGCGGGGACTTAGTCTCCTCCCTGGGTTTGGACACTGGCATCCTGCTTTATGTGTG
ACACCACTGCACCCCTCTGAGCCTCGGTTTCCCCATCTGTAAAATAGAAGCGATCTA
CCCTCACAGGTCAGTTGTAGGGATGAACCATGAAAATACTAGAGTCTCTGTTTTTTG
ACAGGAACTCAAAAAACAGATCCTAAATGTACATTTAAAGAGGGTGTGAGGAGGCAA
GCAGTCAGCAGAGGATTCCAGCAGGTGACATTTTAGGGAGCTGGAGACAGCAGAGCC
TGGGGTTGCTAAGTTCCTGATGTTGCCCACCAGGCTATTGCTCTGAGCAGCGCTGCC
TCCCAGCTTTCTGGAACCTTCTGGGACGCCTGGGGTGCATCAAGTCCCAAGGGGACA
GGGAGCAGAAGGGGGGGCTCTGGAAGGAGCAAAATCACACCCAGAGCCTGCAGCTTC
TCAGATTTCCTTAAAGGTTTTGTGTGTGTGTGTGTGTGTGTGTGTGTGTGTATGTGT
GTGTGTGTGTGTGTGTGTGTGTGTTTTCTCTAAAAGTCCTATGGCCAGACTTTGTTT
CCCAAGGGTCATATGACTGCTCCTCTCACCCCACACTGGCCCGGGGCGGGCTGGGC
GCGGGCCCCTGCGGGTGTTGCAACGCCCGGCCAGAAAGTGGGCATCAGCTGTTCCGC
CTGGCCCACGTGACCCGCCGAGCATAAATGTGACCGGCCGCGGCTCCGGCAGTCAAC
GCCTGCCTCCTCTCGAGCGTCCTCAGCGCAGCCGCCGCCCGCGGAGCCAGCACGAAC
GAGCCCAGCACCGGCCGGATGGAGCGTCCGCAACCCGACAGGCAAGCGCGGGGCGCG
GGACGCGGGACGGGCGCCTTTCTCTCCCAACCCTGCTTGCGTCCTAGCCCCACCCCG
GGACACTGCCACACAGCGACAGAGCCCAGGAGCCAGAAACTTGGGCTCTGGAGTCAG
GAGGTGCGGGGTTCTGATCCTGCCTGTGCCCGTAGGGTAGTTGGAGGGAGGAACGGT
AATTTACATGCCTGGCACCCTGGTATGCGGTTGGTGACCAAGATGGGAGTGTCCCTA
GAGTATCCAGTCTTTGAGGTAGCCAATTTTTTTTTTTAATCCTACTTTCGAGGTGTGT
```

Fig. 3 *Top, HMOX1* exon-intron structure and position of the CpG island. *Bottom, HMOX1* partial genomic sequence showing exons 1 and 2 (*gray background*) and introns (*white background*). Also shown, the five CpG (in *white* on *black* background) where we measured DNA methylation, the major "A" allele of SNP rs2071746 (*white on black background*) and the polymorphic $(GT)_n$ microsatellite (*underlined*) in exon 2

and gene expression were analyzed in samples collected from the same individuals. *HMOX1* was not differentially expressed in CF patients compared with controls (Wilcoxon $p = 0.11$) or in patients stratified according to the lung disease severity (Kruskal-Wallis $p = 0.39$) (Fig. 5a). Also, expression and DNA methylation levels (mean methylation of the amplicon and methylation at CpG#2) were not correlated, be it in the whole cohort (Spearman's $r = 0.09$, $p = 0.48$) or separately in the control (Spearman's $r = -0.14$, $p = 0.62$) and in the CF (Spearman's $r = 0.08$, $p = 0.58$) populations (Fig. 5b).

RNA could not be extracted from the NEC samples of the METHYLCF cohort because the whole amount of cells had to be used to isolate genomic DNA. Therefore, to determine the expression levels in NEC samples, we inspected data from three publicly available transcriptomic studies [5, 17, 18]. *HMOX1* was not differentially expressed in CF compared with control NEC. The only study that compared mild versus severe CF patients was not informative for this gene [5].

EDNRA was differentially methylated but not differentially expressed in CF blood samples

EDNRA encodes a G protein-coupled receptor that, following binding to endothelin, triggers cellular proliferation and contraction of smooth muscle cells. In CF airways, higher level of endothelin may contribute to the pulmonary phenotype [19]. In the METHYLCF cohort,

EDNRA was less methylated in CF than in control blood samples (Wilcoxon $p = 0.017$) and DNA methylation level correlated with the lung disease severity (Kruskal-Wallis $p = 0.028$) (Fig. 6a, b). The DNA methylation at individual CpG sites was homogeneous, close to the mean methylation in the region (Fig. 6c).

Gene expression was not detectable in blood cells, even in CF samples where *EDNRA* was less methylated. Thus, we concluded that loss of DNA methylation at *EDNRA* was a consequence rather than a cause of lung disease severity.

DNA methylation levels at *GSTM3* were associated with lung disease severity and correlated with the *GSTM3*B* allele

The mean DNA methylation at *GSTM3* was not significantly different in CF and control samples; however, it was associated with CF lung disease severity in NEC samples (ANOVA $p = 0.016$) (Fig. 7a). DNA methylation at individual CpG sites was homogeneous (Fig. 7b).

Previous studies showed that various polymorphisms of the *GST(M)* genes contribute to lung disease severity in CF patients [20] and that GST activity may modulate *P. aeruginosa* lung infection [21]. Of note, the *GSTM3*B* allele, a 3-bp deletion that has a protective effect in CF patients, is 6.1 kb downstream of the region analyzed in this study. To determine whether this polymorphic sequence affected DNA methylation levels, we

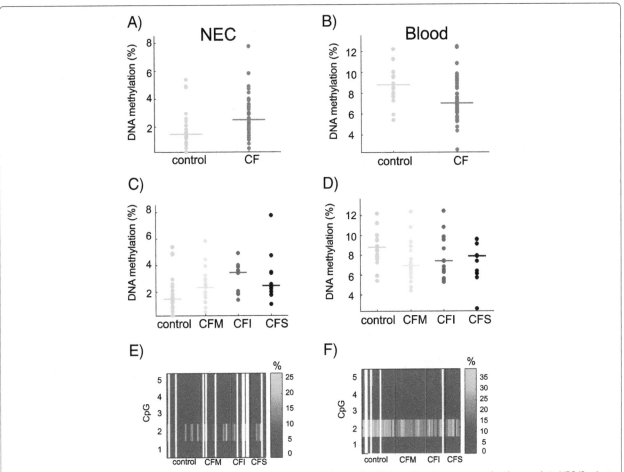

Fig. 4 DNA methylation at *HMOX1*. The *dot plots* represent the mean DNA methylation of *HMOX1* in CF patients compared with controls in NEC (Student $p = 0.018$) (**a**) and in blood (Wilcoxon $p = 0.009$) samples (**b**). DNA methylation levels depended on pulmonary severity (NEC, ANOVA $p = 0.052$ (**c**); blood, Kruskal-Wallis $p = 0.035$ (**d**)). The *horizontal line* indicates the median in each group. The heat maps represent the DNA methylation at five CpG dinucleotides in NEC (**e**) and blood (**f**) samples. *White lines* represent missing data. *CFM* mild CF patient, *CFI* intermediary CF patient, *CFS* severe CF patient

genotyped patients and controls for the micro-deletion (Additional file 3: Table S2). Interestingly, in both NEC and blood samples, DNA methylation levels at *GSTM3* correlated with the presence of the *GSTM3*B* allele (Spearman's NEC $r = -0.43$ $p = 5\ 10^{-4}$; blood $r = -0.42$ $p = 2.8\ 10^{-4}$). DNA methylation levels in homozygous *GSTM3*B* carriers were lower than in heterozygous carriers, where they were lower than in homozygous *GSTM3*A* carriers (Fig. 7c, d).

Replication of DNA methylation analysis in an independent set of CF patients

To replicate data obtained in the METHYLCF cohort, we selected 30 additional p.Phe508del homozygous patients with severe ($n = 12$) or mild lung disease ($n = 18$) from an independent CF cohort enrolled by the French CF Gene Modifier Consortium (FrGMC) [22]. Of note,

the phenotype of this set of patients was more extreme than that of the METHYLCF cohort (Table 1). Genomic DNA being available for blood and not for NEC cells, we decided to replicate blood differentially methylated regions (*EDNRA* and *HMOX1*), leaving replication of NEC regions for future studies. DNA methylation was measured by locus-specific pyrosequencing. To analyze *EDNRA*, we used a pyrosequencing assay located 350 bp downstream of the region that was targeted by BS-NGS. In the replication set of patients, DNA methylation at *EDNRA* was significantly associated with lung disease severity (Kruskal-Wallis $p = 0.047$) (Additional file 4: Figure S3). DNA methylation in mild CF patients was higher than in controls (Wilcoxon $p = 0.023$) and slightly higher than in severe patients (not significant). Overall, *EDNRA* DNA methylation levels by pyrosequencing were higher than those obtained by BS-NGS: this is consistent with previous results by

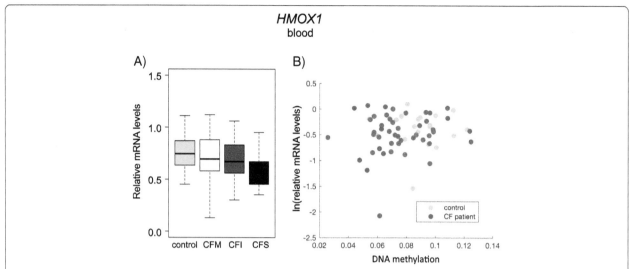

Fig. 5 *HMOX1* gene expression in blood samples. **a** The *box plots* represent the relative expression of *HMOX1* in CF patients with different lung disease and controls (Kruskal-Wallis $p = 0.39$). *CFM* mild CF patient, *CFI* intermediary CF patient, *CFS* severe CF patient. **b** Correlation between gene expression and mean DNA methylation levels in blood samples from CF patients and controls. Whole cohort (Spearman's $r = 0.09$, $p = 0.48$); controls (Spearman's $r = -0.14$, $p = 0.62$); CF patients (Spearman's $r = 0.08$, $p = 0.58$)

Potapova et al. [23] who compared the two methods and showed a trend towards higher values in the range between 0 and 20% DNA methylation.

For *HMOX1*, all tested primers failed to provide a linear pyrosequencing signal in the region of interest.

Discussion

In this study, we provide the first DNA methylation profile using tissue samples collected from CF patients. We measured DNA methylation at CpG islands associated with *CFTR* and 13 CF modifier genes. DNA methylation

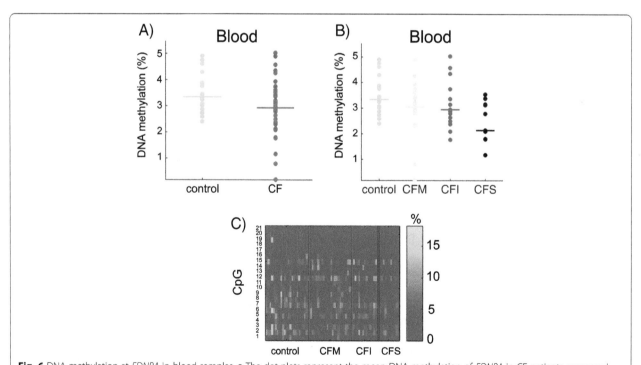

Fig. 6 DNA methylation at *EDNRA* in blood samples. **a** The dot plots represent the mean DNA methylation of *EDNRA* in CF patients compared with controls (Wilcoxon $p = 0.017$). **b** DNA methylation at *EDNRA* correlated with pulmonary severity (Kruskal-Wallis $p = 0.028$). The *horizontal line* indicates the median in each group. **c** The heat map represents DNA methylation at 21 CpG dinucleotides. *CFM* mild CF patient, *CFI* intermediary CF patient, *CFS* severe CF patient

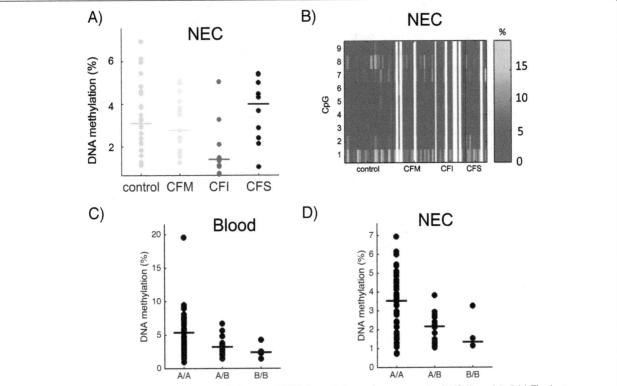

Fig. 7 DNA methylation at *GSTM3*. The mean DNA methylation at *GSTM3* depended on pulmonary severity (ANOVA *p* = 0.016) (**a**). The *heat maps* represent DNA methylation at nine CpG dinucleotides in NEC samples; *white lines* represent missing data (**b**). Low methylation levels correlated with the *GSTM3*B* allele both in blood (Spearman's *r* = −0.42, *p* = 3 10^{-4}) (**c**) and NEC samples (Spearman's *r* = −0.43, *p* = 5 10^{-4}) (**d**). *CFM* mild CF patient, *CFI* intermediary CF patient, *CFS* severe CF patient. The *horizontal line* in the *dot plots* indicates the median in each group. *A/A* homozygous *GSTM3*A*, *A/B* heterozygous, *B/B* homozygous *GSTM3*B*

levels were altered not only in NEC, which are directly affected by the disease (CF patients often have rhinitis and nasal polyposis), but also in blood cells where *CFTR* is not expressed. By combining the DNA methylation data obtained in NEC and blood cells, we correctly classified 75% of the subjects, distinguishing homozygous p.Phe508del CF patients from controls. This finding suggests that DNA methylation variations in specific genes may provide a CF-specific molecular signature.

Our study has also disclosed a number of genes whose methylation seemed to be co-regulated in CF samples. Concomitant DNA methylation changes in two or more genes have been already described in solid tumors, including in lung adenocarcinomas [24] and in sputum samples of asthmatic smokers [25]. More recently, van Eijk *et al.* identified networks of co-methylation and co-expression modules in blood samples collected from healthy individuals [26]. In this genome-wide analysis, co-methylation and co-expression modules contained few overlapping genes, but several pairs of methylation and expression modules were significantly correlated [26]. Moreover, because they were enriched in gene ontology categories, these modules were considered biologically relevant. The actual mechanism responsible for

their generation is unknown, however, the existence of factors that affect DNA methylation and gene expression acting in *trans* at the module level was hypothesized [26]. In our study, using stringent conditions, we observed gene co-methylation exclusively in patient samples. Therefore, we suggest the involvement of *trans*-acting factors that are specifically activated by the disease, namely by the oxidative stress and the inflammatory and immune responses. A genome-wide DNA methylation analysis of CF samples is required to better understand this phenomenon.

By comparing patients and controls, we found significant DNA methylation variations at two CF modifier genes: *HMOX1* (in NEC and blood cells) and *EDNRA* (in blood cells). Moreover, the DNA methylation level at three genes (*GSTM3* in NEC and *HMOX1* and *EDNRA* in blood samples) was associated with lung disease severity. The association between pulmonary severity and DNA methylation at *EDNRA* was replicated using blood samples from an independent set of CF patients. The magnitude of the methylation changes in lung severity modifier genes was small. Three lines of evidence show that small epigenomic changes can be biological meaningful. First, many epidemiological studies showed that the environment induces small epigenetic changes associated with a clinical outcome. In patients affected by chronic obstructive pulmonary disease and exposed to fine particulate matter (PM$_{2.5}$) constituents, hypomethylation of the NOS2A gene (about -1.5%) was associated with a higher (about $+18\%$) fractional concentration of exhaled nitric oxide (FeNO), a biomarker of airway inflammation [27]. In patients with type 2 diabetes mellitus (T2DM), a CpG dinucleotide in the first intron of the FTO gene was hypomethylated (-3.35%) and the odds of belonging to the T2DM group increased by 6.1% for every 1% decrease in DNA methylation [28]. Second, experimental studies in animals showed the impact of small methylation changes on gene expression. In the offspring of rat fed with a protein-restricted diet during pregnancy, a small decrease of DNA methylation in the promoter of PPARα was associated with an increase of gene expression [29]. Third, a genome-wide expression analysis in patients affected by type 2 diabetes mellitus showed that small expression changes in multiple genes belonging to the same pathway had a bigger impact than a high-fold change in a single gene [30, 31]. Collectively, these findings lead us to suggest that small DNA methylation variations in lung modifier genes can impact cystic fibrosis severity.

HMOX1 encodes a protein that is important for iron homeostasis and cell protection from oxidative damage during stress. Activating and repressive factors regulate the HMOX1 basal expression by interacting with the promoter and various stimuli (i.e., heme, cadmium, and oxidative stress) switch on its induced expression via binding to responsive elements [32]. Of note, the CpG island targeted in our DNA methylation analysis contains an HMOX1 hydrogen peroxide-responsive element [14]. CF tissues are exposed to continuous stress by the immune and inflammatory responses. Here, we found that HMOX1 was differentially methylated both in blood and NEC samples from CF patients compared with controls, but the direction of the methylation change was not the same in the two tissue models. One possible explanation is that DNA methylation levels result from a balance between the burden of halogenic compounds produced by the inflammatory response (especially by neutrophils) that favors methylation gain [10, 11] and other oxidative products responsible for methylation loss [9]. The contribution of these opposing factors is likely to be different in blood and NEC because NEC are directly affected by cystic fibrosis. In addition, in NEC samples, the increase of DNA methylation at the promoter of HMOX1 was non-monotonic in CF patients stratified according to the lung disease severity. The intensity of the inflammatory response and of the oxidative stress in the airway tissues varies among patients and correlates with the lung disease [33, 34]. The proportion of oxidative products changing DNA methylation in opposite

directions may be variable in stratified CF patients so that the final ratio results in a U-shaped curve. The possible effect of DNA methylation on HMOX1 transcription deserves further analysis. In NEC samples, the small amount of cells did not allow us to carry DNA methylation and gene expression analysis on the same samples. In blood samples, we failed to demonstrate a significant impact of DNA methylation on expression, possibly due to the lack of statistical power of the present cohort.

EDNRA encodes a G protein-coupled receptor that, following ligation to endothelin, causes contraction of smooth cells. Previous genetic studies showed an association between EDNRA DNA polymorphisms and pulmonary disease in four independent cohorts of CF patients [19]. Also, a functional study showed that an allele that is deleterious for the lung function resulted in higher EDNRA mRNA levels in human tracheal smooth muscle cells [19]. Our study shows that EDNRA was hypomethylated in CF patients and DNA methylation levels were associated with pulmonary disease severity in blood cells. Because EDNRA transcripts were not detected in control nor in CF samples, we conclude that loss of DNA methylation had no impact on gene expression and was probably a consequence rather than a cause of lung disease severity.

Compelling evidence shows that DNA methylation is affected not only by environmental but also by genetic factors. Of note, methylation levels at 2–7% of CpG sites are associated with cis-DNA variants and may provide the molecular mechanisms for the associated quantitative trait locus [26]. In the present study, we realized that two differentially methylated regions mapped close to polymorphic sequences that have been previously shown to be associated with the pulmonary function in airway diseases: two DNA variants were in the 5′ untranslated region of HMOX1 and the third one was in the body of GSTM3. Since we found no correlation between DNA methylation levels and two polymorphic sequences in HMOX1, we suggest that DNA methylation and the two polymorphisms are independently associated with lung function. This result should be validated in an independent cohort. Conversely, two findings in our study suggest that DNA methylation in the GSTM3 gene is under genetic control. First, DNA methylation at GSTM3 was highly correlated with the presence of the GSTM3*B allele both in NEC and blood samples. Second, we found a high positive correlation between GSTM3 DNA methylation levels in the blood and NEC samples from the same individuals. These results are consistent with a previous study showing that diplotypes in the GSTM3 gene predicted DNA methylation levels at five CpG dinucleotides scattered in the gene, outside the region we analyzed [35]. The GSTM3*B allele, a 3-bp deletion in intron 6, is associated with higher level of GSTM3 mRNA and protein expression [36]. To explain this association, it was proposed that the 3-bp deletion generates a binding site for the transcription factor YY1 [37]. We hypothesize that upon activation by YY1 or another transcription factor, the *GSTM3*B* intronic sequence binds to the gene promoter via a chromatin loop and causes a reduction in the DNA methylation level in the same region. The GSTM3 protein conjugates various toxic compounds to glutathione, thus, similarly to HMOX1, has a protective effect in cells, and is particularly beneficial to CF damaged tissues.

The present study has limitations. We analyzed 48 CF patients and 24 healthy controls. Confirmatory studies should be carried out on a larger number of patients. DNA methylation was analyzed in 14 lung modifier genes and restricted to the promoter regions. Future studies should cover the whole genome including other genic and intergenic regulatory regions (enhancers, insulators, etc). We could not analyze gene expression in NEC samples because the whole amount of cells had to be used for DNA extraction.

Conclusions

In summary, we showed that DNA methylation was altered in nasal epithelial and blood samples from CF patients and, using stringent conditions, we observed modules of gene co-methylation exclusively in patient samples. Through the analysis of 13 lung disease-modifiers genes, we found DNA methylation changes of small magnitude in two genes (*HMOX1* and *EDNRA*). DNA methylation was associated with pulmonary severity in three genes (*HMOX1*, *GSTM3*, and *EDNRA*) and with a polymorphic deletion that has a protective effect in cystic fibrosis at one gene (*GSTM3*). Some of these small DNA methylation changes are a consequence of the disease. Other changes may result in small expression variations that collectively and over time modulate the lung disease severity. Genome-wide epigenomic, transcriptomic and genomic analyses are needed to further understand how genetic and epigenetic factors contribute to the large spectrum of lung disease severity in cystic fibrosis.

Methods
Study cohorts
The study was approved by the local Institutional Review Board (CPP Sud Méditerranée III, Nîmes #2013.02.01bis). Informed written consent was obtained from all participants. Table 1 lists the demographic and relevant clinical features of two cohorts. CF patients were homozygous for the p.Phe508del mutation and ≥18-year-old. Exclusion criteria for CF patients included lung transplantation and pulmonary exacerbation during sample collection.

The *METHYLCF* cohort includes 48 CF patients and 24 healthy controls with no history of airway diseases or allergy. It was enrolled in four CF centers in the South

of France. CF patients were stratified into three groups based on the severity of the lung disease and mainly using the FEV_1% predicted: mild (48% of patients), intermediary (27%), and severe (25%). Patients with FEV_1% predicted values that corresponded to the top and bottom quartiles were classified as mild and severe, respectively [38]. CF patients of age ≥34 years were considered mild because of their long survival. The age distribution did not differ between patients and controls (Wilcoxon $p = 0.30$). The male-to-female ratio was slightly, but not significantly, higher in CF patients than in controls (χ^2 $p = 0.22$).

From the already available *FrGMC* cohort (French Ethical Board, CPP #2004/15) [22], a replication set of CF patients (12 patients with severe and 18 patients with mild pulmonary disease) was selected. They were stratified using the same criteria as for the METHYLCF cohort.

Biological samples
Biological samples were collected from the METHYLCF cohort, whereas blood genomic DNA was already available for the replication FrGMC cohort.

Nasal epithelial cells were collected from the inferior turbinate using nasal curettes (Rhino-probe, Arlington) after nebulization with 5% xylocaine (Astrazeneca, France). NEC were collected from both nostrils, pooled together in 1 ml RNA protect Cell Reagent (#76526 Qiagen), and then shipped to the handling center at room temperature.

Whole blood samples were collected in EDTA (5 ml) and in PAXgene (2.5 ml) tubes (#762165, Becton Dickinsen) for DNA and RNA extraction, respectively.

DNA extraction
NEC collected in RNAprotect Cell Reagent (#76526 QIAGEN) were treated with 1 mg/mL RNAse. Genomic DNA was extracted using the QIAamp DNA Micro Kit (#56304, QIAGEN) as previously described [39]. The mean DNA yield was 5.1 ± 2.8 µg in controls and 3.9 ± 3.1 µg in CF patients (range 0 to 12.4 µg). DNA yield was not significantly different between groups (Wilcoxon $p = 0.19$).

Genomic DNA was extracted from whole blood samples using the Flexigene DNA kit (#51206, QIAGEN) according to the manufacturer's recommendations.

RNA extraction
RNA was extracted from whole blood samples using the PAXgene Blood RNA kit (#762124, PreAnalytix), according to the manufacturer's recommendations.

Bisulfite conversion
NEC and blood DNA samples were treated with sodium bisulfite as previously described [40].

DNA methylation analysis by amplicon sequencing

Fusion primers were designed to amplify 133 to 264 bp-long amplicons in the region of interest (Additional file 5: Table S1). Each forward primer contained a MID (Multiplex Identifiers, Roche) to allow computational screening of each sample. PCR products were obtained using the PyroMark PCR kit (#978703, QIAGEN), and 10 μM forward and reverse primers in a 25-μl final volume. PCR conditions were 95 °C for 15 min, followed by 94 °C for 30 s, the annealing temperature for 30s, 72 °C for 30 s for 45 cycles, and then 72 °C for 10 min. Amplicons were purified with the QIAquick PCR Purification Kit (#28106 QIAGEN) and quantified using a NanoDrop 2000 Spectrophotometer (Thermo Scientific) and a Qubit 2.0 fluorometer (Life Technologies). In each sequencing run, 112 purified amplicons were pooled in equimolar amounts. Emulsion PCR and subsequent bidirectional sequencing were done according to the GS Junior emPCR Amplification Method Manual-Lib-A (#05996520001, Roche) and GS Junior Sequencing Method Manual (#05996554001, Roche), respectively.

Sequence analysis

We measured DNA methylation using bisulfite and next-generation sequencing (BS-NGS). To filter and order the raw sequencing data, we developed a pipeline. The script works in a Galaxy environment and includes four steps: (i) a barcode splitter to separate sequences per sample; (ii) a sequence trimming to remove all the MID (multiplex identifiers, Roche) and adaptor sequences; (iii) a barcode splitter to separate sequences per gene; and (iv) analysis of fasta/bam files with BiQAnalyzer HT [41]. BiQAnalyzer HT removes non-fully converted sequences and determines the methylation status of each CpG site within amplicons. It provides a text file where each CpG site is either 1 (methylated) or 0 (unmethylated). A minimal conversion rate of 0.97 was used. Before filtering, the number of reads per analyzed amplicon ranged from 9 to 2704. We retained only the BS-NGS measurements for which the number of sequences was large enough as to have either a coefficient of variation of the mean methylation percentage smaller than 5% or a standard deviation not higher than 1% (the first condition is too stringent for very small methylation percentages). After filtering, 95% of the reads were in the interval [98; 1460].

DNA methylation analysis by pyrosequencing

PCR products were amplified using the PyroMark PCR Kit ((#978703, QIAGEN) in 25 μL reaction volume. For *EDNRA*, the pool of forward and reverse primers (one of which was biotin-labeled at the 5′) as well as the sequencing primer were from the Hs_EDNRA_02_PM PyroMark CpG Assay (#978746, QIAGEN, Hilden, Germany). The PCR program was 94 °C for 15 min, followed by 94 °C 30 s, 56 °C for 30 s, 72 °C for 30 s during 45 cycles, and 72 °C for 10 min. PCR products were purified using 1 μL Streptavidin Sepharose HP™ (#17-5113-01, GE Healthcare) and a PyroMark Q24 Workstation. Pyrosequencing reactions were performed in a PyroMark Q24 (QIAGEN) using the PyroMark Gold Q24 reagents (#970802, QIAGEN) according to the manufacturer's instructions. Before the assays, we tested the signal linearity using mixtures of methylated and unmethylated genomic DNA (0, 20, 40, 60, 80, and 100%); standard errors were from three replicates.

Genotyping

HMOX1 (GT)$_n$ microsatellite

Using blood genomic DNA, we amplified a 113–135-bp DNA fragment spanning the $(GT)_n$ microsatellite with a FAM-labeled sense primer (5′-AGAGCCTGCAGCTTCT-CAGA-3′) and an unlabeled reverse primer (5′-ACAAA GTCTGGCCATAGGAC-3′). The PCR program was 94 °C 30 s, 57 °C 90 s, 72 °C 90 s for 30 cycles. PCR products were analyzed using an ABI 3130xl Genetic Analyzer (Applied Biosystem), and the microsatellite size was measured with the Gene Mapper software (Applied Biosystem).

HMOX1 SNP rs2071746

A 139-bp PCR fragment surrounding the A(-413)T SNP (rs2071746) was amplified with the following program: 95 °C 30 s, 64 °C 30 s, 72 °C 30 s for 35 cycles. Primers were forward 5′-GCAGAGGATTCCAGCAGGTG-3′ and reverse 5′-CAGGCGTCCCAGAAGGTTCC-3′. After purification with the QIAquick kit (QIAGEN) and labeling with the Big Dye Terminator (Life Technologies), DNA was sequenced using an ABI 3130xl Genetic Analyzer (Applied Biosystem).

GSTM3 *A and GSTM3*B alleles

A 202-bp PCR fragment was amplified using primers 5′-GCTACCTGGACAACTGAAAC-3′ and 5′-CGGTTC TGATCCAAGATATC-3′ and the following program: 95 °C 5 min, then (95 °C 30 s, 56 °C 30 s, 72 °C 1 min) for 25 cycles and 72 °C 15 min. PCR products were analyzed using an ABI 3130xl Genetic Analyzer (Applied Biosystem) and their size measured with the Gene Mapper software (Applied Biosystem).

Gene expression

For reverse transcription, 500 ng of total blood RNA from each sample was added to Rnase-free water (final volume 8 μl) followed by DNase I treatment for 15 min at room temperature. Samples were then added to a mix containing 4 μl of first strand 5× buffer, 2 μl of 10× dithiothreitol, 1 μl of 10 mM dNTP mix, 300 ng/μl of hexaprimer (random primers), 20–40 U/μl of RNasin® enzyme (Promega), and 200 U/μl of MMLV-RT enzyme

(Life Technology). The reverse transcription reaction program consisted of three steps: 10 min at 25 °C, 50 min at 37 °C, and 15 min at 70 °C. mRNA expression was measured using a LightCycler 480 real-time PCR system and SYBR Green I Master mix® (Roche Diagnostics) (primers are listed in Additional file 5: Table S1). Standard curves were generated for each run by serial dilution of control cDNA. Gene expression levels were expressed as ratios relative to that of reference genes (GAPDH for HMOX1 and TBP for EDNRA). Real-time PCR reactions were done in duplicate in two independent reverse transcriptions.

Statistical analysis

For a given gene, the mean methylation of each individual site as well as the mean methylation percentage over all sites were left for statistical analysis. To homogenize the variance of the mean methylation percentage (which is maximal at 50% and zero at 0 or 100%), we worked with its logit transformation.

To evaluate the repeatability of the BS-NGS methylation analyses, we duplicated the measurements corresponding to the $n_g = 14$ genes of interest for 4 CF patients in the $n_t = 2$ tissues (blood and NEC) with 106 degrees of freedom (instead of $4 \times n_g \times n_t = 112$ due to few missing values).

To compare the mean methylation level of a given gene in a given tissue between controls and CF patients, and across the whole cohort stratified according to the severity of the lung disease (i.e., controls, mild, intermediary, and severe CF patients), depending on the statistical features of the data (normality or not, homoscedasticity or not), we used either parametric tests (i.e., Student, Welch, and analysis of variance tests) or non-parametric tests (i.e., Wilcoxon and Kruskal-Wallis tests). P values <0.05 were considered statistically significant. To compare the methylation status of the individual CpG sites between controls and CF patients, we used Fisher's exact test. To take the multiplicity of the hypotheses into account, we used Bonferroni's correction and a family-wise error rate (FWER) of 5% was considered significant.

The ability of the 14 genes in both tissues to discriminate between controls and CF patients was further evaluated using a partial least square discriminant analysis. The descriptors were the normalized mean methylation levels in one of the tissues or both. The PLS response was discrete with two levels, −1 for controls and +1 for CF patients: positive PLS estimates correspond to a classification into the control class and negative ones to a classification into the CF patients class, hence a percentage of correct classification.

We studied the correlations of the mean methylation levels of the genes in both tissues using Spearman's non-parametric correlation coefficient. To take the multiplicity of the hypotheses into account, we used

Bonferroni's correction and a FWER of 10% was considered significant.

The expression ratios of HMOX1 in blood obtained with PCR were log transformed before their mean was taken. Because the resulting values were non-Gaussian, the expression levels between controls and stratified or unstratified CF patients were compared with Kruskal-Wallis' and Wilcoxon's tests. The correlation with the mean methylation level and the methylation status of the individual CpG sites was analyzed with Spearman's coefficient.

Spearman's coefficient was used to test the correlation of lung function (characterized by degree of severity, FEV1% predicted and FVC) with CF patient genotypes at GSTM3, (homozygous GSTM3*A, GSTM3*A/GSTM3*B and homozygous GSTM3*B) and at HMOX1 (rs2071746 A/A, A/T and T/T; and the $(GT)_n$ microsatellite length where we considered both the largest or the smallest n of the two alleles). Multivariate regression models were also used to correct for factors such as demographic and clinical data (Table 1).

Additional files

Additional file 1: Figure S1. Correlation between DNA methylation data in two independent experiments (Spearman's $r = 0.97$ $p = 0$).

Additional file 2: Figure S2. DNA methylation distribution at 14 analyzed genes in CF and control samples.

Additional file 3: Table S2. Distribution of HMOX1 and GSTM3 genotypes in CF patients and controls

Additional file 4: Figure S3. DNA methylation analysis at EDNRA in an independent set of CF patients. DNA methylation at EDNRA was associated with pulmonary severity in blood samples collected from this independent set of severe (CFS) and mild (CFM) CF patients (Kruskal-Wallis $p = 0.047$). CF patients were from the FrGMC cohort; controls were from the METHYLCF cohort. The mean DNA methylation was measured by pyrosequencing.

Additional file 5: Table S1. Primers.

Abbreviations

BMI: Body mass index; CF: Cystic fibrosis; COPD: Chronic obstructive pulmonary disease; FEV1%: Percentage of forced expiratory volume in 1 second; FVC%: Percentage of forced vital capacity; FWER: Family- wise error rate; MRSA: Methicillin-resistant Staphylococcus aureus; NEC: Nasal epithelial cells; NGS: Next-generation sequencing; PI: Pancreatic insufficiency; PLS: Partial least square; SNP: Single-nucleotide polymorphism

Acknowledgements
We are greatly indebt to cystic fibrosis patients and to the medical and paramedical staff of Montpellier, Nice, Hyères, and Toulouse CF Centers for their contribution to the METHYLCF cohort and to CF Centers throughout France for their contribution to the FrGMC cohort. We thank Florin Grigorescu (Montpellier, France) for the helpful discussion and anonymous reviewers for their comments.

Funding
The project was funded by VLM, INSERM, and Montpellier Hospital. MM was supported by the Ciência Sem Fronteiras Program (CNPq, Brazil) and EB by CHU Montpellier. The FrGMC cohort was supported by INSERM, APHP, UPMC Univ Paris 06, ANR (R09186DS), DGS, VLM, and AICM. The funders had no role in the method design, data analysis, decision to publish, or preparation of the manuscript.

Authors' contributions
MM, JV, MT, and AB carried out the molecular analyses. IR performed the statistical analysis and contributed to manuscript writing. RC, LM, SL, MMu, HC, LG, IV, and DC enrolled patients and controls, recorded clinical parameters, collected biological samples, and stratified patients. EB developed bioinformatic pipelines. RC and MC participated in the study design. AD conceived, designed, and coordinated the study and wrote the manuscript. All authors read and approved the final manuscript.

Competing interests
The authors declare that they have no competing interests.

Consent for publication
Not applicable.

Author details
[1]Laboratoire de Génétique de Maladies Rares, EA7402 Montpellier University, Montpellier, France. [2]Equipe de Statistique Appliquée—ESPCI ParisTech, PSL Research University—UMRS1158, Paris, France. [3]Laboratoire de Génétique Moléculaire—CHU Montpellier, Montpellier, France. [4]CRCM, Renée Sabran Hospital—CHU Lyon, Hyères, France. [5]CRCM, Pasteur Hospital—CHU Nice, Nice, France. [6]Sorbonne Universités, UPMC Univ Paris 06, Paris, France. [7]INSERM U938—CRSA, Paris, France. [8]APHP, Trousseau Hospital, Paris, France. [9]CRCM, Larrey Hospital—CHU Toulouse, Toulouse, France. [10]CRCM, Arnaud de Villeneuve Hospital—CHU Montpellier, Montpellier, France.

References
1. Jirtle RL, Skinner MK. Environmental epigenomics and disease susceptibility. Nat Rev Genet. 2007;8(4):253–62.
2. Bjornsson HT, Fallin MD, Feinberg AP. An integrated epigenetic and genetic approach to common human disease. Trends Genet. 2004;20(8):350–8.
3. Cutting GR. Modifier genes in Mendelian disorders: the example of cystic fibrosis. Ann N Y Acad Sci. 2010;1214:57–69.
4. Dequeker E, Stuhrmann M, Morris MA, Casals T, Castellani C, Claustres M, Cuppens H, des Georges M, Ferec C, Macek M, Pignatti PF, Scheffer H, Schwartz M, Witt M, Schwarz M, Girodon E. Best practice guidelines for molecular genetic diagnosis of cystic fibrosis and CFTR-related disorders—updated European recommendations. Eur J Hum Genet. 2009;17(1):51–65.
5. Wright JM, Merlo CA, Reynolds JB, Zeitlin PL, Garcia JG, Guggino WB, Boyle MP. Respiratory epithelial gene expression in patients with mild and severe cystic fibrosis lung disease. Am J Respir Cell Mol Biol. 2006;35(3):327–36.
6. Guillot L, Beucher J, Tabary O, Le Rouzic P, Clement A, Corvol H. Lung disease modifier genes in cystic fibrosis. Int J Biochem Cell Biol. 2014;52:83–93.
7. Gallati S. Disease-modifying genes and monogenic disorders: experience in cystic fibrosis. Appl Clin Genet. 2014;7:133–46.
8. Collaco JM, Blackman SM, McGready J, Naughton KM, Cutting GR. Quantification of the relative contribution of environmental and genetic factors to variation in cystic fibrosis lung function. J Pediatr. 2010;157(5): 802–7.e13.
9. Valinluck V, Tsai HH, Rogstad DK, Burdzy A, Bird A, Sowers LC. Oxidative damage to methyl-CpG sequences inhibits the binding of the methyl-CpG binding domain (MBD) of methyl-CpG binding protein 2 (MeCP2). Nucleic Acids Res. 2004;32(14):4100–8.
10. Henderson JP, Byun J, Williams MV, Mueller DM, McCormick ML, Heinecke JW. Production of brominating intermediates by myeloperoxidase. A transhalogenation pathway for generating mutagenic nucleobases during inflammation. J Biol Chem. 2001;276(11):7867–75.
11. Valinluck V, Sowers LC. Inflammation-mediated cytosine damage: a mechanistic link between inflammation and the epigenetic alterations in human cancers. Cancer Res. 2007;67(12):5583–6.
12. Bergougnoux A, Claustres M, De Sario A. Nasal epithelial cells: a tool to study DNA methylation in airway diseases. Epigenomics. 2015;7(1):119–26.
13. Park JE, Yung R, Stefanowicz D, Shumansky K, Akhabir L, Durie PR, Corey M, Zielenski J, Dorfman R, Daley D, Sandford AJ. Cystic fibrosis modifier genes related to Pseudomonas aeruginosa infection. Genes Immun. 2011;12(5):370–7.
14. Yamada N, Yamaya M, Okinaga S, Nakayama K, Sekizawa K, Shibahara S, Sasaki H. Microsatellite polymorphism in the heme oxygenase-1 gene promoter is associated with susceptibility to emphysema. Am J Hum Genet. 2000;66(1):187–95. Erratum in: Am J Hum Genet 2001; 68(6):1542.
15. Pechlaner R, Willeit P, Summerer M, Santer P, Egger G, Kronenberg F, Demetz E, Weiss G, Tsimikas S, Witztum JL, Willeit K, Iglseder B, Paulweber B, Kedenko L, Haun M, Meisinger C, Gieger C, Müller-Nurasyid M, Peters A, Willeit J, Kiechl S. Heme oxygenase-1 gene promoter microsatellite polymorphism is associated with progressive atherosclerosis and incident cardiovascular disease. Arterioscler Thromb Vasc Biol. 2015;35(1):229–36.
16. Alam J, Igarashi K, Immenschuh S, Shibahara S, Tyrrell RM. Regulation of heme oxygenase-1 gene transcription: recent advances and highlights from the International Conference (Uppsala, 2003) on Heme Oxygenase. Antioxid Redox Signal. 2004;6(5):924–33.
17. Ogilvie V, Passmore M, Hyndman L, Jones L, Stevenson B, Wilson A, Davidson H, Kitchen RR, Gray RD, Shah P, Alton EW, Davies JC, Porteous DJ, Boyd AC. Differential global gene expression in cystic fibrosis nasal and bronchial epithelium. Genomics. 2011;98(5):327–36.
18. Clarke LA, Sousa L, Barreto C, Amaral MD. Changes in transcriptome of native nasal epithelium expressing F508del-CFTR and intersecting data from comparable studies. Respir Res. 2013;14:13.
19. Darrah R, McKone E, O'Connor C, Rodgers C, Genatossio A, McNamara S, Gibson R, Stuart Elborn J, Ennis M, Gallagher CG, Kalsheker N, Aitken M, Wiese D, Dunn J, Smith P, Pace R, Londono D, Goddard KA, Knowles MR, Drumm ML. EDNRA variants associate with smooth muscle mRNA levels, cell proliferation rates, and cystic fibrosis pulmonary disease severity. Physiol Genomics. 2010;41(1):71–7.
20. Flamant C, Henrion-Caude A, Boëlle PY, Brémont F, Brouard J, Delaisi B, Duhamel JF, Marguet C, Roussey M, Miesch MC, Boulé M, Strange RC, Clement A. Glutathione-S-transferase M1, M3, P1 and T1 polymorphisms and severity of lung disease in children with cystic fibrosis. Pharmacogenetics. 2004;14(5):295–301.
21. Feuillet-Fieux MN, Nguyen-Khoa T, Loriot MA, Kelly M, de Villartay P, Sermet I, Verrier P, Bonnefont JP, Beaune P, Lenoir G, Lacour B. Glutathione S-transferases related to P. aeruginosa lung infection in cystic fibrosis children: preliminary study. Clin Biochem. 2009;42(1-2):57–63.
22. Corvol H, Blackman SM, Boëlle PY, Gallins PJ, Pace RG, Stonebraker JR, Accurso FJ, Clement A, Collaco JM, Dang H, Dang AT, Franca A, Gong J, Guillot L, Keenan K, Li W, Lin F, Patrone MV, Raraigh KS, Sun L, Zhou YH, O'Neal WK, Sontag MK, Levy H, Durie PR, Rommens JM, Drumm ML, Wright FA, Strug LJ, Cutting GR, Knowles MR. Genome-wide association meta-analysis identifies five modifier loci of lung disease severity in cystic fibrosis. Nat Commun. 2015;6:8382. doi:10.1038/ncomms9382.
23. Potapova A, Albat C, Hasemeier B, Haeussler K, Lamprecht S, Suerbaum S, Kreipe H, Lehmann U. Systematic cross-validation of 454 sequencing and pyrosequencing for the exact quantification of DNA methylation patterns with single CpG resolution. BMC Biotechnol. 2011;11:6. doi:10.1186/1472-6750-11-6.
24. Tessema M, Yu YY, Stidley CA, Machida EO, Schuebel KE, Baylin SB, Belinsky SA. Concomitant promoter methylation of multiple genes in lung adenocarcinomas from current, former and never smokers. Carcinogenesis. 2009;30(7):1132–8.
25. Sood A, Petersen H, Blanchette CM, Meek P, Picchi MA, Belinsky SA, Tesfaigzi Y. Methylated genes in sputum among older smokers with asthma. Chest. 2012;142(2):425–31.
26. van Eijk KR, de Jong S, Boks MP, Langeveld T, Colas F, Veldink JH, de Kovel CG, Janson E, Strengman E, Langfelder P, Kahn RS, van den Berg LH, Horvath S, Ophoff RA. Genetic analysis of DNA methylation and gene expression levels in whole blood of healthy human subjects. BMC Genomics. 2012;13:636. doi:10.1186/1471-2164-13-636.
27. Chen R, Qiao L, Li H, Zhao Y, Zhang Y, Xu W, Wang C, Wang H, Zhao Z, Xu X, Hu H, Kan H. Fine particulate matter constituents, nitric oxide synthase DNA methylation and exhaled nitric oxide. Environ Sci Technol. 2015;49(19):11859–65.
28. Toperoff G, Aran D, Kark JD, Rosenberg M, Dubnikov T, Nissan B, Wainstein J, Friedlander Y, Levy-Lahad E, Glaser B, Hellman A. Genome-wide survey reveals predisposing diabetes type 2-related DNA methylation variations in human peripheral blood. Hum Mol Genet. 2012;21(2):371–83.
29. Lillycrop KA, Phillips ES, Torrens C, Hanson MA, Jackson AA, Burdge GC. Feeding pregnant rats a protein-restricted diet persistently alters the methylation of specific cytosines in the hepatic PPAR alpha promoter of the offspring. Br J Nutr. 2008;100(2):278–82.

30. Mootha VK, Lindgren CM, Eriksson KF, Subramanian A, Sihag S, Lehar J, Puigserver P, Carlsson E, Ridderstråle M, Laurila E, Houstis N, Daly MJ, Patterson N, Mesirov JP, Golub TR, Tamayo P, Spiegelman B, Lander ES, Hirschhorn JN, Altshuler D, Groop LC. PGC-1alpha-responsive genes involved in oxidative phosphorylation are coordinately downregulated in human diabetes. Nat Genet. 2003;34(3):267–73.

31. Subramanian A, Tamayo P, Mootha VK, Mukherjee S, Ebert BL, Gillette MA, Paulovich A, Pomeroy SL, Golub TR, Lander ES, Mesirov JP. Gene set enrichment analysis: a knowledge-based approach for interpreting genome-wide expression profiles. Proc Natl Acad Sci U S A. 2005;102(43):15545–50.

32. Kim J, Zarjou A, Traylor AM, Bolisetty S, Jaimes EA, Hull TD, George JF, Mikhail FM, Agarwal A. In vivo regulation of the heme oxygenase-1 gene in humanized transgenic mice. Kidney Int. 2012;82(3):278–91.

33. Paredi P, Kharitonov SA, Barnes PJ. Analysis of expired air for oxidation products. Am J Respir Crit Care Med. 2002;166(12 Pt 2):S31–7.

34. Lagrange-Puget M, Durieu I, Ecochard R, Abbas-Chorfa F, Drai J, Steghens JP, Pacheco Y, Vital-Durand D, Bellon G. Longitudinal study of oxidative status in 312 cystic fibrosis patients in stable state and during bronchial exacerbation. Pediatr Pulmonol. 2004;38(1):43–9.

35. Alexander M, Karmaus W, Holloway JW, Zhang H, Roberts G, Kurukulaaratchy RJ, Arshad SH, Ewart S. Effect of GSTM2-5 polymorphisms in relation to tobacco smoke exposures on lung function growth: a birth cohort study. BMC Pulm Med. 2013;13:56. doi:10.1186/1471-2466-13-56.

36. Yengi L, Inskip A, Gilford J, Alldersea J, Bailey L, Smith A, Lear JT, Heagerty AH, Bowers B, Hand P, Hayes JD, Jones PW, Strange RC, Fryer AA. Polymorphism at the glutathione S-transferase locus GSTM3: interactions with cytochrome P450 and glutathione S-transferase genotypes as risk factors for multiple cutaneous basal cell carcinoma. Cancer Res. 1996;56(9):1974–7.

37. Kim J, Kim JD. In vivo YY1 knockdown effects on genomic imprinting. Hum Mol Genet. 2008;17(3):391–401.

38. Schluchter MD, Konstan MW, Drumm ML, Yankaskas JR, Knowles MR. Classifying severity of cystic fibrosis lung disease using longitudinal pulmonary function data. Am J Respir Crit Care Med. 2006;174(7):780–6.

39. Bergougnoux A, Rivals I, Liquori A, Raynal C, Varilh J, Magalhães M, Perez MJ, Bigi N, Des Georges M, Chiron R, Squalli-Houssaini AS, Claustres M, De Sario A. A balance between activating and repressive histone modifications regulates cystic fibrosis transmembrane conductance regulator (CFTR) expression in vivo. Epigenetics. 2014;9(7):1007–17.

40. Grunau C, Buard J, Brun ME, De Sario A. Mapping of the juxtacentromeric heterochromatin-euchromatin frontier of human chromosome 21. Genome Res. 2006;16(10):1198–207.

41. Lutsik P, Feuerbach L, Arand J, Lengauer T, Walter J, Bock C. BiQ Analyzer HT: locus-specific analysis of DNA methylation by high-throughput bisulphite sequencing. Acids Res. 2011;39(Web Server issue):W551–6.

2

SMYD2 promoter DNA methylation is associated with abdominal aortic aneurysm (AAA) and *SMYD2* expression in vascular smooth muscle cells

Bradley J. Toghill[1*], Athanasios Saratzis[1], Peter J. Freeman[2], Nicolas Sylvius[2], UKAGS collaborators and Matthew J. Bown[1]

Abstract

Background: Abdominal aortic aneurysm (AAA) is a deadly cardiovascular disease characterised by the gradual, irreversible dilation of the abdominal aorta. AAA is a complex genetic disease but little is known about the role of epigenetics. Our objective was to determine if global DNA methylation and CpG-specific methylation at known AAA risk loci is associated with AAA, and the functional effects of methylation changes.

Results: We assessed global methylation in peripheral blood mononuclear cell DNA from 92 individuals with AAA and 93 controls using enzyme-linked immunosorbent assays, identifying hyper-methylation in those with large AAA and a positive linear association with AAA diameter ($P < 0.0001$, $R^2 = 0.3175$).
We then determined CpG methylation status of regulatory regions in genes located at AAA risk loci identified in genome-wide association studies, using bisulphite next-generation sequencing (NGS) in vascular smooth muscle cells (VSMCs) taken from aortic tissues of 44 individuals (24 AAAs and 20 controls). In *IL6R*, 2 CpGs were hyper-methylated ($P = 0.0145$); in *ERG*, 13 CpGs were hyper-methylated ($P = 0.0005$); in *SERPINB9*, 6 CpGs were hypo-methylated ($P = 0.0037$) and 1 CpG was hyper-methylated ($P = 0.0098$); and in *SMYD2*, 4 CpGs were hypo-methylated ($P = 0.0012$).
RT-qPCR was performed for each differentially methylated gene on mRNA from the same VSMCs and compared with methylation. This analysis revealed downregulation of *SMYD2* and *SERPINB9* in AAA, and a direct linear relationship between *SMYD2* promoter methylation and *SMYD2* expression ($P = 0.038$). Furthermore, downregulation of *SMYD2* at the site of aneurysm in the aortic wall was further corroborated in 6 of the same samples used for methylation and gene expression analysis with immunohistochemistry.

Conclusions: This study is the first to assess DNA methylation in VSMCs from individuals with AAA using NGS, and provides further evidence there is an epigenetic basis to AAA. Our study shows that methylation status of the *SMYD2* promoter may be linked with decreased *SMYD2* expression in disease pathobiology. In support of our work, downregulated *SMYD2* has previously been associated with adverse cardiovascular physiology and inflammation, which are both hallmarks of AAA. The identification of such adverse epigenetic modifications could potentially contribute towards the development of epigenetic treatment strategies in the future.

Keywords: Epigenetics, Vascular disease, Aneurysm, Inflammation

* Correspondence: bt96@le.ac.uk
[1]Department of Cardiovascular Sciences and the NIHR Leicester Biomedical Research Centre, University of Leicester, Leicester LE2 7LX, UK
Full list of author information is available at the end of the article

Background

Abdominal aortic aneurysm (AAA) is a degenerative cardiovascular disease and a global health concern. AAA is responsible for between 2 and 4% of deaths in white males over the age of 65 [1–3]. Progressive aneurysm growth may lead to rupture, characterised by internal aortic haemorrhage, which results in death in approximately 80% of cases. Surgical repair is the only proven therapeutic option in order to prevent or treat AAA rupture. The exact etiological basis of AAA is still not fully understood, making it difficult to develop a successful pharmaco-therapeutic strategy [4]. However, it is well known that AAA is multifactorial with known risk factors that contribute towards aneurysm development (smoking, male sex, increased age, white European ancestry, atherosclerosis and hyperlipidaemia) [5, 6]. AAA also demonstrates strong heritability [7, 8] and is polygenic. To date, 10 genomic risk loci have been identified through the conduct of genome-wide association studies (GWASs) [9–16].

Epigenetics refers to modifications of the genome that are not exclusively a result of change in the primary DNA sequence and includes DNA methylation, histone modifications and non-coding RNAs [6, 17]. These mechanisms are essential for cell, tissue and organ development and directly interact with DNA sequence/structure to manipulate gene expression at the transcriptional and post-transcriptional levels. However, aberrant epigenetic modifications have been implicated in many complex diseases [18–20]. DNA methylation is the most extensively studied epigenetic modification to DNA and involves the addition of a methyl group to a cytosine base 5′ to a guanine (CpG dinucleotide) by DNA methyltransferase enzymes, which can inhibit gene transcription; however, increased methylation has also been associated with increased expression, since methylation can be necessary for transcriptional binding [21–23]. DNA methylation patterns are long-term inherited signatures that are passed through generations and can be affected by both genetic and environmental factors [24, 25].

There is currently very limited investigation surrounding the role of DNA methylation in AAA, and the only published study to date was conducted by Ryer et al. [26], who performed genome-wide methylation analysis in peripheral blood DNA from 20 AAA patients (11 smokers and 9 non-smokers) and 21 control samples (10 smokers and 11 non-smokers) using Illumina 450k micro-arrays. They identified differentially methylated regions in *ADCY10P1*, *CNN2*, *KLHL35* and *SERPINB9*, and differential expression in *SERPINB9* and *CNN2*.

Variations in DNA sequence at polymorphic loci can result in variable patterns of DNA methylation, known as methylation quantitative trait loci (meQTL) [27–29]. Disease variants can alter transcription factor levels and methylation of their binding sites [30]. The direct mechanisms of the variants associated with AAA are not fully known, but many are intronic and it is likely that their effects are regulatory in nature as opposed to being directly functional. It is therefore feasible that disease-specific meQTLs exist for AAA, and considering many meQTL act in cis [28], their likely effects on methylation are in the genes surrounding the risk loci identified in previous GWASs.

In this study, targeted bisulphite next-generation sequencing (NGS) was performed in vascular smooth muscle cell (VSMC) DNA, isolated from whole aortic tissues of AAA and controls, to investigate the methylation status of CpG islands in regulatory regions of genes located near AAA genomic risk loci. This approach offers a more refined way to assess the methylation status of genes in proximity to AAA risk loci than other methodologies such as the 450k microarray. The identification of adverse epigenetic modifications directly linked to patients with AAA could offer a more comprehensive understanding of AAA pathobiology, and an alternative research avenue in the search for a future treatment strategy [31, 32].

Results

Global DNA methylation and homocysteine analysis in AAA patients and controls

Global genomic DNA methylation was assessed in peripheral blood DNA of 185 individuals, and circulating homocysteine (HCY) was assessed in blood plasma from 137 of the same individuals using ELISAs. Global DNA methylation was significantly higher in men with large AAA (> 55 mm, $n = 48$, global DNA methylation 1.86% (± 0.6%)) compared to men with small AAA (30–55 mm, $n = 45$, global DNA methylation 0.93% (± 0.52)) and controls (< 25 mm, $n = 92$, global DNA methylation 0.79% (± 0.43%)) (Fig. 1a). There was a linear relationship between AAA size and global DNA methylation which was adjusted for patient age (Fig. 1b) (Pearson coefficient R^2 value = 0.3175, $P < 0.0001$). Circulating blood plasma HCY levels were slightly higher in men with AAA compared to controls (9.6 μmol/L ± 0.62, $n = 70$, vs 7.94 μmol/L ± 0.52, $n = 67$, $P = 0.0433$) (Additional file 1: Figure S1a). However, there was no association between global methylation and circulating HCY ($P = 0.095$) (Additional file 1: Figure S1b).

Bisulphite NGS of vascular smooth muscle cell DNA in AAA patients and controls

We isolated CpG islands within regulatory regions of genes associated with AAA (*LRP1*, *ERG*, *MMP9*, *LDLR*, *IL6R*, *SORT1*, *SERPINB9*, *SMYD2* and *DAB2IP*—Table 1) in DNA extracted from the VSMCs of 20 controls and 24 large AAAs (Additional file 1: Table S2). Using targeted bisulphite NGS, we determined the methylation status of these regions and subsequently compared AAA to controls. We identified significant differences in methylation

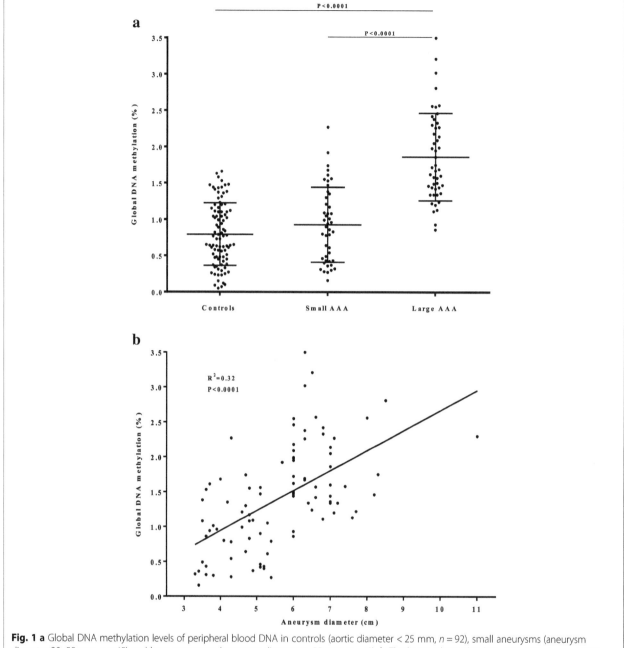

Fig. 1 a Global DNA methylation levels of peripheral blood DNA in controls (aortic diameter < 25 mm, n = 92), small aneurysms (aneurysm diameter 30–55 mm, n = 45) and large aneurysms (aneurysm diameter > 55 mm, n = 48). **b** The linear relationship between AAA size and DNA methylation (n = 93)

between AAA and controls in four genes: *ERG, IL6R, SERPINB9* and *SMYD2* (Fig. 2 and Table 2). There were no significant differences in methylation status between AAA and controls in *DAB2IP, LDLR, LRP1, MMP9, SORT1* and an alternative promoter in *ERG* (Additional file 1: Figure S2). A consistently high mapped sequencing depth was observed for each gene, and there were no significant differences in sequencing coverage between AAA and controls in any gene. Mean mapped sequencing coverage for AAA and controls for each gene is shown in Additional file 1: Figure S3.

The greatest differences in methylation were seen in the *ERG* gene, where 13 CpGs were consecutively hypermethylated (Fig. 2a) in AAAs compared to controls. The mean methylation of these sites was 18.67% ± 3.92 in controls versus 23.8% ± 4.11 in cases with a mean increase in methylation of 5.13% ± 0.43 (P = 0.0005, Q = 0.0014). The 13 CpGs in *ERG* span a region of 118 bp (NG_029732.1:g.5389_5507|gom) which is located within an intronic region directly downstream from the first *ERG* exon.

Table 1 Candidate genes and amplicons for VSMC bisulphite NGS: Genomic locations, primer sequences, annealing temperatures and CpG coverage are displayed for each amplicon targeting candidate genes

Gene symbol	Chromosome and RefSeq ID	RefSeq ID coordinates	Primer sequence 5'-3'	Annealing temp (°C)	CpG coverage
LRP1		3875–4240	F: TAGAAGGGGGTAGTGATTAAAAGTA	54	13
			R: AAAACAAACCCTAATTTAAAAAAAA		
	Chr: 12	4274–4640	F: GGGTATTAAGGTGGGTTTTATTTT	57	25
	NG_016444.1		R: TACTCTAAAATTTCAAACTCCCTCC		
		4680–5036	F: GTATTAGGGAGGAGGGTTTAGTTAG	54	24
			R: TCCTCAATACATAAACCTAAAACTC		
ERG		282787–283319	F:TTTTATTAGGTAGTGGTTAGATTTAGTTTT	54	22
	Chr: 21		R: TACCCCCAATTAATAAATTCCAATATA		
	NG_029732.1	283331–283680	F: TTGGAATTTATTAATTGGGGGTATATAT	57	6
			R: ACACTATCTTTTACAAAATCAATCCAC		
		5162–5540	F: AAAATTTTTGGAAGGGGTTTAGTT	56	37
			R: ATAATATTTTTCCAACCTCATTAAAAAC		
MMP-9		3980–4579	F: TTGGGTTTAAGTAATTTTTTTATTT	52	7
	Chr: 20		R: TAACCCATCCTTAACCTTTTACAAC		
	NG_011468.1	4640–5600	F: GATGGGGGATTTTTTTAGTTTTATT	57	8
			R: TACCCATTTCTAACCATCACTACTC		
LDLR		4376–4725	F: TTTTTTTAAGGGGAGAAATTAATATTTA	54	5
			R: CACAAAAAAATAACAACAACCTTTC		
		4776–5355	F: GAAAGGTTGTTGTTATTTTTTTGTG	57	28
	Chr: 19		R: AAACTCCCTCTCAACCTATTCTAAC		
	NG_009060.1	5464–5972	F: TTTAAGTTTTTTATAGGGTGAGGGAT	56	31
			R: ACACCCAACTCAAAATAACAATAAC		
		6054–6785	F: AATTTTATTGGGTGTAGTTTAATAGGTTAT	54	49
			R: CTCAAAATCATACACTAACCAACCTC		
IL6R		2781–3061	F: TTTTTGTTTAGGTTGGAGTGTAGTG	52	8
	Chr: 1		R: CAAAATTTTAATATTATAATTCACATAAAA		
	NG_012087.1	3095–3729	F: GTTTGTTTTGGTGTAGAGATAGGTG	58	7
			R: ATAAAACTCCCAAATAAACAAAACC		
SORT1	Chr: 1	4363–4804	F: AGATTATTTTTTAGGTTTGTAGGAGTTA	55	24
	NG_028280.1		R: AACAAAAACTACTAAAATCAACCCC		
SMYD2		214280141–214280630	F: TTTTATTTTGAAGTAGTGGTTTTTGTTAA	55	13
			R: TTTTCCAAAATTAAAAATTTTTAAACCT		
	Chr: 1	214280631–214281330	F: AAGGTTTAAAAATTTTTAATTTTGGAAA	58	82
	NC_000001.11		R: AAATCCCCCACCTAAAAAAACTAC		
		214281331–214281733	F: TTTTTTTAGGTGGGGGATTT	55	46
			R: CCCACATTTAAAAACAAAAACCTAC		
SERPINB9		2891544–2892079	F: GGTGTTGAAAATATTTTTGGAGGTA	57	26
			R: AAAATCTACCATTCATCAAACTAACAAA		
		2890564–2891263	F: TTAGAGGGTGGGATTAGAGGTAGTT	54	9
	Chr: 6		R: TTTCCACCTAAAAAACCAAAATTAA		
	NC_000006.12	2889934–2890423	F: GTATTTGGGAATTGTTGATGTTTTT	55	11
			R: ACAACAAAAAATCATTATAATCTATTTTCT		
		2889425–2889933	F: AGAATTTTATATGTAATTTATTTTGG	54	31

Table 1 Candidate genes and amplicons for VSMC bisulphite NGS: Genomic locations, primer sequences, annealing temperatures and CpG coverage are displayed for each amplicon targeting candidate genes *(Continued)*

Gene symbol	Chromosome and RefSeq ID	RefSeq ID coordinates	Primer sequence 5'-3'	Annealing temp (°C)	CpG coverage
DAB2IP			R: TATAATCCCAACACTTTAAAAAACC		
		121689899–121690502	F: TTGGAGTAGTTTGTTTGTTGTGTT	57	10
			R: TAAAAAATTAAATACCCAAAACCTC		
		121690573–121691062	F: TTGTTATATTTTAAGTTGAGATTTTGGG	57	6
	Chr: 9		R: TAACCACATAAAAACATTCCAAACA		
	NC_000009.12	121767430–121767872	F: GGAGAGGAGATAGGAAAGTTTTTAG	57	6
			R: CTAACCTTATCCCTTACAACCACTAC		
		121768273–121768692	F: AATGGAGTGGGAGTTTGTTATAGTG	57	11
			R: TAATAACAACTTTTAACCCCACCTC		

We observed a more conservative site of potential differential methylation in the *IL6R* gene promoter upstream of the transcriptional start site, where we identified significant hyper-methylation of 2 CpGs (NG_012087.1:g.3570_3676|gom) in AAA (89% ± 5.83) vs controls (74.85% ± 2.62) (Fig. 2b). The mean increase in methylation was 14.2% ± 3.22 ($P = 0.0145$, $Q = 0.076$).

In the *SERPINB9* gene, seven CpGs located in the 3' untranslated region downstream of exon 7 were found differentially methylated in AAA compared to controls. Overall, hypo-methylation of the region was observed (NC_000006.12:g.2889674_2889564|lom) where six CpGs were hypo-methylated ($P = 0.0037$, $Q = 0.032$) and one CpG was hyper-methylated ($P = 0.0098$, $Q = 0.0478$)

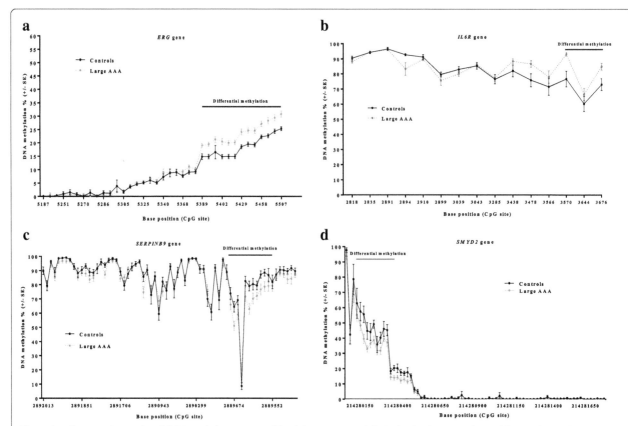

Fig. 2 Vascular smooth muscle cell DNA methylation status of bisulphite-sequenced CpG islands where significant differential methylation was observed in 20 controls vs 24 AAAs (*ERG* (**a**), *IL6R* (**b**), *SERPINB9* (**c**) and *SMYD2* (**d**)). See Tables 2 and 3 for descriptive statistics of differentially methylated CpGs

Table 2 Mean observed CpG site methylation in vascular smooth muscle cell DNA after bisulphite sequencing of candidate genes (total n = 24 AAAs and 20 controls—Additional file 1: Table S2)

Gene, chromosome and RefSeq ID	RefSeq ID coordinate (CpG site)	Overall mean meth % ± SD (AAA n = 24)	Overall mean meth % ± SD (controls n = 20)	Overall meth difference (% ± SE)	P value	Q value	Age-adjusted odds ratio
ERG	5389	19.00 (3.48)	14.83 (4.42)	4.17 (1.23)	0.0019	0.0037	NS
Chr: 21	5393	19.43 (3.72)	14.83 (3.618)	4.60 (1.16)	0.0004	0.0010	NS
NG_029732.1	5400	21.26 (5.76)	16.50 (10.57)	4.76 (2.58)	0.0002	0.0010	6.70
	5402	20.43 (5.45)	14.83 (3.50)	5.60 (1.48)	0.0001	0.0009	NS
	5405	19.96 (4.15)	14.89 (3.43)	5.07 (1.21)	0.0001	0.0008	6.18
	5408	20.13 (4.20)	14.89 (3.51)	5.24 (1.23)	0.0001	0.0008	6.19
	5429	24.13 (5.26)	18.56 ((3.22)	5.58 (1.41)	0.0003	0.0010	NS
	5440	24.65 (4.49)	19.50 (3.40)	5.15 (1.28)	0.0001	0.0008	NS
	5442	24.57 (5.12)	19.28 (3.25)	5.29 (1.38)	0.0002	0.0009	NS
	5458	27.22 (4.88)	22.22 (3.15)	5.00 (1.33)	0.0003	0.0010	NS
	5468	28.35 (6.01)	22.67 (3.16)	5.68 (1.56)	0.0015	0.0032	NS
	5474	29.39 (5.55)	24.39 (3.48)	5.00 (1.50)	0.0012	0.0027	NS
	5507	30.83 (5.57)	25.33 (3.09)	5.49 (1.46)	0.0004	0.0010	NS
IL6R Chr: 1	3570	93.17 (5.62)	76.70 (23.02)	16.47 (4.85)	0.0113	0.0743	3.11
NG_012087.1	3676	84.92 (10.89)	73.00 (17.49)	11.92 (4.31)	0.0176	0.0773	NS
SERPINB9	2889674	51.00 (16.19)	64.47 (16.19)	−13.47 (4.97)	0.0063	0.0375	NS
Chr: 6	2889644	20.46 (20.29)	8.579 (10.64)	11.88 (5.14)	0.0098	0.0478	3.29
NC_000006.12	2889627	71.67 (18.37)	82.53 (14.94)	−10.86 (5.20)	0.0058	0.0375	NS
	2889620	62.88 (19.54)	79.11 (15.15)	−16.23 (5.47)	0.0017	0.0273	NS
	2889612	70.33 (19.08)	80.37 (17.96)	−10.04 (5.71)	0.0064	0.0375	NS
	2889568	76.58 (19.21)	87.21 (11.26)	−10.63 (4.97)	0.0012	0.0273	0.33
	2889564	78.67 16.34)	88.58 (10.65)	−9.91 (4.35)	0.0010	0.0273	NS
SMYD2	214280412	39.50 (14.74)	55.67 (16.24)	−16.17 (6.00)	0.0004	0.0200	NS
Chr: 1	214280441	32.91 (10.53)	44.80 (17.29)	−11.89 (4.87)	0.0014	0.0299	0.09
NC_000001.11	214280507	38.78 (12.67)	49.22 (7.93)	−10.44 (4.56)	0.0016	0.0299	0.06
	214280600	37.04 (11.02)	45.30 (11.43)	−8.26 (4.22)	0.0014	0.0299	0.03

NS not significant

(NC_000006.12:g.2889644|gom) (Fig. 2c). This is in contrast to the overall AAA hyper-methylation reported by Ryer et al., [26], which was conducted on peripheral blood DNA.

Finally in the *SMYD2* gene, we identified four significantly hypo-methylated CpG sites in AAA compared to controls (Fig. 2d) in the gene promoter upstream of the transcriptional start site (NC_000001.11:g 214280412_214280600|lom). Mean control methylation was 48.75% ± 5.02, mean methylation in the cases was 37.06% ± 2.95, and the mean overall decrease in methylation in AAA compared to controls was 11.69% ± 8.2 ($P = 0.0012$, $Q = 0.027$). Figure 3a displays a more detailed, CpG-specific illustration of the differentially methylated sites in the *SMYD2* gene promoter.

Logistic regression was conducted on statistically significant differentially methylated CpGs in each gene to assess the potential effects of age (controls (56 years ±35) and AAA (68 years ±25)) on disease status. Some sites lost significance when applying age as an additional variable for differential methylation; however, there still remained CpGs in each differentially methylated gene that were significant independently of age. In addition, the effects of smoking and sex were assessed in our data by excluding all non-smokers and females respectively (Table 3).

Functional corroboration of differential methylation in AAA patients and controls

Gene expression analysis of differentially methylated genes
In genes where differentially methylated CpGs were identified between AAA and controls (*ERG, IL6R, SERPINB9* and *SMYD2)*, we performed mRNA expression analysis to assess the relationship between DNA methylation and

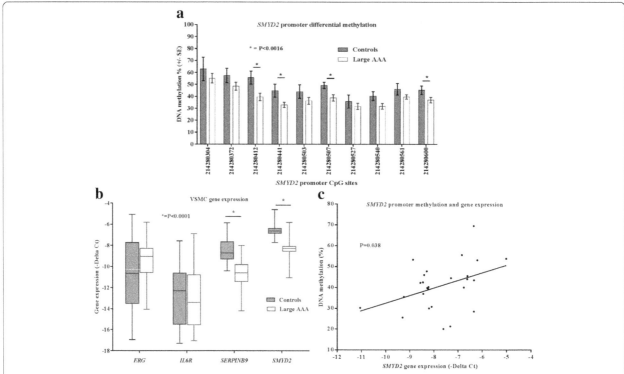

Fig. 3 a Differentially methylated CpGs in the *SMYD2* gene promoter identified after bisulphite sequencing of DNA from 24 AAAs and 20 controls. **b** Relative gene expression levels of differentially methylated candidate genes (*SMYD2, SERPINB9, IL6R* and *ERG*) in vascular smooth muscle cells from 20 controls and 24 AAAs. **c** Linear relationship between gene expression and mean DNA methylation status of the differentially methylated CpGs (NC_000001.11: 214280412, 214280441, 214280507 and 214280600) in *SMYD2* (*n* = 26 where sufficient sequencing coverage and expression data was acquired)

gene expression. We identified a significant overall reduction in the expression of *SMYD2* (*P* < 0.0001) and *SERPINB9* mRNA (*P* < 0.0001) in people with AAA compared to controls. However, there were no differences in expression in the *IL6R* and *ERG* mRNA (Fig. 3b). Subsequently, the methylation values of significantly associated CpG regions were correlated with expression levels in their respective genes using linear regression. In the *SERPINB9* gene, where significant expression and methylation differences were observed, neither the hypo- nor hypermethylated CpG sites were associated with *SERPINB9* expression levels (*P* = 0.243 and *P* = 0.31 respectively). Finally, the mean DNA methylation percentage of differentially methylated CpGs (214280412, 214280441, 214280507 and 214280600) in the *SMYD2* gene promoter was significantly associated with *SMYD2* gene expression (*P* = 0.0383, R^2 = 0.17, (Fig. 3c). However *SMYD2* gene expression did not correlate with aortic diameter.

Immunohistochemistry to assess Smyd2 in aortic tissues

To further corroborate the *SMYD2* methylation/expression relationship and to ensure that the differential gene activity seen between AAA and controls was applicable to the site of aneurysm in the aortic wall, histological

staining of the Smyd2 protein was conducted in whole aortic tissues from six of the same samples used in methylation and gene expression analysis (three AAAs and three controls).

For each of the six samples, three individual stains were conducted (total = 18 slides) on frozen tissue cuts from the same aortic sections. The first was for Smyd2 staining, the second was a negative Smyd2 control with no primary antibody, and the last was for smooth muscle actin (SMA) staining. The controls were checked for the presence of Smyd2, and all were absent of brown colouring, which represents the presence of primary antibody. The SMA slides were then visualised, and where the densest section of smooth muscle fibres was seen, the Smyd2 slides were visualised in the same regions, where photographs were taken (Fig. 4).

The histological technique used for this assay is hard to quantify and is not as objective as other methodologies. However, Fig. 4 subjectively shows a reduced abundance of the Smyd2 protein (brown colouring) in the tissues of those with AAA compared to controls. Whilst a statistical analysis is not available, it appears that the same pattern of downregulated Smyd2 activity (from the mRNA gene expression analysis) is evident in AAA compared to controls. This demonstrates that transcript

Table 3 Mean observed CpG site methylation in vascular smooth muscle cell DNA after bisulphite sequencing of candidate genes after consideration of sex and smoking status as co-variates of methylation (total $n = 24$ AAAs, 16 male controls and 16 smoking controls—Additional file 1: Table S2)

Gene, chromosome and RefSeq ID	RefSeq ID coordinate	Meth % ± SD—all male smokers	Smokers only meth % ± SD	Smokers only meth difference (% ± SE)	Men only meth % ± SD	Men only meth difference (% ± SE)	Smokers only P value	Men only P value
	(CpG site)	(AAA— $n = 24$)	(Controls— $n = 16$)	$n = 16$	(Controls— $n = 16$)	$n = 16$	(24 AAAs vs 16 controls)	(24 AAAs vs 16 controls)
ERG	5389	19.00 (3.48)	14.43 (3.99)	4.57 (1.25)	15.07 (4.65)	3.93 (1.32)	0.0011	0.0059
Chr: 21	5393	19.43 (3.72)	14.86 (3.65)	4.58 (1.25)	14.80 (3.82)	4.63 (1.25)	0.0013	0.0009
NG_029732.1	5400	21.26 (5.76)	17.36 (11.73)	3.90 (2.88)	16.87 (11.58)	4.39 (2.82)	0.0011	0.0007
	5402	20.43 (5.45)	15.14 (3.55)	5.29 (1.64)	14.80 (3.73)	5.63 (1.61)	0.0009	0.0004
	5405	19.96 (4.15)	15.50 (3.54)	4.46 (1.33)	14.93 (3.65)	5.02 (1.32)	0.0014	0.0003
	5408	20.13 (4.20)	15.36 (3.69)	4.77 (1.36)	14.87 (11.57)	5.26 (1.33)	0.0007	0.0002
	5429	24.13 (5.26)	19.21 (2.83)	4.92 (1.53)	18.73 (3.26)	5.40 (1.52)	0.0028	0.0010
	5440	24.65 (4.49)	20.07 (2.92)	4.58 (1.35)	19.60 (3.48)	5.05 (1.37)	0.0010	0.0003
	5442	24.57 (5.12)	19.86 (2.51)	4.71 (1.47)	19.40 (3.36)	5.17 (1.50)	0.0009	0.0006
	5458	27.22 (4.88)	22.93 (2.49)	4.29 (1.41)	22.40 (3.31)	4.82 (1.44)	0.0030	0.0010
	5468	28.35 (6.01)	23.50 (2.34)	4.85 (1.68)	22.80 (3.39)	5.55 (1.71)	0.0127	0.0045
	5474	29.39 (5.55)	25.36 (2.56)	4.03 (1.58)	24.53 (3.70)	4.86 (1.63)	0.0125	0.0037
	5507	30.83 (5.57)	26.21 (1.85)	4.61 (1.54)	25.20 (3.36)	5.63 (1.60)	0.0040	0.0008
IL6R Chr: 1	3570	93.17 (5.62)	78.31 (22.41)	14.85 (4.76)	75.37 (23.77)	17.79 (5.02)	NS	0.0037
NG_012087.1	3676	84.92 (10.89)	75.56 (16.55)	9.35 (4.33)	73.63 (16.97)	11.29 (4.39)	NS	NS
SERPINB9	2889674	51.00 (16.19)	66.27 (15.98)	−15.27 (5.30)	64.07 (16.84)	−13.07 (5.41)	0.0037	0.0120
Chr: 6	2889644	20.46 (20.29)	8.13 (9.93)	12.33 (5.63)	9.27 (11.59)	11.19 (5.76)	0.0139	0.0230
NC_000006.12	2889627	71.67 (18.37)	84.93 (12.93)	−13.27 (5.44)	82.13 (15.91)	−10.47 (5.76)	0.0015	0.0100
	2889620	62.88 (19.54)	80.00 (15.17)	−17.13 (5.95)	81.00 (14.06)	−18.13 (5.84)	0.0018	0.0007
	2889612	70.33 (19.08)	80.27 (20.23)	−9.933 (6.43)	78.93 (19.58)	−8.60 (6.34)	0.0081	0.0168
	2889568	76.58 (19.21)	88.33 (12.19)	−11.75 (5.56)	87.67 (9.31)	−11.08 (5.33)	0.0003	0.0018
	2889564	78.67 16.34)	87.80 (11.82)	−9.13 (4.88)	90.47 (8.57)	−11.80 (4.59)	0.0039	0.0002
SMYD2	214280412	39.50 (14.74)	59.00 (17.12)	−19.50 (6.64)	55.38 (17.33)	−15.88 (6.37)	0.0002	0.0015
Chr: 1	214280441	32.91 (10.53)	48.57 (17.80)	−15.66 (5.37)	41.56 (14.78)	−8.64 (4.64)	0.0004	NS
NC_000001.11	214280507	38.78 (12.67)	50.14 (8.61)	−11.36 (5.14)	50.13 (7.97)	−11.34 (4.81)	0.0012	0.0009
	214280600	37.04 (11.02)	47.57 (10.63)	−10.53 (4.72)	45.78 (12.02)	−8.73 (4.44)	0.0001	0.0020

NS not significant

expression differences are also reflected from a translational perspective at the site of aneurysm, which is in turn associated with *SMYD2* promoter methylation.

Discussion

Peripheral blood global DNA hyper-methylation is commonly a hallmark of chronic inflammation and has been observed as a potential pathological marker of disease [19, 33–37], which supports our findings. In turn, inflammation is a pathological hallmark of AAA, yet up until now, no work has investigated global methylation in AAA. Our study has identified global DNA hypermethylation in patients with a large AAA, and a clear linear relationship with AAA diameter in peripheral blood DNA. These results were adjusted for age, and both groups (cases and controls) consisted exclusively of white males with at least a 10-year smoking history. Age, smoking and inflammation, as stated, can have considerable impact upon global and Cp-specific DNA methylation patterns, consisting of global hypo-methylation (ageing), hyper- and hypo-methylation (smoking) and global hyper-methylation (inflammation). Each of these factors is also significant a risk factor for AAA [6].

In addition, circulating HCY has previously been associated with global methylation [38], is a proposed biomarker for cardiovascular disease [39] and has also been suggested to have a role in AAA [40]. However, the differences in HCY between AAA and controls in this

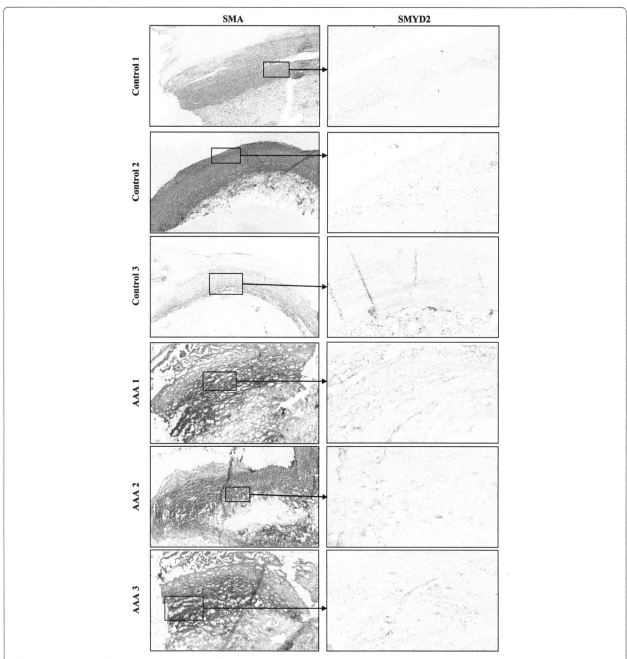

Fig. 4 Immuno-histochemical staining: staining of aneurysmal (*n* = 3) and non-aneurysmal (*n* = 3) abdominal aortic frozen tissue sections for smooth muscle actin (SMA) and Smyd2. Brown = presence of primary specific antibody

study were not clinically different (hyper-homocysteinemia is described above 15 μmol/L) and support the idea that HCY is not an appropriate biomarker for AAA, which is also suggested by Lindqvist et al., [41]. In addition, there was no significant relationship between HCY and global DNA methylation, suggesting that global DNA hyper-methylation is associated with AAA independently of circulating blood HCY. Similarly to other inflammatory diseases, increased global DNA methylation may be a factor in the pathobiology of AAA

and provides a solid rationale for further study in to the role of DNA methylation in AAA.

This study is the first to adopt bisulphite NGS, a gold standard technique in studies of DNA methylation [42], to identify differentially methylated CpG sites associated with AAA in VSMCs. VSMCs are the main constituents of the tunica media, where cellular apoptosis occurs as a hallmark of AAA during aneurysm development and growth [43]. Ryer et al. [26] previously conducted a methylation study on AAA in peripheral blood DNA using Illumina 450k micro-arrays and found differential methylation and gene expression in *CNN2* and *SERPINB9*. In VSMCs, we also identified differential methylation in *SERPINB9*, but the relationship was converse to that seen in DNA isolated from peripheral blood. In our study, after assessing the association between differential methylation and expression in *SERPINB9* with linear regression analysis, it was determined that there was not a direct relationship between the two. *SERPINB9* could however still be important in AAA, as reduced *SERPINB9* expression is associated with atherosclerotic disease progression and is inversely related to the extent of apoptosis within the intima [44].

Most importantly in this study, we identified significant DNA hypo-methylation in the *SYMD2* gene promoter in those with AAA (NC_000001.11:g214280412_214280600|lom), and the mean methylation percentage of the differentially methylated CpGs were significantly correlated with gene expression. *SMYD2* gene expression however was not correlated with aortic diameter, indicating that downregulated *SMYD2* expression, a potential result of *SMYD2* promoter methylation, may be associated with the development, and not progression, of AAA. Our analysis is limited by the fact that all VSMC samples in this study were from individuals with AAA greater than 55 mm. The lack of data from those with smaller AAA limits our ability to make conclusions on whether the level of *SMYD2* expression is linked to AAA size. Future studies should therefore address this. Further to this analysis, the decrease in expression of *SMYD2* in AAA was corroborated in aortic tissues using immunohistochemistry. *SMYD2* codes for an important lysine methyltransferase and is well known for its role in transcriptional regulation [45]. Our results are particularly relevant when considering the role of Smyd2 in inflammation and cardio-physiology, as previous studies have identified a role in cardio-protection and suppression of inflammation. In one study, protein levels of Smyd2 were decreased in cardiomyocytes after cellular apoptosis and after myocardial infarction. In addition, *SMYD2* deletion in cardiomyocytes in vivo promoted apoptotic cell death upon myocardial infarction [46]. It has also been reported that Smyd2 is a negative regulator for macrophage activation. Elevated *SMYD2* expression suppresses the production of pro-inflammatory cytokines, including IL-6 and TNF. In addition, macrophages with high *SMYD2* expression promote regulatory T cell differentiation as a result of increased TGF-β production and decreased IL-6 secretion [47]. In summary, previous work has demonstrated that down-regulation of *SMYD2* is linked with adverse cardio-physiology and an increase in inflammation, both of which are key hallmarks of AAA pathobiology.

Looking more closely at the relationship between *SMYD2* expression and DNA methylation in our study, it is clear that the observed association is opposite to the traditional notion of promoter hyper-methylation inhibiting gene transcription. However, it is becoming better known that this is not definitively correct, as multiple pieces of evidence go against this paradigm [48, 49]. In some cases, methylation is required for activation of transcription and is therefore positively correlated with gene expression [49], which could be appropriate to our findings. Due to this, it must be stated that the exact mechanisms by which methylation controls gene expression is still not definitively understood and is not always the same.

The results presented in this article are preliminary. Future work will concentrate on proving up the *SMYD2* relationship with large-scale replication, whilst also establishing whether there is a causal genetic basis to differential methylation patterns in AAA (meQTL analysis). There is some knowledge of meQTLs, but this is still very scarce and currently there is no comprehensive database highlighting phenotype-specific associations between genotype and methylation. It is therefore likely that existing disease-specific meQTLs are unreported, and in our case, genes that are known AAA risk loci represent potentially insightful targets for further genetic and methylation analysis. Finally, after replication, it is suggested that mechanistic analysis to assess the causal role of *SMYD2* promoter methylation on gene function would be insightful.

There were limitations to this work, including limited sample size which was hindered due to the difficulty of obtaining aortic tissues from individuals. This problem in turn had an effect on sample group demographics, and we could not control for gender, smoking and age as effectively as we would have liked—as we did in the peripheral blood global methylation assay.

Conclusion

This pilot study is the first to assess DNA methylation in VSMCs of people with AAA, and the first to assess methylation in AAA using bisulphite NGS. The results presented in this article demonstrate further evidence that there is an epigenetic basis to AAA and is proof of concept for further large-scale analysis. Global peripheral blood DNA hyper-methylation is associated with large AAA, increasing AAA size, and this relationship appeared independently of circulating HCY. Gene-specific changes in DNA methylation were also identified in VSMC DNA (*IL6R*, *SMYD2*, *SERPINB9* and *ERG*). In particular, there was a significant association between the methylation status of CpGs in the *SMYD2* promoter and *SMYD2* gene expression, which was further corroborated in whole aortic tissues using immunohistochemistry. In summary, this article demonstrates that global and gene-specific DNA methylation changes exist in AAA and are potentially involved in the pathobiology of the disease and could be useful in the future development of epigenetic therapies.

Methods
Population
Patients and controls were recruited from two sources: The UK aneurysm growth study (UKAGS) (http://www2.le.ac.uk/projects/ukags) and our local AAA research programme based at the NIHR Leicester Cardiovascular Biomedical Research Centre. The UKAGS collects peripheral blood samples from men with AAA (aortic diameter 30 mm or greater) and healthy controls (all screened for AAA) recruited from the English NHS AAA Screening Programme. Furthermore, AAA biopsies (aortic wall) are taken from patients undergoing open AAA repair in our regional vascular unit, where we have also established an aortic tissue collection from cadaveric organ donors. Ethical approval by an NHS Research Ethics Committee was obtained for both studies. Risk factors associated with AAA can be confounders of DNA methylation and include age, smoking, gender and ethnicity. Where possible, to help prevent any confounding effects of these factors, white males over the age of 65 with at least a 10-year history of smoking were selected for inclusion.

Vascular smooth muscle cell culture from aortic biopsies
Aortic tissues were collected by vascular surgeons within the vascular surgery group at the University of Leicester. There were no signs of AAA or atherosclerosis in the abdominal aortas obtained from organ donors, which were therefore used as controls. For each aortic tissue sample collected from those undergoing open AAA surgery ($n = 24$) and from cadaveric organ donors ($n = 20$), explant culture was performed to isolate and grow VSMCs in vitro. It was important to isolate individual cells given that independent cell lines have different epigenetic profiles. VSMCs are a good surrogate for such analysis considering aneurysmal formation is characterised by inflammation and VSMC apoptosis in the tunica media of the aortic wall.

Thin sections of the tunica media were carved from the whole aortic tissues no longer than 48 h after surgery (AAA samples), or death (cadaveric organ donor samples), and placed in T80 cell culture flasks containing 10 ml smooth muscle cell Medium 231 (ThermoFisher Scientific - M2315005) with added Smooth Muscle Growth Supplement (ThermoFisher Scientific - S00725). A 25-ml bottle of growth serum was added to each 500 ml 231 media prior to use. The flasks containing the media, growth supplement and aortic tissues were left to incubate at 37 °C until visible primary cell growth was observed (~ 2 weeks). After primary growth of VSMCs, confluent cells were detached from the flask with the addition of 2× trypsin-EDTA

solution and incubated at 37 °C for 3 min. Five-millilitre sterile phosphate buffered saline (PBS) and fetal bovine serum solution (20:1 ratio) was added to neutralise the trypsin. Solutions containing detached cells were aspirated from the flasks into sterile universal tubes and centrifuged at 500×g for 6 min at 20 °C. Supernatants were discarded and cell pellets were re-suspended in 3 ml media. One millilitre of the 3 ml suspensions were added to three new T80 cell culture flasks, each containing 10 ml of media. This process was repeated until sufficient amounts of isolated VSMCs (two clearly visible cell pellets) were available for DNA and RNA extraction, however, never beyond passage 3 to attenuate any potential changes in cellular gene expression or DNA methylation patterns because of the artificial media/culture environment. Isolated cultured cells stored by cryogenic freezing at − 80 °C until needed for experimental analysis. In the earliest stages of establishing our explant culture work flows, α-smooth muscle immunofluorescent cell type staining (internal work not published) was conducted to assess for fibroblast contamination in our standard protocol, and this protocol was used in this study.

DNA extraction
DNA was extracted from 185 peripheral blood mononuclear cell (PBMC) samples from the UKAGS for global methylation analysis (Additional file 1: Table S1), and 44 VSMC samples (Additional file 1: Table S2) isolated from whole aortic tissues for bisulphite NGS analysis using the DNeasy Blood & Tissue Kit (Qiagen) and according to the manufacturer's standard protocol. DNA concentrations were measured using a NanoDrop Spectrophotometer. Each sample was diluted to 25 ng/µl using sterilised distilled water in a 100 µl final volume working solution.

Global methylation analysis in peripheral blood DNA
Genome-wide global DNA methylation was assessed in DNA derived from the 185 PBMC samples from our UKAGS resource (Additional file 1: Table S1) using a colorimetric enzyme linked immunosorbent-assay (ELISA) following the manufacturers' protocol (Epigentek - Methyl flash DNA quantification (colorimetric)). All reactions were performed in duplicate, and for each new 96-well plate, new standardisation controls were conducted. Mixed plate repeats were also performed to assess for potential batch effects. Absorbance (optical density (OD)) was measured with a micro-plate colorimetric spectrophotometer (Bio-Tek ELX808IU Ultra Microplate Reader), and ODs were converted to genome methylation percentage with the use of positive (5 ng of 50% methylated DNA) and negative control (20 ng of un-methylated DNA) reactions, which were also conducted in duplicate.

Homocysteine analysis in circulating blood plasma

A total of 137 of the blood plasma samples from the global methylation assay were available (67 controls and 70 AAAs) for homocysteine (HCY) analysis (Additional file 1: Table S3). Colorimetric ELISAs were performed following the manufacturers' protocol (Cell Biolabs Inc. – Homocysteine ELISA kit – STA-670). All reactions were conducted in duplicate with new controls and mixed plate repeats for each new 96-well ELISA plate. ODs were measured with a micro-plate colorimetric spectrophotometer and HCY levels were determined using the standard curve method.

Bisulphite conversion of DNA and targeted Illumina NGS

Each of the 44 VSMC DNA samples isolated from aortic biopsies was bisulphite-converted using the MOD50 kit (Sigma–Aldrich) according to the manufacturer's standard protocol. Eluted bisulphite-treated DNA was concentrated at 25 ng/µl in sterilised distilled water.

Genes for this study were chosen based on the results from the recent AAA GWAS meta-analysis conducted by Jones et al., [15], and included the following: low-density lipoprotein receptor-related protein 1 (*LRP1*), ETS-related gene (*ERG*), matrix metallopeptidase 9 (*MMP9*), low-density lipoprotein receptor (*LDLR*), interleukin 6 receptor *(IL6R)*, sortilin 1, (*SORT1*), SET and MYND domain containing 2 (*SMYD2*) and DAB2-interacting protein (*DAB2IP*). In addition, *SERPINB9* was included, which was differentially methylated in PBMCs from patients with AAA in the Ryer et al. study previously described. We also tried to target the differentially methylated CpG Island in the *CNN2* gene from the Ryer et al. study but bisulphite-specific PCR primers could not be designed due to the repetitive nature of the target sequence.

Within all candidate genes (with the exception of the *SERPINB9* CpG Island which is located in the gene body), promoters and transcriptional start sites were identified using the transcriptional regulatory element promoter database (http://rulai.cshl.edu/TRED) and the NCBI gene database (https://www.ncbi.nlm.nih.gov/gene). All candidate gene sequences (GRCh38) were compiled in a single file for use as an in-house reference sequence for bioinformatics analysis.

Bisulphite-specific PCR primers were designed to isolate genomic regions of interest within candidate genes using methprimer [50] (Table 1). Each 20 µl bisulphite-specific PCR reaction consisted of 8 µl sterilised distilled water, 10 µl 2× jumpstart red-taq polymerase ready mix (SIGMA), 1 µl mixed forward and reverse primers (5 µM, SIGMA) and 1 µl bisulphite converted DNA (25 ng/µl). Each bisulphite-specific primer pair was optimised via PCR:

94 °C for 2 min—1 cycle

94 °C for 15 s—40 cycles
52 °C - 58 °C for 35 s—40 cycles
72 °C for 35 s—40 cycles
72 °C for 5 min—1 cycle

Amplicons were checked by fractionation using agarose gel electrophoresis (10 µl per well, 2% agarose gel).

All bisulphite-specific PCR reactions (total $n = 1146$) were cleaned with ExoSAP-IT PCR Clean-up (Affymetrix). On ice, 5 µl of the PCR reactions were mixed with 2 µl ExoSAP-IT solution and incubated at 37 °C for 15 min, then at 80 °C for 15 min. The cleaned products were pooled together (2 µl of each reaction) to create a single pooling of each separate DNA sample (44 individual samples containing 26 amplified genomic regions spanning 9 genes).

The NEBNext Ultra™ DNA Library Prep Kit for Illumina (E7370) and NEBNext Mutiplex Oligos for Illumina dual index kit (E7600) were used for end repair and adaptor ligation and quantification (bar coding for multiplex sequencing) using the standard manufacturer protocols. Clean-up of adaptor-ligated DNA was performed using Agencourt AMPure XP - PCR Purification (A63880) following the manufacturer's standard protocol.

Each of the adaptor ligated, bar-coded DNA samples was ran on an Agilent 2100 Bioanalyzer Instrument (expert High-Sensitivity DNA chip) to assess sample DNA concentration and the presence and distribution of correct fragment sizes [51]. These readings were used to pool the individual samples prior to sequencing at 4 nM. Indexed libraries were pooled, and paired-end multiplexed sequencing (2 × 310 bp) was performed on an Illumina Miseq platform using MiSeq Reagent Kit v3. Sequencing was conducted in the NUCLEUS Genomics Centre, University of Leicester.

Bioinformatics workflow and data analysis

A provisional data quality filter (Phred quality score (Q) = 15) was applied to fastq files with Trimmomatic [52] to remove all low-quality raw sequencing reads, remove contaminating Illumina adaptor sequences, and to clip and discard partially poor-quality sequencing reads four bases at a time using the sliding window function.

BWA-meth (https://github.com/brentp/bwa-meth) was used to align the provisionally filtered sequencing reads to the in house reference sequence. BWA-meth is specifically designed for targeted bisulphite NGS analysis.

PCR duplicates for all BAM files were marked and excluded with the Picard utility MarkDuplicates (http://broadinstitute.github.io/picard/>), and the final processed files were sorted and indexed with Samtools [53].

Finally, for each sample, the methylation values (normalised as methylation %) of each sequenced CpG in each individual were extracted to a bedGraph file at a threshold of Q50 (read alignment filter) and Q20 (base quality filter) with a minimum read depth of 5 using PileOMeth (https://github.com/dpryan79/PileOMeth). Sequencing data that did not meet these criteria were excluded from the study.

The data extracted using PileOMeth was taken forward for statistical analysis using IBM SPSS Statistics 24 and GraphPad prism 7. To correct for non-normal distribution, each methylation value at CpG sites in each individual were ranked with rank-based inverse normal transformations (Blom normal scores in SPSS). Unpaired multiple t tests were conducted on the ranked normal scores at each sequenced CpG site between cases vs controls for each gene separately. For multiple comparison testing, the false discovery rate approach was adopted. Discovery was determined using the two-stage linear step-up procedure of Benjamini, Krieger and Yekutieli, with $Q = 10\%$. CpG sites with significant P values and Q values were then manually checked to ensure only sites with a noticeably visual difference in methylation between cases, and controls were included to investigate further. The effects of smoking and sex were assessed in our original data by excluding all non-smokers and females respectively, where we then reported P values. Logistic regression was performed using SPSS at each significant CpG site to adjust for age differences between cases vs controls. Age-adjusted odds ratios were reported for each CpG unless significance was lost.

RNA extraction, genomic DNA removal and cDNA synthesis

VSMCs from the same samples as used in the NGS assay ($n = 44$) were used for gene expression analysis. RNA isolation was performed following the manufacturer's protocol (QIAGEN RNeasy Mini Kit). Total RNA was eluted in 30 µl RNase-free water and taken forward for contaminating genomic DNA digestion using the manufacturer's standard protocol (DNase I - RNase-Free: New England BioLabs M0303). CDNA was then synthesised from the DNase digested RNA using the ThermoFisher High-Capacity cDNA Reverse Transcription Kit (4368814) according to the manufacturer's protocol. CDNA was eluted in a final volume of 40 µl and stored at − 20 °C until further use.

TaqMan gene expression analysis

For each gene where differential methylation was observed after VSMC bisulphite sequencing (*SMYD2*, *IL6R*, *SERPINB9* and *ERG*), pre-designed TaqMan gene expression assays were purchased from Fisher Scientific UK. *GAPDH* was used as the normalisation reference gene:

Hs01554629_m1 (*ERG* gene FAM-MGB dye)

Hs00220210_m1 (*SYMD2* gene: FAM-MGB dye)
Hs01075666_m1 (*IL6R* gene: FAM-MGB dye)
Hs00394497_m1 (*SERPINB9* gene: FAM-MGB dye)
Hs02758991_g1 (*GAPDH* gene: FAM-MGB dye)

The TaqMan qPCR assays were all performed on an Applied Biosystems Step One Plus qPCR system in duplicate and the qPCR constituents were as follows in a 20 µl total volume: 10 µl 2× TaqMan Gene Expression Master Mix (ThermoFisher), 1 µl 20× TaqMan Gene Expression primers/probes (ThermoFisher), 8 µl sterilised distilled water and 1 µl cDNA (25 ng/µl).

The TaqMan qPCR cycling conditions were as follows:

50 °C for 2 min − 1 cycle
95 °C for 20 s − 1 cycle
95 °C for 1 s − 40 cycles
60 °C for 20 s − 40 cycles

Mean Ct values of the target gene were subtracted from the mean Ct values of the housekeeper gene to acquire inverse delta Ct values (higher value represents higher expression).

Immunohistochemistry to assess protein expression in aortic tissues

Histological staining of the Smyd2 protein was conducted on frozen sections from aortic tissues to corroborate differential *SMYD2* gene expression and promoter methylation in three AAAs and three controls. This work was conducted in the Histology Facility, Core Biotechnology Services, University of Leicester.

Three assays were conducted for each individual aortic tissue sample, meaning a total of 18 slides were prepared for microscopy. Six slides (three controls and three AAAs) were prepared with a Smyd2 polyclonal rabbit antibody at 50: 1 (ThermoFisher - PA5-51339), six control slides (three controls and three AAAs) were prepared with no primary antibody, and six slides (three controls and three AAAs) were prepared with a smooth muscle actin (SMA) polyclonal rabbit antibody at 100: 1 (ThermoFisher - PA5-19465). Visualisation of Smyd2 staining was concentrated on the regions where the highest density of smooth muscle fibres were seen from the SMA. The primary antibodies were detected using the NOVOLINK polymer detection system (Novocastra RE7140-K) using the standard protocol, and visualisation of the slides was performed on the Hamamatsu Nanozoomer 2.0HT Slide Scanner with the use of NDP VIEW2 software (U12388-01).

Summary of statistical analysis

Statistical analyses were conducted using GraphPad Prism 7 (GraphPad Software, Inc., CA, USA) and IBM

SPSS Statistics 24 (IBM, NY, USA). Continuous para-metric data are presented as mean value ± standard error (SE) and non-parametric data are presented as median value and range; categorical data are presented as absolute value or percentage. The chi-square test was used to compare categorical data. Student's t test was used to compare continuous parametric data. Pearson's correlation coefficient was calculated to assess linear dependence between two variables. A P value of < 0.05 was considered statistically significant. Where applicable, a Q value has been reported ($Q < 0.1$), adjusted for false discovery rate (FDR).

Abbreviations
AAA: Abdominal aortic aneurysm; ELISA: Enzyme-linked immunosorbent assay; HCY: Homocysteine; MeQTL: Methylation quantitative trait loci; NGS: Next-generation sequencing; PBMC: Peripheral blood mononuclear cell; UKAGS: UK aneurysm growth study; VSMC: Vascular smooth muscle cell

Acknowledgements
Not applicable. List of UKAGS collaborators: Mr. Rajiv Pathak: rajiv.pathak@dgh.nhs.uk, Mr. Marcus J Brooks: marcus.brooks@uhbristol.nhs.uk, Mr. Paul Hayes: paul.hayes@addenbrookes.nhs.uk, Prof Chris H Imray: Christopher.Imray@uhcw.nhs.uk, Mr. John Quarmby: John.Quarmby@derbyhospitals.nhs.uk, Mr. Sohail A Choksy: sohailchoksy@nhs.net, Mr. Jonothon J Earnshaw: Jonothan.Earnshaw@glos.nhs.uk, Prof Cliff P Shearman: C.P.Shearman@soton.ac.uk, Mr. Eric Grocott: eric.grocott@googlemail.com, Mr. Thomas Rix: thomasrix@nhs.net, Prof Ian C Chetter: Ian.chetter@hey.nhs.uk, Mr. William Tennant: william.tennant@nuh.nhs.uk, Mr. Gabor Libertiny: gabor.libertiny@ngh.nhs.uk, Mr. Tim Sykes: Tim.Sykes@sath.nhs.uk, Mr. Mark Dayer: mark.dayer@tst.nhs.uk, Ms. Lynda Pike: lynda.pike@nhs.net, Mr. Arun Pherwani: Arun.Pherwani@uhns.nhs.uk, Mr. Colin Nice: colin.nice@ghnt.nhs.uk, Mr. Neil Browning: neil.browning@asph.nhs.uk, Prof Charles N McCollum: charles.mccollum@manchester.ac.uk, Mr. Syed W Yusuf: Syed.Yusuf@bsuh.nhs.uk, Mr. Mark Gannon: mark.gannon@heartofengland.nhs.uk, Mr. Jamie Barwell: Jamie.Barwell@nhs.net, Mrs. Sara Baker: Sara.Baker@rbch.nhs.uk, Mr. Srinivasa R Vallabhaneni: fempop@liverpool.ac.uk, Mr. JV Smyth: JV.Smyth@cmft.nhs.uk, Prof Alun H Davies: a.h.davies@imperial.ac.uk, Mr. Tim Lees: Tim.Lees@nuth.nhs.uk, Mr. Louis Fligelstone: Louis.Fligelstone@wales.nhs.uk, Prof Rob Sayers: rs152@leicester.ac.uk, Prof Nilesh J Samani: njs@leicester.ac.uk, Dr. Mike J Sweeting: mjs212@medschl.cam.ac.uk, Prof John Thompson: trj@le.ac.uk.

Funding
Sample collection for this study was funded by the British Heart Foundation (UK Aneurysm Growth Study, CS/14/2/30841) and the National Institute for Health Research (NIHR Leicester Cardiovascular Biomedical Research Unit Biomedical Research Informatics Centre for Cardiovascular Science, IS_BRU_0211_20033). BJT held a PhD fellowship (2014–2017) awarded by the University Of Leicester, Department Of Cardiovascular Sciences, which funded this study.

Authors' contributions
BJT conceived the study, designed the experiments, carried out the experiments, conducted the bioinformatics and statistical analysis, and wrote the manuscript. MJB conceived the study, helped design the experiments, supervised the study and critically reviewed the manuscript. AS conceived the study, helped supervise the study and critically reviewed the manuscript. PJF and NS assisted with next generation sequencing preparations and reactions, and critically reviewed the manuscript. All authors approved the final manuscript for submission.

Consent for publication
Not applicable.

Competing interests
The authors declare that they have no competing interests.

Author details
[1]Department of Cardiovascular Sciences and the NIHR Leicester Biomedical Research Centre, University of Leicester, Leicester LE2 7LX, UK. [2]Department of Genetics and Genome Biology, University of Leicester, Leicester LE1 7RH, UK.

References
1. Sidloff D, Stather P, Dattani N, Bown M, Thompson J, Sayers R, Choke E. Aneurysm global epidemiology study: public health measures can further reduce abdominal aortic aneurysm mortality. Circulation. 2014;129:747–53.
2. Lozano R, Naghavi M, Foreman K, Lim S, Shibuya K, Aboyans V, Abraham J, Adair T, Aggarwal R, Ahn SY, et al. Global and regional mortality from 235 causes of death for 20 age groups in 1990 and 2010: a systematic analysis for the global burden of disease study 2010. Lancet. 2012;380:2095–128.
3. Earnshaw JJ. Triumphs and tribulations in a new national screening programme for abdominal aortic aneurysm. Acta Chir Belg. 2012;112: 108–10.
4. Meijer CA, Stijnen T, Wasser MN, Hamming JF, van Bockel JH, Lindeman JH. Doxycycline for stabilization of abdominal aortic aneurysms: a randomized trial. Ann Intern Med. 2013;159:815–23.
5. Toghill BJ, Saratzis A, Bown MJ. Abdominal aortic aneurysm—an independent disease to atherosclerosis? Cardiovasc Pathol. 2017;27:71–5.
6. Toghill BJ, Saratzis A, Harrison SC, Verissimo AR, Mallon EB, Bown MJ. The potential role of DNA methylation in the pathogenesis of abdominal aortic aneurysm. Atherosclerosis. 2015;241:121–9.
7. Wahlgren CM, Larsson E, Magnusson PK, Hultgren R, Swedenborg J. Genetic and environmental contributions to abdominal aortic aneurysm development in a twin population. J Vasc Surg. 2010;51:3–7. discussion 7
8. Joergensen TM, Christensen K, Lindholt JS, Larsen LA, Green A, Houlind K. High heritability of liability to abdominal aortic aneurysms: a population based twin study. Eur J Vasc Endovasc Surg. 2016;52:41–6. Editor's choice
9. Bown MJ, Jones GT, Harrison SC, Wright BJ, Bumpstead S, Baas AF, Gretarsdottir S, Badger SA, Bradley DT, Burnand K, et al. Abdominal aortic aneurysm is associated with a variant in low-density lipoprotein receptor-related protein 1. Am J Hum Genet. 2011;89:619–27.
10. Bradley DT, Hughes AE, Badger SA, Jones GT, Harrison SC, Wright BJ, Bumpstead S, Baas AF, Gretarsdottir S, Burnand K, et al. A variant in LDLR is associated with abdominal aortic aneurysm. Circ Cardiovasc Genet. 2013;6: 498–504.
11. Gretarsdottir S, Baas AF, Thorleifsson G, Holm H, den Heijer M, de Vries JP, Kranendonk SE, Zeebregts CJ, van Sterkenburg SM, Geelkerken RH, et al. Genome-wide association study identifies a sequence variant within the DAB2IP gene conferring susceptibility to abdominal aortic aneurysm. Nat Genet. 2010;42:692–7.
12. Harrison SC, Smith AJ, Jones GT, Swerdlow DI, Rampuri R, Bown MJ, Folkersen L, Baas AF, de Borst GJ, Blankensteijn JD, et al. Interleukin-6 receptor pathways in abdominal aortic aneurysm. Eur Heart J. 2013;34:3707–16.
13. Helgadottir A, Thorleifsson G, Magnusson KP, Gretarsdottir S, Steinthorsdottir V, Manolescu A, Jones GT, Rinkel GJ, Blankensteijn JD, Ronkainen A, et al. The same sequence variant on 9p21 associates with myocardial infarction, abdominal aortic aneurysm and intracranial aneurysm. Nat Genet. 2008;40:217–24.
14. Jones GT, Bown MJ, Gretarsdottir S, Romaine SP, Helgadottir A, Yu G, Tromp G, Norman PE, Jin C, Baas AF, et al. A sequence variant associated with sortilin-1 (SORT1) on 1p13.3 is independently associated with abdominal aortic aneurysm. Hum Mol Genet. 2013;22:2941–7.
15. Jones GT, Tromp G, Kuivaniemi H, Gretarsdottir S, Baas AF, Giusti B, Strauss E, van't Hof FN, Webb T, Erdman R, et al. Meta-analysis of genome-wide association studies for abdominal aortic aneurysm identifies four new disease-specific risk loci. Circ Res. 2016;120: 341–53.
16. Toghill BJ, Saratzis A, Liyanage L. S, Sidloff D, Bown MJ. Genetics of Aortic Aneurysmal Disease. 2016. eLS. 1–10. https://doi.org/10.1002/9780470015902.a0026851.
17. Verma M, Banerjee HN. Epigenetic inhibitors. Methods Mol Biol. 2015;1238: 469–85.

18. Egger G, Liang G, Aparicio A, Jones PA. Epigenetics in human disease and prospects for epigenetic therapy. Nature. 2004;429:457–63.

19. Zaina S, Heyn H, Carmona FJ, Varol N, Sayols S, Condom E, Ramirez-Ruz J, Gomez A, Goncalves I, Moran S, Esteller M. A DNA methylation map of human atherosclerosis. Circ Cardiovasc Genet. 2014;7(5):692–700.

20. Wongtrakoongate P. Epigenetic therapy of cancer stem and progenitor cells by targeting DNA methylation machineries. World J Stem Cells. 2015;7:137–48.

21. Bestor TH. The DNA methyltransferases of mammals. Hum Mol Genet. 2000; 9:2395–402.

22. Goldberg AD, Allis CD, Bernstein E. Epigenetics: a landscape takes shape. Cell. 2007;128:635–8.

23. Yang X, Han H, De Carvalho DD, Lay FD, Jones PA, Liang G. Gene body methylation can alter gene expression and is a therapeutic target in cancer. Cancer Cell. 2014;26:577–90.

24. Aslibekyan S, Claas SA, Arnett DK. Clinical applications of epigenetics in cardiovascular disease: the long road ahead. Transl Res. 2014;165(1):143–53.

25. Cornuz J, Sidoti Pinto C, Tevaearai H, Egger M. Risk factors for asymptomatic abdominal aortic aneurysm: systematic review and meta-analysis of population-based screening studies. Eur J Pub Health. 2004;14:343–9.

26. Ryer EJ, Ronning KE, Erdman R, Schworer CM, Elmore JR, Peeler TC, Nevius CD, Lillvis JH, Garvin RP, Franklin DP, et al. The potential role of DNA methylation in abdominal aortic aneurysms. Int J Mol Sci. 2015;16:11259–75.

27. Drong AW, Nicholson G, Hedman AK, Meduri E, Grundberg E, Small KS, Shin SY, Bell JT, Karpe F, Soranzo N, et al. The presence of methylation quantitative trait loci indicates a direct genetic influence on the level of DNA methylation in adipose tissue. PLoS One. 2013;8:e55923.

28. Richmond RC, Sharp GC, Ward ME, Fraser A, Lyttleton O, McArdle WL, Ring SM, Gaunt TR, Lawlor DA, Davey Smith G, Relton CL. DNA methylation and BMI: investigating identified methylation sites at HIF3A in a causal framework. Diabetes. 2016;65:1231–44.

29. Zhi D, Aslibekyan S, Irvin MR, Claas SA, Borecki IB, Ordovas JM, Absher DM, Arnett DK. SNPs located at CpG sites modulate genome-epigenome interaction. Epigenetics. 2013;8:802–6.

30. Bonder MJ, Luijk R, Zhernakova DV, Moed M, Deelen P. Disease variants alter transcription factor levels and methylation of their binding sites. Nature Genetics. 2017;49:131–138.

31. Valdespino V, Valdespino PM. Potential of epigenetic therapies in the management of solid tumors. Cancer Manag Res. 2015;7:241–51.

32. Napoli C, Grimaldi V, De Pascale MR, Sommese L, Infante T, Soricelli A. Novel epigenetic-based therapies useful in cardiovascular medicine. World J Cardiol. 2016;8:211–9.

33. Fuggle NR, Howe FA, Allen RL, Sofat N. New insights into the impact of neuro-inflammation in rheumatoid arthritis. Front Neurosci. 2014;8:357.

34. Stenvinkel P, Karimi M, Johansson S, Axelsson J, Suliman M, Lindholm B, Heimburger O, Barany P, Alvestrand A, Nordfors L, et al. Impact of inflammation on epigenetic DNA methylation—a novel risk factor for cardiovascular disease? J Intern Med. 2007;261:488–99.

35. Zhang S, Zhong B, Chen M, Yang L, Yang G, Li Y, Wang H, Wang G, Li W, Cui J, et al. Epigenetic reprogramming reverses the malignant epigenotype of the MMP/TIMP axis genes in tumor cells. Int J Cancer. 2014;134:1583–94.

36. Liyanage VR, Jarmasz JS, Murugeshan N, Del Bigio MR, Rastegar M, Davie JR. DNA modifications: function and applications in normal and disease states. Biology (Basel). 2014;3:670–723.

37. Kim M, Long TI, Arakawa K, Wang R, Yu MC, Laird PW. DNA methylation as a biomarker for cardiovascular disease risk. PLoS One. 2010;5:e9692.

38. Pushpakumar S, Kundu S, Narayanan N, Sen U. DNA hypermethylation in hyperhomocysteinemia contributes to abnormal extracellular matrix metabolism in the kidney. FASEB J. 2015;29:4713–25.

39. Antoniades C, Antonopoulos AS, Tousoulis D, Marinou K, Stefanadis C. Homocysteine and coronary atherosclerosis: from folate fortification to the recent clinical trials. Eur Heart J. 2009;30:6–15.

40. Krishna SM, Dear A, Craig JM, Norman PE, Golledge J. The potential role of homocysteine mediated DNA methylation and associated epigenetic changes in abdominal aortic aneurysm formation. Atherosclerosis. 2013;228: 295–305.

41. Lindqvist M, Hellstrom A, Henriksson AE. Abdominal aortic aneurysm and the association with serum levels of homocysteine, vitamins B6, B12 and folate. Am J Cardiovasc Dis. 2012;2:318–22.

42. Patterson K, Molloy L, Qu W, Clark S. DNA methylation: bisulphite modification and analysis. J Vis Exp. 2011;56:e3170.

43. Thompson RW, Liao S, Curci JA. Vascular smooth muscle cell apoptosis in abdominal aortic aneurysms. Coron Artery Dis. 1997;8:623–31.

44. Hendel A, Cooper D, Abraham T, Zhao H, Allard MF, Granville DJ. Proteinase inhibitor 9 is reduced in human atherosclerotic lesion development. Cardiovasc Pathol. 2012;21:28–38.

45. Huang J, Perez-Burgos L, Placek BJ, Sengupta R, Richter M, Dorsey JA, Kubicek S, Opravil S, Jenuwein T, Berger SL. Repression of p53 activity by Smyd2-mediated methylation. Nature. 2006;444:629–32.

46. Sajjad A, Novoyatleva T, Vergarajauregui S, Troidl C, Schermuly RT, Tucker HO, Engel FB. Lysine methyltransferase Smyd2 suppresses p53-dependent cardiomyocyte apoptosis. Biochim Biophys Acta. 1843;2014:2556–62.

47. Xu G, Liu G, Xiong S, Liu H, Chen X, Zheng B. The histone methyltransferase Smyd2 is a negative regulator of macrophage activation by suppressing interleukin 6 (IL-6) and tumor necrosis factor alpha (TNF-alpha) production. J Biol Chem. 2015;290:5414–23.

48. Medvedeva YA, Khamis AM, Kulakovskiy IV, Ba-Alawi W, Bhuyan MSI, Kawaji H, Lassmann T, Harbers M, Forrest AR, Bajic VB. Effects of cytosine methylation on transcription factor binding sites. BMC Genomics. 2014;15:119.

49. Rishi V, Bhattacharya P, Chatterjee R, Rozenberg J, Zhao J, Glass K, Fitzgerald P, Vinson C. CpG methylation of half-CRE sequences creates C/EBPalpha binding sites that activate some tissue-specific genes. Proc Natl Acad Sci U S A. 2010;107:20311–6.

50. Li LC, Dahiya R. MethPrimer: designing primers for methylation PCRs. Bioinformatics. 2002;18:1427–31.

51. Mueller O, Hahnenberger K, Dittmann M, Yee H, Dubrow R, Nagle R, Ilsley D. A microfluidic system for high-speed reproducible DNA sizing and quantitation. Electrophoresis. 2000;21:128–34.

52. Bolger AM, Lohse M, Usadel B. Trimmomatic: a flexible trimmer for Illumina sequence data. Bioinformatics. 2014;30:2114–20.

53. Li H, Handsaker B, Wysoker A, Fennell T, Ruan J, Homer N, Marth G, Abecasis G, Durbin R. The sequence alignment/map format and SAMtools. Bioinformatics. 2009;25:2078–9.

3

Cord blood hematopoietic cells from preterm infants display altered DNA methylation patterns

Olivia M. de Goede[1,2], Pascal M. Lavoie[1,3] and Wendy P. Robinson[1,2]*

Abstract

Background: Premature infants are highly vulnerable to infection. This is partly attributable to the preterm immune system, which differs from that of the term neonate in cell composition and function. Multiple studies have found differential DNA methylation (DNAm) between preterm and term infants' cord blood; however, interpretation of these studies is limited by the confounding factor of blood cell composition. This study evaluates the epigenetic impact of preterm birth in isolated hematopoietic cell populations, reducing the concern of cell composition differences.

Methods: Genome-wide DNAm was measured using the Illumina 450K array in T cells, monocytes, granulocytes, and nucleated red blood cells (nRBCs) isolated from cord blood of 5 term and 5 preterm (<31 weeks gestational age) newborns. DNAm of hematopoietic cells was compared globally across the 450K array and through site-specific linear modeling.

Results: Nucleated red blood cells (nRBCs) showed the most extensive changes in DNAm, with 9258 differentially methylated (DM) sites (FDR < 5%, |Δβ| > 0.10) discovered between preterm and term infants compared to the <1000 prematurity-DM sites identified in white blood cell populations. The direction of DNAm change with gestational age
at these prematurity-DM sites followed known patterns of hematopoietic differentiation, suggesting that term hematopoietic cell populations are more epigenetically mature than their preterm counterparts. Consistent shifts in DNAm between preterm and term cells were observed at 25 CpG sites, with many of these sites located in genes involved in growth and proliferation, hematopoietic lineage commitment, and the cytoskeleton. DNAm in preterm and term hematopoietic cells conformed to previously identified DNAm signatures of fetal liver and bone marrow, respectively.

Conclusions: This study presents the first genome-wide mapping of epigenetic differences in hematopoietic cells across the late gestational period. DNAm differences in hematopoietic cells between term and <31 weeks were consistent with the hematopoietic origin of these cells during ontogeny, reflecting an important role of DNAm in their regulation. Due to the limited sample size and the high coincidence of prematurity and multiple births, the relationship between cause of preterm birth and DNAm could not be evaluated. These findings highlight gene regulatory mechanisms at both cell-specific and systemic levels that may be involved in fetal immune system maturation.

Keywords: DNA methylation, Cord blood, Preterm birth, Illumina 450K array, Epigenetics, Nucleated red blood cells, Epigenetic clock, Gestational age

* Correspondence: wrobinson@cfri.ca
[1]BC Children's Hospital Research Institute, Room 2082, 950W 28th Avenue, Vancouver, BC V5Z 4H4, Canada
[2]Department of Medical Genetics, University of British Columbia, Vancouver, BC V6T 1Z3, Canada
Full list of author information is available at the end of the article

Background

Preterm birth (PTB), defined as birth prior to 37 weeks gestational age (GA), occurs in approximately 11% of live births and accounts for over half of infant mortality cases worldwide [1]. If a premature infant survives the immediate postnatal period, they face increased risk of developing major short- and long-term health problems including cerebral palsy, chronic lung disease, visual and hearing impairments, and adult metabolic diseases [2–6]. This elevated risk is attributable to organ immaturity, as well as an increased risk of medical complications linked to oxidative stress and inflammation during the neonatal period [7–9].

The immune system is not spared from the effects of PTB. The composition and function of hematopoietic cell populations change dramatically throughout gestation as the embryonic and fetal immune system mature. Premature interruption of the immunologically protected intrauterine environment results in an extremely fragile infant whose immune system is unprepared for the microbe-ridden external environment. A variety of systemic and cell-specific alterations in immune function have been identified in preterm infants that greatly increase their vulnerability to infection [10–12].

The importance of DNA methylation (DNAm) in the process of hematopoietic cell lineage commitment is well established [13, 14], and multiple studies have found differential methylation between cord blood of preterm and term infants [15–17]. However, these studies have used whole blood, which is a mixed-cell sample in which overall DNAm levels are influenced by cell composition [18, 19]. As a result, these studies cannot distinguish prematurity-associated DNAm patterns due to differences in cell composition from DNAm patterns reflecting developmental changes in immune function.

Using the Illumina Infinium Human Methylation 450 Bead Chip (450K array), we provide genome-wide DNAm profiles of T cells, monocytes, granulocytes, and nucleated red blood cells (nRBCs) collected from cord blood of infants born at term or highly preterm (<31 weeks GA). These DNAm profiles were compared between cell types and across GA to evaluate an epigenetic basis for altered neonatal immune function with prematurity. This study provides important insights into the role of DNAm in early hematopoietic system maturation in humans.

Methods
Study participants and sample collection
Ethics approval for this study was obtained from the University of British Columbia Children's and Women's (C&W) Research Ethics Board (certificate numbers H07-02681 and H04-70488). Written informed parental consent to participate was obtained. Individual patient data is not reported. Cord blood was collected from neonates delivered by caesarean section at the C&W Health Centre of BC (Vancouver, Canada). A total of 10 infants were involved in the study: 5 preterm (GA range 26–30 weeks) and 5 term (GA 38 weeks) (Table 1). None of the subjects had histological evidence of chorioamnionitis in the placenta. All term births were singleton, and the caesarean section was performed in the absence of labor. The preterm births had more variable clinical characteristics, including one case of preeclampsia, four births from multiple pregnancies, and a case of labor preceding the caesarean section (Table 1). In the cases of multiple pregnancies, only one subject was used and other siblings were excluded. Since the preterm births were all <31 weeks, immune function is expected to be significantly altered compared to term births regardless of the cause of prematurity.

T cells, monocytes, and nRBCs were collected from cord blood by fluorescence-activated cell sorting (FACS). These sorting methods were designed to prevent erythrocyte-white blood cell (WBC) cross-contamination, a common occurrence in cord blood [20] and are described in detail

Table 1 Subject characteristics and cell types collected from each subject

	Sex	GA (weeks)	Multiple birth	Presence of labor	Indication for PTB	Cells collected
term_1	M	38	No	No	n/a	all
term_2	M	38	No	No	n/a	all
term_3	F	38	No	No	n/a	all
term_4	M	38	No	No	n/a	all
term_5	M	38	No	No	n/a	all
preterm_A	M	26	No	No	Preeclampsia	T cells, nRBCs
preterm_B	F	29	Yes	No	Placental insufficiency	T cells, gran., mono.
preterm_C	M	30	Yes	No	Placental insufficiency	all
preterm_D	F	30	Yes	No	Placental insufficiency	T cells, mono., nRBCs
preterm_E	M	30	Yes	Yes	Twin-to-twin transfusion syndrome	all

For the column "Cells collected": *all* T cells, granulocytes, monocytes, and nRBCs; *gran.* granulocytes; *mono.* monocytes; *n/a* not applicable

in the Additional file 1. Granulocytes were collected by density gradient centrifugation and hypotonic red blood cell lysis. All cell populations were collected from all term subjects; however, due to small sample volumes and variability in blood cell counts, some cell populations could not be collected from some preterm subjects (Table 1).

DNA extraction and DNA methylation data collection

DNA was extracted from all samples using standard protocols and purified with the DNeasy Blood & Tissue Kit (Qiagen, MD, USA). DNA was bisulphite-converted using the EZ DNA Methylation Kit (Zymo Research, CA, USA) before amplification and hybridization to the 450K array following manufacturer's protocols (Illumina, CA, USA). Samples were randomly distributed across four 450K array chips, as shown in Additional file 1: Figure S1. 450K array chips were scanned with a HiScan reader (Illumina).

Raw intensity data for all hematopoietic cells were background corrected in GenomeStudio (Illumina). Quality control was performed using the 835 control probes included in the array. The intensity data were then exported from GenomeStudio and were converted into M values using the lumi package [21] in R software [22]. Sample identity and quality were evaluated as described in Additional file 1. The 450K array targets 485577 DNAm sites, but probe filtering was performed as described in Additional file 1 to produce a final dataset of 429765 sites. Red-green color bias was corrected for using the lumi package [21], and the data were normalized by subset within-array quantile normalization [23].

DNA methylation data analysis

Unsupervised Euclidean clustering of the samples based on DNAm β values and principal component analysis based on DNAm M values were performed as exploratory global analysis steps. DNAm was then evaluated at subsets of the 450K array based on surrounding CpG density. These subsets are detailed in Additional file 1. Median DNAm (β values) of these CpG site groups were compared between all cell types using ANOVA followed by Tukey's honest significant difference test, using a multiple comparison-adjusted p value threshold of 0.005. DNAm-based estimates of GA for the samples were calculated using a method developed by Knight et al. [24] in cord blood.

Differential methylation based on cell type and birth group (preterm or term) was assessed by linear modeling using the R package limma [25]. The same model was used to assess both PTB-associated and cell type-specific DNAm: the interaction of cell type

and birth group was the variable of interest, and sex was included in the model as a covariate. Since each cell type was collected from the same set of individuals and the sample size was small, DNAm may have been influenced by inter-individual differences. To adjust for this, the model included a within-individual consensus correlation estimated using the *duplicateCorrelation()* function in limma [25]. Resulting p values were adjusted for multiple comparisons by the Benjamini and Hochberg [26] false detection rate (FDR) method.

For the comparison between preterm and term samples, statistically significant sites ("prematurity-associated DM sites") were limited to those with an FDR < 5% and a $|\Delta\beta| > 0.10$. Prematurity-associated DM sites were identified separately for each cell population. For cell type-specific DNAm, statistically significant sites ("cell type-DM sites") were limited to those with an FDR < 5% and a $|\Delta\beta| > 0.20$. Cell-type DM sites were identified separately within the two birth groups. ErmineJ was used to evaluate enrichment of gene ontology (GO) terms in genes associated with the cell-type and prematurity-associated DM sites [27].

Several other studies have performed similar evaluations of DNAm differences between preterm and term births, using whole cord blood instead of isolated cell populations [15–17]. The PTB-associated CpG sites discovered in those studies (29 CpG sites from Parets et al. [17]; 1347 CpG sites from Fernando et al. [16]; and 1555 CpG sites from Cruickshank et al. [15]) were overlapped with the prematurity-associated and the cell-type DM CpG sites identified in this study. Several subject characteristics varied between these studies, including cause of prematurity, ethnicity, and maternal age range. Overlapping sites are thus likely to be those present in multiple cell populations and also be unrelated to ethnicity or to PTB etiology. The specifics of these overlaps are described in Additional file 1.

To assess how prematurity-associated DNAm might reflect hematopoietic origin, DNAm patterns were compared between the birth groups and the cell types at a set of previously identified CpG sites that showed differential methylation between erythroblasts derived from fetal liver and erythroblasts derived from adult bone marrow ("source-DM sites") [28]. These source-DM sites were divided into two groups: the top 100 CpG sites hypomethylated in adult BM erythroblasts ("BM-hypomethylated sites") and the top 100 CpG sites hypomethylated in FL erythroblasts ("FL-hypomethylated sites"), with ranking and selection based on Lessard et al.'s β values. Median DNAm (β value) at these source-DM sites was compared between cell types and birth groups by ANOVA followed by Tukey's honest significant difference test, with a multiple comparison-adjusted *p* value threshold of 0.005.

Results

Global DNA methylation profiles of preterm and term hematopoietic cells

Cell type was the dominant influence when DNAm profiles of term and preterm cell populations were compared by array-wide Euclidean clustering (Fig. 1a). Prematurity also had an observable impact on epigenetic relationships between the samples, with some samples clustering by birth group (preterm or term) within each cell type. However, these GA subgroups were not perfect, with some preterm samples clustering more closely with their term counterparts. Evaluating genome-wide DNAm by β value density distributions suggests that the effect of prematurity is largest in nRBCs. Term nRBCs were hypomethylated

relative to preterm nRBCs, whereas all of the WBC populations showed similar distributions between term and preterm samples (Fig. 1b).

None of the cord blood hematopoietic cell populations differed in median array-wide DNAm between preterm and term infants; although term nRBCs were notably hypomethylated compared to preterm nRBCs, this difference was not significant (Fig. 1c). To identify genomic regions where the association between DNAm and prematurity is strongest, subsets of the 450K array based on CpG density were evaluated. In WBCs, no significant differences were observed between preterm and term samples at any of the CpG density subgroups. In nRBCs, term cells displayed hypomethylation relative to their preterm counterparts at the intermediate and low CpG density regions, however, these differences did not pass the multiple-test corrected significance threshold ($p > 0.005$).

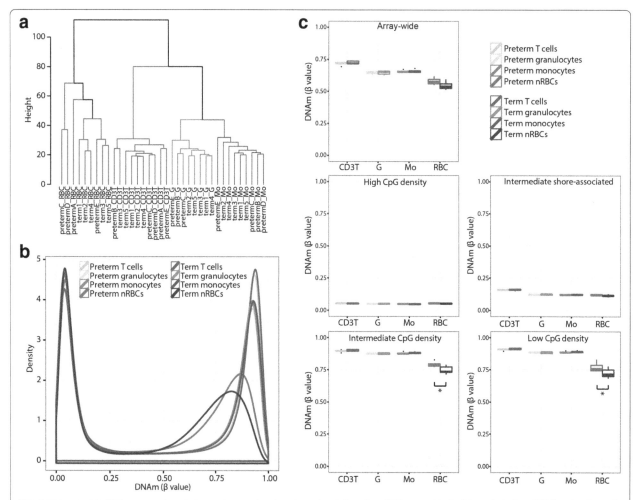

Fig. 1 Genome-wide DNAm comparisons between term and preterm hematopoietic cells. **a** 450K array-wide Euclidean clustering of DNAm data for each preterm and term subject. **b** DNAm β value density distributions; each *line* represents the mean of that birth group/cell type combination. **c** Comparison of median DNAm between GA groups and cell types array-wide, and in CpG sites grouped by CpG density. *$p < 0.05$, **$p < 0.005$

Prematurity-associated differentially methylated CpG sites
Linear modeling was performed within each cell type to identify cell-specific prematurity-associated DM sites (FDR < 5%, |Δβ| > 0.10). nRBCs showed the greatest difference between preterm and term samples, with 9258 prematurity-associated DM sites; more than tenfold greater than observed in granulocytes, monocytes, and T cells (Table 2; Additional file 2). The majority of prematurity-associated DM sites were specific to a single cell type, making it unlikely that these changes were driven by chance genetic differences between the samples (Additional file 1: Figure S2). Twenty-five of the prematurity-associated DM sites were identified across all cell types, 17 of which increased in DNAm and 8 of which decreased in DNAm with GA (Additional file 2). All cell populations had the highest number of their prematurity-DM sites in the gene body and intergenic regions (Fig. 2a), which is consistent with the representation of these gene regions on the 450K array (33 and 24% of CpG sites assayed, respectively). In nRBCs, the TSS-upstream and 5′ UTR gene regions were also highly represented in prematurity-DM sites. This likely reflects the global nature of erythrocyte demethylation with maturity [29, 30].

The direction of DNAm change at prematurity-associated DM sites in each cell type paralleled their patterns of DNAm upon terminal differentiation (Table 2). For T cells, the high percentage of prematurity-associated DM sites with increased DNAm in term samples (72%) is consistent with increased DNAm with terminal differentiation of these cells [13, 14, 31]. For granulocytes and monocytes, the majority of prematurity-associated DM sites were hypomethylated in term samples (69 and 61%, respectively) in keeping with the documented loss of DNAm in myeloid cells [13, 14, 31]. For nRBCs, the vast majority of prematurity-associated DM sites (94%) showed reduced DNAm in term samples. Given that terminal erythroid differentiation is associated with global demethylation [29, 30], this change likely reflects an increasing proportion of mature erythroblasts in the nRBC population at term. Overall, these data suggest an epigenetic basis for the increased maturity of cord blood hematopoietic cell populations during fetal development.

GO pathway analysis of the prematurity-associated DM sites revealed enrichment of distinct sets of genes for each cell type (Additional file 3). The two significant GO terms in granulocytes (FDR < 10%) related to negative regulation of the ERK1 and ERK2 cascades, and Ras guanyl-nucleotide

exchange factor activity. In T cells, the only significant GO term (FDR < 10%) was embryonic placenta development. Monocyte prematurity-associated DM sites were associated with eight significant GO terms (FDR < 10%), all of which were related to epidermal and hair growth and development. Evaluating the nRBC prematurity-associated DM sites revealed 152 significantly enriched GO terms (FDR < 10%); recurring themes in this list included Ras- and Rho-related activity, the cytoskeleton, and terms related to renal, muscle, and neuronal processes.

Cell-specific DNA methylation patterns differ between preterm and term infants
After establishing the cell-specific DNAm differences between preterm and term births, we next investigated whether prematurity affects cell-type differences in DNAm (Additional file 4). Linear modeling revealed that nRBCs were the most distinct cell type in term samples, consistent with our previous findings [20, 32], but interestingly, T cells were the most distinct cell type of the preterm samples (Table 3). The relatively low number of monocyte- and granulocyte-DM sites in both GA groups was likely because these cell types are both of the myeloid lineage and thus epigenetically similar, in contrast to T cells and nRBCs, which are the only representatives of their respective hematopoietic lineages. In the WBC populations, the number of cell type-DM sites did not change drastically between preterm and term samples (Table 3). In contrast, the number of nRBC-DM sites nearly tripled between the preterm and term samples. This large change coincides with the increased hypomethylation in term nRBCs relative to their preterm counterparts (Fig. 1b), which made term nRBCs more distinct from term WBCs.

When the cell-type DM sites were compared between preterm and term samples based on gene region and the cell type of interest's relative DNAm—that is, whether the unique cell population has DNAm that is higher, lower, or in between the DNAm levels of the other cell types—there was little difference in genomic representation or direction of DNAm change, particularly within WBCs (Fig. 2b). In nRBCs, the increase in the number of CpG sites hypomethylated relative to WBCs occurred in all gene regions, but most dramatically in the gene body and intergenic regions. Overall, this indicates that prematurity does not have a major impact on WBCs' epigenetic relationships to each other. In contrast,

Table 2 Number of prematurity-associated DM sites for each cell type (FDR < 5%, |Δβ| > 0.10)

	T cells	Granulocytes	Monocytes	nRBCs
Total	273	987	692	9258
DNAm decreases with GA	76 (28%)	679 (69%)	425 (61%)	8731 (94%)
DNAm increases with GA	197 (72%)	308 (31%)	267 (39%)	527 (6%)

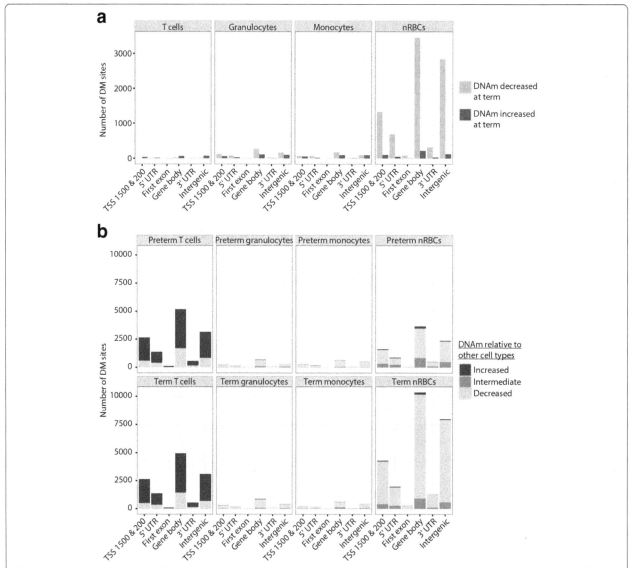

Fig. 2 Prematurity-associated and cell type-DM sites grouped by gene region and changes in DNAm. **a** Prematurity-associated DM sites (FDR < 5%, |Δβ| > 0.10). **b** Cell type-DM sites (FDR < 5%, |Δβ| > 0.20). *TSS 1500 & 200* 1500 or 200 bp upstream from transcriptional start site, *UTR* untranslated region

nRBCs become more epigenetically distinct from WBCs as gestation progresses, adopting their uniquely hypomethylated profile [20, 32, 33]. The representation of these changes across all gene regions is likely reflective of nRBC demethylation being a global and largely passive process [29, 30].

Comparison to other epigenetic studies of preterm birth
Previous studies have also identified distinctive DNAm patterns between preterm and term infants [15–17]. However, these studies were performed on either whole cord blood samples or the buffy coat and were not able to distinguish systematic prematurity-associated changes

Table 3 Number of cell type-DM sites (FDR < 5%, | Δβ| > 0.20)

	T cells	Granulocytes	Monocytes	nRBCs
Preterm	12974	1410	1665	9056
Term	12662	1900	1508	26176
Common	10991 (85%, 87%)	1201 (85%, 63%)	1221 (73%, 81%)	7645 (84%, 29%)

Percentages of cell type-DM sites in common between the two GA groups are reported relative to the number of preterm DM sites first, then number of term DM sites

from those caused by shifts in cell composition across gestation. We compared our prematurity-associated and cell type-DM sites with the CpG sites found to be significantly associated with PTB (FDR < 5%) by Parets et al. [17] (29 CpG sites), Fernando et al. [16] (1347 CpG sites), and Cruickshank et al. [15] (1555 CpG sites).

For all three of the comparison studies, approximately 30% of their DM sites were also discovered in at least one of our sets of prematurity-associated DM sites (9/29, 369/1347, and 427/1555 replicated DM sites) (Fig. 3a). When our identified cell type-DM sites were overlapped with the three comparison studies, the overlap was much lower than that with the prematurity-associated DM sites (Fig. 3b). The greatest overlap by number of sites occurred at T cell-specific and nRBC-specific DM sites (Fig. 3b). However, a notable proportion of Parets et al.'s [17] 29 differentially methylated CpG sites in PTB were associated with monocyte-specific DNAm in our data, a trend not seen in comparison with the other two

studies. This could be due to a chance difference in average monocyte proportions between their preterm and term subjects or it could have come from Parets et al.'s [17] use of the buffy coat rather than the whole blood. Additionally, a subset of 196 of Fernando et al.'s [16] prematurity-associated CpG sites that were associated only with the state of being premature, and not with GA, showed almost no overlap with our cell type-DM sites (Fig. 3b).

A CpG site in *MYL4*, encoding myosin light chain 4, was the only DM site identified by Fernando et al. [16], Cruickshank et al. [15], and Schroeder et al. [34]; however, it was not observed in any of our cell populations. We also did not find any prematurity-associated DM sites in *ESR1*, encoding the estrogen receptor, in any of our cell populations, despite this gene being identified by both Fernando et al. [16] and Schroeder et al. [34]. However, we did replicate some of Fernando et al.'s [16] top findings of differential methylation in *NCOR2, DNAJC17, PYCR2, ATP6V0A1, RARA, FBLN7, IGF2BP1,*

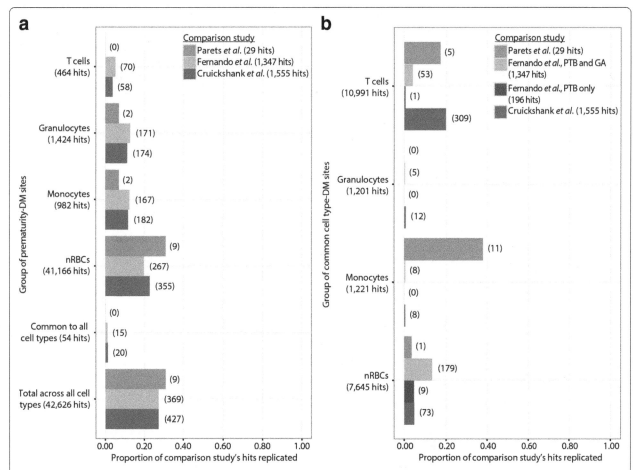

Fig. 3 Overlap of prematurity-associated and cell type-DM sites with prematurity-associated CpG sites identified in previous studies. Proportion of prematurity-associated CpG sites found by Cruickshank et al., Fernando et al., and Parets et al. [15–17] also represented in (**a**) the prematurity-associated DM sites (FDR < 5%); and (**b**) the cell type-DM sites (FDR < 5%, |Δβ| > 0.20) identified in this study. The *numbers* beside *bars* are the number of overlapping CpG sites between the two lists

and *ATP2B2*, as well as differential methylation observed by Cruickshank et al. [15] in *NFIX, OXT, DNMT3A, RUNX1*, and *AIRE*. We also found prematurity-associated DNAm in *ADORA2A* and *GABBR1*, which was identified by both of these studies. Of the 54 prematurity-associated DM sites we observed across all cell types, seven were also identified by both Fernando et al. [16] and Cruickshank et al. [15]. Of these shared CpG sites, two are located in the gene body of *WWTR1*, and two are located in the 5′UTR of *CLIP2*; the other three are intergenic.

To further explore how the cell-specific DNAm changes we observed compared to trends in whole cord blood, we applied the recently published epigenetic clock for GA to our data [24]. This GA-epigenetic clock was designed using cord blood samples and, unlike the epigenetic clock designed for adult samples [35], was only validated in cord blood. This is unsurprising, since cord blood is the most frequently studied tissue in studies of the fetus or neonate. However, we were curious to see how this whole blood-based algorithm would perform on its constituent cell types. In all preterm cell populations, estimated DNAm GA was an overestimate of actual GA (Additional file 5). In term samples, the GA estimates were more accurate: when estimated GA was averaged across all cell types within an individual, none

of the term individuals had estimates over 1 week different than their actual GA (Additional file 5). There were also some intriguing cell type-specific trends in GA estimates: for example, T cells had the highest GA estimates in preterm individuals, but one of the lowest in term individuals (Fig. 4). Additionally, monocytes in term individuals were consistently estimated as the "oldest" cell population, whereas nRBCs had low GA estimates regardless of birth group.

DNA methylation associated with prematurity may reflect hematopoietic origin

While changes in DNAm may in part reflect an aging "clock" [24], our cell-specific GA-epigenetic clock analyses above suggest that other factors can modify this trend. One such factor may be the predominant hematopoietic organ, which shifts from the liver to the bone marrow early in the third trimester of gestation [36]. We hypothesized that the preterm samples used in this study, which range from 26 to 30 weeks GA, have a greater proportion of liver-derived cells than the term samples. Hematopoietic source-related methylation differences with PTB were evaluated using CpG sites previously associated with liver- or bone marrow-specific DNAm in ex vivo-derived nRBCs [28]. From these 5937 "source-DM sites", two groups of CpG sites were

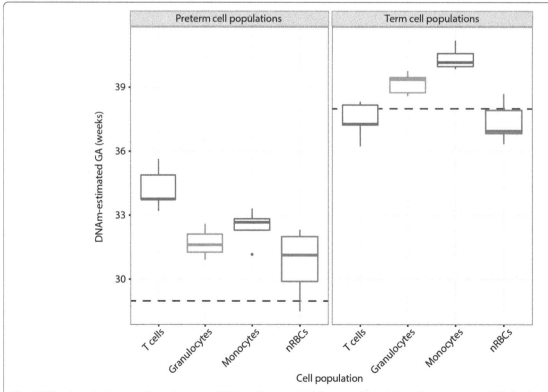

Fig. 4 DNAm-based estimates of gestational age (GA) by cell type and birth group. *Dashed lines* reflect mean actual GA for the birth group. Estimates calculated using methods published by Knight et al. [24]

assessed: the top 100 sites that were hypomethylated in adult bone marrow-derived nRBCs relative to fetal liver-derived nRBCs ("BM-hypomethylated sites") and the top 100 sites that displayed the opposite pattern ("FL-hypomethylated sites"). Only one of these 200 CpG sites overlapped with the 148 CpG sites used in the GA-epigenetic clock [24]. In our samples, all cell types displayed the same trend, with preterm samples less methylated at FL-hypomethylated sites and term samples less methylated at BM-hypomethylated sites (Fig. 5). This difference at BM-hypomethylated sites was significant in all cell types except T cells.

Overlapping our prematurity-associated DM sites with Lessard et al.'s hematopoietic source-DM sites provided further support for a relationship between DNAm and hematopoietic origin of cord blood cell types (Table 4). CpG sites that increased in DNAm with increasing GA overlapped almost exclusively with fetal liver-hypomethylated CpG sites, likely reflecting the reduced contribution of the liver to hematopoiesis as gestation progresses. In contrast, many of the CpG sites that decreased in DNAm with increasing GA were also associated with hypomethylation in bone marrow-derived hematopoietic cells, corresponding with this organ becoming the primary source of hematopoietic cells towards the end of the third trimester.

Discussion

Previous DNAm studies using cord blood have identified significant differences between preterm and term infants [15–17]; however, interpretation of these studies is limited by the confounding factor of cord blood cell composition. Granulocytes and T cells are the two most abundant cell types in whole blood and thus are the most likely to influence overall DNAm, but cell type proportions show considerable inter-individual variability and also change with gestational age [12]. Some DNAm changes previously associated with prematurity may simply reflect these changes in cell composition

with GA. This study is the first evaluation of the epigenetic impact of PTB in hematopoietic cell populations isolated from the same individuals.

An important question is the functional role of these DNAm changes in hematopoietic cell populations during ontogeny. There was a notable difference in the number of cell-specific prematurity-associated DM sites in each cell type, ranging from 273 in T cells to 9258 in nRBCs (Table 2). The number of prematurity-associated DM sites in a given cell type may relate to the magnitude of phenotypic differences between preterm and term populations. For example, we and others have reported major functional differences in dendritic cells and macrophages between preterm and term infants [37–40]. In contrast, fewer functional differences have been observed between preterm and term T cells [41, 42]. For granulocytes, much less is known regarding gestational differences. The high number of prematurity-associated DM sites we observed in granulocytes (987) suggests a more dynamic maturation across late gestation than for either monocytes or T cells. Alternatively, it is possible that these DNAm changes may reflect differences in the composition of granulocyte subsets, including a mixture of eosinophils, basophils, and mast cells, between age groups. However, this is less likely given that our granulocytes were overwhelmingly represented by neutrophils in both preterm and term samples (>95%; data not shown). Given the extent of DNAm differences between preterm and term granulocytes, functional studies may provide new insight into the limitations of the preterm immune system.

Our findings showed moderate overlap with previous studies of prematurity-associated DNAm in whole cord blood, with approximately 30% of the DM sites from each comparison study also discovered in at least one of our cell types (Fig. 3a). This is a greater amount of overlap than the 161 of 1347 CpG sites Fernando et al. [16] found in common with Cruickshank et al. [15] and another study not evaluated in this paper [34]. This

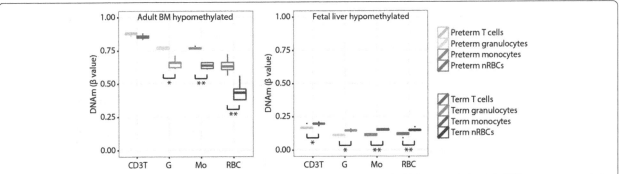

Fig. 5 Comparison of median DNAm between GA groups and cell types at CpG sites associated with hematopoietic origin. DNAm was compared at the top 100 CpG sites hypomethylated in erythroblasts derived from adult bone marrow stem cells (*left*) and the top 100 CpG sites hypomethylated in erythroblasts derived from fetal liver stem cells, as identified by Lessard et al. [28]. *$p < 0.05$, **$p < 0.005$

Table 4 Overlap between cell-specific prematurity-associated DM sites (FDR < 5%, |Δβ| > 0.10) and Lessard et al.'s source-DM sites [25]

	DNAm decreases with GA				DNAm increases with GA			
	T cell	Gran.	Mono.	nRBC	T cell	Gran.	Mono.	nRBC
Total	76	679	425	8731	197	308	267	527
Overlap with BM-hypo. sites	25, 32.9%	197, 29.0%	213, 50.1%	895, 10.3%	1, 0.5%	0, 0.0%	1, 0.4%	1, .02%
Overlap with FL-hypo. sites	0, 0.0%	1, 0.1%	0, 0.0%	2, 0.0%	22, 11.2%	74, 24.0%	70, 26.2%	89, 16.9%

Gran. granulocytes, *mono.* monocytes

increased overlap with other studies may reflect reduced noise our data due to eliminating variation due to cell composition differences. Alternatively, since we compared four sets of prematurity-associated DM sites (one per cell type), all of which were of fairly large size, we may have increased our chance of overlap just by having a greater number of hits.

When the cell type-DM sites discovered in this study are overlapped with the three comparison studies, the highest numbers of overlapping sites were observed with T cell- and nRBC-specific DM sites, the two cell types with the strongest cell-specific DNAm patterns (Fig. 3b). We additionally compared our cell type-DM sites to a subgroup of 196 CpG sites identified by Fernando et al. to be associated with PTB but not GA—and thus thought to reflect systematic differences due to prematurity rather than cell composition—and found almost no overlap (Fig. 3b). This supports their assertion that those 196 CpG sites are more likely related to the molecular mechanisms of PTB than cell composition, compared to the 1151 CpG sites they identified as associated with both PTB and GA. Thus, subsequent studies of PTB may be able to work around concerns of variability in cell composition by considering prematurity as a separate variable from GA.

Two CpG sites highlighted by Fernando et al. as being of potential interest for preterm delivery were not replicated in our study. One site, in *MYL4* (myosin light chain 4), was the only DM site identified by Fernando et al. [16], Cruickshank et al. [15], and Schroeder et al. [34]; the other, *ESR1* (estrogen receptor) was observed in both Fernando et al. [16] and Schroeder et al. [34]. Fernando et al. suggested that these sites may be related to the labor process, since *MYL4* activity is involved in the myometrial contraction pathway [43], and upregulation of *ESR1* leads to the increase in estrogen activity required for contractions [44]. The lack of replication in our data may be a consequence of all of our subjects being born by caesarean section, whereas all of the comparison studies included at least some subjects born by vaginal delivery. Notable genes in which we replicated the differential methylation found by other studies include *ADORA2A*, which has been associated with the inflammatory pathway in the myometrium [45], and *GABBR1*, which encodes a gamma-aminobutyric

acid receptor and has been associated with chemotaxis in cord blood-derived stem cells [46]. These two genes were identified as differentially methylated in both Fernando et al. [16] and Cruickshank et al. [15].

GO pathway analyses of the prematurity-associated DM sites highlighted potentially important differences in gene regulation that are unique to each cell type (Additional file 3). For instance in granulocytes, prematurity-associated DM sites were significantly enriched for genes associated with the Ras-Raf-MEK-ERK cascade. Defects in this pathway have been associated with impaired neutrophil extracellular trap formation and with respiratory burst in neutrophils [47, 48], both of which are also deficient in preterm infants [49, 50]. The prevalence of prematurity-associated DM sites in genes associated with these functions could reflect either reduced functional ability in preterm neutrophils or a low proportion of neutrophils within the preterm granulocyte population. Some of our findings point towards novel pathways potentially involved in the maturation of hematopoietic cells, such as the enrichment for prematurity-associated DM sites in genes associated with placental development in T cells and with dermal development in monocytes. In nRBCs, prematurity-associated DNAm changes were widespread and associated with GO terms related to the cytoskeleton, membrane composition and cell-cell junctions, and motility. This may reflect the large-scale structural changes that occur in erythroblasts as they mature and prepare to extrude their nucleus.

Based on gestational age differences, DNAm conformed to the epigenetic profile of the dominant hematopoietic organ when evaluated in source-DM sites [28]: the liver in mid-gestation, and the bone marrow in late gestation (Fig. 5). Despite these candidate sites being identified exclusively in nRBCs derived ex vivo from hematopoietic stem cells, the DNAm trends in this study were consistent across both nRBCs and WBCs. Thus, our findings indicate that hematopoietic sources have epigenetic signatures that are shared across multiple cell lineages derived from that organ. Additionally, our analysis of these hematopoietic source-DM sites revealed that nRBCs actually gain DNAm with increasing GA in functionally relevant regions of the genome, specifically the fetal liver-hypomethylated sites. This is a rare occurrence, as the overwhelming trend is for nRBCs to become demethylated both during erythropoiesis [29, 30] and as the fetus approaches term. This novel observation is important to our understanding of hematopoiesis during ontogeny since it indicates that although nRBC demethylation is largely global and passive [29, 30], it also has some selectivity, with certain CpG sites protected from the widespread DNAm loss.

The main limitation in our study is the small sample size. Other studies evaluating prematurity-associated DNAm had sample sizes ranging from 22 [16] to 50 [17]. With only ten subjects, our study had reduced power to detect changes in DNAm. Considering the large epigenetic differences between cell lineages, we expect that our study was sufficiently powered to compare cell types. However, differential methylation associated with prematurity is expected to be of a smaller scale than cell type differences, so this may have led to an underestimation of prematurity-associated DM sites. There was also an increased chance that genetic factors impacted our findings, since some CpG sites are methylation quantitative trait loci (mQTLs), or sites where DNAm is more strongly associated with individuals than cell type [20, 51]. We mitigated this concern by performing an additional probe filtering step to remove suspected mQTLs, as described in the Supplementary Methods (Additional file 1).

It is possible that heterogeneity in our subjects' clinical characteristics reduced our ability to detect prematurity-associated differential methylation. All births were caesarean sections with no indications of infection; however, one preterm case was attributed to preeclampsia, and four of the five preterm births were multiples (Table 1). This raises the concern that multiplicity in the preterm subjects may have confounded our results. There is limited information on how the immune system differs with multiple births: it has been shown that intrauterine infection occurs more often in preterm births with dizygotic twins compared to monozygotic twins or singletons, but no differences in postnatal outcome have been associated with zygosity [52]. Additionally, CD4+ T cell activity has been observed to be significantly lower in preterm dizygotic twins than in preterm singletons [53]. For DNAm studies, the effect of twin births has only been assessed within twin pairs [54], not between twins and singletons. For our study, considering the extreme difference in GA of our preterm and term cases (<31 versus >38 weeks), we expect that prematurity will have a much greater effect on DNAm than differences due to multiplicity. This is in keeping with Fernando et al.'s [16] observation of more distinct clustering of DNAm in extreme PTB cases compared to intermediate PTB cases. Despite these limitations, we identified prematurity-DM sites that showed reasonable overlap with prior studies [15–17] and cell-specific DNAm patterns that were consistent with our previous findings in cord blood cell populations [20, 32].

Conclusions
The preterm immune system differs from that of the term neonate in both cell composition and function, resulting in heightened vulnerability to infection in preterm infants. We identified epigenetic markers of immune system differences with prematurity by comparing DNAm of major cord blood hematopoietic cell populations across gestation. Changes in DNAm between preterm and term hematopoietic cells in our study likely reflect a shift from the liver to the bone marrow as the predominant hematopoietic source with advancing gestational age. Granulocytes were identified as a candidate cell population of particular interest in preterm infants' susceptibility to infection, due to their relatively high number of prematurity-associated DM sites and the enrichment of these sites for GO terms related to the Ras-Raf-MEK-ERK cascade. Our findings provide important insights into the epigenetic regulation of hematopoietic cell-specific functions during fetal development. These data may have clinical implications, as they highlight gene regulatory mechanisms on both cell-specific and systemic levels that are involved in neonatal immune system maturity. Larger samples will be required to determine the potential impact of cause of PTB (such multiple gestations or preeclampsia) on these epigenetic profiles.

Additional files

Additional file 1: Supplementary methods and figures.

Additional file 2: "Location and genomic context of prematurity-DM sites". Each sheet is a table summarizing prematurity-DM sites (FDR < 5%, $|\Delta\beta| > 0.10$) for each cell type, as well as one for the DM sites common across all cell types.

Additional file 3: Significantly enriched GO terms (corrected p value <0.10) from ErmineJ analysis of prematurity-DM sites for each cell type, ordered by corrected p value.

Additional file 4: "Location and genomic context of cell type-DM sites". Each sheet is a table summarizing cell type-DM sites (FDR < 5%, $|\Delta\beta| > 0.20$) for each birth group/cell type combination.

Additional file 5: DNAm-based estimates of gestational age (GA) using Knight et al.'s [24] methods.

Abbreviations
450K array: Illumina Infinium Human Methylation 450 BeadChip; BM: Bone marrow; DM: Differentially methylated; DNAm: DNA methylation; FDR: False detection rate; FL: Fetal liver; GA: Gestational age; GO: Gene ontology; nRBC: Nucleated red blood cell; PTB: Preterm birth; WBC: White blood cell

Acknowledgements
We thank the BC Children's and Women's Hospital staff for their help with subject recruitment. We also thank Paul Villeneuve for his contributions to the cell sorting method; Kristi Finlay for obtaining consent for preterm subjects in this study; Ruby Jiang, Mihoko Ladd, Dr. Maria Peñaherrera for their work in sample processing; and Dr. Lisa Xu for flow cytometer operation.

Funding
This research was funded by grants from the Canadian Institutes of Health Research (CIHR; MOP-123478 to PML and MOP-49520 to WPR). OMdG is supported by a CIHR Frederick Banting and Charles Best Graduate

Scholarship–Master's Award. PML is supported by a Clinician-Scientist Award from the BC Children's Hospital Research Institute and a Career Investigator Award from the Michael Smith Foundation for Health Research. WPR is supported by a Scientist Award from the BC Children's Hospital Research Institute.

Authors' contributions

OMdG sorted cord blood cells for DNA methylation analyses, performed DNA methylation data analyses, and wrote the manuscript. PML and WPR supervised and designed the research, interpreted the data, and co-wrote the manuscript. All authors read and approved the final manuscript.

Competing interests

The authors declare that they have no competing interests.

Author details

[1]BC Children's Hospital Research Institute, Room 2082, 950W 28th Avenue, Vancouver, BC V5Z 4H4, Canada. [2]Department of Medical Genetics, University of British Columbia, Vancouver, BC V6T 1Z3, Canada. [3]Department of Pediatrics, University of British Columbia, Vancouver, BC V6T 1Z3, Canada.

References

1. Blencowe H, Cousens S, Chou D, Oestergaard M, Say L, Moller AB, et al. Born too soon: the global epidemiology of 15 million preterm births. Reprod Health. 2013;10 Suppl 1:S2.
2. Barker DJ, Gelow J, Thornburg K, Osmond C, Kajantie E, Eriksson JG. The early origins of chronic heart failure: Impaired placental growth and initiation of insulin resistance in childhood. Eur J Heart Fail. 2010;12(8):819–25.
3. Koupil I, Leon DA, Lithell HO. Length of gestation is associated with mortality from cerebrovascular disease. J Epidemiol Community Health. 2005;59(6):473–4.
4. O'Connor AR, Wilson CM, Fielder AR. Opthalmological problems associated with preterm birth. Eye (Lond). 2007;21(10):1254–60.
5. O'Reilly M, Sozo F, Harding R. Impact of preterm birth and bronchopulmonary dysplasia on the developing lung: long-term consequences for respiratory health. Clin Exp Pharmacol Physiol. 2013;40(11):765–73.
6. Saigal S, Doyle LW. An overview of mortality and sequelae of preterm birth from infancy to adulthood. Lancet. 2008;371(9608):261–9.
7. Lavoie PM, Lavoie JC, Watson C, Rouleau T, Chang BA, Chessex P. Inflammatory response in preterm infants is induced early in life by oxygen and modulated by total parenteral nutrition. Pediatr Res. 2010;68(3):248–51.
8. Strunk T, Currie A, Richmond P, Simmer K, Burgner D. Innate immunity in human newborn infants: prematurity means more than immaturity. J Matern Fetal Neonatal Med. 2011;24(1):25–31.
9. Takala TI, Makela E, Suominen P, Matomaki J, Lapinleimu H, Lehtonen L, et al. Blood cell and iron status analytes of preterm and full-term infants from 20 weeks onwards during the first year of life. Clin Chem Lab Med. 2010; 48(9):1295–301.
10. Dowling DJ, Levy O. Ontogeny of early life immunity. Trends Immunol. 2014;35(7):299–310.
11. Kan B, Razzaghian HR, Lavoie PM. An immunological perspective on neonatal sepsis. Trends Mol Med. 2016;22(4):290–302.
12. Sharma AA, Jen R, Butler A, Lavoie PM. The developing human preterm neonatal immune system: a case for more research in this area. Clin Immunol. 2012;145(1):61–8.
13. Cedar H, Bergman Y. Epigenetics of haematopoietic cell development. Nat Rev Immunol. 2011;11(7):478–88.
14. Ji H, Ehrlich LI, Seita J, Murakami P, Doi A, Lindau P, et al. Comprehensive methylome map of lineage commitment from haematopoietic progenitors. Nature. 2010;467(7313):338–42.
15. Cruickshank MN, Oshlack A, Theda C, Davis PG, Martino D, Sheehan P, et al. Analysis of epigenetic changes in survivors of preterm birth reveals the effect of gestational age and evidence for a long term legacy. Genome Med. 2013;5(10):96.
16. Fernando F, Keijser R, Henneman P, van der Kevie-Kersemaekers AMF, Mannens MM, van der Post JAM, et al. The idiopathic preterm delivery methylation profile in umbilical cord blood DNA. BMC Genomics. 2015;16:736.
17. Parets SE, Conneely KN, Kilaru V, Fortunato SJ, Syed TA, Saade G, et al. Fetal DNA methylation associates with early spontaneous preterm birth and gestational age. PLoS One. 2013;8(6), e67489.
18. Lam LL, Emberly E, Fraser HB, Neumann SM, Chen E, Miller GE, et al. Factors underlying variable DNA methylation in a human community cohort. Proc Natl Acad Sci U S A. 2012;109 Suppl 2:17253–60.
19. Reinius LE, Acevedo N, Joerink M, Pershagen G, Dahlen SE, Greco D, et al. Differential DNA methylation in purified human blood cells: implications for cell lineage and studies on disease susceptibility. PLoS One. 2012;7(7), e41361.
20. de Goede OM, Razzaghian HR, Price EM, Jones MJ, Kobor MS, Robinson WP, et al. Nucleated red blood cells impact DNA methylation and expression analyses of cord blood hematopoietic cells. Clin Epigenetics. 2015;7(1):95.
21. Du P, Kibbe WA, Lin SM. lumi: a pipeline for processing Illumina microarray. Bioinformatics. 2008;24(13):1547–8.
22. R Core Team. R: A language and environment for statistical computing. Vienna: the R Foundation for Statistical Computing; 2014.
23. Maksimovic J, Gordon L, Oshlack A. SWAN: Subset-quantile within array normalization for Illumina Infinium HumanMethylation450 BeadChips. Genome Biol. 2012;13(6):R44. -2012-13-6-r44.
24. Knight AK, Craig JM, Theda C, Bækvad-Hansen M, Bybjerg-Grauholm J, Hansen CS, et al. An epigenetic clock for gestational age at birth based on blood methylation data. Genome Biol. 2016;17:206.
25. Ritchie ME, Phipson B, Wu D, Hu Y, Law CW, Shi W, et al. limma powers differential expression analyses for RNA-sequencing and microarray studies. Nucleic Acids Res. 2015;43(7):e47.
26. Benjamini Y, Hochberg Y. Controlling the false discovery rate: a practical and powerful approach to multiple testing. J R Stat Soc. 1995;57:289–300.
27. Gillis J, Mistry M, Pavlidis P. Gene function analysis in complex data sets using ErmineJ. Nat Protoc. 2010;5(6):1148–59.
28. Lessard S, Beaudoin M, Benkirane K, Lettre G. Comparison of DNA methylation profiles in human fetal and adult red blood cell progenitors. Genome Med. 2015;7(1):1. -014-0122-2 . eCollection 2015.
29. Shearstone JR, Pop R, Bock C, Boyle P, Meissner A, Socolovsky M. Global DNA demethylation during mouse erythropoiesis in vivo. Science. 2011; 334(6057):799–802.
30. Yu Y, Mo Y, Ebenezer D, Bhattacharyya S, Liu H, Sundaravel S, et al. High resolution methylome analysis reveals widespread functional hypomethylation during adult human erythropoiesis. J Biol Chem. 2013;288(13):8805–14.
31. Alvarez-Errico D, Vento-Tormo R, Sieweke M, Ballestar E. Epigenetic control of myeloid cell differentiation, identity and function. Nat Rev Immunol. 2015;15(1):7–17.
32. de Goede OM, Lavoie PM, Robinson WR. Characterizing the hypomethylated DNA methylation profile of nucleated red blood cells from cord blood. Epigenomics. 2016;8(11):1481–94.
33. Bakulski KM, Feinberg JI, Andrews SV, Yang J, Brown S, McKenney S L, et al. DNA methylation of cord blood cell types: applications for mixed cell birth studies. Epigenetics. 2016;11(5):354–62.
34. Schroeder JW, Conneely KN, Cubells JC, Kilaru V, Newport DJ, Knight BT, et al. Neonatal DNA methylation patterns associate with gestational age. Epigenetics. 2011;6(12):1498–504.
35. Horvath S. DNA methylation age of human tissues and cell types. Genome Biol. 2013;14:R115.
36. Holt PG, Jones CA. The development of the immune system during pregnancy and early life. Allergy. 2000;55(8):688–97.
37. Sharma AA, Jen R, Kan B, Sharma A, Marchant E, Tang A, et al. Impaired NLRP3 inflammasome activity during fetal development regulates IL-1beta production in human monocytes. Eur J Immunol. 2015;45(1):238–49.
38. Weatherstone KB, Rich EA. Tumor necrosis factor/cachectin and interleukin-1 secretion by cord blood monocytes from premature and term neonates. Pediatr Res. 1989;25(4):342–6.
39. Strunk T, Prosser A, Levy O, Philbin V, Simmer K, Doherty D, et al. Responsiveness of human monocytes to the commensal bacterium Staphylococcus epidermidis develops late in gestation. Pediatr Res. 2012;72(1):10–8.

40. Lavoie PM, Huang Q, Jolette E, Whalen M, Nuyt AM, Audibert F, et al. Profound lack of interleukin (IL)-12/IL-23p40 in neonates born early in gestation is associated with an increased risk of sepsis. J Infect Dis. 2010;202(11):1754–63.

41. Currie AJ, Curtis S, Strunk T, Riley K, Liyanage K, Prescott S, et al. Preterm infants have deficient monocyte and lymphocyte cytokine responses to group B streptococcus. Infect Immun. 2011;79(4):1588–96.

42. Melville JM, Moss TJ. The immune consequences of preterm birth. Front Neurosci. 2013;7:79.

43. Salomonis N, Cotte N, Zambon AC, Pollard KS, Vranizan K, Doniger SW, et al. Identifying genetic networks underlying myometrial transition to labor. Genome Biol. 2005;6(2):R12.

44. Hirota Y, Cha J, Dey SK. Prematurity and the puzzle of progesterone resistance. Nat Med. 2010;16(5):529–31.

45. Lee Y, Sooranna SR, Terzidou V, Christian M, Brosens J, Huhtinen K, et al. Interactions between inflammatory signals and the progesterone receptor in regulating gene expression in pregnant human uterine myocytes. J Cell Mol Med. 2012;16(10):2487–503.

46. Zangiacomi V, Balon N, Maddens S, Tiberghien P, Versaux-Botteri C, Deschaseaux F. Human cord blood-derived hematopoietic and neural-like stem/progenitor cells are attracted by the neurotransmitter GABA. Stem Cells Dev. 2009;18(9):1369–78.

47. Tortorella C, Stella I, Piazzolla G, Simone O, Cappiello V, Antonaci S. Role of defective ERK phosphorylation in the impaired GM-CSF-induced oxidative response of neutrophils in elderly humans. Mech Ageing Dev. 2004;125(8):539–46.

48. Hakkim A, Fuchs TA, Martinez NE, Hess S, Prinz H, Zychlinsky A, et al. Activation of the Raf-MEK-ERK pathway is required for neutrophil extracellular trap formation. Nat Chem Biol. 2011;7(2):75–7.

49. Yost CC, Cody MJ, Harris ES, Thornton NL, McInturff AM, Martinez ML, et al. Impaired neutrophil extracellular trap (NET) formation: a novel innate immune deficiency of human neonates. Blood. 2009;113(25):6419–27.

50. Peden DB, VanDyke K, Ardekani A, Mullett MD, Myerberg DZ, VanDyke C. Diminished chemiluminescent responses of polymorphonuclear leukocytes in severely and moderately preterm neonates. J Pediatr. 1987;111(6 Pt 1):904–6.

51. Yuen RK, Avila L, Penaherrera MS, von Dadelszen P, Lefebvre L, Kobor MS, et al. Human placental-specific epipolymorphism and its association with adverse pregnancy outcomes. PLoS One. 2009;4(10), e7389.

52. Spiegler J, Hartel C, Schulz L, von Wurmb-Schwark N, Hoehn T, Kribs A, et al. Causes of delivery and outcomes of very preterm twins stratified by zygosity. Twin Res Hum Genet. 2012;15(4):532–6.

53. Aquilano G, Capretti MG, Nanni F, Corvaglia L, Aceti A, Gabrielli L, et al. Altered intracellular ATP production by activated CD4+ T-cells in very preterm infants. J Immunol Res. 2016;2016:8374328.

54. Gordon L, Joo JE, Powell JE, Ollikainen M, Novakovic B, Li X, et al. Neonatal DNA methylation profile in human twins is specified by a complex interplay between intrauterine environmental and genetic factors, subject to tissue-specific influence. Genome Res. 2012;22(8):1395–406.

4

Genome-wide DNA methylation analysis in blood cells from patients with Werner syndrome

T. Guastafierro[1,2] ⓘ, M. G. Bacalini[3], A. Marcoccia[2,4], D. Gentilini[5], S. Pisoni[5], A. M. Di Blasio[5], A. Corsi[6], C. Franceschi[3,7,8], D. Raimondo[6*], A. Spanò[1], P. Garagnani[7,8,9,10,11,12] and F. Bondanini[2,13]

Abstract

Background: Werner syndrome is a progeroid disorder characterized by premature age-related phenotypes. Although it is well established that autosomal recessive mutations in the WRN gene is responsible for Werner syndrome, the molecular alterations that lead to disease phenotype remain still unidentified.

Results: To address whether epigenetic changes can be associated with Werner syndrome phenotype, we analysed genome-wide DNA methylation profile using the Infinium MethylationEPIC BeadChip in the whole blood from three patients affected by Werner syndrome compared with three age- and sex-matched healthy controls. Hypermethylated probes were enriched in glycosphingolipid biosynthesis, FoxO signalling and insulin signalling pathways, while hypomethylated probes were enriched in PI3K-Akt signalling and focal adhesion pathways. Twenty-two out of 47 of the differentially methylated genes belonging to the enriched pathways resulted differentially expressed in a publicly available dataset on Werner syndrome fibroblasts. Interestingly, differentially methylated regions identified CERS1 and CERS3, two members of the ceramide synthase family. Moreover, we found differentially methylated probes within ITGA9 and ADAM12 genes, whose methylation is altered in systemic sclerosis, and within the PRDM8 gene, whose methylation is affected in dyskeratosis congenita and Down syndrome.

Conclusions: DNA methylation changes in the peripheral blood from Werner syndrome patients provide new insight in the pathogenesis of the disease, highlighting in some cases a functional correlation of gene expression and methylation status.

Keywords: Werner syndrome, DNA methylation, Systemic sclerosis

Introduction

Werner syndrome (WS) is a rare adult premature ageing disease. Individuals affected by WS generally have a normal development until the third decade of life, when premature ageing phenotypes and symptoms begin to manifest including premature greying or loss of hair, bird-like faces, cataracts, sclerodermiform skin atrophy [1] and skin ulcers [2]. Because of apparent similar skin changes, WS is often misdiagnosed as systemic sclerosis (SSc) [3].

Particularly, WS represents an important part of the differential diagnosis in patients who present with scleroderma-like skin changes, as they share with SSc patients the main histological changes of skin. These include replacement of the subcutaneous tissue by hyalinized connective, hyalinized connective tissue in the lower dermis and hyalinization and formation of aneurysms in the dermal blood vessels [4, 5]. Otherwise, progeria, premature cataract, type 2 diabetes mellitus, sensorineural hearing loss, premature atherosclerosis and dyslipidemia in WS besides the absence of anti-scleroderma antibodies may be helpful for the differential diagnosis of WS [6]. WS is associated to an

---

* Correspondence: domenico.raimondo@uniroma1.it
T. Guastafierro and MG Bacalini have shared first name.
P. Garagnani and F. Bondanini have co-senior authorship.
[6]Department of Molecular Medicine, Sapienza University of Rome, Rome, Italy
Full list of author information is available at the end of the article

autosomal recessive mutation of Werner syndrome gene (*WRN*) [7, 8]; up today, more than 50 different disease-causing mutations in the *WRN* gene have been identified [8]. The *WRN* gene encodes a 180-kDa nuclear protein member of the RecQ subfamily of DNA and RNA helicases [8, 9] with an intrinsic 3′ to 5′ DNA helicase activity [10]. DNA helicases are involved in many aspects of DNA metabolism including transcription, replication and recombination [9]. Given its role as a 3′ to 5′ exonuclease and based on interactions between this protein and Ku70/80 heterodimer in DNA end processing, WRN plays a critical role in the repair of DNA double-strand breaks [11]. Additionally, recent studies suggest a role for WRN in maintaining DNA telomere stability [10]. It is well known that defects in telomere structure and/or function may have a strong impact on human health, leading to premature ageing and a variety of diseases [12, 13]. Overall, WRN protein is crucial in maintaining genome structure and integrity, and accordingly, WS patients exhibit early age-associated biomarkers like DNA damage accumulation and chromosomal instability [14]. Despite these evidences, the molecular bases of WS phenotype are still largely unknown.

Alterations in epigenetic patterns, in particular in DNA methylation profiles, could contribute to WS phenotype. DNA methylation, which consists in the addition of a methyl group to the cytosine in a CpG dinucleotide, is a key mechanism in development and differentiation, and profound changes in DNA methylation patterns occur during ageing and age-related diseases [15, 16]. At present, few studies assessed DNA methylation changes in WS. Heyn and colleagues evaluated genome-wide DNA methylation in lymphoblastoid cell lines (LCLs) derived from Epstein-Barr virus (EBV)-immortalized B cells of four WS patients and three healthy controls [7], observing profound DNA methylation changes in WS. More recently, a WRN promoter was found hypermethylated in naïve B lymphocytes and lymphoblastoid cell lines from two brothers carrying a novel homozygous WRN mutation. Finally, Maierhofer et al. applied Horvath's epigenetic clock to a dataset including the whole blood from WS patients and age-matched controls, demonstrating that the disease is associated with an increase in epigenetic age [17].

In the present study, we analysed genome-wide DNA methylation in the peripheral whole blood from three WS patients and their age- and sex-matched controls using the Infinium MethylationEPIC BeadChip. Our results indicate that specific genome-wide alterations occur in WS, which partially overlap those that occur in diffuse and limited SSc [18].

Methods

Samples

Subjects were enrolled in the study "Early diagnosis of scleroderma and identification of predictor factors of disease development". The study was approved by the Ethic Committee of ASL RM/B, and all participants signed the informed consent forms. WS was diagnosed according to clinical appearance and genetic testing. Two patients were siblings and carried the mutation c.2630+1G>A in the *WRN* coding sequence; the third patient carried the homozygous g.nt 77177 a>g mutation (GenBank acc. no. AY44237) in the *WRN* gene. They suffer of skin ulcers of suspected SSc derivation as they showed typical SSc skin signs being bound tightly to the underlying tissues and around joints causing an inability to pinch or lift the skin up (Fig. 1a) and calcinosis cutis, a type of calcinosis which characterized scleroderma skin ulcers [19–21] (Fig. 1b). Serum was collected, and anti-nuclear antibodies (ANAs) were analysed using indirect immunofluorescence (IIF) assays on ANA-HEp-2 cells (A. Menarini Diagnostics) according to the manufacturer's protocol. Anti-extractable nuclear antigen (ENA) antibodies [anti-Sjögren's syndrome A (SS-A), anti-Sjögren's syndrome B (SS-B), anti-Smith (SM), anti-ribonucleoprotein (RNP), anti-topoisomerase I (SCL-70) and anti-histidyl-tRNA synthetase (JO-1)] were also dosed using chemiluminescence assay (Zenit RA autoimmunity, A. Menarini Diagnostics) according to the manufacturer's protocol. These autoantibodies, considered as SSc diagnostic markers, were undetectable in selected patients (data not shown). SSc patients were also subjected to nailfold video capillaroscopy (NVC). Capillaroscopic alterations and ectasias compatible with diagnosis of acrocyanosis were detected in all three patients (Fig. 1c). Peripheral blood was collected from the three patients.

Genome-wide DNA methylation analysis

Genomic DNA was extracted from whole blood using the Duplica˙ Blood DNA Kit (EuroClone) on Duplica˙ Prep Platform (EuroClone). DNA was bisulfite-converted using the EZ DNA Methylation-Gold Kit (Zymo Research) and analysed using the Infinium HumanMethylationEPIC BeadChip (Illumina) following the manufacturer's instructions. Signal intensity files were extracted using the *minfi* Bioconductor package. Normalization was performed using the functional normalization function implemented in *minfi*. DNA methylation data are available at the NCBI Gene Expression Omnibus (GEO) database (http://www.ncbi.nlm.nih.gov/geo/) under accession number GSE100825.

DNA methylation differences between WS patients and healthy controls were assessed using two different approaches: (1) we used analysis of variance (ANOVA) to find out whether there was any statistically significant difference between WS patients and healthy controls when we consider differentially methylated positions (DMPs), that is single-CpG sites whose methylation differs between the groups under investigation. DMPs with a non-adjusted P value < 0.001 were retained as significantly differentially methylated. (2) We applied multivariate analysis

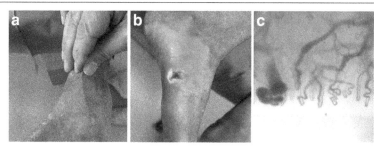

Fig. 1 SSc typical skin signs in WS patients. **a** Skin unable to be lifted in pliers and/or pinched. **b** Cutaneous ulcer with calcinosis. **c** Capillaroscopic alterations and ectasias

of variance (MANOVA) on sliding windows including three adjacent CpG sites as described in Bacalini et al. [15] in order to identify statistically significantly differentially methylated regions (DMRs), that is regions in which multiple adjacent CpG sites differ between the two groups under investigation. This approach was applied only to the CpG probes mapping in CpG-rich regions (CpG islands, shores and shelves) associated to a gene. DMRs with a non-adjusted P value < 0.001 and in which at least two adjacent CpG sites had a minimum absolute DNA methylation difference of 0.3 were retained as significant. Although a 0.15 threshold for mean DNA methylation difference is often suggested [22], here we preferred to use a more stringent criterion, given the small sample size, in order to provide a short, but more likely reproducible, list of DMRs.

Infinium450k datasets GSE42865 and GSE75310 were downloaded from the GEO database. GSE42865 includes DNA methylation data from the LCL of three Hutchinson-Gilford progeria syndrome patients, four WS patients and three healthy patients, together with peripheral blood mononuclear cells from three healthy subjects and naïve B cells from three healthy subjects. We compared the four WS patients and the three LCL controls by ANOVA. GSE75310 includes DNA methylation data from four dyskeratosis congenita (DKC) patients. To investigate DNA methylation differences in DKC, we performed ANOVA using as controls the same four GEO samples (GSM796678, GSM796674, GSM796667 and GSM796671) used in the original paper [23].

Pathway enrichment
Pathway analysis of differentially methylated genes and gene clusters was performed with the publicly available tool Enrichr (http://amp.pharm.mssm.edu/Enrichr) that provides access to various gene set libraries [24, 25]. Enrichr currently contains annotated gene sets from 102 gene set libraries organized in eight categories. Details of the gene set libraries in Enrichr can be found in previously published studies [24, 25]. Enrichment analysis checks whether the input set of genes significantly

overlaps with annotated gene sets. Each gene set within the Enrichr database is associated with a functional term or an enrichment term such as a pathway, cell line or disease. The output of Enrichr is ranked lists of terms, one list for each gene set library. The most highly ranked enrichment terms for the user's input gene list provide knowledge about the input list. We used the publicly Kyoto Encyclopedia of Genes and Genomes (KEGG) pathway gene sets to identify the relatively enriched biological pathways and then extracted the enriched gene sets for each pathway. We considered pathways as enriched if their P value was lower than 0.05 and ranked them using the combined score (CS). CS is a multiplication of the P value computed using Fisher's exact test with the z-score of the deviation from the expected rank. In particular, c is equal to log $(p) \times z$ where c is the combined score, p is the P value computed using Fisher's exact test and z is the z-score computed to assess the deviation from the expected rank.

Differential gene expression analysis
In order to perform differential expression analysis, we used shinyGEO (http://gdancik.github.io/shinyGEO/) [26], a Web-based tool that allows a user to download the expression and sample data from a GEO dataset, select a gene of interest and perform a survival or differential expression analysis using the available data [27]. For both analyses, shinyGEO is able to generate R code, ensuring that all analyses are reproducible. The tool is developed using shiny, a Web-based application framework for R, a language for statistical computing and graphics. Fold change (FC) and P value are calculated by two-sample Student's t test for differential expression.

Results
We used the Infinium MethylationEPIC BeadChip to compare whole blood genome-wide DNA methylation profiles between three WS patients and three age- and sex-matched controls (CTRs). We conducted principal component analysis to identify major traits of variation between samples (Additional file 1: Figure S1). No clear

separation between WS patients and CTRs was observed along the first component (which explained 32% of the variance) or the second component (which explained 51% of the remaining variance).

ANOVA identified 1125 DMPs that distinguished WS patients from CTRs (non-adjusted P value < 0.001) (Fig. 2a, Additional file 2, Table 1). Of these, 511 probes (mapping in 382 genic regions) were hypermethylated and 614 probes (mapping in 416 genic regions) were hypomethylated in WS patients compared to CTRs. Several DMPs showed large DNA methylation differences between the two groups, with 87/511 hypermethylated and 110/614 hypomethylated probes having mean methylation differences larger than 0.15 [22]. The volcano plot in Fig. 2a shows many CpG sites with non-adjusted P values lower than 0.001 but having low DNA methylation differences between the groups. This behaviour might be related to the small number of analysed WS samples.

Next, we performed a gene set enrichment analysis in order to examine in silico the biological functions of DMPs in WS patients compared to CTR samples. To this aim, we searched for the overrepresented biological pathways associated with the differentially methylated genes, using as input for the Enrichr tool the list of genes associated with DMPs. Enrichr's combined score, a combination of the P value and z-score, was considered in order to prioritize enriched pathways (see "Methods" for more details). As detailed in the Enrichr manuscript [24], the combined score provides a compromise between both methods (P value and z-score) and, in several benchmarks, has reported the best ranking when compared with the other scoring schemes.

We took into account the most highly ranked enriched pathways according to the combined score while we considered significantly overrepresented pathways, those

showing a P value < 0.05. Results for enriched biological pathways are shown in Table 2. The hypermethylated genes in KEGG pathway enrichment analysis results were associated with the enriched pathways with P value < 0.05, including "glycosphingolipid biosynthesis (HSA-00603)", "FoxO signalling pathway (HSA-04068)" and "insulin signalling pathway (HSA-04910)". The hypomethylated genes were associated with the enriched pathways with P value < 0.05, including "PI3K-Akt signalling (HSA-04151)" and "focal adhesion (HSA-04510)".

Finally, we also used the pipeline reported by Bacalini et al. [15] (see "Methods" for details) to identify differentially methylated regions (DMRs) between WS patients and CTRs. We found 27 DMRs with a non-adjusted P value < 0.001 (Additional file 3, Additional file 4, Table 3), 20 of which were clearly hypermethylated in WS patients compared to CTRs. Even these 27 DMRs (corresponding to 37 genes) were employed in the Enrichr pathway analysis approach, revealing "sphingolipid metabolism (HSA-00600)" and "sphingolipid signalling pathway (HSA-04071)" among the most enriched pathways with P value < 0.05. Both pathways contain *CERS3* and *CERS1* genes, which are two members of the ceramide synthase family.

To assess if the epigenetic pattern that we described in our WS cohort was reproducible in other WS DNA methylation datasets, we compared our results with those available from a publicly available dataset (GEO accession ID: GSE42865). GSE42865 includes DNA methylation data from LCL from four WS patients and three CTRs, measured using the previous version of the microarray, the Infinium450k BeadChip [7]. Of the 1125 DMPs identified in our analysis, 581 are included in the Infinium450k design. Of these, only two probes were differentially methylated between LCLs of WS patients and

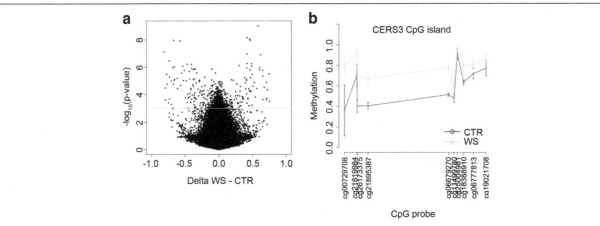

Fig. 2 DNA methylation alterations in Werner syndrome. **a** Volcano plot of DMPs between WS patients and CTRs. The difference between mean DNA methylation values in WS patients and in CTRs is plotted on the *x*-axis, while the non-adjusted *P* value for ANOVA between the two groups is on the *y*-axis (− 1 × log10 scale). The green line corresponds to a non-adjusted *P* value of 0.001. **b** DNA methylation profile of the CpG island located in the *CERS3* gene

Table 1 List of the top 20 DMPs

Probe	CHR	MAPINFO	Gene	CpG island name	Relation with respect to the CpG island	Non-adjusted P value
cg15294279	3	174,842,010	NAALADL2	–	–	9.48E−10
cg26845082	3	13,555,664	–	–	–	6.75E−09
cg23432430	12	125,538,377	–	chr12:125534060-125534527	S_Shelf	8.75E−09
cg00597723	5	158,691,793	UBLCP1	chr5:158690013-158690541	S_Shore	1.05E−08
cg13956086	7	2,434,521	–	–	–	1.08E−08
cg16995742	2	237,992,612	COPS8	chr2:237994004-237994876	N_Shore	1.19E−08
cg17779733	22	49,589,242	–	–	–	7.55E−08
cg06052372	16	83,967,808	–	–	–	1.04E−07
cg23928292	12	21,815,474	–	–	–	1.19E−07
cg10360725	8	144,139,316	–	–	–	3.31E−07
cg14782559	6	33,131,893	COL11A2	chr6:33129291-33129718	S_Shelf	5.19E−07
cg18673341	7	22,481,962	STEAP1B	–	–	5.28E−07
cg26822175	22	27,018,010	CRYBA4	–	–	5.71E−07
cg08161337	22	45,814,116	RIBC2	–	–	5.86E−07
cg22664298	5	128,795,827	ADAMTS19	chr5:128795503-128797417	Island	7.16E−07
cg13885829	1	17,482,041	–	–	–	7.98E−07
cg07584620	1	2,265,881	MORN1	chr1:2266007-2266432	N_Shore	8.00E−07
cg20757478	6	31,012,262	–	–	–	8.99E−07
cg17628377	5	180,121,337	–	–	–	1.03E−06
cg15865722	11	68,860,657	–	–	–	1.03E−06

CTRs with a comparable statistical threshold (non-adjusted P value < 0.001), while 30 probes were found differentially methylated, considering a less stringent threshold (non-adjusted P value < 0.05) (Additional file 2). To investigate the reason of this low reproducibility of the data, we focused on the DMPs mapping in the genes resulting from the pathway enrichment analysis that are present in both the microarray versions. We included in this analysis also the probes mapping in *CERS1* and *CERS3* DMRs. As shown in Additional file 5, DNA methylation in LCL showed a larger variation in WS and/or CTR subjects, possibly as a consequence of

Table 2 Gene set enrichment analysis details are reported for DPMs (hypo- and hypermethylated genes) and DMRs

	Enriched pathways	Pathway ID	Overlap	P value	Combined score	Overlapping genes
Hypermethylated probes	Glycosphingolipid biosynthesis	HSA-00603	3/14	0.001	2.29	ST3GAL1, GBGT1, ST3GAL2
	FoxO signal pathway	HSA-04068	8/133	0.003	2.60	MAPK10, USP7, AKT2, STAT3, PTEN, FOXO3, SKP2, GABARAP
	Insulin signalling pathway	HSA-04910	8/139	0.004	2.49	MAPK10, PTPN1, SHC2, AKT2, PRKAK1B, FASN, TSC2, CRKL
Hypomethylated probes	PI3K-Akt signalling	HSA_04151	16/341	0.006	3.46	CSF-1R, CDKN1B, TNXB, VWF, LAMA1, FLT4, THBS1, PTK2, LPAR5, PPP2R2B, PPP2R2D, MAPK1, COL6A6, FGFR1, BCL2L1, ITGA9
	Focal adhesion	HSA_04510	12/202	0.01	3.26	MAPK10, TNXB, VWF, ROCK2, LAMA1, FLT4, PXN, MAPK1, COL6A6, THBS1, PTK2, ITGA9
DMRs	Sphingolipid metabolism	HSA_00600	2/120	0.003	4.54	CERS3, CERS1
	Sphingolipid signalling pathway	HSA_047071	1/47	0.001	5.36	CERS3

Overlap indicates the number of hits from the differentially methylated gene sets compared to the KEGG gene set library, while "overlapping genes" column contains names of these hits. Differentially expressed genes between normal and WS fibroblasts in Cheung HH dataset [30] are reported. Enriched pathways were selected based on the P value. Combined score is a multiplication of the P value computed using Fisher's exact test with the z-score of the deviation from the expected rank (see "Methods" for details)

Table 3 List of the 27 DMRs

CHR	CpG island name	Relation with respect to the CpG island	Gene	Non-adjusted P value
18	chr18:30349690-30352302	S_Shore	KLHL14	1.70E−05
12	chr12:50297580-50297988	S_Shore	FAIM2	0.00013152
6	chr6:33396050-33396296	S_Shelf	SYNGAP1	0.000168391
5	chr5:110074605-110075223	N_Shore	SLC25A46	0.000215085
22	chr22:46366726-46368726	S_Shore	WNT7B	0.000233192
17	chr17:46620367-46621373	S_Shore	HOXB2; HOXB-AS1	0.000244681
10	chr10:5930914-5932389	N_Shore	ANKRD16; FBXO18	0.000261161
3	chr3:9811466-9811736	S_Shore	CAMK1	0.000281325
2	chr2:129075197-129077639	Island	HS6ST1	0.000292046
15	chr15:71145995-71146820	N_Shore	LARP6; LRRC49	0.000302033
15	chr15:101084428-101085178	Island	CERS3	0.000341282
6	chr6:143999154-143999667	N_Shore	PHACTR2	0.000393635
6	chr6:166755812-166756510	S_Shore	LOC100289495; SFT2D1	0.000423645
4	chr4:146540053-146540656	N_Shore	MMAA	0.000441415
17	chr17:20059028-20060060	N_Shore	SPECC1	0.000444431
6	chr6:31939730-31940559	S_Shore	DXO; STK19	0.000488824
15	chr15:99791328-99792042	N_Shore	TTC23; LRRC28	0.0004949
19	chr19:19006031-19007546	S_Shore	GDF1; CERS1	0.000509929
22	chr22:20134462-20134705	S_Shelf	LOC388849	0.000585452
19	chr19:19624954-19627258	Island	NDUFA13; TSSK6	0.000650774
1	chr1:160990718-160991225	N_Shore	F11R	0.000652612
6	chr6:33266302-33267582	N_Shore	RGL2	0.000759464
8	chr8:117886284-117887319	S_Shore	RAD21-AS1; RAD21; MIR3610	0.000773036
19	chr19:35491151-35492020	N_Shore	GRAMD1A	0.000798561
17	chr17:4692249-4693977	N_Shore	GLTPD2	0.000853965
4	chr4:9783035-9784960	Island	DRD5	0.000911924
4	chr4:1396291-1401730	Island	NKX1-1	0.000928038

the immortalization process [28]. Accordingly, DNA methylation differences were observed between LCL and naïve B cells (Additional file 5). It is therefore plausible that low overlapping between the GSE42865 dataset and ours is related to the different sources of genomic DNA (LCL vs whole blood). Despite these differences, we noted that several CpG sites showed the same trend in DNA methylation changes between WS patients and CTRs, in particular within the *CERS3* DMR and in *VWF* and *FOXO3* genes.

Furthermore, we compared WS-related epigenetic changes with those occurring in DKC, a premature ageing disease associated to impaired telomere maintenance. We exploited a publicly available Infinium450k dataset (GEO accession ID: GSE75310) performed on the whole blood from four DKC subjects [23]. Of the 581 WS-DMPs included in the Infinium450k design, 14 were differentially methylated in DKC with a comparable threshold (non-

adjusted P value < 0.001), but only three of them showed a concordant DNA methylation change in the two diseases (Additional file 2). Despite this, we noted that the list of WS-DMPs included three CpG sites (cg27111250, cg10129063 and cg27639662) mapping in the S_Shore of the chr4:81109887-81110460 CpG island in the PR domain containing eight (*PRDM8*) genes and hypermethylated in WS patients compared to CTRs. This is of particular interest, as *PRDM8* is hypermethylated in DKC, aplastic anaemia (AA) [23] and Down syndrome [29], although in a different region (spanning from the CpG island chr4:81118137-81118603 to the CpG island chr4:81119095-81119391).

Unfortunately, RNA was not available from the samples included in our cohort and it was not possible to assess the expression of the genes identified as differentially methylated. To overcome this limitation, we exploited the GEO database in order to get more insight into the relationship between DNA methylation changes and WS features. We selected the genes emerged from the pathway enrichment analysis, and we checked their expression in a dataset including ten WS and ten CTR skin fibroblasts (GEO accession ID: GSE48761) [30]. Enrichment analysis checks whether an input set of genes significantly overlaps with annotated gene sets. We found 47 genes overlapping with the annotated gene set from KEGG pathways (see the overlap column in Table 1), and 22 out of 47 of these genes were differentially expressed between WR patients and CTRs (P value < 0.05), as reported in Fig. 3 and Additional file 6: Figure S2. The relationship between DNA methylation and expression changes is reported in Additional file 7. Focusing on the DMRs, *HS6ST1* was hypermethylated in the whole blood from WS patients and downregulated in WS fibroblasts, while both *CERS1* and *CERS3* were hypermethylated in the whole blood from WS patients and upregulated in WS fibroblasts. Interestingly, we also noted that *PRDM8* gene expression is downregulated in WS fibroblasts (data not shown), like it was previously reported in the whole blood from DKC patients [30]. Although further investigation is needed, possibly on the same biological specimen from the same subject, these data underline an association between deregulation of DNA methylation and gene expression in healthy donors and patients.

Discussion

In this study, we analysed genome-wide DNA methylation in the peripheral blood from three WS subjects and three age- and sex-matched controls, identifying 1125 DMPs and 27 DMRs. Pathway enrichment analysis on the list of hypermethylated and hypomethylated DMPs and on the list of DMR provided interesting hints on the epigenetic alterations in WS patients, possibly related with the disease phenotype. Bioactive sphingolipids have been suggested to play a role in ageing and cellular senescence, and in WS, lipid profile is often abnormal [31]. Moreover, glycosphingolipids are elevated during ageing in the mouse brain and liver as well as in human fibroblasts obtained

from elderly individuals [32]. On the other hand, longevity is associated with genetic variation in insulin-FOXO pathways [33] and the signalling pathway connecting insulin and FoxO transcription factors provides the most compelling example for a conserved genetic pathway at the interface between ageing and cancer. FoxO proteins are a subgroup of the forkhead family of transcription factors, which regulate the expression of genes involved in several physiological events including apoptosis, cell cycle control, glucose metabolism, oxidative stress resistance and longevity [34, 35]. In response to insulin or growth factors, the protein is phosphorylated by Akt, downstream of PI3K. This constitutes a signal to export FoxO from the nucleus to the cytoplasm, thereby decreasing the expression of FoxO target genes [34, 36]. Dysregulation of the insulin/PI3K/Akt pathway is implicated in several human diseases including cancer, diabetes, cardiovascular diseases and neurological diseases [37, 38]. Insulin signalling pathway regulates ageing in many organisms, ranging from simple invertebrates to mammals, including human [36]. Notably, the enrichment of hypermethylated probes in insulin and FoxO signalling pathways is also consistent with the clinical and metabolic evidences that relate WS to disorders of lipid and liver function [39].

The hypomethylated probes were enriched in PI3K-Akt signalling (HSA-04151) and focal adhesion (HSA-04510) pathways. Even these results underline the importance of deregulated methylation positions we detected. In fact, the role of the PI3K/Akt/FoxO signalling in longevity appears to be well conserved across species [40] and senescence-related morphological alteration is one of the main features of the senescent phenotype and it is deeply dependent to changes of cellular structural determinants in terms of their levels and activities. These determinants included integrins, focal adhesion complexes and small Rho GTPases [41].

Importantly, the alteration of the above-mentioned pathways in WS was fully supported when we correlated gene sets detected as differentially methylated and a publicly available gene expression dataset on WS fibroblasts

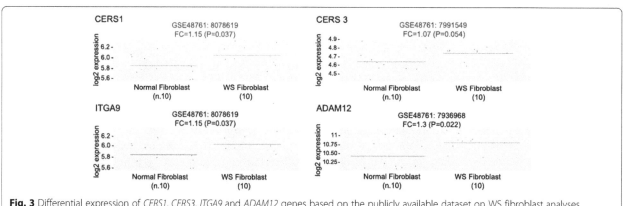

Fig. 3 Differential expression of *CERS1*, *CERS3*, *ITGA9* and *ADAM12* genes based on the publicly available dataset on WS fibroblast analyses

including ten WS and ten CTR [30]. Indeed, several of the differentially methylated genes belonging to the enriched pathways showed altered RNA expression in fibroblasts from WS patients compared to healthy controls (Fig. 3, Additional file 6: Figure S2).

The small overlap between WS and DKC DMPs suggests that the two progeroid diseases show distinct patterns of epigenetic alterations. However, the *PRDM8* gene resulted hypermethylated in both WS and DKC, although in different CpG sites, suggesting that epigenetic remodelling in different premature ageing syndromes can converge on the same gene.

Interestingly, the gene *ITGA9*, which encodes for an alpha integrin, was previously reported to be hypomethylated and overexpressed in fibroblasts from patients with diffuse and limited SSc compared to fibroblasts from healthy controls [18]. Although in our dataset WS affected the methylation of a distinct CpG site in the body of *ITGA9* (cg22345769, which is not assessed in the study by Altorok et al. [18], as included in the Infinium MethylationEPIC and not in the Infinium450k microarray), it is noteworthy that the methylation and mRNA level of the same gene is altered in both the diseases. In fact, in our bioinformatics analysis, we detected a fold change equal to 1.15 for the differential expression of *ITGA9* (P value = 0.037). Furthermore, we found that another gene is implicated in cell-cell and cell-matrix interactions, *ADAM12*, that resulted hypomethylated both in WS and in diffuse and limited SSc [18] and overexpressed (fold change equal to 1.3, P value = 0.022) in WS with respect to CTR when we analysed the GSE48761 dataset [30] (Fig. 3). *ADAM12* is involved in the process of fibrosis through enhancing TGF-β signalling pathway [42, 43]. Of note, the peripheral blood from WS patients shared the same hypomethylated CpG site in the body of *ADAM12* with SSc fibroblasts.

Finally, *CERS3* was shown to exhibit differentially methylated status between WS patients and CTRs. A distinct expression profile was also found between normal and WS fibroblasts in the GSE48761 dataset (Fig. 3) [30]. This gene regulates sphingolipid synthesis and is involved in the synthesis of ceramides with ultra-long-chain acyl moieties (ULC-Cers), playing an important role in creating a barrier for the epidermis from the environment [44]. A mutation in CERS3 is responsible of autosomal recessive congenital ichthyosis [45], and interestingly, scleroderma-like changes have been described in different clinical variants of ichthyosis [46]. All these data are of particular interest as WS has some features that are typical SSc signs like skin sclerosis and calcification and ankle ulcerations [47, 48]. Indeed, skin sclerosis is a hallmark of SSc and skin thickening represents the definitive diagnostic criterion of SSc in the vast majority of cases [19–21, 49]. Common to SSc subjects, WS patients show SSc-like skin involvement, calcinosis and skin ulcers; in particular, ulcers are present in 50% of WS patients and

are generally the main cutaneous symptom [50, 51]. Anyway, the possibility for WS patients to be diagnosed as SSc was avoided as, based on 2013 ACR/EULAR classification criteria for SSc disease [52], serological analyses performed on subjects enrolled in the present study for anti-ANA and anti-ENA (anti-SS-A, anti-SS-B, anti-SM, anti-RNP, anti-SCL-70 and anti-JO-1) antibodies resulted negative. The presence of serum autoantibodies directed to multiple intracellular antigens is considered the serological hallmark of SSc [53].

The emerging connection in WS and SSc epigenetic alterations that we observed in this study could be related to the fact that the WS patients that we analysed for genome-wide DNA methylation were recruited in the framework of a study on SSc. Although we cannot exclude a priori this hypothesis, it is worth to note that (1) the CpG sites cg17287034 in the *ADAM12* gene and cg06679270 in the *CERS3* gene showed the same DNA methylation trend in the GSE42865 dataset and in ours (we cannot check *ITGA9* because the probe is missing in the Infinium450k design) (Additional file 5) and (2) *ITGA9*, *ADAM12* and *CERS3* showed gene expression changes between WS and normal fibroblasts according to the GSE48761 dataset. As in both the publicly available studies WS patients were not selected on the basis of a SSc-like phenotype, we can suggest that SSc-related epigenetic changes are an intrinsic characteristic of WS. Although further studies should confirm our findings, they could be particularly interesting since recent studies highlight that another progeroid disease, the Hutchinson-Gilford progeria syndrome, is associated with SSc-like skin changes [54, 55] and that SSc-associated fibrosis is considered as an accelerated ageing phenotype [56].

Conclusions

In summary, we identified for the first time genome-wide DNA methylation changes in the peripheral blood from WS patients, providing important new insight in the pathogenesis of the diseases and emphasizing the potential role of DNA methylation changes in Werner disorder.

Additional files

Additional file 1: Figure S1. Principal Component Analysis for the DNA methylation levels of the probes included in the InfiniumEPIC beadchip in WS and CTR.

Additional file 2: List of DMPs between WS and CTR. (XLS 531 kb)

Additional file 3: DNA methylation profiles of the 27 DMRs between WS and CTR. For each DMR, the title of the plot reports the name of the gene/genes in which the DMR maps, the not-adjusted P value of the ANOVA comparison between WS and CTR, and the length of the DMR in base pairs. Below the plot, the name of the CpG island and the position of the DMR respect to the CpG island are reported.

Additional file 4: List of DMRs between WS and CTR.

Additional file 5: Boxplots of DNA methylation values of the CpG probes in the genes belonging to the enriched pathways in naïve B cells,

CTR LCL, WS LCL from the GSE42865 dataset and in CTR and WS whole blood samples assessed in this study. The brown dots correspond to the individual samples. For each CpG probe, the title of the plot reports the name of the CpG probe, the name of the gene/genes in which the probe maps and, if present, the name of the CpG island and the position of the probe respect to the CpG island.

Additional file 6: Figure S2. Differential expression of several genes belonging to the enriched pathways resulted based on publicly available dataset on WS fibroblasts analyses.

Additional file 7: Relationship between DNA methylation of DMPs and DMRs in the present dataset and RNA expression in the GSE48761 dataset.

Abbreviations
DKC: Dyskeratosis congenita; DMPs: Differentially methylated positions; DMRs: Differentially methylated regions; LCLs: Lymphoblastoid cell lines; NVC: Nailfold video capillaroscopy; SSc: Systemic sclerosis; WRN: Werner syndrome gene; WS: Werner syndrome

Acknowledgements
Not applicable.

Funding
This study was supported by the European Union's Seventh Framework Programme [grant number 602757 ("HUMAN: Health and the understanding of Metabolism, Aging and Nutrition")] and by the European Union's H2020 Project [grant number 634821 ("PROPAG-AGEING: The continuum between healthy ageing and idiopathic Parkinson Disease within a propagation perspective of inflammation and damage: the search for new diagnostic, prognostic and therapeutic targets") and JPco-fuND ("ADAGE: Alzheimer's Disease pathology within the ageing physiology")].

Authors' contributions
TG, MGB, AM, CF, PG, DR and FB designed the research. TG managed the biobank and performed the clinical laboratory analyses. MGB performed the statistical analysis of the data. DR performed the bioinformatics analysis and wrote the paper. AM recruited the patients, performed the NVC and wrote the paper. TG, MGB, DG, SP and DBAM performed the experiments. PG, CF, AS, FB, TG, AC and MGB analysed and discussed the data. TG and MGB were the major contributors in writing the manuscript. All authors read and approved the final manuscript.

Competing interests
The authors declare that they have no competing interests.

Author details
[1]UOC of Clinical Biochemistry, Sandro Pertini Hospital, Rome, Italy. [2]CRIIS (Interdisciplinary, Interdepartmental and Specialistic Reference Center for Early Diagnosis of Scleroderma, Treatment of Sclerodermic Ulcers and Videocapillaroscopy), Sandro Pertini Hospital, Rome, Italy. [3]IRCCS Institute of Neurological Sciences, Bologna, Italy. [4]UOSD Ischemic Microangiopathy and Sclerodermic Ulcers, Sandro Pertini Hospital, Rome, Italy. [5]Centre for Biomedical Research and Technologies, Italian Auxologic Institute, IRCCS, Milan, Italy. [6]Department of Molecular Medicine, Sapienza University of Rome, Rome, Italy. [7]Department of Experimental, Diagnostic and Specialty Medicine, University of Bologna, Bologna, Italy. [8]Interdepartmental Center "L. Galvani", University of Bologna, Bologna, Italy. [9]Center for Applied Biomedical Research (CRBA), St. Orsola-Malpighi University Hospital, Bologna, Italy. [10]Clinical Chemistry, Department of Laboratory Medicine, Karolinska Institute at Huddinge University Hospital, S-141 86 Stockholm, Sweden. [11]CNR Institute for Molecular Genetics, Unit of Bologna, Bologna, Italy. [12]Laboratory of Musculoskeletal Cell Biology, Rizzoli Orthopedic Institute, Bologna, Italy. [13]UOC of Clinical Pathology, Saint' Eugenio Hospital, Rome, Italy.

References
1. Takemoto M, Mori S, Kuzuya M, Yoshimoto S, Shimamoto A, Igarashi M, Tanaka Y, Miki T, Yokote K. Diagnostic criteria for Werner syndrome based on Japanese nationwide epidemiological survey. Geriatr Gerontol Int. 2013;13:475–81.
2. Coppede F. The epidemiology of premature aging and associated comorbidities. Clin Interv Aging. 2013;8:1023–32.
3. Goto M, Okawa-Takatsuji M, Aotsuka S, Nakai H, Shimizu M, Goto H, Shimamoto A, Furuichi Y. Significant elevation of IgG anti-WRN (RecQ3 RNA/DNA helicase) antibody in systemic sclerosis. Mod Rheumatol. 2006;16:229–34.
4. Bes C, Vardi S, Guven M, Soy M. Werner's syndrome: a quite rare disease for differential diagnosis of scleroderma. Rheumatol Int. 2010;30:695–8.
5. Capell BC, Tlougan BE, Orlow SJ. From the rarest to the most common: insights from progeroid syndromes into skin cancer and aging. J Invest Dermatol. 2009;129:2340–50.
6. Hettema ME, Zhang D, de Leeuw K, Stienstra Y, Smit AJ, Kallenberg CG, Bootsma H. Early atherosclerosis in systemic sclerosis and its relation to disease or traditional risk factors. Arthritis Res Ther. 2008;10:R49.
7. Heyn H, Moran S, Esteller M. Aberrant DNA methylation profiles in the premature aging disorders Hutchinson-Gilford progeria and Werner syndrome. Epigenetics. 2013;8:28–33.
8. Muftuoglu M, Oshima J, von Kobbe C, Cheng WH, Leistritz DF, Bohr VA. The clinical characteristics of Werner syndrome: molecular and biochemical diagnosis. Hum Genet. 2008;124:369–77.
9. Zhu X, Zhang G, Kang L, Guan H. Epigenetic regulation of werner syndrome gene in age-related cataract. J Ophthalmol. 2015;2015:579695.
10. Edwards DN, Machwe A, Chen L, Bohr VA, Orren DK. The DNA structure and sequence preferences of WRN underlie its function in telomeric recombination events. Nat Commun. 2015;6:8331.
11. Rossi ML, Ghosh AK, Bohr VA. Roles of Werner syndrome protein in protection of genome integrity. DNA Repair (Amst). 2010;9:331–44.
12. Burla R, Carcuro M, Raffa GD, Galati A, Raimondo D, Rizzo A, La Torre M, Micheli E, Ciapponi L, Cenci G, et al. AKTIP/Ft1, a new shelterin-interacting factor required for telomere maintenance. PLoS Genet. 2015; 11:e1005167.
13. Burla R, Carcuro M, Torre ML, Fratini F, Crescenzi M, D'Apice MR, Spitalieri P, Raffa GD, Astrologo L, Lattanzi G, et al. The telomeric protein AKTIP interacts with A- and B-type lamins and is involved in regulation of cellular senescence. Open Biol. 2016;6
14. Jovanovic SV, Clements D, MacLeod K. Biomarkers of oxidative stress are significantly elevated in Down syndrome. Free Radic Biol Med. 1998;25:1044–8.
15. Bacalini MG, Boattini A, Gentilini D, Giampieri E, Pirazzini C, Giuliani C, Fontanesi E, Remondini D, Capri M, Del Rio A, et al. Erratum: A meta-analysis on age-associated changes in blood DNA methylation: results from an original analysis pipeline for Infinium 450k data. Aging (Albany NY). 2016;8: 831.
16. Bacalini MG, D'Aquila P, Marasco E, Nardini C, Montesanto A, Franceschi C, Passarino G, Garagnani P, Bellizzi D. The methylation of nuclear and mitochondrial DNA in ageing phenotypes and longevity. Mech Ageing Dev. 2017. doi:10.1016/j.mad.2017.01.006.
17. Maierhofer A, Flunkert J, Oshima J, Martin GM, Haaf T, Horvath S. Accelerated epigenetic aging in Werner syndrome. Aging (Albany NY). 2017;9:1143–52.
18. Altorok N, Tsou PS, Coit P, Khanna D, Sawalha AH. Genome-wide DNA methylation analysis in dermal fibroblasts from patients with diffuse and limited systemic sclerosis reveals common and subset-specific DNA methylation aberrancies. Ann Rheum Dis. 2015;74:1612–20.
19. Silver RM. Clinical aspects of systemic sclerosis (scleroderma). Ann Rheum Dis. 1991;50(Suppl 4):854–61.
20. Haustein UF. Systemic sclerosis-scleroderma. Dermatol Online J. 2002;8:3.
21. van den Hoogen F, Khanna D, Fransen J, Johnson SR, Baron M, Tyndall A, Matucci-Cerinic M, Naden RP, Medsger TA Jr, Carreira PE, et al. classification criteria for systemic sclerosis: an American College of Rheumatology/European League Against Rheumatism Collaborative Initiative. Ann Rheum Dis. 2013;72:1747–55.
22. Du P, Zhang X, Huang CC, Jafari N, Kibbe WA, Hou L, Lin SM. Comparison of beta-value and M-value methods for quantifying methylation levels by microarray analysis. BMC Bioinformatics. 2010;11:587.
23. Weidner CI, Lin Q, Birkhofer C, Gerstenmaier U, Kaifie A, Kirschner M, Bruns H, Balabanov S, Trummer A, Stockklausner C, et al. DNA methylation in PRDM8 is indicative for dyskeratosis congenita. Oncotarget. 2016;7:10765 72.

24. Chen EY, Tan CM, Kou Y, Duan Q, Wang Z, Meirelles GV, Clark NR, Ma'ayan A. Enrichr: interactive and collaborative HTML5 gene list enrichment analysis tool. BMC Bioinformatics. 2013;14:128.

25. Kuleshov MV, Jones MR, Rouillard AD, Fernandez NF, Duan Q, Wang Z, Koplev S, Jenkins SL, Jagodnik KM, Lachmann A, et al. Enrichr: a comprehensive gene set enrichment analysis web server 2016 update. Nucleic Acids Res. 2016;44:W90-7.

26. Dumas J, Gargano MA, Dancik GM. shinyGEO: a web-based application for analyzing gene expression omnibus datasets. Bioinformatics. 2016;32:3679-81.

27. Barrett T, Troup DB, Wilhite SE, Ledoux P, Evangelista C, Kim IF, Tomashevsky M, Marshall KA, Phillippy KH, Sherman PM, et al. NCBI GEO: archive for functional genomics data sets—10 years on. Nucleic Acids Res. 2011;39:D1005-10.

28. Grafodatskaya D, Choufani S, Ferreira JC, Butcher DT, Lou Y, Zhao C, Scherer SW, Weksberg R. EBV transformation and cell culturing destabilizes DNA methylation in human lymphoblastoid cell lines. Genomics. 2010;95:73-83.

29. Bacalini MG, Gentilini D, Boattini A, Giampieri E, Pirazzini C, Giuliani C, Fontanesi E, Scurti M, Remondini D, Capri M, et al. Identification of a DNA methylation signature in blood cells from persons with Down syndrome. Aging (Albany NY). 2015;7:82-96.

30. Cheung HH, Liu X, Canterel-Thouennon L, Li L, Edmonson C, Rennert OM. Telomerase protects werner syndrome lineage-specific stem cells from premature aging. Stem Cell Reports. 2014;2:534-46.

31. Oshima J, Martin GM, Hisama FM. Werner syndrome. 1993.

32. Hernandez-Corbacho MJ, Jenkins RW, Clarke CJ, Hannun YA, Obeid LM, Snider AJ, Siskind LJ. Accumulation of long-chain glycosphingolipids during aging is prevented by caloric restriction. PLoS One. 2011;6:e20411.

33. Brunet A. Aging and the control of the insulin-FOXO signaling pathway. Med Sci (Paris). 2012;28:316-20.

34. Greer EL, Brunet A. FOXO transcription factors at the interface between longevity and tumor suppression. Oncogene. 2005;24:7410-25.

35. Eijkelenboom A, Burgering BM. FOXOs: signalling integrators for homeostasis maintenance. Nat Rev Mol Cell Biol. 2013;14:83-97.

36. Greer EL, Brunet A. FOXO transcription factors in ageing and cancer. Acta Physiol (Oxf). 2008;192:19-28.

37. Peltier J, O'Neill A, Schaffer DV. PI3K/Akt and CREB regulate adult neural hippocampal progenitor proliferation and differentiation. Dev Neurobiol. 2007;67:1348-61.

38. Adams HH, Hibar DP, Chouraki V, Stein JL, Nyquist PA, Renteria ME, Trompet S, Arias-Vasquez A, Seshadri S, Desrivieres S, et al. Novel genetic loci underlying human intracranial volume identified through genome-wide association. Nat Neurosci. 2016;19:1569-82.

39. Murata K, Nakashima H. Clinical and metabolic studies on Werner's syndrome: with special reference to disorders of lipid and liver function. Adv Exp Med Biol. 1985;190:285-304.

40. Hay N. Interplay between FOXO, TOR, and Akt. Biochim Biophys Acta. 2011;1813:1965-70.

41. Cho KA, Ryu SJ, Oh YS, Park JH, Lee JW, Kim HP, Kim KT, Jang IS, Park SC. Morphological adjustment of senescent cells by modulating caveolin-1 status. J Biol Chem. 2004;279:42270-8.

42. Atfi A, Dumont E, Colland F, Bonnier D, L'Helgoualc'h A, Prunier C, Ferrand N, Clement B, Wewer UM, Theret N. The disintegrin and metalloproteinase ADAM12 contributes to TGF-beta signaling through interaction with the type II receptor. J Cell Biol. 2007;178:201-8.

43. Taniguchi T, Asano Y, Akamata K, Aozasa N, Noda S, Takahashi T, Ichimura Y, Toyama T, Sumida H, Kuwano Y, et al. Serum levels of ADAM12-S: possible association with the initiation and progression of dermal fibrosis and interstitial lung disease in patients with systemic sclerosis. J Eur Acad Dermatol Venereol. 2013;27:747-53.

44. Jennemann R, Rabionet M, Gorgas K, Epstein S, Dalpke A, Rothermel U, Bayerle A, van der Hoeven F, Imgrund S, Kirsch J, et al. Loss of ceramide synthase 3 causes lethal skin barrier disruption. Hum Mol Genet. 2012;21:586-608.

45. Radner FP, Marrakchi S, Kirchmeier P, Kim GJ, Ribierre F, Kamoun B, Abid L, Leipoldt M, Turki H, Schempp W, et al. Mutations in CERS3 cause autosomal recessive congenital ichthyosis in humans. PLoS Genet. 2013, 9:e1003536.

46. Giacomin MF, Franca CM, Oliveira ZN, Machado MC, Sallum AM, Silva CA. Generalized morphea in a child with harlequin ichthyosis: a rare association. Rev Bras Reumatol Engl Ed. 2016;56:82-5.

47. Brown WT, Kieras FJ, Houck GE Jr, Dutkowski R, Jenkins EC. A comparison of adult and childhood progerias: Werner syndrome and Hutchinson-Gilford progeria syndrome. Adv Exp Med Biol. 1985;190:229-44.

48. Imura H, Nakao Y, Kuzuya H, Okamoto M, Yamada K. Clinical, endocrine and metabolic aspects of the Werner syndrome compared with those of normal aging. Adv Exp Med Biol. 1985;190:171-85.

49. Valentini G, D'Angelo S, Della Rossa A, Bencivelli W, Bombardieri S. European Scleroderma Study Group to define disease activity criteria for systemic sclerosis. IV. Assessment of skin thickening by modified Rodnan skin score. Ann Rheum Dis. 2003;62:904-5.

50. Lok C, Ruto F, Labeille B, Pietri J, Denoeux JP. Leg ulcers in Werner's syndrome. Report of one case. J Mal Vasc. 1991;16:381-2.

51. Yeong EK, Yang CC. Chronic leg ulcers in Werner's syndrome. Br J Plast Surg. 2004;57:86-8.

52. Mohan C, Assassi S. Biomarkers in rheumatic diseases: how can they facilitate diagnosis and assessment of disease activity? BMJ. 2015, 351:h5079.

53. Kayser C, Fritzler MJ. Autoantibodies in systemic sclerosis: unanswered questions. Front Immunol. 2015;6:167.

54. Madej-Pilarczyk A, Rosinska-Borkowska D, Rekawek J, Marchel M, Szalus E, Jablonska S, Hausmanowa-Petrusewicz I. Progeroid syndrome with scleroderma-like skin changes associated with homozygous R435C LMNA mutation. Am J Med Genet A. 2009;149A:2387-92.

55. Zhang S, Zhang K, Jiang M, Zhao J. Hutchinson-Gilford progeria syndrome with scleroderma-like skin changes due to a homozygous missense LMNA mutation. J Eur Acad Dermatol Venereol. 2016;30:463-5.

56. Luckhardt TR, Thannickal VJ. Systemic sclerosis-associated fibrosis: an accelerated aging phenotype? Curr Opin Rheumatol. 2015;27:571-6.

5

DNA methylation as a predictor of fetal alcohol spectrum disorder

Alexandre A. Lussier[1,2], Alexander M. Morin[1], Julia L. MacIsaac[1], Jenny Salmon[3,4], Joanne Weinberg[2], James N. Reynolds[5], Paul Pavlidis[6,8], Albert E. Chudley[3,4] and Michael S. Kobor[1,7*]

Abstract

Background: Fetal alcohol spectrum disorder (FASD) is a developmental disorder that manifests through a range of cognitive, adaptive, physiological, and neurobiological deficits resulting from prenatal alcohol exposure. Although the North American prevalence is currently estimated at 2–5%, FASD has proven difficult to identify in the absence of the overt physical features characteristic of fetal alcohol syndrome. As interventions may have the greatest impact at an early age, accurate biomarkers are needed to identify children at risk for FASD. Building on our previous work identifying distinct DNA methylation patterns in children and adolescents with FASD, we have attempted to validate these associations in a different clinical cohort and to use our DNA methylation signature to develop a possible epigenetic predictor of FASD.

Methods: Genome-wide DNA methylation patterns were analyzed using the Illumina HumanMethylation450 array in the buccal epithelial cells of a cohort of 48 individuals aged 3.5–18 (24 FASD cases, 24 controls). The DNA methylation predictor of FASD was built using a stochastic gradient boosting model on our previously published dataset FASD cases and controls (GSE80261). The predictor was tested on the current dataset and an independent dataset of 48 autism spectrum disorder cases and 48 controls (GSE50759).

Results: We validated findings from our previous study that identified a DNA methylation signature of FASD, replicating the altered DNA methylation levels of 161/648 CpGs in this independent cohort, which may represent a robust signature of FASD in the epigenome. We also generated a predictive model of FASD using machine learning in a subset of our previously published cohort of 179 samples (83 FASD cases, 96 controls), which was tested in this novel cohort of 48 samples and resulted in a moderately accurate predictor of FASD status. Upon testing the algorithm in an independent cohort of individuals with autism spectrum disorder, we did not detect any bias towards autism, sex, age, or ethnicity.

Conclusion: These findings further support the association of FASD with distinct DNA methylation patterns, while providing a possible entry point towards the development of epigenetic biomarkers of FASD.

Keywords: Fetal alcohol spectrum disorder, Epigenetics, DNA methylation, Biomarkers, Neurodevelopmental disorders

* Correspondence: msk@cmmt.ubc.ca
[1]Department of Medical Genetics, Centre for Molecular Medicine and Therapeutics, British Columbia Children's Hospital Research Institute, University of British Columbia, Vancouver, British Columbia, Canada
[7]Human Early Learning Partnership, University of British Columbia, Vancouver, British Columbia, Canada
Full list of author information is available at the end of the article

Background

Fetal alcohol spectrum disorder (FASD) is a leading preventable cause of developmental disability, with a North American prevalence currently estimated at 2–5% [1–3]. FASD presents through a wide spectrum of phenotypes, ranging from growth deficits and physical abnormalities to cognitive and behavioral deficits, as well as motor and sensory impairments, immune dysfunction, and increased vulnerability to mental health problems in adulthood [4–6]. On the most severe end of the spectrum lies fetal alcohol syndrome (FAS), which is characterized by growth retardation, a distinct set of facial dysmorphisms, and central nervous system abnormalities [7, 8]. By contrast, Alcohol-Related Neurodevelopmental Disorder (ARND) describes the less visible and largest group within the spectrum, where individuals with confirmed alcohol exposure during pregnancy show primarily behavioral, adaptive, and/or cognitive abnormalities without obvious facial dysmorphisms [9]. Of note, individuals across the spectrum show cognitive and behavioral deficits, which can be as serious in those without any physical features as in those with full FAS [10].

Although children with FAS are often diagnosed in infancy or in early life, FASD in general has proven difficult to identify, particularly in the absence of the overt facial features characteristic of FAS. As such, many individuals with FASD are not identified until they reach school age, where they begin to struggle with increased social pressure and cognitive challenges [11]. However, early cognitive and behavioral interventions may potentially attenuate some of the deficits associated with FASD and improve the long-term outcomes of these individuals [12]. As early diagnosis is a strong predictor of positive outcome, early screening tools are necessary to help identify at-risk children at a young age and potentially buffer some of the deficits associated with prenatal alcohol exposure (PAE) [13, 14].

While self-report methods are most commonly used for assessing PAE and a child's risk for FASD, these are not always accurate and can underestimate alcohol consumption during pregnancy [15–17]. Over the past decades, various biomarkers of alcohol exposure have been developed to complement self-report measures, focusing primarily on the direct or indirect products of ethanol metabolism, which can be measured in biological specimens from both the mother and infant [18]. Although these biomarkers are very sensitive to alcohol exposure, they present a number of limitations when attempting to determine whether prenatal alcohol exposure has occurred or to gain insight into the biological underpinnings of alcohol-induced deficits and the developmental profiles associated with FASD. For example, many of these biomarkers have short windows of detection (e.g., urine, blood, plasma) or are limited by specimen availability (e.g., placenta, meconium), making them useful for identification of alcohol exposure around the time of parturition, but not in infants and children over the course of development [19]. As such, objective and persistent measures are needed to aid in the screening and diagnosis of children at risk for FASD.

Epigenetic marks are now emerging as potential biomarkers or signatures of early-life exposures. Broadly defined, epigenetics refers to modifications of DNA and its regulatory components, including chromatin and non-coding RNA, that potentially modulate gene transcription without changing underlying DNA sequences [20–22]. In addition to their role in the regulation of cellular processes, these may also bridge environmental factors and genetic regulation to capture a lasting signature of early life exposures. In particular, DNA methylation is emerging as a candidate biomarker for environmental exposures and disease. Typically found on the cytosine residues of cytosine-guanine dinucleotides (CpG), this epigenetic mark is both stable over time and dynamic in response to environmental factors [23]. Several pre- and postnatal environmental influences have been associated with altered DNA methylation patterns, hinting at possible malleability by early-life environments and suggesting a potential utility as biomarkers [24, 25]. For example, prenatal exposure to cigarette smoke is associated with lasting alterations to DNA methylation patterns, which are now being used as biomarkers of cigarette smoke exposure in infants [26].

While in its infancy in relation to FASD, epigenetic biomarkers show promise for early screening of at-risk individuals, as the DNA methylome retains a lasting signature of prenatal alcohol exposure in both the central nervous system and peripheral tissues (reviewed in [27]). Numerous studies performed in animal and cell culture models have identified both short-term and persistent alterations to DNA methylation patterns following PAE. Although some of these models reflect supra-physiological levels of alcohol exposure or display modest effect sizes in response to PAE, the findings from these pre-clinical models suggest the possibility that PAE may directly influence epigenetic patterns and that these may play a role in PAE-induced deficits [27–33]. By contrast, fewer studies have investigated DNA methylation patterns in individuals with FASD. More targeted methods identified differences in DNA methylation levels in the promoter region of *Drd4* in a large Australian cohort of children exposed to alcohol during breastfeeding [34]. Others have employed discovery-driven approaches, assessing genome-wide DNA methylation patterns in case-control studies of FASD. The first of these came from a small cohort of children, where slight differences in DNA methylation patterns within the protocadherin (PCDH) gene clusters reported with a rather modest significance threshold [35]. Recently, we analyzed DNA methylation profiles in a large cohort of children with FASD recruited by NeuroDevNet (NDN), a Canadian Networks of Centres of Excellence, where we identified a signature of 658 differentially methylated CpGs [36]. Although few results have been validated across different cohorts, these findings set the stage for broader applications of DNA methylation in the context of FASD, creating a framework upon which to build future epigenomic studies of FASD.

To validate the findings from our previous DNA methylation study, we assessed the genome-wide DNA methylation profiles of buccal epithelial cells (BEC) from an independent cohort of 24 individuals with FASD, aged 3.5–18, and 24 typically developing controls, aged 5–17. Given that our initial study provided a framework for genome-wide assessment of DNA methylation patterns in individuals with FASD, we used the findings from the NDN study as a foundation for the identification of replicable epigenetic differences associated with FASD. Notably, nearly 25% of statistically significant associations from the NDN cohort were validated in this new cohort at a false-discovery rate (FDR) < 0.05 [37]. In addition to the validation analyses, we also assessed whether DNA methylation profiles could be used to identify individuals with FASD, generating a classification algorithm that uses DNA methylation levels to accurately predict FASD status. Taken together, these results suggested that there were replicable differences in DNA methylation patterns between individuals with FASD and controls, which could potentially contribute to the development of a screening tool for at-risk children.

Methods

The Kids Brain Health Network cohort of children with FASD

The present cohort was collected as a replication study by Kids Brain Health Network (KBHN), formerly NeuroDevNet, and is hereby referred to as the KBHN cohort [38]. Ethics for this study were reviewed and approved by the "Children's and Women's Research Ethics Board – Clinical" (H10-01149). All experimental procedures were reviewed and approved by the University of Manitoba and the University of British Columbia. Written informed consent was obtained from a parent or legal guardian, and assent was obtained from each child before study participation. The clinics used previously described guidelines for the diagnosis of FASD [39]. Children with FASD and typically developing controls were recruited from the Manitoba FASD diagnostic clinic in Winnipeg, Manitoba, Canada. Briefly, buccal epithelial cell (BEC) samples were collected for DNA methylation analysis from 25 FASD and 26 age-

and sex-matched control children aged between 3.5 and 18, prior to pre-processing (Table 1). BECs were collected using the Isohelix buccal swabs and Dri-Capsule (Cell Projects Ltd., Kent, UK). To collect buccal cells, the swab was inserted into the participants' mouth and rubbed firmly against the inside of the left cheek for 1 min. The swab was then placed into a sterile tube with a Dri-Capsule and the tube sealed. An identical procedure was followed for the right cheek. Participants did not have any dental work performed 48 h prior to collection, and no food was consumed less than 60 min prior to collection to avoid contamination.

DNA methylation 450K assay

DNA was extracted from BECs using the Isohelix DNA isolation kit (Cell Projects, Kent, UK). Seven hundred fifty nanograms of genomic DNA was subjected to bisulfite conversion using the Zymo EZ DNA Methylation Kit (Zymo Research, Irvine, California), which converts DNA methylation information into sequence base differences by deaminating unmethylated cytosines to uracil while leaving methylated cytosines unchanged. One hundred sixty nanograms of converted DNA was applied to the HumanMethylation450 BeadChip array from Illumina (450K array), which enables the simultaneous

Table 1 Characteristics of the NeuroDevNet II FASD cohort

	FASD cases	Controls
N	24	24
ARND	18	
Partial FAS	6	
FAS	1	
FASD	1	
Age (years)		
Range	3.5–18	5–17
Mean	9.1	11.6
Sex		
Female	9	13
Male	15	11
Self-declared ethnicity		
Caucasian	4 (2)[a]	22
First Nations	17 (20)[a]	1
Asian	1 (0)[a]	1
Not reported	2	0
Caregiver status		
Biological parents	7	24
Biological grandparents	3	0
Adopted/legal guardian	8	0
Foster care	6	0

[a]Including mixed lineage First Nations

quantitative measurement of 485,512 CpG sites across the human genome, following the manufacturer's instructions. Chips were scanned on an Illumina HiScan, with the 51 samples run in two batches and each containing a similar number of FASD and control samples, randomly distributed across the chips. Two pairs of technical replicates were also included and showed a Pearson correlation coefficient $r > 0.994$ in both cases, highlighting the technology's reproducibility on our in-house platform. Inter-sample correlations ranged from 0.926–0.99.

DNA methylation data quality control and normalization
The raw DNA methylation data were subjected to a rigorous set of quality controls, first of the samples, and then of the probes. Of the 51 initial samples, 3 were removed from the final dataset based on poor quality data, which was identified through skewed internal controls and/or > = 5% of probes with a detection p value > 0.05 (2 controls and 1 FASD). Next, probes were removed from the dataset according to the following criteria: (1) probes on X and Y chromosomes ($n = 11,648$), (2) SNP probes ($n = 65$), (3) probes with bead count < 3 in 10% of samples ($n = 726$), (4) probes with 10% of samples with a detection p value > 0.01 ($n = 11,864$), and (5) probes with a polymorphic CpG and non-specific probes ($N = 19,337$ SNP-CpG and 10,484 non-specific probes) [40]. A final filtering step was performed to set the methylation values to NA for any remaining probe-sample pair where bead count < 3 or detection p value > 0.01. Data normalization was performed using the SWAN method on the final dataset, composed of 48 samples (24 FASD and 24 controls) and 431,544 probes [41]. Finally, batch effects (chip number and chip position) were removed using the ComBat function from the *SVA* package in R [42]. Statistical analyses were performed using on ComBat-corrected M values, which represent the log2 ratio of methylated/unmethylated, where negative values indicate less than 50% methylation and positive values indicate more than 50% methylation [43]. Percent methylation differences (beta-values) were used in graphical representations of the data and indicate the percentage of methylation calculated by methylated/(methylated + unmethylated), ranging from 0 (fully unmethylated) to 1 (fully methylated).

Differential methylation analysis and validation of NeuroDevNet (NDN) findings
Cell type deconvolution was performed to assess the proportions of CD14, CD34, and buccal epithelial cells in each sample using DNA methylation levels at CpGs highly correlated with these cell types [44]. Surrogate variable analysis (SVA) was also performed on ComBat-corrected, normalized data using the *SVA* package in R

to identify surrogate variables (SVs) representative of unwanted heterogeneity [42]. Using DNA methylation data from all 48 samples, SVA identified 6 SVs not associated with clinical status (FASD vs control). As these were partially associated with known covariates, such as cell type proportions and age, the SVs were included in the linear regression analysis to account for their effects. More specifically, linear modeling was performed on the 648 differentially methylated probes identified in the initial NDN study and found in the present dataset using the *limma* package in R and a model that included clinical status and all identified SVs as covariates [36, 45]. Significant differentially methylated probes between groups were identified at a false-discovery rate (FDR) < 0.05 following multiple test correction by the Benjamini-Hochberg method and were required to show the same direction of change as the NDN cohort's findings [46]. Further evaluation of potential biological significance was performed using an arbitrary threshold of > 5% mean percent DNA methylation difference between FASD and controls.

DNA methylation pyrosequencing assay
The bisulfite pyrosequencing assay was designed with PyroMark Assay Design 2.0 (Qiagen; Additional file 1: Table S1). The region of interest was amplified by PCR using the HotstarTaq DNA polymerase kit (Qiagen) as follows: 15 min at 95 °C, 45 cycles of 95 °C for 30s, 58 °C for 30s, and 72 °C for 30s, and a 5-min 72 °C final extension step. For pyrosequencing, single-stranded DNA was prepared from the PCR product with the Pyromark™ Vacuum Prep Workstation (Qiagen) and the sequencing was performed using sequencing primers on a Pyromark™ Q96 MD pyrosequencer (Qiagen). The quantitative levels of methylation for each CpG dinucleotide were calculated with Pyro Q-CpG software (Qiagen).

The NDN cohort of children with FASD
DNA methylation data from the previous cohort of children with FASD were obtained from GEO (GSE80261) and normalized as described in our original publication [36]. This cohort was collected by NeuroDevNet, a Canadian Network of Centres for Excellence, and is hereby referred to as the NDN cohort [36]. Briefly, we selected the individuals with a confirmed diagnosis of FASD from this dataset, as well as age- and sex-matched typically developing controls, resulting in dataset composed of 83 children with FASD (55 ARND, 18 partial FAS, 10 FAS) and 96 typically developing controls. The mean age (in years) for individuals with FASD was 11.88 and 11.28 for controls, both ranging from 5 to 18 years old. The proportions of males and female differed slightly between groups, with 42 females and 41 males in

the FASD cases and 57 females and 39 males in the control group. A skew in self-declared ethnicity was present between the groups, as the majority of controls identified as Caucasian, while the majority of children in the FASD group identified as First Nations. This skew was addressed in the initial epigenome-wide association study through the use of a more ethnically homogeneous subset of the cohort. DNA methylation data were obtained from BEC using the Illumina 450K array and were normalized using the beta-mixture quantile normalization method.

Cohort of individuals with autism spectrum disorder

Processed DNA methylation data from a publically available dataset of individuals with autism spectrum disorder (ASD) were obtained from GEO (GSE50759). Briefly, this dataset was composed of 48 individuals with ASD and 48 typically developing controls. As per the authors' description of the GEO data, these were preprocessed using the R packages *minfi* and *sva* to obtain normalized M values [47]. The samples consisted of 58 males and 38 females, consistent with the skew towards males in ASD. The mean age (8.84) and range (1–28 years old) differed from the NDN and KBHN studies, and the genetic ancestry of most individuals was Caucasian (European), though a proportion of the cohort was of Nigerian ancestry. DNA methylation data of these samples were obtained from BEC using the Illumina 450K array.

DNA methylation as a predictor of FASD status

A predictive model of FASD status was created using DNA methylation data and the *caret* package in R [48]. First, a predictive model was created using stochastic gradient boosting on the NDN cohort (83 FASD cases, 96 controls) using the beta-values of the differentially methylated probes identified in the NDN study (648 probes) [36]. The parameters of the modeling were optimized for area under the receiver operating characteristic (ROC) curve by grid tuning for repeated cross-validation (number of trees 50–1500; 1, 5, or 9 interaction depth; 0.1 shrinkage). The optimal model for predicting clinical FASD status using 648 probes was 550 trees, 1 of interaction depth, and 20 minimum observations per node. The KBHN cohort (24 FASD, 24 controls) was then used to verify the predictive sensitivity and specificity of the model. In parallel, 450K data from a cohort of children with ASD were tested to verify the predictive specificity of the model for FASD. The predictor was tested using normalized data that was uncorrected for batch effects to better mimic the potential use of the predictive model by independent groups.

Results

The KBHN cohort of children with FASD

As noted, we analyzed genome-wide DNA methylation patterns from 24 children with FASD and 24 typically developing controls, matched for sex and age, ranging from 3.5 to 18 years of age (Table 1). We found that self-declared ethnicity, primary caregiver, and mean age were significantly different between the FASD and control participants (Student's t test; $p < 0.05$). We corrected for the potential effects of age on DNA methylation through the statistical methods outlined below. However, given the confound in self-declared ethnicity and caregiver status, we could not correct for these effects and relied on the previous correction of ethnic bias in the initial NDN study (see below) [36]. Furthermore, we could not account for the different life experiences of individuals with FASD, including potential exposure to adverse early life events at considerably higher levels than those in the general population. It is possible that these distinct experiences in themselves may potentially be associated with DNA methylation patterns.

Children with FASD and typically developing controls showed differential DNA methylation patterns

Following quality control and normalization, 431,544 sites of the 485,512 sites remained in the final dataset of 48 samples, which were corrected for batch effects using ComBat. While BECs are mostly a homogeneous population of cells, they contain small proportions of CD34- and CD14-positive white blood cells, which can potentially skew DNA methylation analyses. As such, cell type deconvolution was performed to identify any blood contamination in the samples, identifying a trend towards significance in the proportions of different cells types between groups (CD34+, $p = 0.115$; CD14+, $p = 0.224$; BEC, $p = 0.068$). To account for this factor in addition to other potential confounding variables within the dataset, we performed SVA to identify patterns of variation, identifying 6 surrogate variables when protecting the effects of group (FASD vs control). These were correlated with known sources of variation within the data, including cell type proportions and age (Additional file 2: Figure S1).

To identify DNA methylation patterns specific to the FASD group, we coupled differential DNA methylation analysis using a two-group design with the surrogate variables to correct for undesirable variation in the data. Given that we already accounted for ethnicity-related probes as much as possible in the NDN study, it was concluded that the effects of ethnic background would be lessened by using the final 658 differentially methylated CpGs [36]. As such, we performed linear modeling on the probes that were differentially methylated in the first study and remained in the dataset after pre-processing (648 CpGs of 658 from NDN). Of these, 161 CpGs displayed statistically significant differential methylation in the same direction as the initial cohort in the KBHN FASD group compared to the controls at an FDR < 0.05 (Fig. 1a; Additional file 1: Table S2). To assess the probability of validating this many probes, random group subsampling was performed 10,000 times. As none showed more differentially methylated probes than the original replication cohort (maximum = 31 differentially methylated probes), the probability of validating 161/648 probes was < 1e−4 (Additional file 2: Figure S2). Of the 161 validated probes, 82 were up-methylated, while 79 were down-methylated in FASD compared to control samples. Several genes contained multiple differentially methylated CpGs across both co-horts, including *Hla-dpb1* (5), *Fam59b* (4), *Capn10* (3), *Des* (3), *Slc6a3* (3), *Slc38a2* (3), *Fam24a* (2), *H19* (2), and *Tgfb1i1* (2) (Table 2). Moreover, 53 CpGs showed > 5% difference in methylation, an arbitrary cutoff often used to gauge potential biological significance [49]. Three genes contained 2 or more differentially methylated (DM) probes that showed both an FDR < 0.05 and > 5% difference in percent methylation, including *Fam59b* (4 probes), *Hla-dpb1* (2 probes), and *Slc6a3* (2 probes). In particular, the *Fam59b* CpGs were located

Table 2 Genes containing multiple differentially methylated CpGs in FASD

Gene	No. of CpGs	Direction of change
Hla-dpb1	5	Up
Fam59b	4	Down
Des	3	Down
Slc6a3	3	Up
Slc38a2	3	Down
Capn10	3	Up
Fam24a	2	Up
H19	2	Down
Tgfb1i1	2	Down

within a CpG island and showed substantial differences in DNA methylation levels between FASD and control groups, with an average 13% methylation difference across the array probes in the CpG island (Fig. 2). Three additional sites located in intergenic regions showed > 10% percent DNA methylation difference between groups.

Overall, the percent methylation differences between groups of the 648 analyzed probes were highly correlated between the NDN and KBHN cohorts, as determined by linear modeling (r = 0.638, p < 2.2e−16; Fig. 1b). We also compared the ranking of probes by p value from linear

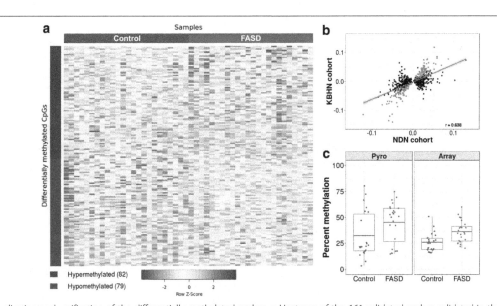

Fig. 1 Visualization and verification of the differentially methylated probes. **a** Heatmap of the 161 validated probes validated in the KBHN cohort at an FDR < 0.05 (79 hypermethylated in FASD; 82 hypomethylated in FASD). The percent methylation values (ranging from 0 to 100) were centered, scaled, and trimmed, resulting in a standardized DNA methylation level ranging from − 2 to + 2 (blue-red scale). **b** Scatter plot of the differences in percent methylation between FASD and controls for the 648 differentially probes identified in the NDN cohort. The mean differences between groups were highly correlated between both the NDN and KBHN cohorts (r = 0.638, p < 2.2e-16). The red points show the probes that were statistically significant (FDR < 0.05) and showed the same direction of change across both studies **c** Verification by bisulfite pyrosequencing in FASD (*blue*) and control (*gray*) samples. The left panel shows the DNA methylation levels from the pyrosequencing assay, while the right panel shows the results from the 450K array. The CpG assayed was located in the CACNA1A gene body (cg24800175) and showed statistically significant differences between groups (p = 0.04)

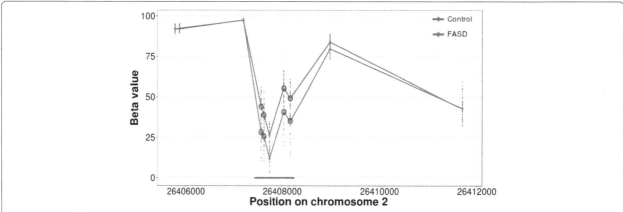

Fig. 2 Several differentially methylated CpGs were located in the *Fam59b* gene body. DNA methylation levels for FASD (*blue*) and controls (*gray*) are shown for 10 CpGs within the gene, with the red circles representing the validated hits in KBHN (FDR < 0.05). These were located in a CpG island, illustrated by the green bar at the bottom, which showed an average 13% difference in DNA methylation levels in individuals with FASD versus controls across all five CpGs covered by the 450K array

modeling between the NDN and KBHN cohorts; no significant similarities were identified ($p = 0.91$). Of note, 21 of the significant probes with > 5% methylation difference between groups from the NDN study were validated in the present analysis (39 of 41 were present in the filtered KBHN dataset). This proportion (54%) was much higher than all validated probes (25%), suggesting that these potentially represented more robust associations with FASD.

Bisulfite pyrosequencing verified the differential DNA methylation of CACNA1A

To verify that the differential DNA methylation results did not depend on the method used to measure them, we assessed DNA methylation levels of the cg24800175 probe in CACNA1A. We selected this probe as it was also verified in the initial NDN study, where it similarly showed a > 5% difference in DNA methylation between individuals with FASD and controls. Pyrosequencing results confirmed the DNA methylation levels observed on the 450K array, showing similar DNA methylation levels and differences between groups for CpGs located in *CACNA1A* (Fig. 1c). The Pearson correlation between these two methods was 0.826 and the Bland–Altman plot showed little difference when comparing the 450K array to pyrosequencing, suggesting good concordance between DNA methylation data from the two methods (Additional file 2: Figure S3). Linear regression analysis of pyrosequencing data between FASD cases and controls confirmed differential DNA methylation in this site, even without correcting for covariates ($p = 0.04$).

DNA methylation patterns classified individuals with FASD versus controls

To assess whether DNA methylation data could be used to predict FASD status, we created a predictive algorithm of FASD using machine learning approaches. First,

we selected the normalized DNA methylation data (beta-values) of 179 samples from the NDN cohort (83 FASD; 96 control) in the 648 initial probes that were also found in the KBHN data. Our strategy was to build the predictor using an initial training cohort (NDN), followed by subsequent evaluation in the test cohort (KBHN). See Fig. 3 for an overview of steps used to build the FASD predictor.

Using a gradient boosting model in the *caret* package to optimize both sensitivity and specificity (area under the receiver operating characteristic (ROC) curve), we

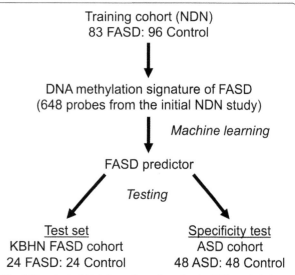

Fig. 3 Flowchart of bioinformatic analyses for the DNA methylation predictor of FASD. Briefly, samples from the NDN cohort were used as the training set, and machine learning was performed on the DNA methylation signature of FASD identified in the initial NDN study. The resulting FASD predictor was tested on the KBHN test set, as well as an independent cohort composed of individuals with autism spectrum disorder and typically developing controls to test the specificity of the predictor for FASD

created a predictive model to assess the probability of FASD based on DNA methylation patterns [48]. This method provided weighted values for the different features (CpGs) of the model to determine their importance in classifying the samples. Of the 648 initial features, 183 had non-zero influence on the predictive model and could be used to predict FASD status (Additional file 1: Table S3). As the number of non-zero features was similar to the total number of samples, concerns of model over-fitting were reduced.

Through this approach, the predicted sensitivity and specificity for the training cohort were 0.879 and 0.944, respectively, for an area under the curve of 0.977 (95% confidence intervals, 0.972–0.982; Fig. 4a). The performance of the predictor on the training data indicated that DNA methylation could be used to distinguish FASD cases and controls, although these results will need to be carefully assessed in independent test sets or clinical settings.

To get a better understanding of the utility of this tool, we next assessed the predictive model using the normalized, batch-corrected DNA methylation data of the KBHN cohort as a test set. Of note, these data were not corrected for any covariates or surrogate variables other than batch correction. As expected for analysis of an independent test set, the model performed at a lower level in this cohort, displaying 0.917 sensitivity, 0.75 specificity, and 0.920 area under the ROC curve (Table 3; Fig. 4a). The balanced accuracy of the model in this cohort was 0.833 (95% CI 0.698–0.925), and the ROC curve was not significantly different from the one obtained in the training cohort (p = 0.192). Overall, 2 controls were misclassified as FASD and 6 children with FASD were misclassified as controls, giving a negative predictive value (NPV) of 78.6% and a positive predictive value (PPV) of 90%. Given the discrepancies in ethnic backgrounds between FASD and control groups, the misclassified samples were assessed for differences in self-reported ethnicity, caregiver status, age, and buccal cell-type proportions in the classification. We did not identify any skew of these data in the misclassified controls, which were both Caucasian males aged 15 and 16, respectively. Although every misclassified individual with FASD had a previous diagnosis of ARND, a category that was present in high proportion within this cohort, no other patterns emerged between the correctly and incorrectly classified individuals with FASD (3 females/3 males; 3 First Nations/1 Métis/2 Caucasian; aged 6–18). Taken together, these findings suggested that differences in these demographic variables between the groups did not drive their classification.

The DNA methylation predictors were not biased by ASD in an independent cohort

BEC samples from an independent published autism spectrum disorder (ASD) cohort were used to assess the specificity of the model in the FASD cohorts. To this end, we used a publically available dataset of 450K array data from the BECs of 48 individuals with ASD and 48 typically developing controls from the gene expression omnibus (GSE50759) [47]. Using processed GEO data from this cohort, the predictor correctly identified the vast majority of individuals in the cohort as non-FASD. The model only misclassified 1 individual as FASD, for a specificity of 0.990 (95% CI 0.943–0.9997), higher than the predicted specificity in the training set. This sample, a 3-year-old female with ASD (51% African ancestry, 41% European ancestry) did not have any particular distinguishing features compared to the correctly classified samples, suggesting that the predictive model was not biased for ASD, sex, age, or African ancestry in this independent cohort.

Discussion

Epigenetic marks are emerging as potential biomarkers and mediators of environmental exposures, and a growing body of literature suggests that epigenetic factors may be involved in the etiology of FASD. In particular, our recent study using a large cohort of children with FASD to date identified a signature of 658 differentially methylated CpGs in the BEC of individuals with FASD compared to typically developing controls [36]. Here, we present a validation of genome-wide DNA methylation data in a small cohort of individuals with FASD, where we successfully replicated 161 of the 658 differentially methylated CpGs identified in the initial NDN cohort. Furthermore, we demonstrated that DNA methylation data could be utilized to generate a predictive algorithm to classify individuals as FASD or control with high accuracy. These results indicated that DNA methylation in BECs could potentially be used towards developing a screening tool for children at risk for FASD.

Our present findings represent the initial validation of genome-wide DNA methylation differences in individuals with FASD. Of the 161 validated CpGs at an FDR < 0.05, 53 had > 5% difference in DNA methylation levels between groups. This 5% threshold is often used for assessing potential biological relevance in epigenetic studies of psychiatric and neurodevelopmental disorders, and therefore, we confined our interpretation of possible functional implications to CpGs with this effect size [47, 49–51]. Importantly, the biological significance of a 5% difference in DNA methylation is poorly understood, and its functional relevance may be limited in relation to gene expression or cellular physiology. Nevertheless, 21 CpGs showed a > 5% difference between FASD cases and controls at an FDR < 0.05 in both the KBHN and NDN cohorts, suggesting that these may represent the strongest associations with FASD. Although the DNA methylation differences between FASD and controls

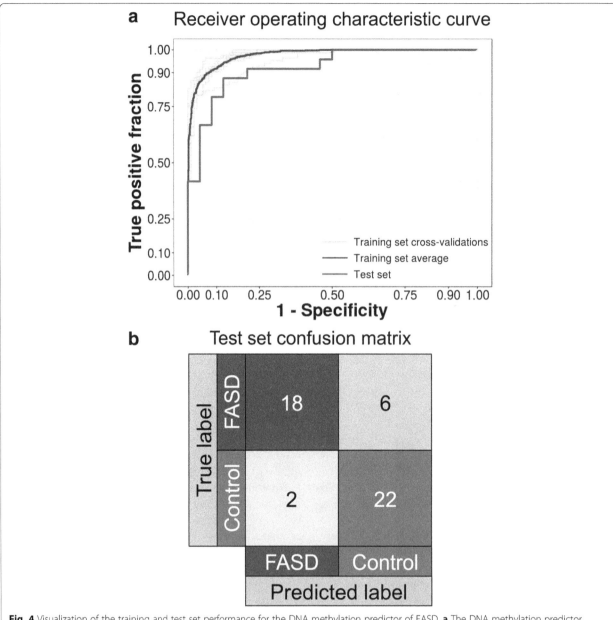

Fig. 4 Visualization of the training and test set performance for the DNA methylation predictor of FASD. **a** The DNA methylation predictor created using the 648 probes identified in NDN showed high accuracy in the training cohort (*dark gray*; area under the curve = 0.977) and slightly poorer accuracy in the KBHN test set (*blue*; area under the curve = 0.920). These curves were not significantly different (*p* = 0.192). **b** The confusion matrix displays number of samples classified correctly or incorrectly. Of note, six individuals with FASD in the test set were classified as controls, while only two control samples were misclassified as FASD

were highly correlated between the NDN and KBHN cohorts (*p* < 2.2e−16), the majority of CpGs showing the same direction of change did not achieve statistical significance (301/648), potentially due to the replication cohort's small size or the absence of individuals with only PAE in this cohort. In addition, we verified the results from the 450K array by bisulfite pyrosequencing, confirming the differential DNA methylation results for a CpG located in CACNA1A and supporting that our findings were not an artifact of array technology. As

discussed in a recent comprehensive review, we note that the functional relevance of these differences is highly dependent on multiple factors, including subcellular differences, transcription factor binding regulation, density and cooperativity of DNA methylation, or other *cis*-regulatory elements [52].

Several genes previously associated with PAE or FASD contained multiple differentially methylated CpGs with > 5% difference in DNA methylation between groups, including *Fam59b*, *Hla-dpb1*, and *Slc6a3*. The *Hla-dpb1*

Table 3 Summarized results from the classification algorithm

Training set (NDN)	
AUC	0.977
Accuracy	0.914
Sensitivity	0.879
Specificity	0.944
Test set (KBHN)	
AUC	0.920
Accuracy	0.833
Sensitivity	0.75
Specificity	0.917
False positives	2
False negatives	6
PPV	0.900
NPV	0.786
Negative control (ASD)	
Accuracy	0.990
Sensitivity	NA
Specificity	0.990
False positives	1

locus, a member of the major histocompatibility complex proteins, contained several differentially methylated CpGs, which overlapped with a differentially methylated region identified in the NDN study. Given its important function in immune regulation and potential role in rheumatoid arthritis, these differences could potentially reflect some of the immune deficits associated with FASD [53]. Furthermore, the *Fam59b* gene contained several CpGs with substantial (> 10%) differences in DNA methylation levels between individuals with FASD and controls, potentially representing a particularly sensitive locus with regard to FASD. Of note, only one validated CpG was located in one of the protocadherin gene clusters (*Pcdhb18*), which were considerably enriched in previous genome-wide studies of DNA methylation in individuals with FASD [35, 36]. Given that these different studies only showed one overlapping probe, this could indicate higher variability within these gene clusters that may be associated with other variables not present in the current dataset, such as differences in age, body mass index, ethnicity, and socio-economic status.

Of particular interest, we replicated the differential DNA methylation patterns of the two genes involved in dopamine signaling from the NDN cohort, the dopamine transporter *Slc6a3* and the dopamine receptor D4 (*Drd4*). Given the important role of the dopaminergic system in brain development and its interactions with neuroendocrine and immune systems, these differences could potentially reflect broader alterations to signaling

pathways in the organism. Of note, the BEC of children exposed to alcohol during prenatal life and breastfeeding also display altered DNA methylation patterns in the promoter region of *Drd4* [34]. Furthermore, several disorders previously associated with allelic variation and DNA methylation in this gene show either overlaps or co-morbidities with FASD, including attention deficit hyperactivity disorder, bipolar disorder, anxiety disorder, schizophrenia, and substance abuse [54–64]. Although it is tempting to interpret these findings in the context of PAE-induced deficits, DNA methylation differences in BEC likely do not fully reflect alterations in the central nervous system. Nevertheless, it has been suggested that BECs may act as a suitable surrogate tissue in human studies of DNA methylation, as they are also derived from the ectoderm [65]. While we did not measure these genes in additional tissues, evidence from animal models suggests that PAE can cause lasting alterations to the epigenome of central nervous system tissues, and as such, these results could potentially represent broader associations with epigenomic patterns in the brain [27].

Although these findings represent the initial validation of genome-wide DNA methylation data in children with FASD, a few particularities of the KBHN cohort limit the interpretability and generalizability of these results. Similar to the original cohort, the KBHN replication cohort was confounded by ethnicity, as the vast majority of FASD cases were from First Nations communities, while controls were mainly Caucasian. Given that genetic background influences DNA methylation patterns, differences between groups may have been, at least partly, due to ethnicity. Unfortunately, the KBHN cohort was too small to separate the groups into more ethnically homogeneous subsets, a method we had previously used to account for ethnicity-related differences in DNA methylation [37]. As such, we performed linear modeling on the sites that had been previously identified in the NDN study, which were partially filtered for ethnicity-related differences during the analysis of the first cohort. However, some of the top differentially methylated genes could potentially be influenced by ethnicity differences between groups in spite of our best efforts. For instance, three known polymorphisms are located within the *Fam59b* locus (dbSNP minor allele frequencies: rs774397935, 1.04%; rs4665833, 5.1%; rs181971256, 21.4%). Although, as of now, none of these are known methylation quantitative trait loci (mQTL), the *Fam59b* gene body contains several mQTLs in the developing human brain, and genetic variation outside the region could potentially influence DNA methylation levels [66]. In addition, nearby genetic variation can also influence DNA methylation patterns in the promoter of *Drd4*, which may be reflected in this cohort through the skew in ethnicity between groups [59]. Although the frequencies of these alleles in First Nation populations have not been assessed, genetic differences between groups could potentially influence DNA methylation levels within this differentially methylated region.

In addition to self-declared ethnicity, significant differences in the primary caregiver were present between groups, as all controls lived with their biological families, while the majority of children with FASD were generally in foster care or adoptive families. While the effects of this disparity on the epigenome are unclear, they could influence DNA methylation patterns through a number of factors, including nutrition, early-life adversity, and socio-economic status [67]. Individuals with FASD also tend to have life experiences different from those of typically developing children, which include early life adversity (e.g., maltreatment or neglect), separation from the biological family/placement in foster care (as occurred in our cohorts), poverty, and familial stress [13, 68]. Importantly, both pre- and postnatal experiences are known to play a role in early programming and thus may also influence DNA methylation patterns. As such, it may be difficult to separate the impacts of PAE and environmental adversity, and studies evaluating FASD may in many instances assess a combination of different factors and exposures, which is often the reality in this population. Nevertheless, our findings demonstrated clear and replicable associations between FASD and DNA methylation patterns across two independent cohorts. We believe that our use of SVA to partially account for unknown factors that could influence DNA methylation reduced some of the potential confounds associated with the cohort design. Future studies with larger groups that are balanced for ethnicity, age, and additional variables, including a focus on environmental stress/adversity, will be necessary to tease out these differences and further validate our findings.

Finally, we show here that DNA methylation patterns can be utilized as predictive variables for FASD in clinical populations. These findings complement and extend previous studies that investigated different molecular and physiological markers to help screen children for potential prenatal alcohol exposure, including alcohol metabolites in mothers and children, circulating miRNA in mothers, and cardiac orienting response in children [69–72]. In particular, eye tracking measures have been used in a small cohort of children to distinguish children with FASD, ADHD, or typically developing controls with relatively good accuracy [73]. In contrast to these studies, we selected only individuals with confirmed FASD from the initial NDN training cohort to create a DNA methylation-based predictor that was specific to individuals with an FASD diagnosis. The classification model was tuned to screen children at a higher risk for FASD with high sensitivity and specificity in an attempt to

balance the false-positive and false-negative rates. Importantly, our results suggest that DNA methylation predictors can achieve high accuracy in the classification of individuals with FASD versus controls across multiple cohorts. Moreover, the predictive algorithm appeared to be largely independent of typical confounding factors, such as age, sex, ethnicity, and cell type composition of the samples. The predictor was also unbiased towards individuals with ASD and although there was no report of FASD in this independent cohort, it is possible that our reported false-positive could be due to an undiagnosed FASD case [74]. Collectively, these results support the use of DNA methylation as a potential screening tool for FASD.

Conclusions
Given the broad spectrum of cognitive, behavioral, and biological deficits associated with PAE, FASD places an important strain on both societal resources and the affected individuals and families. As such, accurate screening methods are necessary to identify children at risk for FASD at an early age, when interventions are most effective. Our findings provide an initial stepping-stone towards epigenetic biomarkers of FASD and could potentially be adapted for the development of related screening tools for neurodevelopmental disorders. Validation of these tools across different cohorts, with increased sample sizes, varying ages, ethnicities, and better documented environmental exposures will be essential to parse out the strongest associations and to develop reliable epigenetic screening tools for FASD.

Abbreviations
450K array: Illumina HumanMethylation450 BeadChip array; ARND: Alcohol-related neurodevelopmental disorder; ASD: Autism spectrum disorder; BEC: Buccal epithelial cell; CpG: Cytosine-guanine dinucleotide; DNA: Deoxyribonucleic acid; FAS: Fetal alcohol syndrome; FASD: Fetal alcohol spectrum disorder; FDR: False discovery rate; GEO: Gene Expression Omnibus; KBHN: Kids Brain Health Network; NDN: NeuroDevNet; PAE: Prenatal alcohol exposure; PCDH: Protocadherin; RNA: Ribonucleic acid; ROC: Receiver operator characteristic; SV: Surrogate variable; SVA: Surrogate variable analysis

Acknowledgements
MSK is the Canada Research Chair in Social Epigenetics, the Sunny Hill BC Leadership Chair in Child Development, and a Senior Fellow of the Canadian Institute For Advanced Research (CIFAR).

Funding
This research is supported by the Kids Brain Health Network (formerly NeuroDevNet), a Canadian Network of Center's for Excellence and a program of the federal government to advance science and technology. AAL is supported by a Developmental Neurosciences Research Training award from Brain Canada & NeuroDevNet. JW is supported by grants from the US

National Institutes of Health/National Institute on Alcohol Abuse and Alcoholism (R37 AA007789 and RO1 AA022460); and the Canadian Foundation for Fetal Alcohol Research. PP is supported by an NSERC Discovery Grant.

Authors' contributions
AAL all bioinformatic analyses and wrote the manuscript. AMM performed the pyrosequencing. JS and AEC collected the samples. AEC, JW, PP, JNR, and MSK helped with the study design, interpretation, and writing. All authors read and approved the final manuscript.

Consent for publication
Written informed consent was obtained from a parent or legal guardian and assent was obtained from each child before study participation.

Competing interests
The authors declare that they have no competing interests.

Author details
[1]Department of Medical Genetics, Centre for Molecular Medicine and Therapeutics, British Columbia Children's Hospital Research Institute, University of British Columbia, Vancouver, British Columbia, Canada. [2]Department of Cellular and Physiological Sciences, Life Sciences Institute, University of British Columbia, Vancouver, British Columbia, Canada. [3]Department of Pediatrics and Child Health, Faculty of Medicine, University of Manitoba, Winnipeg, Manitoba, Canada. [4]Department of Biochemistry and Medical Genetics, Faculty of Medicine, University of Manitoba, Winnipeg, Manitoba, Canada. [5]Department of Biomedical and Molecular Sciences, Centre for Neuroscience Studies, Queen's University, Kingston, Ontario, Canada. [6]Michael Smith Laboratories, University of British Columbia, Vancouver, British Columnbia, Canada. [7]Human Early Learning Partnership, University of British Columbia, Vancouver, British Columbia, Canada. [8]Department of Psychiatry, University of British Columbia, Vancouver, British Columbia, Canada.

References
1. May PA, Gossage JP, Kalberg WO, Robinson LK, Buckley D, Manning M, et al. Prevalence and epidemiologic characteristics of FASD from various research methods with an emphasis on recent in-school studies. Dev Disabil Res Rev. 2009;15:176–92.
2. May PA, Baete A, Russo J, Elliott AJ, Blankenship J, Kalberg WO, et al. Prevalence and characteristics of fetal alcohol spectrum disorders. Pediatrics. 2014;134:855–66. https://doi.org/10.1542/peds.2013-3319.
3. May PA, Keaster C, Bozeman R, Goodover J, Blankenship J, Kalberg WO, et al. Prevalence and characteristics of fetal alcohol syndrome and partial fetal alcohol syndrome in a Rocky Mountain Region City. Drug Alcohol Depend. 2015;155:118–27. https://doi.org/10.1016/j.drugalcdep.2015.08.006.
4. Zhang X, Sliwowska JH, Weinberg J. Prenatal alcohol exposure and fetal programming: effects on neuroendocrine and immune function. Exp Biol Med. 2005;230:376–88.
5. Pei J, Denys K, Hughes J, Rasmussen C. Mental health issues in fetal alcohol spectrum disorder. J Ment Health. 2011;20:473–83.
6. Mattson SN, Crocker N, Nguyen TT. Fetal alcohol spectrum disorders: neuropsychological and behavioral features. Neuropsychol Rev. 2011;21:81–101.
7. Jones KL, Smith DW. Recognition of the fetal alcohol syndrome in early infancy. Lancet. 1973;302:999–1001. https://doi.org/10.1016/S0140-6736(73)91092-1.
8. Astley SJ, Clarren SK. Diagnosing the full spectrum of fetal alcohol-exposed individuals: introducing the 4-digit diagnostic code. Alcohol Alcohol. 2000;35:400–10.
9. Jacobson SW, Jacobson JL, Stanton ME, Meintjes EM, Molteno CD. Biobehavioral markers of adverse effect in fetal alcohol spectrum disorders. Neuropsychol Rev. 2011;21:148–66.
10. Pollard I. Neuropharmacology of drugs and alcohol in mother and fetus. Semin Fetal Neonatal Med. 2007;12:106–13.
11. Senturias Y, Baldonado M. Fetal spectrum disorders: an overview of ethical and legal issues for healthcare providers. Curr Probl Pediatr Adolesc Health Care. 2014;44:102–4. doi:https://doi.org/10.1016/j.cppeds.2013.12.010
12. Paley B, O'Connor MJ. Behavioral interventions for children and adolescents with fetal alcohol spectrum disorders. Alcohol Res Heal. 2011;34:64–75. http://www.ncbi.nlm.nih.gov/pmc/articles/PMC3860556/
13. Streissguth AP, Bookstein F, Barr H, Sampson P, O'Malley K, Young J. Risk factors for adverse life outcomes in fetal alcohol syndrome and fetal alcohol effects. J Dev Behav Pediatr. 2004;25:228–38. https://doi.org/10.1097/00004703-200408000-00002.
14. Fox SE, Levitt P, Nelson CA III. How the timing and quality of early experiences influence the development of brain architecture. Child Dev. 2010;81:28–40. https://doi.org/10.1111/j.1467-8624.2009.01380.x.
15. Russell M, Martier SS, Sokol RJ, Mudar P, Jacobson S, Jacobson J. Detecting risk drinking during pregnancy: a comparison of four screening questionnaires. Am J Public Health. 1996;86:1435–9. http://www.ncbi.nlm.nih.gov/pmc/articles/PMC1380656/
16. Jones TB, Bailey BA, Sokol RJ. Alcohol use in pregnancy: insights in screening and intervention for the clinician. Clin Obstet Gynecol. 2013;56:114–23. https://doi.org/10.1097/GRF.0b013e31827957c0.
17. Burns E, Gray R, Smith LA. Brief screening questionnaires to identify problem drinking during pregnancy: a systematic review. Addiction. 2010;105:601–14. https://doi.org/10.1111/j.1360-0443.2009.02842.x.
18. Concheiro-Guisan A, Concheiro M. Bioanalysis during pregnancy: recent advances and novel sampling strategies. Bioanalysis. 2014;6:3133–53. https://doi.org/10.4155/bio.14.278.
19. Cabarcos P, Álvarez I, Tabernero MJ, Bermejo AM. Determination of direct alcohol markers: a review. Anal Bioanal Chem. 2015;407:4907–25. https://doi.org/10.1007/s00216-015-8701-7.
20. Bird A. Perceptions of epigenetics. Nature. 2007;447:396–8.
21. Meaney MJ. Epigenetics and the biological definition of gene X environment interactions. Child Dev. 2010;81:41–79.
22. Henikoff S, Greally JM. Epigenetics, cellular memory and gene regulation. Curr Biol. 2016;26:R644–8. http://dx.doi.org/10.1016/j.cub.2016.06.011
23. Boyce WT, Kobor MS. Development and the epigenome: the "synapse" of gene-environment interplay. Dev Sci. 2015;18:1–23. https://doi.org/10.1111/desc.12282.
24. Joubert BR, Håberg SE, Nilsen RM, Wang X, Vollset SE, Murphy SK, et al. 450K epigenome-wide scan identifies differential DNA methylation in newborns related to maternal smoking during pregnancy. Environ Health Perspect. 2012;120:1425–31. https://doi.org/10.1289/ehp.1205412.
25. Heijmans BT, Tobi EW, Stein AD, Putter H, Blauw GJ, Susser ES, et al. Persistent epigenetic differences associated with prenatal exposure to famine in humans. Proc Natl Acad Sci U S A. 2008;105:17046–9. https://doi.org/10.1073/pnas.0806560105.
26. Reese SE, Zhao S, Wu MC, Joubert BR, Parr CL, Håberg SE, et al. DNA Methylation score as a biomarker in newborns for sustained maternal smoking during pregnancy. Environ Health Perspect. 2017;125:760–6. https://doi.org/10.1289/EHP333.
27. Lussier AA, Weinberg J, Kobor MS. Epigenetics studies of fetal alcohol spectrum disorder: where are we now? Epigenomics. 2017;9:291–311. https://doi.org/10.2217/epi-2016-0163.
28. Chater-Diehl EJ, Laufer BI, Castellani CA, Alberry BL, Singh SM. Alteration of gene expression, DNA methylation, and histone methylation in free radical scavenging networks in adult mouse hippocampus following fetal alcohol exposure. PLoS One. 2016;11:e0154836.
29. Laufer BI, Mantha K, Kleiber ML, Diehl EJ, Addison SMF, Singh SM. Long-lasting alterations to DNA methylation and ncRNAs could underlie the effects of fetal alcohol exposure in mice. Dis Model Mech. 2013;6:977–92. https://doi.org/10.1242/dmm.010975.
30. Liu Y, Balaraman Y, Wang G, Nephew KP, Zhou FC. Alcohol exposure alters DNA methylation profiles in mouse embryos at early neurulation. Epigenetics. 2009;4:500–11.
31. Hicks SD, Middleton FA, Miller MW. Ethanol-induced methylation of cell cycle genes in neural stem cells. J Neurochem. 2010;114:1767–80.
32. Zhou FC, Chen Y, Love A. Cellular DNA methylation program during neurulation and its alteration by alcohol exposure. Birth Defects Res Part A - Clin Mol Teratol. 2011;91:703–15.

33. Otero NKH, Thomas JD, Saski CA, Xia X, Kelly SJ. Choline supplementation and DNA methylation in the hippocampus and prefrontal cortex of rats exposed to alcohol during development. Alcohol Clin Exp Res. 2012;36:1701–9.

34. Fransquet PD, Hutchinson D, Olsson CA, Wilson J, Allsop S, Najman J, et al. Perinatal maternal alcohol consumption and methylation of the dopamine receptor DRD4 in the offspring: the triple B study. Environ Epigenetics. 2016;2:dvw023. http://dx.doi.org/10.1093/eep/dvw023

35. Laufer BI, Kapalanga J, Castellani CA, Diehl EJ, Yan L, Singh SM. Associative DNA methylation changes in children with prenatal alcohol exposure. Epigenomics. 2015;7 August:1–16. https://doi.org/10.2217/epi.15.60.

36. Portales-Casamar E, Lussier AA, Jones MJ, MacIsaac JL, Edgar RD, Mah SM, et al. DNA methylation signature of human fetal alcohol spectrum disorder. Epigenetics Chromatin. 2016;9:81–101.

37. Portales-Casamar E, Lussier AA, Jones MJ, MacIsaac JL, Edgar RD, Mah SM, et al. DNA methylation signature of human fetal alcohol spectrum disorder. Epigenetics Chromatin. 2016;9:25. https://doi.org/10.1186/s13072-016-0074-4.

38. Reynolds JN, Weinberg J, Clarren S, Beaulieu C, Rasmussen C, Kobor M, et al. Fetal alcohol spectrum disorders: gene-environment interactions, predictive biomarkers, and the relationship between structural alterations in the brain and functional outcomes. Semin Pediatr Neurol. 2011;18:49–55.

39. Chudley AE, Conry J, Cook JL, Loock C, Rosales T, LeBlanc N. Fetal alcohol spectrum disorder: Canadian guidelines for diagnosis. Can Med Assoc J. 2005;172(5 Suppl):S1–21.

40. Price ME, Cotton AM, Lam LL, Farré P, Emberly E, Brown CJ, et al. Additional annotation enhances potential for biologically-relevant analysis of the Illumina Infinium HumanMethylation450 BeadChip array. Epigenetics Chromatin. 2013;6:4. https://doi.org/10.1186/1756-8935-6-4.

41. Teschendorff AE, Marabita F, Lechner M, Bartlett T, Tegner J, Gomez-Cabrero D, et al. A Beta-mixture quantile normalization method for correcting probe design bias in Illumina Infinium 450k DNA methylation data. Bioinformatics. 2012;29:189–96.

42. Leek JT, Johnson WE, Parker HS, Jaffe AE, Storey JD. The sva package for removing batch effects and other unwanted variation in high-throughput experiments. Bioinformatics. 2012;28:882–3.

43. Du P, Zhang X, Huang C-C, Jafari N, Kibbe WA, Hou L, et al. Comparison of Beta-value and M-value methods for quantifying methylation levels by microarray analysis. BMC Bioinformatics. 2010;11:587. https://doi.org/10.1186/1471-2105-11-587.

44. Smith AK, Kilaru V, Klengel T, Mercer KB, Bradley B, Conneely KN, et al. DNA extracted from saliva for methylation studies of psychiatric traits: evidence tissue specificity and relatedness to brain. Am J Med Genet Part B Neuropsychiatr Genet. 2015;168:36–44.

45. Smyth GK. Linear models and empirical bayes methods for assessing differential expression in microarray experiments. Stat Appl Genet Mol Biol. 2004;3:Article3.

46. Benjamini Y, Hochberg Y. Controlling the false discovery rate: a practical and powerful approach to multiple testing. J R Stat Soc Ser B. 1995;57:289–300. https://doi.org/10.2307/2346101.

47. Berko ER, Suzuki M, Beren F, Lemetre C, Alaimo CM, Calder RB, et al. Mosaic epigenetic dysregulation of ectodermal cells in autism spectrum disorder. PLoS Genet. 2014;10:e1004402. https://doi.org/10.1371/journal.pgen.1004402.

48. Kuhn M. Building predictive models in R using the caret package. J Stat Software. 2008;1(5) https://doi.org/10.18637/jss.v028.i05.

49. Breton CV, Marsit CJ, Faustman E, Nadeau K, Goodrich JM, Dolinoy DC, et al. Small-magnitude effect sizes in epigenetic end points are important in children's environmental health studies: the Children's environmental health and disease prevention research center's epigenetics working group. Environ Health Perspect. 2017;125:511–26. https://doi.org/10.1289/EHP595.

50. Ladd-Acosta C, Hansen KD, Briem E, Fallin MD, Kaufmann WE, Feinberg AP. Common DNA methylation alterations in multiple brain regions in autism. Mol Psychiatry. 2014;19:862–71.

51. Rakyan VK, Down TA, Balding DJ, Beck S. Epigenome-wide association studies for common human diseases. Nat Rev Genet. 2011;12:529–41.

52. Lappalainen T, Greally JM. Associating cellular epigenetic models with human phenotypes. Nat Rev Genet. 2017;18:441–51. http://dx.doi.org/10.1038/nrg.2017.32

53. Liu Y, Aryee MJ, Padyukov L, Fallin MD, Hesselberg E, Runarsson A, et al. Epigenome-wide association data implicate DNA methylation as an intermediary of genetic risk in rheumatoid arthritis. Nat Biotechnol. 2013;31:142–7. https://doi.org/10.1038/nbt.2487.

54. Sánchez-Mora C, Ribasés M, Casas M, Bayés M, Bosch R, Fernàndez-Castillo N, et al. Exploring DRD4 and its interaction with SLC6A3 as possible risk factors for adult ADHD: a meta-analysis in four European populations. Am J Med Genet Part B, Neuropsychiatr Genet. 2011;156B:600–12.

55. Dadds MR, Schollar-Root O, Lenroot R, Moul C, Hawes DJ. Epigenetic regulation of the DRD4 gene and dimensions of attention-deficit/hyperactivity disorder in children. Eur Child Adolesc Psychiatry. 2016;25:1081–9. https://doi.org/10.1007/s00787-016-0828-3.

56. Ji H, Wang Y, Jiang D, Liu G, Xu X, Dai D, et al. Elevated DRD4 promoter methylation increases the risk of Alzheimer's disease in males. Mol Med Rep. 2016;14:2732–8.

57. Cheng J, Wang Y, Zhou K, Wang L, Li J, Zhuang Q, et al. Male-specific association between dopamine receptor D4 gene methylation and schizophrenia. PLoS One. 2014;9:e89128. https://doi.org/10.1371/journal.pone.0089128

58. Kordi-Tamandani DM, Sahranavard R, Torkamanzehi A. Analysis of association between dopamine receptor genes' methylation and their expression profile with the risk of schizophrenia. Psychiatr Genet. 2013;23:183–7. http://journals.lww.com/psychgenetics/Abstract/2013/10000/Analysis_of_association_between_dopamine_receptor.1.aspx.

59. Docherty SJ, Davis OSP, Haworth CMA, Plomin R, D'Souza U, Mill J. A genetic association study of DNA methylation levels in the DRD4 gene region finds associations with nearby SNPs. Behav Brain Funct. 2012;8:31. https://doi.org/10.1186/1744-9081-8-31.

60. Ptáček R, Kuželová H, Stefano GB. Dopamine D4 receptor gene DRD4 and its association with psychiatric disorders. Med Sci Monit. 2011;17:RA215–20. https://doi.org/10.12659/MSM.881925.

61. Bau CH, Almeida S, Costa FT, Garcia CE, Elias EP, Ponso AC, et al. DRD4 and DAT1 as modifying genes in alcoholism: interaction with novelty seeking on level of alcohol consumption. Mol Psychiatry. 2001;6:7–9.

62. Zhang H, Herman AI, Kranzler HR, Anton RF, Zhao H, Zheng W, et al. Array-based profiling of DNA methylation changes associated with alcohol dependence. Alcohol Clin Exp Res. 2013;37(Suppl 1):E108–15.

63. Faraone SV, Bonvicini C, Scassellati C. Biomarkers in the diagnosis of ADHD—promising directions. Curr Psychiatry Rep. 2014;16:497. https://doi.org/10.1007/s11920-014-0497-1.

64. Chen D, Liu F, Shang Q, Song X, Miao X, Wang Z. Association between polymorphisms of DRD2 and DRD4 and opioid dependence: evidence from the current studies. Am J Med Genet Part B Neuropsychiatr Genet. 2011;156:661–70. https://doi.org/10.1002/ajmg.b.31208.

65. Lowe R, Gemma C, Beyan H, Hawa MI, Bazeos A, Leslie RD, et al. Buccals are likely to be a more informative surrogate tissue than blood for epigenome-wide association studies. Epigenetics. 2013;8:445–54.

66. Hannon E, Spiers H, Viana J, Pidsley R, Burrage J, Murphy TM, et al. Methylation QTLs in the developing brain and their enrichment in schizophrenia risk loci. Nat Neurosci. 2015;19:48–54. https://doi.org/10.1038/nn.4182.

67. Esposito EA, Jones MJ, Doom JR, MacIsaac JL, Gunnar MR, Kobor MS. Differential DNA methylation in peripheral blood mononuclear cells in adolescents exposed to significant early but not later childhood adversity. Dev Psychopathol. 2016;28 4pt2:1385–99. https://doi.org/10.1017/S0954579416000055.

68. Coggins TE, Timler GR, Olswang LB. A state of double jeopardy: impact of prenatal alcohol exposure and adverse environments on the social communicative abilities of school-age children with fetal alcohol spectrum disorder. Lang Speech Hear Serv Sch. 2007;38:117–27. http://dx.doi.org/10.1044/0161-1461(2007/012)

69. Balaraman S, Schafer JJ, Tseng AM, Wertelecki W, Yevtushok L, Zymak-Zakutnya N, et al. Plasma miRNA profiles in pregnant women predict infant outcomes following prenatal alcohol exposure. PLoS One. 2016;11:e0165081. https://doi.org/10.1371/journal.pone.0165081.

70. Mesa DA, Kable JA, Coles CD, Jones KL, Yevtushok L, Kulikovsky Y, et al. The use of cardiac orienting responses as an early and scalable biomarker of alcohol-related neurodevelopmental impairment. Alcohol Clin Exp Res. 2017;41:128–38. https://doi.org/10.1111/acer.13261.

71. Goh PK, Doyle LR, Glass L, Jones KL, Riley EP, Coles CD, et al. A decision tree to identify children affected by prenatal alcohol exposure. J Pediatr. 2016; 177:121–127.e1. https://doi.org/10.1016/j.jpeds.2016.06.047

72. McQuire C, Paranjothy S, Hurt L, Mann M, Farewell D, Kemp A. Objective measures of prenatal alcohol exposure: a systematic review. Pediatrics. 2016; 138 https://doi.org/10.1542/peds.2016-0517.

73. Tseng P-H, Cameron IGM, Pari G, Reynolds JN, Munoz DP, Itti L. High-throughput classification of clinical populations from natural viewing eye movements. J Neurol. 2013;260:275–84. https://doi.org/10.1007/s00415-012-6631-2.

74. Kelleher E, Corvin A. Overlapping etiology of neurodevelopmental disorders. In: The genetics of Neurodevelopmental disorders: Wiley; 2015. p. 29–48. https://doi.org/10.1002/9781118524947.ch2.

6

Maternal intake of methyl-group donors affects DNA methylation of metabolic genes in infants

Sara Pauwels[1,2*], Manosij Ghosh[1], Radu Corneliu Duca[1], Bram Bekaert[3,4], Kathleen Freson[5], Inge Huybrechts[6], Sabine A. S. Langie[2,7], Gudrun Koppen[2], Roland Devlieger[8,9] and Lode Godderis[1,10]

Abstract

Background: Maternal nutrition during pregnancy and infant nutrition in the early postnatal period (lactation) are critically involved in the development and health of the newborn infant. The Maternal Nutrition and Offspring's Epigenome (MANOE) study was set up to assess the effect of maternal methyl-group donor intake (choline, betaine, folate, methionine) on infant DNA methylation. Maternal intake of dietary methyl-group donors was assessed using a food-frequency questionnaire (FFQ). Before and during pregnancy, we evaluated maternal methyl-group donor intake through diet and supplementation (folic acid) in relation to gene-specific (IGF2 DMR, DNMT1, LEP, RXRA) buccal epithelial cell DNA methylation in 6 months old infants ($n = 114$) via pyrosequencing. In the early postnatal period, we determined the effect of maternal choline intake during lactation (in mothers who breast-fed for at least 3 months) on gene-specific buccal DNA methylation ($n = 65$).

Results: Maternal dietary and supplemental intake of methyl-group donors (folate, betaine, folic acid), only in the periconception period, was associated with buccal cell DNA methylation in genes related to growth (IGF2 DMR), metabolism (RXRA), and appetite control (LEP). A negative association was found between maternal folate and folic acid intake before pregnancy and infant LEP (slope = −1.233, 95% CI −2.342; −0.125, $p = 0.0298$) and IGF2 DMR methylation (slope = −0.706, 95% CI −1.242; −0.107, $p = 0.0101$), respectively. Positive associations were observed for maternal betaine (slope = 0.875, 95% CI 0.118; 1.633, $p = 0.0241$) and folate (slope = 0.685, 95% CI 0.245; 1.125, $p = 0.0027$) intake before pregnancy and RXRA methylation. Buccal DNMT1 methylation in the infant was negatively associated with maternal methyl-group donor intake in the first and second trimester of pregnancy and negatively in the third trimester. We found no clear association between maternal choline intake during lactation and buccal infant DNA methylation.

Conclusions: This study suggests that maternal dietary and supplemental intake of methyl-group donors, especially in the periconception period, can influence infant's buccal DNA methylation in genes related to metabolism, growth, appetite regulation, and maintenance of DNA methylation reactions.

Keywords: Methyl-group donors, DNA methylation, LEP, IGF2 DMR, RXRA, DNMT1, Lactation, Pregnancy

* Correspondence: sara.pauwels@med.kuleuven.be
[1]Department of Public Health and Primary Care, Environment and Health, KU Leuven - University of Leuven, Kapucijnenvoer 35 blok D box 7001, 3000 Leuven, Belgium
[2]Flemish Institute of Technological Research (VITO), Unit Environmental Risk and Health, Boeretang 200, 2400 Mol, Belgium
Full list of author information is available at the end of the article

Background

During pregnancy, environmental exposures can influence the development of the offspring and increase the risk for metabolic diseases, like obesity, later in life. One maternal factor that has consistently been shown to influence later phenotype is maternal nutrition [1]. This has been most clearly shown in studies of the Dutch Hunger Winter (1944–1945), a 5-month period of extreme food shortage in the Netherlands at the end of World War II. Long-term follow-up studies from this cohort found that exposure to famine in early gestation was associated with low birth weight and increased risk of obesity in adulthood, whereas, exposure in late gestation showed decreased glucose tolerance [2]. Studies from this cohort indicate that there are different windows of susceptibility during pregnancy (embryogenesis, organogenesis, and tissue differentiation) where maternal nutrition can influence offspring's health [3]. In addition, the early postnatal period is another critical period in which nutrition can program the infant. Several physiological and metabolic mechanisms are not fully mature at birth and continue to develop in the immediate postnatal period [4].

One of the underlying mechanisms responsible for metabolic programming is epigenetic modifications, such as DNA methylation [5]. The process of methylation and demethylation is a natural process allowing the cell to grow and differentiate. Shortly after fertilization, DNA methylation marks on the maternal and paternal genome are globally demethylated, which is followed by de novo methylation just before implantation. This is a critical window of fetal development during pregnancy where dietary factors can influence the fetal methylome [6]. Methyl-group donors derived from food (choline, betaine, folate, and methionine) and supplements (folic acid), which contains a methyl-group (CH_3), enter the one-carbon (I-C) metabolism at different sites and are, in the end, converted to the universal methyl-group donor S-adenosylmethionine (SAM). SAM will donate a methyl-group for the methylation of the DNA [7]. Choline plays a role in the structural integrity of cell membranes, in the lipid-cholesterol transport and metabolism, and in normal brain development (precursor of acetylcholine) [8, 9]. Betaine is essential in the preimplantation embryo, in which it may play a role as an osmolyte, and for correct neural tube formation [10]. The intake of folate or vitamin B_9 (400 μg per day) is recommended during pregnancy to prevent neural tube defects, placental abruption, preterm birth, and low birth weight [11, 12]. Methionine is an indispensable amino acid required for protein synthesis. Diets with an inappropriate balance of methionine can adversely affect fetal development [13].

Many animal studies examined the effect of maternal methyl-group-supplemented diets on offspring epigenome, health, and longevity. A classic example is the agouti viable yellow mouse, which has a yellow coat color, is obese, hyperinsulinemic, and is more susceptible to cancer. Maternal dietary supplementation with methyl-group donors shifts the coat color of the offspring towards the brown pseudoagouti phenotype and lowers the disease risk [14]. Animal models have confirmed the biological possibility of fetal programming in response to maternal methyl-group donor supplementation and make it reasonable to think that similar processes could happen in humans. Until now, some studies in humans have shown that maternal methyl-group donor intake can influence offspring methylation. For example, Increased *IGF2* methylation and decreased *PEG3* and LINE-1 methylation were observed in cord blood with increased folic acid supplement consumption after 12 weeks of pregnancy [15]. However, the long-term effects on offspring health remain unknown in humans.

Methyl-groups are transferred from SAM to the DNA by DNA methyltransferases (*DNMTs*). *DNMT1* is responsible for maintaining DNA methylation patterns through mitosis [16]. *DNMT3A* and *DNMT3B* are responsible for the establishment of new or "de novo" DNA methylation patterns during early embryogenesis, which is a vulnerable period where nutritional insults can disrupt the correct establishment of epigenetic marks. According to Heijmans et al. [5], a decrease of 5.2% in insulin-like growth factor 2 (*IGF2*) differentially methylated region (DMR) whole blood DNA methylation was observed in 60 adults exposed to periconception famine compared to same-sex siblings who were not exposed. This association was only seen when there was an exposure in early gestation, not in mid or late gestation. *IGF2* is a maternally imprinted gene that is important for fetal growth and development. The *IGF2* DMR is only methylated on the maternal allele, so this region might be vulnerable to nutritional exposures in the pre- and periconception period [17]. Another study from the Dutch Hunger Winter found a significant increase in leptin (*LEP*) whole blood DNA methylation of men exposed to famine in early and late gestation. These results suggest that environmentally induced DNA methylation changes may not be limited to the periconception period (period starting 14 weeks before conception until 10 weeks postconception) [18] but it appears to extend to the whole prenatal period [19]. Leptin is a hormone, produced by adipose tissue, which is implicated in appetite control (inhibits food intake) and fat metabolism. *LEP* promoter methylation differences can influence *LEP* expression [20]. Godfrey et al. [21] observed that lower maternal carbohydrate intake in early pregnancy was associated with higher methylation of the retinoid X receptor-α (*RXRA*) gene in umbilical cord blood. In addition, the authors found that greater methylation levels in *RXRA* were more strongly correlated with greater adiposity (fat mass and percentage fat mass) in later childhood (9 years old) in two independent cohorts. *RXRA* is known to have beneficial effects on insulin sensitivity, adipogenesis, and fat metabolism, through its binding to the transcription factor peroxisome proliferator-activated receptor (*PPAR*) [22].

Early postnatal life has shown to be another critical window for metabolic programming in which nutrition can induce epigenetic changes in the infant. In the early postnatal period, newborns are either breast-fed or formula-fed. The ideal nutrient composition of breast milk and the peculiar feeding behavior associated with breastfeeding seem to have a protective effect against the development of obesity later in life. However, the different epigenetic mechanisms involved remain unclear [23]. One study found that the duration of breastfeeding was negatively associated with *LEP* whole blood methylation in 17 months old children. It was hypothesized that the breast milk content contributes to programming of the neuroendocrine system by changing *LEP* methylation. The decrease in *LEP* methylation could be one of the mechanisms by which breastfeeding contributes to protection against childhood obesity [20]. Some human studies hypothesize that specific breast milk components could possibly induce epigenetic changes and influence the child's health outcome. For example, the high cholesterol content of breast milk may reduce endogenous cholesterol synthesis, probably by down-regulation of hepatic hydroxymethyl glutaryl coenzyme A (HMGCoA) reductase through epigenetic mechanisms [23]. Consequently, it is important to investigate the effect of maternal dietary choline intake during lactation on infant DNA methylation levels. Hence, the methyl-group donor choline can influence choline breast milk composition and infant choline status. Folate breast milk concentration on the other hand is maintained even when the mother is folate deficient and is unaffected by maternal folic acid supplementation [24]. Methionine is present in breast milk in low concentrations. Amino acid composition of breast milk can be influenced by lactation stage but not by maternal dietary protein intake [25, 26].

In this study, we investigated the effect of maternal dietary methyl-group donor intake (choline, betaine, folate, and methionine) and supplemental intake (folic acid) before and during each trimester of pregnancy on gene-specific methylation (*DNMT1*, *IGF2* DMR, *RXRA*, and *LEP)* in buccal epithelial cells of 6 months old infants. Buccal swabs were chosen to collect DNA because the samples are easy to collect and it is a non-invasive technique, which is important to consider when taken DNA samples from infants. Buccal samples mainly exist of exfoliated (dead) epithelial cells but have a more homogenous cell population compared to blood samples

[27]. Next, we determined the effect of maternal choline intake during lactation, in mothers who breast-fed for at least 3 months, on gene-specific buccal DNA methylation. In the gene-specific DNA methylation analysis, we included *DNMT 1*, *IGF2* DMR, *RXRA*, and *LEP*.

Methods
Study subjects
We studied participants enrolled in the MANOE (Maternal Nutrition and Offspring's Epigenome) study, an ongoing prospective, observational cohort study initiated in April 2012. Healthy Caucasian women who desired to become pregnant or who were in the first trimester of pregnancy were recruited at the Department of Obstetrics and Gynecology of the University Hospitals Leuven (Belgium). We enrolled 150 women (34 women before pregnancy and 116 in the first trimester of pregnancy) between April 2012 and January 2015. The last delivery of the cohort took place in September 2015 and the last 6 months postpartum (PP) visit in March 2016. Exclusion criteria were the following: non-Caucasian women, multiple pregnancies (twins, triplets, etc.), and infertility treatment. Of the 150 enrolled women, 36 mother–infant pairs were excluded from analysis due to missing nutritional data ($n = 2$), missing buccal swab samples ($n = 15$), development of pregnancy complications (gestational diabetes ($n = 8$) and preeclampsia ($n = 1$)), preterm delivery ($n = 6$), extreme high intake of folic acid (4 mg/day) ($n = 2$), and birth defects ($n = 2$). This gives us a total of 114 mother–infant pairs for statistical analysis. Further statistical analysis was performed on a subsample of 65 lactating mother–infant pairs. A flowchart of the mother–infant pairs enrolled in the MANOE study and included in the statistical analysis is presented in Fig. 1. The recruitment process has been described in more detail in a previous study [28].

This study was conducted according to the guidelines laid down in the Declaration of Helsinki and all procedures involving human subjects were approved by the UZ Leuven-Committee for Medical Ethics (reference number: ML7975). Written informed consent was obtained from all subjects.

Maternal and neonatal measurements
All 114 women were followed up during pregnancy at their scheduled ultrasounds (11–13 weeks, 18–22 weeks, and 30–34 weeks of gestation), 6 weeks, and 6 months PP. From the women recruited before pregnancy ($n = 34$), extra measurements were taken before conception. To assess maternal intake of dietary methyl-group donors (methionine, folate, betaine, and choline) before pregnancy, during each trimester of pregnancy, and PP, a food-frequency questionnaire (FFQ) was developed, validated [29, 30], and implemented in the MANOE study. The

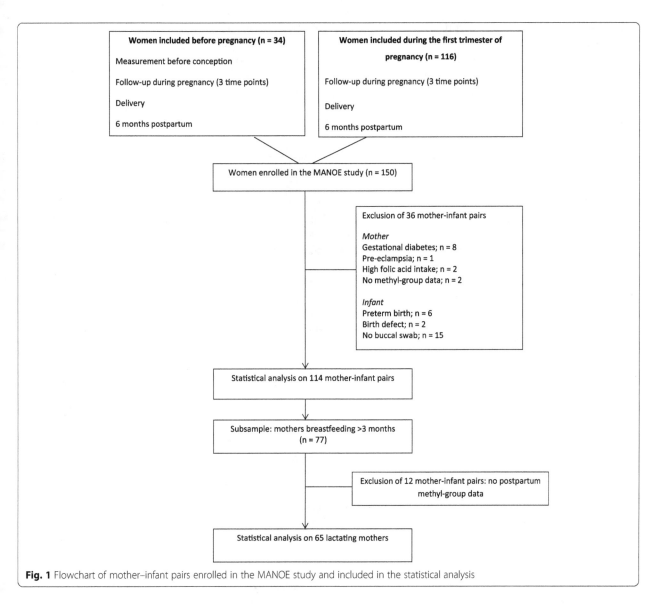

Fig. 1 Flowchart of mother–infant pairs enrolled in the MANOE study and included in the statistical analysis

FFQ contains 51 food items and women were asked to indicate their answers in a list of frequencies and portion sizes to calculate the usual daily intake of the four methyl-group donors (mg or µg/day). Twenty-one FFQs were obtained before pregnancy, 94 FFQs at 11–13 weeks, 85 FFQs at 18–22 weeks, 82 FFQs at 30–34 weeks of pregnancy, 79 FFQs 6–8 weeks PP, and 60 FFQs 6 months PP. To assess the intake of methyl-group donors through supplement use, questions were asked about the use of nutritional supplements (frequency, brand/type, dosage) before, during each trimester of pregnancy, and PP. Only the intake of folic acid (synthetic form of folate) was registered, since there was no report on the supplemental intake of methionine, betaine, and choline. Furthermore, using a combination of questionnaires and interviews, we collected information about a range of socio-demographic factors, life style habits, and physical activity. Information

on mothers' smoking status before and during pregnancy was obtained at each consultation. Questions were asked about smoking before and in each trimester of pregnancy and the number of cigarettes smoked on average per day. From these data, a dichotomous variable for maternal smoking before and during pregnancy was derived (did not smoke/smoked). Height and prepregnancy weight were used to calculate the prepregnancy body mass index (BMI, kg/m^2).

Six months after birth, data on breastfeeding was derived and scores were given ranging from 0–4 (0 = formula feeding; 1 = <1 month of breastfeeding; 2 = 1–3 months of breastfeeding; 3 = 3–6 months of breastfeeding; 4 = >6 months of breastfeeding). We measured infant weight and length at the 6 months PP visit. Maternal and neonatal measurements have been described in more detail in a previous study [28].

Sample collection and DNA extraction

A Cytobrush plus Medscan® was used to brush against the inner cheeks of the infant. The brush handle was cut off and put inside a 15-mL Falcon tube with PBS and stored immediately at −20 °C, until DNA extraction. DNA extraction from cytobrush was performed using the QIAamp DNA Blood Mini Kit (Qiagen Inc., Valencia, CA). The final elution volume obtained was 100 μL. The quantity and purity of DNA were determined by a NanoDrop spectrophotometer.

Gene-specific DNA methylation measurements
Bisulfite conversion and PCR

Genomic DNA (200 ng) was bisulfite converted using the EZ-96 DNA Methylation-Gold™ Kit (#D5008, Zymo Research). Converted DNA was eluted with 30 μL of M-elution buffer. Subsequently, 1 μL of converted DNA was amplified by PCR in a total volume of 25 μL containing 0.2 μM of primers and 2× Qiagen PyroMark PCR Master Mix (#978703, Qiagen). Primers for *DNMT1*, *RXRA*, and *LEP* were ordered from Qiagen (#PM00075761, #PM00144431, #PM00129724 PyroMark CpG Assays). The analyzed sequences are part of the promoter region and lie within a CpG island. For the *RXRA* gene, the analyzed sequence also lies in a transcriptional regulatory site. Primer sequences for *IGF2* DMR were taken from the original paper. Imprinting of the *IGF2* gene is regulated by this differentially methylated region which is located upstream of the imprinted promoters of *IGF2* exon 3 [31].

PCR for *DNMT1*, *RXRA*, and *LEP* consisted of an initial hold at 95 °C for 15 min followed by 45 cycles of 30 s at 94 °C, 30 s at 54 °C, and 30 s at 72 °C. PCR amplification ended with a final extension step at 72 °C for 10 min. PCR for *IGF2* DMR consisted of an initial hold at 5 °C for 15 min followed by 5 cycles of 30 s at 94 °C, 30 s at 68 °C, and 30 s at 72 °C. This was followed by 50 cycles of 30 s at 94 °C, 30 s at 64 °C, and 30 s at 72 °C and ended with a final extension step at 72 °C for 10 min.

Pyrosequencing

In order to assess CpG methylation levels, 20 μL of biotinylated PCR product was immobilized to Streptavidin Sepharose High Performance beads (#17-5113-01, GE Healthcare) followed by annealing to 25 μL of 0.3 μM sequencing primer at 80 °C for 2 min with a subsequent 10 min cooling down period. Pyrosequencing was performed using Pyro Gold reagents (#970802, Qiagen) on the PyroMark Q24 instrument (Qiagen) following the manufacturer's instructions. Pyrosequencing results were analyzed using the PyroMark analysis 2.0.7 software (Qiagen). Five CpGs were analyzed for *DNMT1*, three CpGs for *IGF2* DMR, four CpGs for *LEP*, and five CpGs for *RXRA*. Six samples were randomly selected for technical variation analysis.

Statistical analysis

First, we assessed changes in the intake of maternal methyl-group donors during and after pregnancy using a multivariate regression model for longitudinal measurements with methyl-group donor intake as a response variable and time point as a factor (LSD post hoc test).

Next, we determined the effect of maternal methyl-group donor intake on gene-specific DNA methylation (*IGF2* DMR, *LEP*, *RXRA*, *DNMT1*) using linear mixed models. Linear mixed models were used with gene-specific DNA methylation as a response variable and methyl-group donor intake, CpG site, and their interaction as explanatory variables. Other covariates were included in the multivariable model to correct for possible confounding. Potential confounders were selected based on the association with infant DNA methylation and maternal nutrition: maternal age, maternal prepregnancy BMI, maternal smoking before and during each trimester of pregnancy (0 = did not smoke before and during pregnancy, 1 = smoked before and during pregnancy), gestational weight gain, and duration of breastfeeding (0 = formula feeding; 1 = <1 month of breastfeeding; 2 = 1–3 months of breastfeeding; 3 = 3–6 months of breastfeeding; 4 = > 6 months of breastfeeding). A random intercept was modeled to deal with the clustered nature of the data. Analyses were performed separately per time point (prepregnancy, 11–13 weeks pregnancy, 18–22 weeks pregnancy, 30–34 weeks pregnancy). First, the interaction between maternal methyl-group donor intake and CpG site was tested. A significant interaction test implies that the association between methyl-group donor intake and CpG methylation is different between the individual CpGs. In this case, results were reported per individual CpG. A non-significant interaction test indicates lack of evidence for a differential association between methyl-group donor intake and methylation at different CpGs. In this case, a main effect of methyl-group donor intake over the different CpGs was reported.

Next, an independent *t* test was performed on a subsample of lactating women (*n* = 65) to assess the effect of maternal choline intake during breastfeeding on buccal DNA methylation at 6 months. The mean maternal methyl-group donor intake during the 6 months after delivery was calculated using the two FFQs administrated during this period and using supplement information. Finally, we determined the effect of maternal choline intake during lactation on gene-specific DNA methylation (*IGF2* DMR, *LEP*, *RXRA*, *DNMT1*) using linear mixed models. Linear mixed models were used with gene-specific DNA methylation as a response variable and choline intake, CpG site, and their interaction as explanatory variables. Other covariates were included in the multivariable model to correct for possible confounding. Potential confounders were selected based on the association with infant DNA methylation and maternal nutrition: maternal age, maternal prepregnancy BMI, maternal smoking, and gestational weight gain. A random intercept was modeled to deal with the clustered nature of the data. Analyses were performed separately per time point (0–3 months postpartum, 3–6 months postpartum). First, the interaction between maternal choline intake and CpG site was tested. A significant interaction test implies that the association between choline intake and CpG methylation is different between the individual CpGs. In this case, results were reported per individual CpG. A non-significant interaction test indicates lack of evidence for a differential association between choline intake and methylation at different CpGs. In this case, a main effect of choline intake over the different CpGs was reported.

All tests were two-sided, a 5% significance level was assumed for all tests. Analyses have been performed using SAS software (version 9.4 of the SAS System for Windows).

Results

Characteristics of the 114 mother–infant pairs included in the statistical analysis are presented in Table 1. The mean maternal age and standard deviation (SD) was 31 ± 3.7 years, mean prepregnancy BMI and SD was 23 ± 3.4 kg/m^2, and the mean gestational weight gain and SD was 14.8 ± 4.1 kg. Only four women smoked before and during the first trimester of pregnancy. One woman continued smoking during the second and third trimester. The infants, 54 of which were girls (47.4%), had a mean weight and SD of 7875.4 ± 877.6 g, a mean length and SD of 67.9 ± 2.6 cm, and the mean age and SD was 6.3 ± 2.4 months. Only 7% of the women decided to exclusively use formula feeding, while the biggest group of women (39.5%) breastfed for more than 6 months.

Most of the women in the study had a methionine intake above the daily requirement of 10.4 mg/kg body weight per day [32]. Dietary methionine intake was significantly lower 6 months PP (1533.6 mg/day) than the intake in the third trimester (1659.3 mg/day, $p = 0.043$) of pregnancy and 6–8 weeks PP (1678.5 mg/day, $p = 0.01$). The dietary intake of folate, choline, and betaine was stable and did not change during pregnancy and in the PP period (Table 2). All women took a folic acid supplement in the first trimester of pregnancy to reach an uptake of 400 µg of folate per day. Remarkably, some continued taking the supplement throughout pregnancy and lactation, despite the recommendation of starting 4 weeks before conception until 12 weeks of pregnancy [33].

The supplemental intake of folic acid on the other hand was highest in the first trimester of pregnancy (507.2 µg) and significantly different from the folic acid

Table 1 Maternal and infant characteristics ($n = 114$)

Characteristics	Mean (SD)	Range
Mother		
Maternal age (years)	31 (3.7)	25–41
Prepregnancy BMI (kg/m^2)	23 (3.4)	17.9–33
Gestational weight gain (kg)	14.8 (4.1)	5.3–28.9
Infant		
Weight (g)	7875.4 (877.6)	6240–11,120
Length (cm)	67.9 (2.6)	62–76.5
Age (months)	6.3 (0.4)	4.6–7.2
	%	N
Maternal smoking (yes)		
Before pregnancy	3.5	4
First trimester	3.5	4
Second trimester	0.9	1
Third trimester	0.9	1
Gender		
Boy	52.6	60
Girl	47.4	54
Duration of breastfeeding		
0 months	7	8
<1 month	6.1	7
1–3 months	19.3	22
3–6 months	28.1	32
>6 months	39.5	45

intake in the other four time points ($p = 0.000$). Postpartum, the intake of folic acid was significantly lower as compared to the intake during every trimester of pregnancy. Within the PP period, the folic acid intake at 6 months (68 µg/day) was significantly lower than the intake at 6–8 weeks (204.3 µg/day, $p = 0.000$). For choline, the adequate choline intake was 425 mg for non-pregnant women, 450 mg for pregnant women, and 550 mg for lactating women [34]. Most women had an average intake of about 300 mg choline per day, which lies below the adequate intake. For betaine, no guideline for dietary intake exists.

Gestational methyl-group donor intake and infant buccal DNA methylation

We estimated the association of maternal methyl-group donor intake before pregnancy and during each trimester of pregnancy on infant (6 months old) gene-specific DNA methylation (*DNMT1*, *IGF2* DMR, *RXRA*, and *LEP*) in buccal epithelial cells. The statistically significant associations and trends between maternal methyl-group donor intake and buccal DNA methylation are presented in Table 3.

Table 2 Intake of maternal methyl-group donors through diet and supplements (folic acid) during pregnancy and in the postpartum (PP) period

Methyl-group donors	First trimester (10–13 weeks) Mean (SE) Range N = 94	Second trimester (18–22 weeks) Mean (SE) Range N = 85	Third trimester (30–34 weeks) Mean (SE) Range N = 82	6–8 weeks PP Mean (SE) Range N = 79	6 months PP Mean (SE) Range N = 60	p	Dietary guidelines
Methionine (mg) (mg/kg)	1644.4 (45.9) 792–2932 12.4–45.1	1608.4 (44.3) 746.1–2684.4 9.6–38	1659.3 (48.7) 789–2957 9.2–40.6	1678.5 (52) 786.4–3499.4 12.5–45.8	1562.4[a] (43.5) 710.5–2562.8 11.7–35.1	0.047	Daily requirement 10.4 mg/kg body weight
Folate (μg)	272.1 (8.7) 131–531	263.2 (8.8) 98–519.8	279.8 (10.5) 112–619	264 (9) 97.4–520.7	263.7 (10.7) 98–564.4	0.17	Recommended intake Non-pregnant 200-300 Pregnant 400 Lactation 300
Folic acid (μg)	507.2[b] (14.1) 171–1000	399.9 (23.9) 0–1000	391.3 (25.6) 0–1000	204.3[c] (24.7) 0–800	68[c] (14.9) 0–600	0.000	
Choline (mg)	274.4 (7.4) 137–451	268.1 (7.4) 137.3–469.2	280.3 (8.6) 130–552	278.2 (8.5) 128.3–547.4	268.4 (7.8) 115.3–435.5	0.26	Adequate intake Non-pregnant 400 Pregnant 450 Lactation 550
Betaine (mg)	162.6 (5.7) 63–342	169.2 (6.2) 52.6–354.3	173.2 (6.7) 68–320	170.2 (6.5) 80–349.3	162.6 (7.2) 34.5–326.2	0.53	/

p values were obtained using a multivariate regression model for longitudinal measurements
[a]Methionine intake 6 months PP was significantly lower than the intake in the third trimester of pregnancy and 6–8 weeks PP
[b]Folic acid intake was significantly higher in the first trimester of pregnancy compared to the other time points
[c]Folic acid intake in the PP period was significantly lower than the intake during pregnancy and within the PP period, folic acid intake 6 months PP was lower

Before pregnancy, maternal betaine, folate, and folic acid intakes were associated with buccal epithelial methylation levels of *RXRA*, *LEP*, and *IGF2* DMR. A higher intake of folate and betaine was associated with higher *RXRA* methylation across all CpGs (for folate, 0.685% increase in *RXRA* methylation per 100 μg folate increase; 95% CI 0.245, 1.125; $p = 0.027$; for betaine, 0.875% increase in *RXRA* methylation per 100 mg betaine increase; 95% CI 0.118, 1.633; $p = 0.0241$). In addition, a higher intake of folate was associated with lower *LEP* methylation across all CpGs (−1.233% decrease in *LEP* methylation per 100 μg folate increase; 95% CI −2.342, −0.125; $p = 0.0298$). For folic acid, a higher intake before pregnancy was associated with lower *IGF2* DMR methylation across all CpGs (−0.706% decrease in *IGF2* DMR methylation per 100 μg folic acid increase; 95% CI −1.242, −0.107; $p = 0.0101$).

In the first trimester of pregnancy, only borderline significant results were found between maternal methyl-group donor intake and *IGF2* DMR and *DNMT1* methylation.

In the second trimester of pregnancy, a higher intake of folic acid was associated with lower *DNMT1* methylation across all CpGs (−0.027% decrease in *DNMT1* methylation per 100 μg folic acid increase; 95% CI −0.051, −0.004; $p = 0.0204$).

In the third trimester of pregnancy, a higher intake of choline and folate was associated with higher *DNMT1* CpG1 methylation (0.156% increase in *DNMT1* CpG1 methylation per 100 mg choline increase; 95% CI 0.029, 0.283; $p = 0.0166$) and higher *DNMT1* CpG3 methylation (0.131% increase in *DNMT1* CpG3 methylation per

100 μg folate increase; 95% CI 0.016, 0.246; $p = 0.0256$), respectively.

Choline intake of lactating women and infant methylation
First, we found statistically significant differences in buccal *RXRA* methylation (CpG4 and mean CpG) between low and high (≥275.27 mg/day) maternal dietary intake of choline during lactation. The results are shown in Fig. 2. We observed significantly higher *RXRA* methylation percentages when the mother consumed a diet high in choline during lactation compared to a diet low in choline (for CpG4, 5.7 ± 1.4 vs. 5 ± 1.2%, $p = 0.023$, 95% CI −1.407, −1.083; for mean CpG, 7.3 ± 1.1 vs. 6.7 ± 1.2%, $p = 0.04$, 95% CI −1.180, −0.028).

Next, we estimated the association between choline intake of lactating women during two time points (0–3 months after pregnancy and 3–6 months after pregnancy) and infant gene-specific DNA methylation (*DNMT1*, *IGF2* DMR, *RXRA*, and *LEP*) in buccal epithelial cells. No significant association between maternal choline intake and infant DNA methylation levels was found (Table 4).

Discussion
This research supports the hypothesis that maternal methyl-group donor intake before and during pregnancy could possibly induce epigenetic alterations in offspring genes related to metabolism and genes important to maintain DNA methylation patterns. We first studied the effect of maternal methyl-group donor intake (through diet and supplements) before and during each trimester of pregnancy on gene-specific DNA methylation (*IGF2* DMR,

Table 3 Associations between maternal methyl-group donor intake (before and during pregnancy) and infant *DNMT1*, *IGF2* DMR, *RXRA*, and *LEP* methylation in buccal epithelial cells

Time point	Before pregnancy			First trimester			Second trimester	Third trimester	
	B (95% CI) p value N = 21			B (95% CI) p value N = 94			B (95% CI) p value N = 85	B (95% CI) p value N = 82	
Gene	*RXRA*	*LEP*	*IGF2* DMR	*IGF2* DMR		*DNMT1*	*DNMT1*	*DNMT1*	
Nutrient	All CpG sites[a]	All CpG sites[a]	All CpG sites[a]	CpG1[b]	All CpG sites[a]	All CpG sites[a]	All CpG sites[a]	CpG1[b]	CpG3[b]
Betaine	0.875 (0.118; 1.633) 0.0241			2.341 (−0.138; 4.82) 0.0640					
Choline						−0.092 (−0.191; 0.008) 0.07		0.156 (0.029; 0.283) 0.0166	
Folate	0.685 (0.245; 1.125) 0.0027	−1.233 (−2.342; −0.125) 0.0298							0.131 (0.016; 0.246) 0.0256
Folic acid			−0.706 (−1.242; −0.107) 0.0101	1.013 (−0.095; 2.121) 0.0728		−0.033 (−0.072; 0.006) 0.0987	−0.027 (−0.051; −0.004) 0.0204		
Methionine						−0.013 (−0.029; 0.002) 0.0831			

β-Estimate is an absolute change in percentage of methylation; slope >(<) 0 means positive (negative) association

[a]When there was no evidence for a differential association between maternal methyl-group donor intake and DNA methylation at the different CpG locations, the main effect of maternal methyl-group donor intake over all CpG locations was reported

[b]When there was a significant interaction test, the association between maternal methyl-group donor intake and DNA methylation was different between CpG locations. In this case, the results were reported per CpG location

CI confidence interval

LEP, *RXRA*, *DNMT1*) in buccal epithelial cells of 6 months old infants. We observed that maternal methyl-group donor intake (folate, folic acid, betaine), only in the periconception period, was associated with DNA methylation in genes related to growth (*IGF2* DMR), metabolism (*RXRA*), and appetite control (*LEP*). For *LEP*, only a negative association was observed with prepregnancy folate intake. A similar negative association was found in studies from the Dutch Hunger Winter where adult men exposed to famine (low availability of methyl-group donors) *in utero* had 2.8% higher *LEP* methylation levels than their unexposed siblings [19]. In addition, we observed a negative association between *IGF2* DMR methylation and folic acid use before pregnancy and a borderline positive association in the first trimester of pregnancy. Haggarty et al. [15] observed higher levels of *IGF2* cord blood methylation when folic acid supplements were used after 12 weeks of gestation as compared to use in the periconception period. One other study reported no difference in *IGF2* DMR cord blood methylation between infants born to no, moderate, or high folic acid users before pregnancy [35].

On the other hand, Steegers-Theunissen et al. [36] reported a 4.5% higher *IGF2* DMR methylation in 17 months old children when mother used a folic acid supplement of 400 μg per day in the periconceptional period. The different reported conclusions in relation to the timing of folic acid use could be important since changes in *IGF2* methylation could have unintended consequences for health and disease. For *RXRA*, positive associations were observed with prepregnancy betaine and folate intake. Other studies have also reported both increases and decreases in methylation depending on the timing of the exposure and locus under study [19]. Ours and other results show that both hypo- and hypermethylation can be found in offspring exposed to a higher intake of methyl-group donors. There is no simple correlation between maternal methyl-group donor intake and offspring DNA methylation. It is rather believed that these epigenetic modifications in the offspring are a coordinated adaptive response to environmental challenges, whereby the developing offspring adjusts their physiology to suit the expected postnatal environment [3].

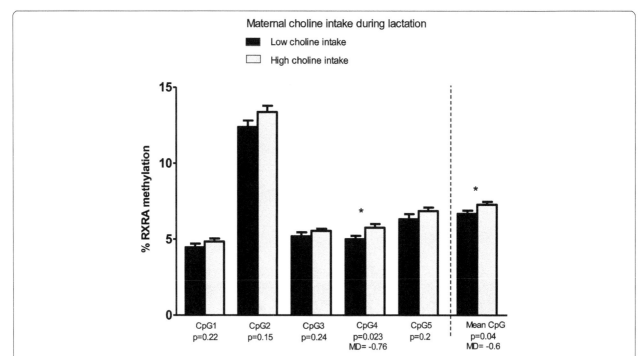

Fig. 2 Buccal *RXRA* methylation by maternal dietary choline intake during lactation. The graphs represent the mean methylation values and standard error of the mean bars of 65 infants. The results are based on a low or high dietary maternal intake of choline during lactation. The overall p values and significant p values with mean differences (MD) are also shown

Table 4 Associations between maternal choline intake during lactation (0–3 months after pregnancy and 3–6 months after pregnancy) and infant DNMT1, IGF2 DMR, RXRA, and LEP methylation in buccal epithelial cells

Time point	0–3 months after pregnancy	3–6 months after pregnancy
Gene[a]	B (95% CI) p value N = 58	B (95% CI) p value N = 42
DNMT1	0.080 (−0.062; 0.223) 0.2675	−0.060 (−0.360; 0.240) 0.6925
IGF2 DMR	−0.984 (−2.416; 0.449) 0.1764	0.479 (−2.157; 3.114) 0.7189
RXRA	0.291 (−0.057; 0.640) 0.1012	0.321 (−0.207; 0.850) 0.2319
LEP	−0.226 (−1.271; 0.818) 0.6695	0.057 (−1.713; 1.828) 0.9490

β-Estimate is an absolute change in percentage of methylation; slope >(<) 0 means positive (negative) association
[a]When there was no evidence for a differential association between maternal choline intake and DNA methylation at the different CpG locations, the main effect of maternal choline intake over all CpG locations was reported, which was the case for all genes
CI confidence interval

The periconception period, during which embryonic development takes place, is a vulnerable period, where nutritional exposures can disrupt the correct establishment of epigenetic marks [1]. Studies from the Dutch Hunger Winter show that adults who had been exposed to famine early in gestation and late gestation had a 5.2 and 0.9% (not significant) lower methylation of the *IGF2* DMR gene as opposed to unexposed same-sex siblings, respectively (timing specific) [2, 5]. It is thought that epigenetic marks set in the periconception period were more stable and could be maintained for many cell divisions (long-term stability), while marks set later in gestation on the other hand seem to be more flexible and can be removed within a few cell divisions (short-term flexibility) [37]. It has been suggested that the stable epigenetic marks represent those determined by genetics, whereas, the variable marks are more likely to be influenced by the environment, for example, nutrition [38].

Our results also indicate that maternal methyl-group donor intake during every trimester of pregnancy could influence *DNMT1* buccal DNA methylation in the infant. In the first and second trimester of pregnancy, lower intakes of maternal methyl-group donors resulted in higher *DNMT1* buccal DNA methylation levels. In the third trimester on the other hand, higher maternal methyl-group donor intake resulted in higher *DNMT1* methylation. Several animal studies have reported altered *DNMT1* expression in offspring due to differences in maternal diet (folic acid supplementation, low-protein diet, low-choline diet) [39–42]. Kovacheva et al. [42] reported that choline deficiency during gestation can modulate the fetal DNA methylation machinery through hypomethylation of the regulatory CpGs within the *DNMT1* gene, leading to its overexpression. It was already shown that hypomethylation of a single CpG (position 101) in the *DNMT1* gene can upregulate its expression [43], which was confirmed in the previous study. Maternal choline deficiency also resulted in an increased global and gene-specific (*IGF2*) DNA methylation in fetal livers [42]. It has already been shown that *DNMT1* is important for the maintenance of *IGF2* methylation patterns [44]. The negative associations found between *DNMT1* methylation and maternal methyl-group donor intakes are in line with the results from animal studies. This could also indicate that the observed changes in DNA methylation in the other genes could result from an increased expression of *DNMT1*. However, in the third trimester of pregnancy, a positive association was observed. This change in direction could be due to a shift in the I-C metabolism during gestation. A higher rate of transsulfuration was previously reported in the first trimester of pregnancy and a higher rate of transmethylation in the third trimester [45].

It is becoming clear that nutrition cannot only affect the epigenome in the prenatal life but that it extends into the early postnatal period. Breast milk has the ideal combination of nutrients, hormones, and other factors essential for the proper development and health of babies and seems to have a protective effect against the development of obesity later in life [23]. Exclusive breastfeeding is recommended by the World Health Organization (WHO) for the first 6 months of life [46]. In this study, we found that 6 months old infants from mothers who breast-fed for at least 3 months and had a high dietary intake of choline (≥275.27 mg/day) had statistically significant higher buccal *RXRA* CpG4 and CpG mean methylation levels. However, we were not able to reproduce these results when dividing the postnatal period in two periods (0–3 months and 3–6 months). In the second analysis, we did not find an association between maternal choline intake during lactation and buccal DNA methylation. Results from other studies show that the mean daily choline intake is about 300 mg/day [47]. These results are in line with the mean daily choline intake of women in the MANOE study (Table 2), but the choline intake is still far below the adequate intake of 450 mg/day during pregnancy and 550 mg/day during lactation. Maternal dietary choline intake can affect choline breast milk composition and infant choline status. Folate breast milk concentration on the other hand is maintained even when the mother is folate deficient and is unaffected by maternal folic acid supplementation [24]. In breast milk, choline is mainly present as phosphocholine (45%)

and glycerophosphocholine (29%), with smaller amounts of free choline (9%), phosphatidylcholine (7%), and sphingomyelin (10%). Ilcol et al. [48] have demonstrated that free choline concentrations in breast milk were influenced by maternal circulating choline status (serum free choline). Also, serum-free choline concentrations in infants were correlated with free choline, phosphocholine, glycerophosphocholine, and total choline contents of breast milk.

A possible explanation for the increase in *RXRA* methylation in infants from breastfeeding mothers with higher dietary choline intake during the first 6 months could be through higher availability of methyl-group donors in the I-C metabolism. Choline is first oxidized into betaine, which contains three methyl-groups. A methyl-group is donated to homocysteine for the formation of methionine, which is further transformed to SAM. SAM will eventually donate the methyl-group to the DNA and is converted into S-adenosylhomocysteine (SAH). Thus, a higher dietary intake of choline could lead to higher DNA methylation levels through the provision of methyl-group donors [49]. Next to the role of choline in the I-C metabolism, it also plays a role in the lipid-cholesterol transport and metabolism (choline is a precursor of phosphatidylcholine) [50]. A specific species of phosphatidylcholine is the peroxisome proliferator-activated receptor alpha (*PPARα*). PPAR and *RXR* form heterodimers that regulate transcription of genes involved in insulin action, adipocyte differentiation, lipid metabolism, and inflammation [22]. According to a study in two independent cohorts, greater cord blood methylation of *RXRA* was strongly correlated with greater adiposity in later childhood (9 years old). A 17, 20, and 6% increase in fat mass and a 10, 12, and 4 increase in percentage fat mass were found per SD change in *RXRA* methylation [21]. A potential mechanistic pathway involved is that the induction of transcription by *RXRA* is dependent on its binding to ligands including the peroxisome proliferator-activated receptors. Our results indicate that a higher maternal choline (≥275.27 mg/day) intake during lactation was linked with higher *RXRA* methylation in infant's buccal cells. We do not know if the positive association found between *RXRA* cord blood methylation and fat mass in 9-year-old children [21] applies for *RXRA* buccal epithelial cell methylation at 6 months of age and childhood fat mass. In this study, we were not able to link our methylation data with infant characteristics. Children of the MANOE study will therefore be followed up at a later age, to link the infant's methylation levels with anthropometric measurements (weight, height, and fat mass).

Many studies have found an association between small DNA methylation changes at single CpGs or over a very limited genomic region (<10% and often only 1–5%) and disease phenotypes. Such small DNA methylation changes are known to be set during epigenetic sensitive periods and play a role in creating a large diversity in phenotypes linked to the onset of many complex diseases. Our study also observed small changes in DNA methylation, but the true biological relevance and how these small changes could give rise to the disease phenotype (mechanism) remains unknown [51].

There are some strengths and limitations in the present study we need to address. The strengths of the present study include a unique study design that allowed us to collect longitudinal maternal data (starting before pregnancy, during each trimester of pregnancy, and in the PP period) and offspring gene-specific DNA methylation data in buccal epithelial cells at the age of 6 months. The use of a validated food-frequency questionnaire designed to assess the intake of the nutrients under study adds to the strength of our study. In addition, at each study time point detailed information about supplement use was obtained. We also have detailed covariate data allowing for adjustment for potential confounding variables. Another advantage is the use of bisulfite pyrosequencing for DNA methylation analysis in candidate genes. It enables the determination of DNA methylation levels at individual CpG sites and the calculation of the average methylation percentage of that region. Single CpG site methylation in the promoter region of a gene can be involved in the regulation of transcription, especially when it lies in a relevant transcription factor binding site, and could be associated with diseases. From example, the loss of DNA methylation in one CpG site in the promoter region of *TET1* was associated with air pollution and childhood asthma and could possibly be a potential biomarker for childhood asthma [52]. CpG methylation within the same CpG island in promoter regions has shown to be highly correlated and these methylation patterns have been shown to differ from methylation patterns elsewhere, indicating that they have a specific biological role [53].

A first limitation is that we measured offspring methylation in buccal epithelial cells and not in the target tissue (adipose tissue). We do not know to what extent the methylation changes found in buccal epithelial cells reflect the changes in the less accessible target tissue (adipose tissue). DNA extracted from buccal swabs mainly stems from exfoliated epithelial cells but has a more homogenous cell population compared to blood samples. Buccal swabs are most often used in epidemiological studies involving young children because it is non-invasive and easy to obtain [27]. It has been shown that buccal samples are more informative than blood samples in DNA methylation studies with non-blood-based diseases/phenotypes (for example obesity) as outcome [54]. We have measured DNA methylation at the age of 6 months and we do not know if these are the methylation patterns set at birth or if it reflects

methylation changes due to early postnatal exposures. Another limitation is the fact that the Belgian food composition database NUBEL [55] does not contain information about the four methyl-group donors under study. Databases of neighboring countries or the USDA database for choline and betaine [56] content were used in the validation of the FFQ [29, 30]. For folate, the Dutch NEVO food composition database was used [57] and the German BLS nutrient database for methionine [58]. The USDA database was also used for the nutrient content of folate and methionine if not found in NEVO and BSL databases, respectively. In addition, in the validation of the FFQ, we found a lower correlation coefficient (which measures the association between the FFQ and the reference method) of 0.42 for choline [29]. The correlation coefficients should be at least 0.40 and optimally in the range of 0.50–0.70 [59]. A last limitation is that we have only performed an analysis for technical variation in the pyrosequencing analysis and did not include control samples.

Conclusions

This study suggests that maternal dietary and supplemental intake of methyl-group donors, especially in early gestation, can influence infant buccal DNA methylation in genes related to metabolism, growth, appetite regulation, and DNA methylation reactions. We have also shown that nutrition in the early postnatal period, lactation, can influence infant DNA methylation levels. Higher maternal choline intake in mothers who breast-fed their children for more than 3 months resulted in higher buccal *RXRA* methylation levels in 6 months old infants.

Acknowledgements

We acknowledge the women who volunteered to take part in this study. Also, the Unit Leuven Biostatistics and Statistical Bioinformatics Centre (L-BioStat) and in particular Annouschka Laenen who did the statistical analysis. Sabine Langie is the beneficiary of the Cefic-LRI Innovative Science Award 2013 and of a postdoctoral fellowship (12L5216N; http://www.fwo.be/) provided by the Research Foundation-Flanders (FWO) and the Flemish Institute of Technological Research (VITO).

Funding

Funding for the present study was provided by a PhD grant (grant number 11B1812N) from the Research Foundation-Flanders (FWO) and the Flemish Institute of Technological Research (VITO).

Authors' contributions

All authors have made substantial contributions to the manuscript: The study was designed by LG. The food-frequency questionnaire was designed by SP and IH. SP was responsible for the field work and data entry/analysis. SP, LG, and RD participated in the conduction and coordination of the study. The paper was written by SP. The samples were collected by SP and analyzed by SP, BB, RD, KF, and MG. GK, SL, and BB provided help with the interpretation of the results. All authors helped in the evaluation of the results and the writing of the manuscript. All authors read and approved the final manuscript.

Competing interests

The authors declare that they have no competing interests.

Consent for publication
Not applicable.

Author details
[1]Department of Public Health and Primary Care, Environment and Health, KU Leuven - University of Leuven, Kapucijnenvoer 35 blok D box 7001, 3000 Leuven, Belgium. [2]Flemish Institute of Technological Research (VITO), Unit Environmental Risk and Health, Boeretang 200, 2400 Mol, Belgium. [3]Department of Imaging & Pathology, KU Leuven - University of Leuven, 3000 Leuven, Belgium. [4]University Hospitals Leuven; Department of Forensic Medicine; Laboratory of Forensic Genetics and Molecular Archeology, KU Leuven - University of Leuven, 3000 Leuven, Belgium. [5]Center for Molecular and Vascular Biology, KU Leuven - University of Leuven, UZ Herestraat 49 - box 911, 3000 Leuven, Belgium. [6]International Agency for Research on Cancer, 150 Cours Albert Thomas, 69372 LyonCEDEX 08, France. [7]Faculty of Sciences, Hasselt University, 3590 Diepenbeek, Belgium. [8]Department of Development and Regeneration, KU Leuven - University of Leuven, 3000 Leuven, Belgium. [9]Department of Obstetrics and Gynecology, University Hospitals of Leuven, 3000 Leuven, Belgium. [10]IDEWE, External Service for Prevention and Protection at Work, Interleuvenlaan 58, 3001 Heverlee, Belgium.

References
1. Lillycrop KA, Burdge GC. Maternal diet as a modifier of offspring epigenetics. J Dev Orig Health Dis. 2015;6(2):88–95.
2. Roseboom TJ, van der Meulen JH, Ravelli AC, Osmond C, Barker DJ, Bleker OP. Effects of prenatal exposure to the Dutch famine on adult disease in later life: an overview. Twin Res. 2001;4(5):293–8.
3. Jiménez-Chillarón JC, Díaz R, Martínez D, Pentinat T, Ramón-Krauel M, Ribó S, et al. The role of nutrition on epigenetic modifications and their implications on health. Biochimie. 2012;94(11):2242–63.
4. Patel MS, Srinivasan M. Metabolic programming due to alterations in nutrition in the immediate postnatal period. J Nutr. 2010;140(3):658–61.
5. Heijmans BT, Tobi EW, Stein AD, Putter H, Blauw GJ, Susser ES, et al. Persistent epigenetic differences associated with prenatal exposure to famine in humans. Proc Natl Acad Sci U S A. 2008;105(44):17046–9.
6. Faulk C, Dolinoy DC. Timing is everything: the when and how of environmentally induced changes in the epigenome of animals. Epigenetics. 2011;6(7):791–7.
7. Anderson OS, Sant KE, Dolinoy DC. Nutrition and epigenetics: an interplay of dietary methyl donors, one-carbon metabolism and DNA methylation. J Nutr Biochem. 2012;23(8):853–9.
8. Ueland PM. Choline and betaine in health and disease. J Inherit Metab Dis. 2011;34(1):3–15.
9. Caudill MA. Pre- and postnatal health: evidence of increased choline needs. J Am Diet Assoc. 2010;110(8):1198–206.
10. Craig SA. Betaine in human nutrition. Am J Clin Nutr. 2004;80(3):539–49.
11. Thaler CJ. Folate metabolism and human reproduction. Geburtshilfe Frauenheilkd. 2014;74(9):845–51.
12. Lamers Y. Folate recommendations for pregnancy, lactation, and infancy. Ann Nutr Metab. 2011;59(1):32–7.
13. Rees WD, Wilson FA, Maloney CA. Sulfur amino acid metabolism in pregnancy: the impact of methionine in the maternal diet. J Nutr. 2006;136(6 Suppl):1701S–5.
14. Dolinoy DC. The agouti mouse model: an epigenetic biosensor for nutritional and environmental alterations on the fetal epigenome. Nutr Rev. 2008;66 Suppl 1:S7–11.
15. Haggarty P, Hoad G, Campbell DM, Horgan GW, Piyathilake C, McNeill G. Folate in pregnancy and imprinted gene and repeat element methylation in the offspring. Am J Clin Nutr. 2013;97(1):94–9.
16. Chen ZX, Riggs AD. DNA methylation and demethylation in mammals. J Biol Chem. 2011;286(21):18347–53.
17. Chao W, D'Amore PA. IGF2: epigenetic regulation and role in development and disease. Cytokine Growth Factor Rev. 2008;19(2):111–20.
18. Steegers-Theunissen RP, Twigt J, Pestinger V, Sinclair KD. The periconceptional period, reproduction and long-term health of offspring: the importance of one-carbon metabolism. Hum Reprod Update. 2013;19(6):640–55.
19. Tobi EW, Lumey LH, Talens RP, Kremer D, Putter H, Stein AD, et al. DNA methylation differences after exposure to prenatal famine are common and timing- and sex-specific. Hum Mol Genet. 2009;18(21):4046–53.
20. Obermann-Borst SA, Eilers PH, Tobi EW, de Jong FH, Slagboom PE, Heijmans BT, et al. Duration of breastfeeding and gender are associated with methylation of the LEPTIN gene in very young children. Pediatr Res. 2013;74(3):344–9.
21. Godfrey KM, Sheppard A, Gluckman PD, Lillycrop KA, Burdge GC, McLean C, et al. Epigenetic gene promoter methylation at birth is associated with child's later adiposity. Diabetes. 2011;60(5):1528–34.
22. Lenhard JM. PPAR gamma/RXR as a molecular target for diabetes. Receptors Channels. 2001;7(4):249–58.
23. Verduci E, Banderali G, Barberi S, Radaelli G, Lops A, Betti F, et al. Epigenetic effects of human breast milk. Nutrients. 2014;6(4):1711–24.
24. Allen LH. B vitamins in breast milk: relative importance of maternal status and intake, and effects on infant status and function. Adv Nutr. 2012;3(3):362–9.
25. Nutten S. Proteins, peptides and amino acids: role in infant nutrition. Nestle Nutr Inst Workshop Ser. 2016;86:1–10.
26. Zhang Z, Adelman AS, Rai D, Boettcher J, Lönnerdal B. Amino acid profiles in term and preterm human milk through lactation: a systematic review. Nutrients. 2013;5(12):4800–21.
27. Vidovic A, Juras DV, Boras VV, Lukac J, Grubisic-Ilic M, Rak D, et al. Determination of leucocyte subsets in human saliva by flow cytometry. Arch Oral Biol. 2012;57(5):577–83.
28. Pauwels S, Duca RC, Devlieger R, Freson K, Straetmans D, Van Herck E, et al. Maternal methyl-group donor intake and global DNA (hydroxy)methylation before and during pregnancy. Nutrients. 2016;8(8):474.
29. Pauwels S, Doperé I, Huybrechts I, Godderis L, Koppen G, Vansant G. Reproducibility and validity of an FFQ to assess usual intake of methyl-group donors. Public Health Nutr. 2015;18(14):1–10.
30. Pauwels S, Doperé I, Huybrechts I, Godderis L, Koppen G, Vansant G. Validation of a food-frequency questionnaire assessment of methyl-group donors using estimated diet records and plasma biomarkers: the method of triads. Int J Food Sci Nutr. 2014;65(6):768–73.
31. Murphy SK, Huang Z, Hoyo C. Differentially methylated regions of imprinted genes in prenatal, perinatal and postnatal human tissues. PLoS One. 2012;7(7):e40924.
32. World Health Organization FaAOotUN, United Nations University. Protein and amino acid requirements in human nutrition. Report of a joint FAO/WHO/UNU expert consultation (WHO Technical Report, Series 935). 2007.
33. Hoge Gezondheidsraad. Voedingsaanbevelingen voor België - 2016. Brussels; 2016.
34. Institute of Medicine (US) Standing Committee on the Scientific Evaluation of Dietary Reference Intakes and its Panel on Folate OBV, and Choline. Dietary reference intakes for thiamin, riboflavin, niacin, vitamin B6, folate, vitamin B12, pantothenic acid, biotin, and choline. Washington (DC): Institute of Medicine (US) Standing Committee on the Scientific Evaluation of Dietary Reference Intakes and its Panel on Folate OBV, and Choline; 1998.
35. Hoyo C, Murtha AP, Schildkraut JM, Jirtle RL, Demark-Wahnefried W, Forman MR, et al. Methylation variation at IGF2 differentially methylated regions and maternal folic acid use before and during pregnancy. Epigenetics. 2011;6(7):928–36.
36. Steegers-Theunissen RP, Obermann-Borst SA, Kremer D, Lindemans J, Siebel C, Steegers EA, et al. Periconceptional maternal folic acid use of 400 microg per day is related to increased methylation of the IGF2 gene in the very young child. PLoS One. 2009;4(11):e7845.
37. Hochberg Z, Feil R, Constancia M, Fraga M, Junien C, Carel JC, et al. Child health, developmental plasticity, and epigenetic programming. Endocr Rev. 2011;32(2):159–224.
38. van Dijk SJ, Molloy PL, Varinli H, Morrison JL, Muhlhausler BS. Epigenetics and human obesity. Int J Obes (Lond). 2014;85–97.
39. Gong L, Pan YX, Chen H. Gestational low protein diet in the rat mediates Igf2 gene expression in male offspring via altered hepatic DNA methylation. Epigenetics. 2010;5(7):619–26.
40. Lillycrop KA, Slater-Jefferies JL, Hanson MA, Godfrey KM, Jackson AA, Burdge GC. Induction of altered epigenetic regulation of the hepatic glucocorticoid receptor in the offspring of rats fed a protein-restricted diet during pregnancy suggests that reduced DNA methyltransferase-1 expression is involved in impaired DNA methylation and changes in histone modifications. Br J Nutr. 2007;97(6):1064–73.
41. Lan X, Cretney EC, Kropp J, Khateeb K, Berg MA, Peñagaricano F, et al. Maternal diet during pregnancy induces gene expression and DNA methylation changes in fetal tissues in sheep. Front Genet. 2013;4:49.

42. Kovacheva VP, Mellott TJ, Davison JM, Wagner N, Lopez-Coviella I, Schnitzler AC, et al. Gestational choline deficiency causes global and Igf2 gene DNA hypermethylation by up-regulation of Dnmt1 expression. J Biol Chem. 2007;282(43):31777–88.

43. Slack A, Cervoni N, Pinard M, Szyf M. Feedback regulation of DNA methyltransferase gene expression by methylation. Eur J Biochem. 1999;264(1):191–9.

44. Li E, Beard C, Jaenisch R. Role for DNA methylation in genomic imprinting. Nature. 1993;366(6453):362–5.

45. Kalhan SC, Marczewski SE. Methionine, homocysteine, one carbon metabolism and fetal growth. Rev Endocr Metab Disord. 2012;13(2):109–19.

46. World Health Organization. The optimal duration of exclusive breastfeeding. Geneva: Report of an expert consultation; 2001.

47. Zeisel SH, da Costa KA. Choline: an essential nutrient for public health. Nutr Rev. 2009;67(11):615–23.

48. Ilcol YO, Ozbek R, Hamurtekin E, Ulus IH. Choline status in newborns, infants, children, breast-feeding women, breast-fed infants and human breast milk. J Nutr Biochem. 2005;16(8):489–99.

49. Obeid R. The metabolic burden of methyl donor deficiency with focus on the betaine homocysteine methyltransferase pathway. Nutrients. 2013;5(9):3481–95.

50. Jiang X, West AA, Caudill MA. Maternal choline supplementation: a nutritional approach for improving offspring health? Trends Endocrinol Metab. 2014;25(5):263–73.

51. Leenen FA, Muller CP, Turner JD. DNA methylation: conducting the orchestra from exposure to phenotype? Clin Epigenetics. 2016;8:92.

52. Somineni HK, Zhang X, Biagini Myers JM, Kovacic MB, Ulm A, Jurcak N, et al. Ten-eleven translocation 1 (TET1) methylation is associated with childhood asthma and traffic-related air pollution. J Allergy Clin Immunol. 2016;137(3):797–805. e5.

53. Zhang W, Spector TD, Deloukas P, Bell JT, Engelhardt BE. Predicting genome-wide DNA methylation using methylation marks, genomic position, and DNA regulatory elements. Genome Biol. 2015;16:14.

54. Lowe R, Gemma C, Beyan H, Hawa MI, Bazeos A, Leslie RD, et al. Buccals are likely to be a more informative surrogate tissue than blood for epigenome-wide association studies. Epigenetics. 2013;8(4):445–54.

55. NUBEL. Belgian food composition table, ministry of public health, 5th. Brussels: NUBEL; 2010.

56. USDA database for the choline content of common foods, U.S. Department of Agriculture, Agricultural Research Service. Release 2 ed. 2008

57. NEVO. Dutch food composition table. Zeist: NEVO Foundation; 2011.

58. Dehne LI, Klemm C, Henseler G, Hermann-Kunz E. The German food code and nutrient data base (BLS II.2). Eur J Epidemiol. 1999;15(4):355–9.

59. Willett W. Nutritional epidemiology. 3 ed. Oxford: Oxford University Press; 2012. p. 529.

DNA methylation levels are associated with CRF$_1$ receptor antagonist treatment outcome in women with post-traumatic stress disorder

Julius C. Pape[1] (iD), Tania Carrillo-Roa[1], Barbara O. Rothbaum[2], Charles B. Nemeroff[3], Darina Czamara[1], Anthony S. Zannas[1,8], Dan Iosifescu[4,9,10], Sanjay J. Mathew[5], Thomas C. Neylan[6,7], Helen S. Mayberg[2], Boadie W. Dunlop[2] and Elisabeth B. Binder[1,2*]

Abstract

Background: We have previously evaluated the efficacy of the CRF$_1$ receptor antagonist GSK561679 in female PTSD patients. While GSK561679 was not superior to placebo overall, it was associated with a significantly stronger symptom reduction in a subset of patients with probable CRF system hyperactivity, i.e., patients with child abuse and *CRHR1* SNP rs110402 GG carriers. Here, we test whether blood-based DNA methylation levels within *CRHR1* and other PTSD-relevant genes would be associated with treatment outcome, either overall or in the high CRF activity subgroup.

Results: Therefore, we measured *CRHR1* genotypes as well as baseline and post-treatment DNA methylation from the peripheral blood in the same cohort of PTSD-diagnosed women treated with GSK561679 ($N = 43$) or placebo ($N = 45$). In the same patients, we assessed DNA methylation at the PTSD-relevant genes *NR3C1* and *FKBP5*, shown to predict or associate with PTSD treatment outcome after psychotherapy. We observed significant differences in *CRHR1* methylation after GSK561679 treatment in the subgroup of patients with high CRF activity. Furthermore, *NR3C1* baseline methylation significantly interacted with child abuse to predict PTSD symptom change following GSK561679 treatment.

Conclusions: Our results support a possible role of *CRHR1* methylation levels as an epigenetic marker to track response to CRF$_1$ antagonist treatment in biologically relevant subgroups. Moreover, pre-treatment *NR3C1* methylation levels may serve as a potential marker to predict PTSD treatment outcome, independent of the type of therapy. However, to establish clinical relevance of these markers, our findings require replication and validation in larger studies.

Keywords: CRF$_1$ receptor antagonist, DNA methylation, Epigenetics, PTSD, CRHR1, NR3C1, FKBP5

* Correspondence: binder@psych.mpg.de; ebinder@emory.edu
[1]Department of Translational Research in Psychiatry, Max Planck Institute of Psychiatry, Munich, Germany
[2]Department of Psychiatry and Behavioral Sciences, Emory University School of Medicine, Atlanta, GA, USA
Full list of author information is available at the end of the article

Background

Post-traumatic stress disorder (PTSD) is a common psychiatric disorder with a prevalence of about 5% in the general population and an overall lifetime prevalence of 7–12%. Key symptoms of the disorder include intrusive memories, avoidance, and numbing as well as hyperarousal. Typically, these symptoms are long lasting and occur after exposure to traumatic life events. Women are twice as likely to develop the disease than men. PTSD therapies include both evidence-based psychotherapies and pharmacology, but only few patients attain remission. Currently, only two medications, paroxetine and sertraline, are approved by the US Food and Drug Administration (FDA). These SSRIs are capable of significantly reducing PTSD symptoms, but with only 20–30% remission rates to these agents, there is a need for additional pharmacologic treatment options [1].

Among pathophysiologic mechanisms that have been investigated for PTSD, disruptions of regulation of the hypothalamic-pituitary-adrenal (HPA) axis are among the most frequently cited hypotheses [2]. A key regulator of the HPA axis is the corticotropin-releasing factor (CRF) and its type 1 receptor (CRF_1 receptor), and many studies have reported alterations in this system in PTSD [3]. Therefore, it represents a promising novel drug target for this disorder. In response to stress, CRF is secreted by nerve terminals of the paraventricular nucleus of the hypothalamus and binds to the CRF_1 receptor in the adenohypophysis to release adrenocorticotropic hormone (ACTH). This process acts as the initial step of HPA axis activation and leads to the release of a number of hormones from the adrenal cortex including cortisol. Numerous studies in laboratory animals as well as in humans indicate that abnormalities of these HPA axis regulators play a crucial role in stress-related disorders such as PTSD [4].

In humans, for example, a number of independent studies report increased cerebrospinal fluid concentrations of corticotropin-releasing factor in PTSD patients [5–7], suggesting hyperactivity of the hypothalamus and extra-hypothalamus CRF system. Moreover, previous investigations have found that genetic variants in the CRF receptor 1 gene (CRHR1) are associated with differences in CRF signaling and may also impact individual responses to environmental stressors [3]. The most studied are variants within a haplotype tagged by the intronic SNP rs110402 that also comprises rs242924 and rs7209436. Interactions with exposure to child abuse and this haplotype were shown to alter risk for major depression, with individuals homozygous for the G-allele of rs110402 and exposed to child abuse being at higher risk in several but not all studies (see [8] for review). This haplotype has also been associated with differences in the neural activation profile with emotional stimulus processing [9], as well as

neuroendocrine responses in psychological and pharmacological challenge tests [10–14], in which individuals who experienced childhood abuse and carry the G-allele display stronger HPA axis disturbances.

These preclinical and clinical results, taken together, support the role of CRF/CRF_1 receptor as a potential drug target in PTSD. However, antagonism of the CRF_1 receptor may only benefit those patients with initial increases in CRF signaling, which according to the above cited endocrine studies are likely to be those with exposure to child abuse and carrying the G-allele of rs110402.

We recently published a study evaluating the efficacy of a novel CRF_1 receptor antagonist (GSK561679) in a cohort of female PTSD patients in a double-blind, placebo-controlled trial. Although the drug was not superior to placebo overall, it was associated with a significantly stronger symptom reduction in a subset of patients with probable CRF_1 receptor hyperactivity, i.e., patients with childhood abuse and carriers of the GG genotype of the CRHR1 SNP rs110402 [15, 16]. These patients may represent a biologically distinct subtype of PTSD and show distinct biomarker profiles. Markers that predict or monitor treatment outcome would represent an important tool to offer targeted treatment for individual patients. Despite great progress in identifying the underpinnings of the pathophysiology of PTSD and some very promising results in the biomarker field [17, 18], there is still no clinically applicable marker in PTSD, neither for diagnosis nor, perhaps even more significantly, to guide treatment selection. This is likely due to the complex pathophysiology of the disease that may include an interplay of genetics, environment, and epigenetic changes. It is therefore likely that not a single but rather a combination of different biological and clinical markers will need to be identified [18].

In addition to gene variants that predispose to PTSD development, epigenetic changes have been implicated in the pathophysiology of PTSD (for review, see [19]). These modifications may also serve as diagnostic marks as well as predicting and monitoring treatment outcome. Several studies highlight the possible use of epigenetic marks in peripheral tissues such as the blood and saliva as diagnostic markers in PTSD [18, 20, 21]. So far, epigenetic marks of only two genes, also within the HPA axis, NR3C1—encoding the glucocorticoid receptor (GR) and FKBP5—a co-chaperone of the GR, have been shown to associate with treatment response. More specifically, NR3C1 baseline promoter methylation in peripheral blood predicted treatment outcome in PTSD, and in the same study, promoter methylation of FKBP5 decreased in association with symptom improvement [22]. These findings were observed after 12 weeks of psychotherapy and have not yet been investigated in the context of pharmacological treatment.

Extending our previous study showing potential effects of a novel CRF_1 receptor antagonist (GSK561679) in a specific subset of women with PTSD (GG homozygous for rs110402 and with a history of childhood abuse) [16], we here use the same cohort to test whether blood-based epigenetic changes of PTSD relevant genes could serve as potential markers for treatment selection and outcome monitoring in biologically defined subgroups of patients. Given that the drug targets the CRF_1 receptor, we focused our analysis on the methylation of the *CRHR1* gene using the previous subgrouping of patients based on genetic and environmental risk factors. In addition, we explored whether methylation levels of two other genes within the stress hormone system (*NR3C1* and *FKBP5*), previously shown to predict and correlate with PTSD symptom improvement after psychotherapy [22], would also be associated with pharmacological treatment response in our study, again with specific focus on patients with probable CRF system hyperactivity (rs110402 GG-carriers and exposure to child abuse).

Results

Subgroup differences in *CRHR1* baseline methylation and change in *CRHR1* methylation from baseline to post-treatment

First, we tested a model with the main effects and interaction effect of child abuse and rs110402 carrier status on mean *CRHR1* baseline methylation. Seventy-nine subjects were included in this analysis due to missing genotype data in three samples. Neither the main effects nor the interaction effect showed significance ($n = 79$; $p > 0.05$). Next, we tested a model including main effects of treatment as well as interaction effects of treatment by child abuse, treatment by rs110402, child abuse by rs110402, and the three-way interaction of treatment by child abuse by rs110402 on changes in mean methylation levels of *CRHR1* from baseline to post-treatment. Due to missing methylation data in two baseline samples and one post-treatment sample, 57 subjects with baseline and post-treatment methylation data remained for this analysis. There was a significant interaction effect of child abuse by rs110402 carrier status ($n = 57$; $F (1, 41) = 9.05$; $p = 0.004$; $\beta = -0.449$; Cohen's $f = 0.47$; $R^2 = 0.38$; adj. $R^2 = 0.153$; post-hoc power = 0.94) on change in methylation. Further, the three-way interaction of treatment by child abuse by rs110402 showed a significant effect on *CRHR1* methylation levels from pre- to post-treatment ($n = 57$; $F (1, 41) = 4.86$; $p = 0.033$; $\beta = -0.297$; Cohen's $f = 0.344$; $R^2 = 0.38$; adj. $R^2 = 0.153$; post-hoc power = 0.72) (Fig. 1a, b).

Genotype by childhood abuse interaction on methylation change stratified by treatment

To further explore the significant three-way interaction on *CRHR1* methylation, we investigated the interaction

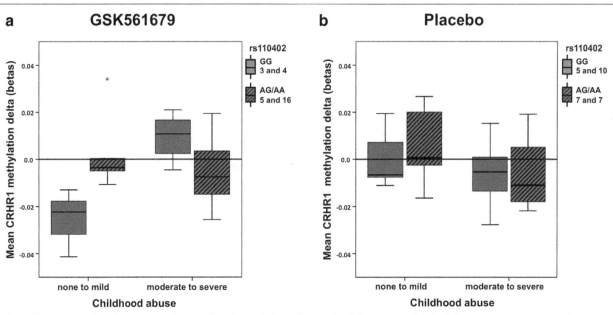

Fig. 1 The boxplots describe the mean change of CRHR1 methylation (top tertile of the most variable CpGs from pre- to post-treatment) in abused and non-abused patients treated with GSK561679 or placebo. GG carriers are shown in blue (plain boxes) and AA/AG in red (striped boxes). Positive values correspond to an increase, whereas negative values correspond to a decrease in methylation from baseline to endpoint. Dots indicate outliers. Three-way interaction of treatment × rs110402 A carrier status × child abuse was significantly associated with mean methylation change ($n = 57$; $p = 0.033$) (**a, b**). After treatment stratification, there was a significant interaction effect of rs110402 A carrier status and child abuse on mean methylation change in subjects treated with GSK561679 ($n = 28$; $p = 0.00005$) (**a**) but not with placebo ($n = 29$; $p > 0.05$) (**b**)

of rs110402 carrier status by child abuse on the change in methylation levels stratified by treatment. The interaction showed a significant effect on pre- to post-treatment CRHR1 methylation change only in patients treated with the CRF_1 receptor antagonist ($n = 28$; F (1, 16) = 29.81; $p = 0.00005$; withstands Bonferroni correction for multiple testing; $ß = - 0.913$; Cohen's $f = 1.366$; $R^2 = 0.73$; adj. $R^2 = 0.55$; post-hoc power = 0.99) (Fig. 1a).

Interestingly, the subset of patients with child abuse and who are also carriers of the GG genotype of rs110402 showed an increase in CRHR1 methylation with GSK561679 treatment. This subgroup was previously described to benefit most from the drug ([16] and Additional file 1: Figure S1). The other three subsets of patients (no abuse and rs110402 GG; no abuse and rs110402 AG/AA; abuse and rs110402 AG/AA) showed no change or decreased methylation after GSK561679 treatment. There was no significant effect in the placebo group ($n = 29$; $p > 0.05$) (Fig. 1b).

Baseline methylation by treatment interaction effects on PTSD symptom change

We next tested whether baseline methylation predicted %-change of PTSD symptoms from pre- to post-treatment. Seventy-nine (CAPS)/78 (PSS) subjects were included in the analysis due to missing genotype data in three samples and missing phenotype data (PSS %-change) in one sample. Neither NR3C1 ($n = 79/78$; $p > 0.05$) nor FKBP5 ($n = 79/78$; $p > 0.05$) showed a significant interaction effect of treatment by baseline methylation on symptom change.

Three-way interaction effects on PTSD symptom change with treatment, baseline methylation, and SNP/child abuse

Next, we included either rs110402 or child abuse in our analysis and tested for two three-way interaction effects (rs110402 × treatment × mean baseline methylation or child abuse × treatment × mean baseline methylation) on symptom reduction measured by change in Clinician-Administered PTSD Scale (CAPS) and PTSD Symptom Scale-Self-Report (PSS-SR) scores. Treatment by baseline methylation by rs110402 carrier status was not significantly associated with differences in PTSD symptom change for neither of the genes (NR3C1: $n = 79/78$, $p > 0.05$; FKBP5: $n = 79/78$, $p > 0.05$).

The three-way interaction that included child abuse was significant for NR3C1 baseline methylation ($n = 78$; F (1, 56) = 4.26; $p = 0.044$; $ß = 0.276$; Cohen's $f = 0.277$; $R^2 = 0.33$; adj. $R^2 = 0.087$; post-hoc power = 0.67) and showed a trend towards significance for FKBP5 baseline methylation ($n = 79$, F (1, 57) = 2.81; $p = 0.099$; $ß = 0.215$; Cohen's $f = 0.222$; $R^2 = 0.28$; adj. $R^2 = 0.017$; post-hoc power = 0.38).

More specifically, CRF_1 receptor antagonist-treated, abused patients with high baseline NR3C1 methylation levels showed the strongest PSS percent change and therefore the best treatment outcome overall (Fig. 2a, b). A post-hoc analysis revealed that the interaction of baseline NR3C1 methylation and child abuse was significantly associated with PSS percent change after CRF_1 receptor antagonist treatment ($n = 38$; F (1, 20) = 4.58; $p = 0.045$; $ß = 0.331$; Cohen's $f = 0.478$; $R^2 = 0.67$; adj. $R^2 = 0.39$; post-hoc power = 0.81) (Fig. 2a) but not placebo ($n = 40$; $p > 0.05$) (Fig. 2b). Results from the same analysis using CAPS score %-change as treatment outcome showed the same direction of effects but did not reach significance (three-way interaction: $n = 79$; $p > 0.05$) (Fig. 2c, d).

For FKBP5, abused patients with high baseline methylation and treated with the CRF_1 receptor antagonist experienced the strongest CAPS percent change ($n = 79$, F (1, 57) = 2.81; $p = 0.099$). The post-hoc analysis, stratifying patients by treatment and testing the interaction effect of baseline methylation by child abuse on PTSD symptom change, did not reach significance in neither one of the treatment groups ($p > 0.05$ for all) (Fig. 3a–d).

Pre- to post-treatment methylation change by treatment interaction effects and three-way interaction effects including SNP or child abuse on PTSD symptom change

To examine the association between FKBP5/NR3C1 methylation change from baseline to post-treatment and symptom improvement, we tested for interaction effects of treatment by pre- to post-methylation change on %-change of PTSD symptoms from pre- to post-treatment. For NR3C1 and FKBP5, 57 subjects were included in the analysis due to missing methylation data in two baseline samples and one post-treatment sample. None of the tested interactions reached significance (FKBP5: $n = 57$, $p > 0.5$; NR3C1: $n = 57$, $p > 0.5$). Further, including either rs110402 or child abuse in our analysis to test for two three-way interactions (rs110402 × treatment × pre- to post-methylation change or child abuse × treatment × pre- to post-methylation change) on symptom reduction also did not show significant effects (FKBP5: $n = 57$, $p > 0.5$; NR3C1: $n = 57$, $p > 0.5$).

Discussion

The objective of this study was to investigate epigenetic marks of PTSD-related genes in association with PTSD symptom changes after CRF_1 receptor antagonist (GSK561679) treatment in female PTSD patients. In a first analysis, we observed significant differences in CRHR1 methylation levels after treatment among patients with probable CRF hyperactivity who previously demonstrated the greatest clinical benefit from the CRF_1 receptor antagonist [16]; this effect was not present among those who received placebo. This subgroup of patients who had experienced child abuse and were

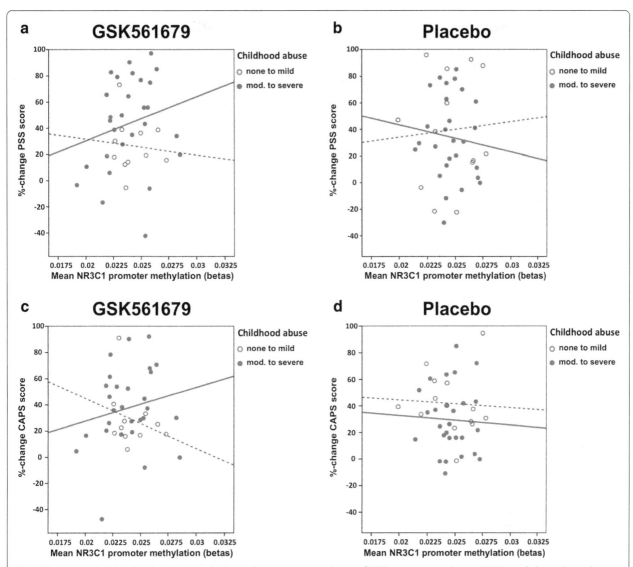

Fig. 2 The scatter plots describe the association between the mean percent change of PTSD symptoms and mean NR3C1 methylation dependent on child abuse in patients treated with GSK561679 (**a**, **c**) or placebo (**b**, **d**). Higher symptom percent change corresponds to improvement (reduction) in PTSD symptoms from baseline to endpoint. Abused patients are shown in red (solid line) and non-abused patients in blue (dashed line). Three-way interaction of NR3C1 baseline methylation × treatment × child abuse was significantly associated with PSS %-change ($n = 79$; $p = 0.044$) (**a**, **b**) but not with CAPS %-change ($n = 78$; $p > 0.05$) (**c**, **d**). After treatment stratification, there was a significant interaction effect of baseline methylation and child abuse on PSS %-change in subjects treated with GSK561679 ($n = 38$; $p = 0.045$) (**a**) but not with placebo ($n = 40$; $p > 0.05$) (**b**). For CAPS %-change, the effect pointed in the same direction without reaching significance (**c**, **d**)

homozygous for the rs110402 GG allele were the only individuals showing a significant increase in *CRHR1* methylation from baseline to the post-GSK561679 treatment time point. All other subjects either showed no change or a reduction in methylation over the time of treatment. On the other hand, baseline *CRHR1* methylation did not predict treatment outcome, suggesting that this epigenetic change may only serve as a potential tracker of symptom changes. The maximum difference in mean *CRHR1* methylation between the subgroups was more than 3%, a change comparable to or even larger than other studies examining

peripheral blood DNA methylation and psychiatric disorders or psychiatric treatment response. In fact, when examining the 11 CpGs composing the *CRHR1* variable methylation score, the maximal effects were observed in CpGs cg27410679 and cg04194664. In the subgroup of patients with child abuse and homozygous for the rs110402 GG allele, these CpGs showed an increase in methylation of up to 3.9% and a maximum methylation difference between the four subgroups of 9.9% (cg04194664) and 7.7% (cg27410679). Future studies should evaluate these optimized markers in larger samples.

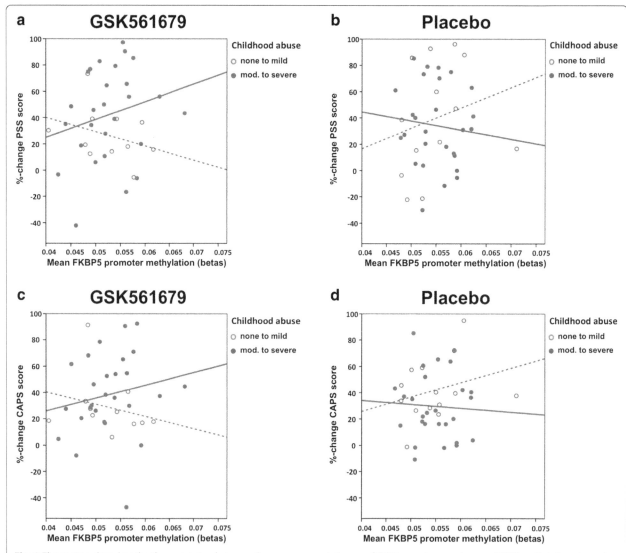

Fig. 3 The scatter plots describe the association between the mean percent change of PTSD symptoms and mean FKBP5 methylation dependent on child abuse in patients treated with GSK561679 (**a**, **c**) or placebo (**b**, **d**). Higher symptom percent change corresponds to improvement (reduction) in PTSD symptoms from baseline to endpoint. Abused patients are shown in red (solid line) and non-abused patients in blue (dashed line). The three-way interaction testing FKBP5 baseline methylation × treatment × child abuse on CAPS %-change had a p value of $p = 0.099$ with an $n = 79$ (**c**, **d**) and $p > 0.05$ with PSS %-change ($n = 78$) (**a**, **b**). After treatment stratification, there was no significant interaction effect of baseline methylation by child abuse on PTSD symptom %-change in neither one of the treatment groups ($p > 0.05$ for all) (**a–d**)

A number of factors can contribute to changes in DNA methylation. In a mixed tissue such as peripheral blood, the most likely contributor is the changes in immune cell subtype composition. Changes in immune responses have been reported in PTSD (reviewed by [23]), and symptom normalization may be associated with a change in immune function and cell type proportion [24–26]. We attempted to account for this using a bioinformatics deconvolution method for blood cell types from genome-wide methylation data [27] and adding the estimated cell type proportions as covariates. In addition, there has been increasing evidence suggesting that dynamic methylation changes, as observed in our study, may

be mediated by certain transcription factors [28–30]. Several studies have reported on the potential role of the glucocorticoid receptor as one of these transcription factors mediating glucocorticoid-induced DNA demethylation [31, 32]. CRF$_1$ receptor antagonists influence the regulation of the HPA axis and by that, ultimately, modulate GR activity. Our previously identified subgroup of patients with rs110402 GG genotype and a history of child abuse displayed a significant increase in *CRHR1* methylation after GSK561679 treatment. Previous studies have shown that this combination of environmental and genetic risk is associated with specific disruptions of HPA axis regulation, including an enhanced cortisol response to the

Trier Social Stress Test and the combined dexamethasone suppression/CRF stimulation test [11–14]. A combination of increased CRF activity and GR activation may exist in this subgroup and normalize with specific CRF_1 receptor antagonist treatment. In fact, a number of studies have also reported GR supersensitivity with PTSD [33, 34] and its normalization with effective treatment [35, 36]. Such a reversal of GR supersensitivity in the subset of patients with response to the antagonist may also lead to changes in GR-mediated DNA methylation. In fact, active GR response elements are shown in the ENCODE project for the *CRHR1* locus [37]. Finally, GSK561679 itself could directly impact *CRHR1* methylation. However, the *CRHR1* expression is low in peripheral blood cells (https://gtex portal.org/), suggesting that the epigenetic regulation of the locus indirectly via receptor blockade and adaptive transcriptional regulation is an unlikely mechanism for inducing this effect.

In our second analysis, we investigated peripheral blood DNA methylation of two genes, for which a previous study had found an association with improvement of PTSD symptoms after prolonged exposure therapy [22]. In a small cohort of combat veterans diagnosed with PTSD, the authors reported that pre-treatment *NR3C1* methylation significantly predicted treatment outcome, with higher *NR3C1* methylation at baseline associated with better response to psychotherapy. The authors also observed a decrease in *FKBP5* promoter methylation over treatment in patients showing clinical improvement [22].

Similar to Yehuda et al. [22], we also find that higher baseline methylation of *NR3C1* is associated with better treatment outcome with the antagonist. However, in our analysis, this is only seen in patients who had also experienced child abuse. No association was found for *FKBP5*, neither for baseline levels predicting treatment outcome nor for change in *FKBP5* methylation being associated with symptom improvement, as reported in Yehuda et al. [22]. While exploratory, our results support the conclusion that peripheral blood DNA methylation of *NR3C1* is associated with PTSD treatment response.

The major limitation of this study is the small sample size, particularly after biological subgrouping. Power calculation for our main hypothesis (change of CRHR1 methylation over treatment and prediction of treatment outcome), however, revealed that power would be sufficient to detect medium to large effect sizes, whereas smaller effect sizes would have been missed. A post-hoc power analysis for the specific effect sizes detected in our study showed that power ranged between 0.673 and 0.999. Further, due to the exploratory nature of our study, we did not apply a systematic correction for multiple testing, increasing the risk for false-positive associations. To identify smaller effects, confirm our results, and reduce the risk of a type I and type II error, much larger sample sizes will be required for future studies.

An additional limitation to this study, which represents a general issue in DNA methylation analyses of mixed tissues, is to rule out cell type composition variation as a potential confounding factor contributing to the observed epigenetic changes. As described, we applied a commonly used bioinformatics cell-type deconvolution method [27] to address this issue. However, this method only accounts for six different cell types in the blood, so that changes in subtypes not covered by this algorithm may still contribute to the observed changes in DNA methylation.

Conclusion

Overall, our results indicate that markers for PTSD likely will need to be an index, comprised of several combination markers. Here, we describe the association of *CRHR1* DNA methylation with treatment response, but only in a specific subset of patients defined by genetic and environmental risk factors. While our association of baseline *NR3C1* methylation with PTSD treatment outcome is supportive of previous findings, both studies are small. Given the exploratory nature of the study and the small sample size, larger studies that stratify patients by potential biomarker status will be needed to fully establish the clinical value of these measures.

Methods
Study overview
Detailed descriptions of the trial design and the study results were published previously [15, 16] and are summarized in the following.

Cohort
Patients were recruited at four academic sites (Emory University, Icahn School of Medicine at Mount Sinai, Baylor College of Medicine, University of California San Francisco/San Francisco Veterans Affairs Medical Center) in the USA. The institutional review boards at each study site approved the study. The cohort used for this study consisted of 88 female patients between 18 and 65 years of age. Males were excluded due to potential reproductive organ toxicity of the investigational medication. All subjects were free of psychotropic medication (except non-benzodiazepine hypnotics) for at least 2 weeks prior to randomization. Subjects had to fulfill criteria for a primary psychiatric diagnosis of DSM-IV-defined PTSD of at least 3 month's duration since the index trauma. PTSD status at the baseline (randomization) visit had to be of at least moderate severity, defined as Clinician-Administered PTSD Scale (CAPS) for DSM-IV [38] past-month and past-week total scores ≥ 50. Important exclusion criteria included current or past diagnosis of a psychotic disorder,

bipolar disorder, or obsessive-compulsive disorder. Subjects with a positive test for drugs of abuse at the screening visit, or who met criteria for substance abuse or dependence within 3 months of the randomization visit, or who presented with significant current suicidal ideation were excluded. Pregnant or lactating women and subjects with an unstable medical condition were also excluded.

Study design

Subjects participated in a parallel-group, double-blind, placebo-controlled randomized clinical trial of a novel CRF_1 receptor antagonist (GSK561679). After randomization, patients were either treated with a nightly dose of 350 mg GSK561679 or placebo over 6 weeks. At the baseline visit (prior to treatment phase), numerous data including demographics, vital signs, and several psychiatric measures were assessed, e.g., level of childhood maltreatment was tested using the Childhood Trauma Questionnaire (CTQ). CAPS score and PTSD Symptom Scale-Self-Report (PSS-SR) [39] were assessed at weeks 1, 2, 4, and 6 after randomization to assess PTSD symptom severity, and the percent change of these scores from pre- to post-treatment were used to determine the degree of improvement in PTSD symptoms. For biological assessments (e.g., methylation levels, genotyping), whole blood was collected at baseline ($n = 88$) as well as after 5 weeks of treatment ($n = 60$ with both baseline and post-treatment) and DNA extraction was performed.

DNA extraction

DNA isolation from whole blood was performed with a *magnetic bead*-based technology on the chemagic 360 extraction robot using the chemagic DNA Blood Kit special (PerkinElmer Inc., Waltham, MA, USA). Quality and quantity of the extracted DNA were assessed using the Epoch Microplate Spectrophotometer (BioTek, Winooski, VT, USA).

Genotyping

Genome-wide SNP genotyping was performed for all subjects using Illumina HumanOmniExpress-24 Bead-Chips according to the manufacturer's protocol. We excluded the relatives of individual subjects from the whole sample ($n = 3$, Pihat ≥ 0.0625) based on mean identity by descent (IBD) in PLINK [40]. Eighty-five subjects remained for further QC. For the genome-wide analyses that were used to correct for population stratification, we only included individuals with a sample-wise call rate ≥ 0.98 and SNPs with call rate ≥ 0.98, Hardy Weinberg equilibrium test (HWE) p value $\geq 1 \times 10^{-5}$ and MAF ≥ 0.05, allowing for a total of 575,455 markers in 85 individuals. To correct for population stratification in an ethnically mixed sample, principal components (PC) for the genetic background were calculated from all

genotypes for each of the individuals using genome-wide complex trait analysis (GCTA) [41].

Methylation analysis

DNA methylation levels were assessed using the Illumina 450k array. After bisulfite conversion with the Zymo EZ-96 DNA Methylation Kit (Zymo Research, Irvine, CA. USA), genome-wide DNA methylation levels were assessed for 84 baseline samples and 60 matching post-treatment samples using Illumina 450K DNA methylation arrays (Illumina, San Diego, CA, USA) as previously published [42].

Quality control of DNA methylation

Minfi Bioconductor R package (version 1.10.2) was used to perform quality control of methylation data including normalization, intensity readouts, cell type composition estimation, and beta and M value calculation. A detection p value larger than 0.01 in at least 75% of the samples led to an exclusion of the probe. Probes that were located close (10 bp from query site) to a SNP which had a minor allele frequency of ≥ 0.05 in any of the populations represented in the sample were removed as well as X chromosome, Y chromosome, and non-specific binding probes. The data were then normalized using functional normalization, which is an extension of quantile normalization included in the minfi R package. The Bioconductor R package shinyMethyl version 0.99.3 was used to identify batch effects by inspecting the association of the first principal component of the methylation levels with plate, sentrix array, and position using linear regression and visual inspection of PCA plots. A linear regression model was fitted in R with the M values for each probe as the dependent variable and plate, sentrix array, and row as the independent variables as factors to remove batch effects. Two baseline samples and one post-treatment sample did not pass quality control, which resulted in 82 baseline samples and 57 matching pairs with 450K methylation data.

Statistical analyses

Statistical analysis was carried out using SPSS v.18.0 (IBM Corp., Armonk, NY, USA) and R software v 3.2 (https://www.r-project.org/). Genotype analysis (SNP rs110402): the intronic SNP rs110402 has been shown to be associated with HPA axis hyperactivity [11, 14, 43]. This may result in a different response to antagonizing the CRF system, depending on a patient's rs110402 genotype. We therefore focused on rs110402 genotype stratification in our analysis. Direct genotypes were taken from the HumanOmniExpress-24 array (rs110402 MAF = 0.401, HWE test p value = 0.52). According to our previous study [16], patients were categorized by rs110402 A-allele carrier status (GG = 33 carriers and 53

A-allele carriers, of which 38 patients had the AG geno-type and 15 were homozygous for the A-allele). Group-ing individuals carrying one or two copies of the minor A-allele of rs110402 has been used in previous studies [9, 11, 44] and helps to preserve power. Additive effects of that SNP have previously been reported [45]. Methy-lation analysis: *CRHR1*: From the *CRHR1* gene locus covered by 33 CpGs on the 450k array, the top tertile (11 CpGs) of the CpGs with the most variable methyla-tion change from pre to post-treatment was selected (Additional file 1: Table S1). The mean methylation of these 11 CpGs was calculated and used for further ana-lysis. *NR3C1*: Mean methylation of 5 CpG sites within the 1F promoter and exon present on the Illumina 450K array was used for the analysis (Additional file 1: Table S2). DNA methylation in the 1F promoter and exon had been shown to predict PTSD treatment outcome [22]. *FKBP5*: Mean methylation level of 3 CpG sites within the exon 1 promoter present on the Illumina 450K array was used for the analysis (Additional file 1: Table S3). DNA methylation of this locus was shown to track with symptom improve-ment [22]. Childhood trauma status was defined as previ-ously described by categorizing individuals as having experienced either no or only mild abuse versus those having experienced at least one type of moderate to severe abuse (emotional abuse ≥ 13, physical abuse ≥ 10, sexual abuse ≥ 8) (57 = abused, 31 = non-abused) using the CTQ [45]. We performed linear regression models adjusted for age, smoking, ancestry PC, and estimated blood cell count to test for main/two-way and three-way interaction effects on methylation changes as well as main/two-way and three-way interactions effects on PTSD symptom %-change. For each of the analysis, only individuals with complete phenotype, methylation data, genotypes, and any additional covariates were included in the model. We calculated power post-hoc using G Power 3.1 [46]. Alpha was set to 0.05, and the number of groups, degrees of free-dom, and eta squares were set according to the test-specific calculations performed in SPSS. Statistical sig-nificance was considered at $p < 0.05$. Due to the explora-tory nature of the study, no correction for multiple testing was applied. As a measure of effect size, Cohen's f was cal-culated and interpreted as follows: $f < 0.25$ = small effect size; $0.25 < f < 0.4$ = medium effect size; $f > 0.4$ = large ef-fect size [47].

Additional file

Additional file 1: Figure S1. The boxplots describe the mean % change of PSS total score in abused and non-abused patients treated with the CRHR1 antagonist or placebo. GG carriers are shown in blue (plain boxes) and AA/AG in red (striped boxes). rs110402 A carrier status by childhood abuse exposure showed a significant interaction effect on PSS score % change over treatment in subjects treated with the CRHR1 antagonist ($n = 43$; F (1, 31) = 4.42; $p = 0.043$) (a) but not in subjects treated with

placebo ($n = 42$, $p > 0.05$) (b). rs110402 GG carriers exposed to child abuse displayed the highest % change of PSS symptoms following CRHR1 treatment. (From Biological Psychiatry; Dunlop et al., 2017). **Table S1.** CRHR1: List of CpGs used for analysis. **Table S2.** NR3C1: List of CpGs used for analysis. **Table S3.** FKBP5: List of CpGs used for analysis.

Abbreviations
ACTH: Adrenocorticotropic hormone; CAPS: Clinician-Administered PTSD Scale; CpG: Cytosine-phosphate-guanine; CRF1: Corticotropin-releasing hormone receptor 1; *CRHR1*: Corticotropin-releasing hormone receptor 1 (gene); CTQ: Childhood Trauma Questionnaire; *FKBP5*: FK506-binding protein 51 kDa gene; GR: Glucocorticoid receptor; HPA: Hypothalamic-pituitary-adrenal; *NR3C1*: Nuclear receptor subfamily 3 group C member 1; PSS: PTSD symptom scale; PTSD: Post-traumatic stress disorder; QC: Quality control; *SKA2*: Spindle and kinetochore-associated complex subunit 2; SNP: Single nucleotide polymorphism

Acknowledgements
We thank Susann Sauer, Anne Löschner, and Maik Ködel for the excellent technical assistance with DNA extraction, genotyping, and DNA methylation assessment.

Funding
Funding for the study was provided from a grant from the National Institute of Mental Health, U19 MH069056 (BWD, HM). Additional support was received from K23 MH086690 (BWD) and VA CSRD Project ID 09S-NIMH-002 (TCN) and the Max Planck Society. The GSK561679 compound was currently licensed by Neurocrine Biosciences. GlaxoSmithKline contributed the study medication and matching placebo, as well as funds to support subject recruitment and laboratory testing, and Neurocrine Biosciences conducted the pharmacokinetic analyses. GlaxoSmithKline and Neurocrine Biosciences were not involved in the data collection, data analysis, or interpretation of findings.

Authors' contributions
EB and JP designed the research and coordinated the experimental work. BD, HM, CN, BR, DI, SM, and TN were responsible for the clinical trial and the different recruitment sites. JP performed the experimental work. JP, TC, and DC performed the statistical analysis. JP and EB prepared the initial manuscript. BR, CN, DC, AZ, DI, SM, TN, HM, and BD revised and edited the manuscript. All authors read and approved the final manuscript.

Competing interests
Dr. Mayberg reports grants from NIMH, grants, and other from GSK, during the conduct of the study; personal fees from Abbott Labs (previously St Jude Medical Inc), outside the submitted work.
Dr. Dunlop reports grants from the National Institute of Mental Health during the conduct of the study; grants from Takeda, grants from Janssen, grants from Acadia, grants from Axsome, outside the submitted work.
Dr. Mathew has served as a consultant to Allergan, Alkermes, and Fortress Biotech. He has received research support from NeuroRx.
Dr. Iosifescu reports grants from the National Institute of Mental Health, during the conduct of the study; personal fees from Axsome, personal fees from Alkermes, grants from Brainsway, grants from LiteCure, personal fees from Lundbeck, grants from Neosync, personal fees from Otsuka, personal fees from Sunovion, outside the submitted work.
Dr. Nemeroff has received research support from the National Institutes of Health (NIH) and Stanley Medical Research Institute. For the last 3 years, he was a consultant for Xhale, Takeda, Taisho Pharmaceutical Inc., Prismic Pharmaceuticals, Bracket (Clintara), Fortress Biotech, Sunovion Pharmaceuticals Inc., Sumitomo Dainippon Pharma, Janssen Research & Development LLC, Magstim, Inc., Navitor Pharmaceuticals, Inc., TC MSO, Inc., Intra-Cellular Therapies, Inc. He is a Stockholder of Xhale, Celgene, Seattle Genetics, Abbvie, OPKO Health, Antares, BI Gen Holdings, Inc., and Corcept Therapeutics Pharmaceuticals Company. Dr. Nemeroff is a member of the scientific advisory boards of the American Foundation for Suicide Prevention

(AFSP), Brain and Behavior Research Foundation (BBRF) (formerly named National Alliance for Research on Schizophrenia and Depression [NARSAD]), Xhale, Anxiety Disorder Association of America (ADAA), Skyland Trail, Bracket (Clintara), Laureate Institute for Brain Research, Inc. and on the board of directors of AFSP, Gratitude America and ADAA. Dr. Nemeroff has income sources or equity of 10.000 USD or more from American Psychiatric Publishing, Xhale, Bracket (Clintara), CME Outfitters, Takeda, Intra-Cellular Therapies, Inc., and Magstim. Dr. Nemeroff holds the following patents: Method and devices for transdermal delivery of lithium (US 6,375,990B1) and Method of assessing antidepressant drug therapy via transport inhibition of monoamine neurotransmitter by ex vivo assay (US 7,148,027B2).
Dr. Binder, Dr. Rothbaum, Dr. Neylan, Dr. Pape, Dr. Carrillo-Roa, Dr. Czamara and Dr. Zannas have nothing to disclose.

Author details

[1]Department of Translational Research in Psychiatry, Max Planck Institute of Psychiatry, Munich, Germany. [2]Department of Psychiatry and Behavioral Sciences, Emory University School of Medicine, Atlanta, GA, USA. [3]Department of Psychiatry and Behavioral Sciences, University of Miami Miller School of Medicine, Miami, FL, USA. [4]Department of Psychiatry, Icahn School of Medicine at Mount Sinai, New York, NY, USA. [5]Menninger Department of Psychiatry & Behavioral Sciences, Baylor College of Medicine & Michael E. Debakey VA Medical Center, Houston, TX, USA. [6]Department of Psychiatry, University of California, San Francisco, San Francisco, CA, USA. [7]The San Francisco Veterans Affairs Medical Center, San Francisco, CA, USA. [8]Department of Psychiatry and Behavioral Sciences, Duke University Medical Center, Durham, NC, USA. [9]New York University School of Medicine, New York, NY, USA. [10]Nathan Kline Institute for Psychiatric Research, Orangeburg, NY, USA.

References

1. Krystal JH, Davis LL, Neylan TC, Raskind MA, Schnurr PP, Stein MB, et al. It is time to address the crisis in the pharmacotherapy of posttraumatic stress disorder: a consensus statement of the PTSD psychopharmacology working group. Biol Psychiatry. 2017. https://doi.org/10.1016/j.biopsych.2017.03.007.
2. Mehta D, Binder EB. Gene × environment vulnerability factors for PTSD: the HPA-axis. Neuropharmacology. 2012;62:654–62.
3. Binder EB, Nemeroff CB. The CRF system, stress, depression and anxiety-insights from human genetic studies. Mol Psychiatry. 2010;15:574–88.
4. Laryea G, Arnett MG, Muglia LJ. Behavioral studies and genetic alterations in corticotropin-releasing hormone (CRH) neurocircuitry: insights into human psychiatric disorders. Behav Sci (Basel, Switzerland). 2012;2:135–71.
5. Baker DG, West SA, Nicholson WE, Ekhator NN, Kasckow JW, Hill KK, et al. Serial CSF corticotropin-releasing hormone levels and adrenocortical activity in combat veterans with posttraumatic stress disorder. Am J Psychiatry. 1999;156:585–8.
6. Bremner JD, Licinio J, Darnell A, Krystal JH, Owens MJ, Southwick SM, et al. Elevated CSF corticotropin-releasing factor concentrations in posttraumatic stress disorder. Am J Psychiatry. 1997;154:624–9.
7. Sautter FJ, Bissette G, Wiley J, Manguno-Mire G, Schoenbachler B, Myers L, et al. Corticotropin-releasing factor in posttraumatic stress disorder (PTSD) with secondary psychotic symptoms, nonpsychotic PTSD, and healthy control subjects. Biol Psychiatry. 2003;54:1382–8.
8. Halldorsdottir T, Binder EB. Gene × environment interactions: from molecular mechanisms to behavior. Annu Rev Psychol. 2017;68:215–41.
9. Glaser YG, Zubieta J-K, Hsu DT, Villafuerte S, Mickey BJ, Trucco EM, et al. Indirect effect of corticotropin-releasing hormone receptor 1 gene variation on negative emotionality and alcohol use via right ventrolateral prefrontal cortex. J Neurosci. 2014;34:4099–107.
10. Cicchetti D, Rogosch FA, Oshri A. Interactive effects of corticotropin releasing hormone receptor 1, serotonin transporter linked polymorphic region, and child maltreatment on diurnal cortisol regulation and internalizing symptomatology. Dev Psychopathol. 2011;23:1125–38.
11. Heim C, Bradley B, Mletzko TC, Deveau TC, Musselman DL, Nemeroff CB, et al. Effect of childhood trauma on adult depression and neuroendocrine function: sex-specific moderation by CRH receptor 1 gene. Front Behav Neurosci. 2009;3:41.
12. Mahon PB, Zandi PP, Potash JB, Nestadt G, Wand GS. Genetic association of FKBP5 and CRHR1 with cortisol response to acute psychosocial stress in healthy adults. Psychopharmacology. 2013;227:231–41.
13. Sumner JA, McLaughlin KA, Walsh K, Sheridan MA, Koenen KC. CRHR1 genotype and history of maltreatment predict cortisol reactivity to stress in adolescents. Psychoneuroendocrinology. 2014;43:71–80.
14. Tyrka AR, Price LH, Gelernter J, Schepker C, Anderson GM, Carpenter LL. Interaction of childhood maltreatment with the corticotropin-releasing hormone receptor gene: effects on hypothalamic-pituitary-adrenal axis reactivity. Biol Psychiatry. 2009;66:681–5.
15. Dunlop BW, Rothbaum BO, Binder EB, Duncan E, Harvey PD, Jovanovic T, et al. Evaluation of a corticotropin releasing hormone type 1 receptor antagonist in women with posttraumatic stress disorder: study protocol for a randomized controlled trial. Trials. 2014;15:240.
16. Dunlop BW, Binder EB, Iosifescu D, Mathew SJ, Neylan TC, Pape JC, et al. Corticotropin-releasing factor type 1 receptor antagonism is ineffective for women with posttraumatic stress disorder. Biol Psychiatry. 2017;23:5295–301.
17. Colvonen PJ, Glassman LH, Crocker LD, Buttner MM, Orff H, Schiehser DM, et al. Pretreatment biomarkers predicting PTSD psychotherapy outcomes: a systematic review. Neurosci Biobehav Rev. 2017;75:140–56.
18. Lehrner A, Yehuda R. Biomarkers of PTSD: military applications and considerations. Eur J Psychotraumatol. 2014;5. https://doi.org/10.3402/ejpt.v5.23797.
19. Klengel T, Pape J, Binder EB, Mehta D. The role of DNA methylation in stress-related psychiatric disorders. Neuropharmacology. 2014;80:115–32.
20. Labonté B, Azoulay N, Yerko V, Turecki G, Brunet A. Epigenetic modulation of glucocorticoid receptors in posttraumatic stress disorder. Transl Psychiatry. 2014;4:e368.
21. Yehuda R, Flory JD, Bierer LM, Henn-Haase C, Lehrner A, Desarnaud F, et al. Lower methylation of glucocorticoid receptor gene promoter 1F in peripheral blood of veterans with posttraumatic stress disorder. Biol Psychiatry. 2015;77:356–64.
22. Yehuda R, Daskalakis NP, Desarnaud F, Makotkine I, Lehrner AL, Koch E, et al. Epigenetic biomarkers as predictors and correlates of symptom improvement following psychotherapy in combat veterans with PTSD. Front Psychiatry. 2013;4:118.
23. Wang Z, Young MRI. PTSD, a disorder with an immunological component. Front Immunol. 2016;7:219.
24. Gill JM, Saligan L, Lee H, Rotolo S, Szanton S. Women in recovery from PTSD have similar inflammation and quality of life as non-traumatized controls. J Psychosom Res. 2013;74:301–6.
25. Gocan AG, Bachg D, Schindler AE, Rohr UD. Balancing steroidal hormone cascade in treatment-resistant veteran soldiers with PTSD using a fermented soy product (FSWW08): a pilot study. Horm Mol Biol Clin Invest. 2012;10:301–14.
26. Morath J, Gola H, Sommershof A, Hamuni G, Kolassa S, Catani C, et al. The effect of trauma-focused therapy on the altered T cell distribution in individuals with PTSD: evidence from a randomized controlled trial. J Psychiatr Res. 2014;54:1–10.
27. Houseman E, Accomando WP, Koestler DC, Christensen BC, Marsit CJ, Nelson HH, et al. DNA methylation arrays as surrogate measures of cell mixture distribution. BMC Bioinformatics. 2012;13:86.
28. Kirillov A, Kistler B, Mostoslavsky R, Cedar H, Wirth T, Bergman Y. A role for nuclear NF-kappaB in B-cell-specific demethylation of the Igkappa locus. Nat Genet. 1996;13:435–41.
29. Feldmann A, Ivanek R, Murr R, Gaidatzis D, Burger L, Schübeler D. Transcription factor occupancy can mediate active turnover of DNA methylation at regulatory regions. PLoS Genet. 2013;9:e1003994.
30. Weaver ICG, D'Alessio AC, Brown SE, Hellstrom IC, Dymov S, Sharma S, et al. The transcription factor nerve growth factor-inducible protein a mediates epigenetic programming: altering epigenetic marks by immediate-early genes. J Neurosci. 2007;27:1756–68.
31. Thomassin H, Flavin M, Espinás ML, Grange T. Glucocorticoid-induced DNA demethylation and gene memory during development. EMBO J. 2001;20:1974–83.
32. Wiench M, John S, Baek S, Johnson TA, Sung M-H, Escobar T, et al. DNA methylation status predicts cell type-specific enhancer activity. EMBO J. 2011;30:3028–39.
33. de Kloet CS, Vermetten E, Heijnen CJ, Geuze E, Lentjes EGWM, Westenberg HGM. Enhanced cortisol suppression in response to dexamethasone administration in traumatized veterans with and without posttraumatic stress disorder. Psychoneuroendocrinology. 2007;32:215–26.
34. Yehuda R. Status of glucocorticoid alterations in post-traumatic stress

disorder. Ann N Y Acad Sci. 2009;1179:56–69.

35. Olff M, de Vries G-J, Güzelcan Y, Assies J, Gersons BPR. Changes in cortisol and DHEA plasma levels after psychotherapy for PTSD. Psychoneuroendocrinology. 2007;32:619–26.

36. Yehuda R, Pratchett LC, Elmes MW, Lehrner A, Daskalakis NP, Koch E, et al. Glucocorticoid-related predictors and correlates of post-traumatic stress disorder treatment response in combat veterans. Interface Focus. 2014;4: 20140048.

37. ENCODE Project Consortium TEP. An integrated encyclopedia of DNA elements in the human genome. Nature. 2012;489:57–74.

38. Blake DD, Weathers FW, Nagy LM, Kaloupek DG, Gusman FD, Charney DS, et al. The development of a clinician-administered PTSD scale. J Trauma Stress. 1995;8:75–90.

39. Foa EB, Riggs DS, Dancu CV, Rothbaum BO. Reliability and validity of a brief instrument for assessing post-traumatic stress disorder. J Trauma Stress. 1993;6:459–73.

40. Purcell S, Neale B, Todd-Brown K, Thomas L, Ferreira MAR, Bender D, et al. PLINK: a tool set for whole-genome association and population-based linkage analyses. Am J Hum Genet. 2007;81:559–75.

41. Yang J, Lee SH, Goddard ME, Visscher PM. GCTA: a tool for genome-wide complex trait analysis. Am J Hum Genet. 2011;88:76–82.

42. Mehta D, Klengel T, Conneely KN, Smith AK, Altmann A, Pace TW, et al. Childhood maltreatment is associated with distinct genomic and epigenetic profiles in posttraumatic stress disorder. Proc Natl Acad Sci. 2013;110:8302–7.

43. Griebel G, Holsboer F. Neuropeptide receptor ligands as drugs for psychiatric diseases: the end of the beginning? Nat Rev Drug Discov. 2012; 11:462–78.

44. Hsu DT, Mickey BJ, Langenecker SA, Heitzeg MM, Love TM, Wang H, et al. Variation in the corticotropin-releasing hormone receptor 1 (CRHR1) gene influences fMRI signal responses during emotional stimulus processing. J Neurosci. 2012;32:3253–60.

45. Bradley RG, Binder EB, Epstein MP, Tang Y, Nair HP, Liu W, et al. Influence of child abuse on adult depression. Arch Gen Psychiatry. 2008;65:190.

46. Faul F, Erdfelder E, Lang A-G, Buchner A. G*power 3: a flexible statistical power analysis program for the social, behavioral, and biomedical sciences. Behav Res Methods. 2007;39:175–91.

47. Cohen J. Statistical power analysis for the behavioral sciences. 2nd ed. Hillsdale: Lawrence Erlbaum Associates, Publishers; 1988.

8

The signature of liver cancer in immune cells DNA methylation

Yonghong Zhang[1†], Sophie Petropoulos[2,3†], Jinhua Liu[1], David Cheishvili[2,4], Rudy Zhou[2], Sergiy Dymov[2], Kang Li[1], Ning Li[1] and Moshe Szyf[2*] (iD)

Abstract

Background: The idea that changes to the host immune system are critical for cancer progression was proposed a century ago and recently regained experimental support.

Results: Herein, the hypothesis that hepatocellular carcinoma (HCC) leaves a molecular signature in the host peripheral immune system was tested by profiling DNA methylation in peripheral blood mononuclear cells (PBMC) and T cells from a discovery cohort (n = 69) of healthy controls, chronic hepatitis, and HCC using Illumina 450K platform and was validated in two validation sets (n = 80 and n = 48) using pyrosequencing.

Conclusions: The study reveals a broad signature of hepatocellular carcinoma in PBMC and T cells DNA methylation which discriminates early HCC stage from chronic hepatitis B and C and healthy controls, intensifies with progression of HCC, and is highly enriched in immune function-related genes such as *PD-1*, a current cancer immunotherapy target. These data also support the feasibility of using these profiles for early detection of HCC.

Keywords: DNA methylation, Hepatocellular carcinoma, Peripheral white blood cells, Immune functions

Backgrounds

The idea that host immuno-surveillance plays an important role in tumorigenesis by eliminating tumor cells and suppressing tumor growth has been proposed by Paul Ehrlich [1, 2] more than a century ago and has fallen out of favor. However, accumulating data from both animal and human clinical studies suggest that the host immune system plays an important role in tumorigenesis through "immuno-editing" which involves three stages: elimination, equilibrium, and escape [3–5]. Presence of tumor infiltrating cytotoxic CD8+ T cells was associated with better prognosis in several clinical studies of human regressive melanoma [6–11], esophageal [12], ovarian [13, 14], and colorectal cancer [15–17]. The immune system is believed to be responsible for the phenomenon of cancer dormancy when circulating cancer cells are detectable in the absence of clinical symptoms [18, 19].

DNA methylation, a covalent modification of DNA, which is a primary mechanism of epigenetic regulation of genome function, is ubiquitously altered in tumors [18, 20–22] including hepatocellular carcinoma (HCC) [23]. Molecular analysis of cancer including DNA methylation is mainly focused on tumors and biomaterial originating in tumor including tumor DNA in plasma [24, 25], circulating tumor cells [26], and the tumor-host microenvironment [27, 28]. The prevailing and widely accepted hypothesis is that molecular changes that drive cancer initiation and progression originate primarily in the tumor itself and that relevant changes in the host occur primarily in the tumor microenvironment [27, 29]. The identity of immune cells in the tumor microenvironment has attracted, therefore, significant attention [30, 31]. Interestingly, recent DNA methylation and transcriptome analysis of tumors revealed tumor stage-specific immune signatures of infiltrating lymphocytes [29, 32]. However, these signatures represent targeted immune cells in the tumor microenvironment, and utilization of such signatures for early diagnosis requires invasive procedures. The tumor-infiltrating immune cells represent only a minor fraction of peripheral blood cells [33–36].

* Correspondence: moshe.szyf@mcgill.ca
†Equal contributors
²Department of Pharmacology and Therapeutics, McGill University, 3655 Sir William Osler Promenade, Montreal, Quebec H3G 1Y6, Canada
Full list of author information is available at the end of the article

Global DNA methylation changes were previously reported in leukocytes, and EWAS (epigenome-wide association studies) studies revealed differences in DNA methylation in leukocytes from bladder, head and neck, and ovarian cancer, and these differences were independent of differences in white blood cell distribution [37]. Differential methylation of 53 CG sites that did not show evidence of association with blood cell composition was found to associate with ovarian cancer risk in blood DNA [38]. A recent study demonstrated association between the state of methylation of multiple CG site in six genes and colorectal cancer in peripheral blood leukocytes as well as an interaction with diet [39]. An EWAS on 48 matched case-controlled pairs in a nested case-control study within a 22-year follow-up cohort of hepatitis B (HepB) carriers revealed methylation variable positions that were associated with progression to HCC and were predictive of the risk of early-onset HCC well before appearance of clinical symptoms [40]. These data provide support for the hypothesis that cancer-specific DNA methylation differences exist in white blood cells; however, it is possible that these just reflect stochastic alterations in DNA methylation in somatic tissues that are associated with cancer.

The question of whether the peripheral host immune system exhibits a distinct DNA methylation response to the cancer state that intensifies in advanced stages of cancer has not been addressed. Addressing this question is essential for understanding the potential importance of epigenetic reprogramming of the immune system in cancer. Does epigenetic reprogramming occur in the immune system during cancer progression and does it play a causal role in HCC progression? If indeed progression of cancer involves epigenetic reprogramming of the host immune system, this has important implications for both therapeutics and diagnostics.

HCC is the fifth most common cancer worldwide [41]. It is particularly prevalent in Asia, and its occurrence is highest in areas where hepatitis B is prevalent, indicating a possible causal relationship [42]. Here, we hypothesize that HCC progression is associated with distinct DNA methylation profiles in the host peripheral immune cells. Since HCC originates in patients with an ongoing inflammatory chronic viral infection, the critical challenge is to delineate DNA methylation markers that differentiate between cancer and the underlying chronic inflammatory liver disease. We hypothesize here that HCC exhibits a DNA methylation profile that is distinct from chronic hepatitis and that there is a clear boundary in the evolution of the DNA methylation profile of white blood cells between the earliest stages of HCC and chronic hepatitis

Our study demonstrates for the first time broad DNA methylation profiles for HCC in peripheral blood mononuclear cells (PBMC) and T cells that are different from controls as well as hepatitis B and C; the differences are intensified during cancer progression. There is a significant overlap between DNA methylation profiles delineated in white blood cells and T cells. Four genes that were differentially methylated in T cells from HCC patients in the discovery cohort were validated by pyrosequencing of T cells DNA in a separate cohort of patients ($n = 79$) and one gene *STAP1* was validated in a third cohort ($n = 48$). HCC DNA methylation profiles are highly enriched in immune functions including genes such as *Programmed cell Death 1 (PD-1)*, a negative regulator of T cell immune response that is an important target in current cancer immunotherapy [43] and show no significant overlap with the DNA methylation profiles of previously described HCC tumors [23]. These data provide proof of principle that there are molecular changes in the host immune cells DNA in HCC. This has important implications for our understanding of the mechanisms of the disease and its treatment as well as for noninvasive diagnostics of cancer in white blood cells DNA.

Results

Correlation between quantitative distribution of site-specific DNA methylation levels and progression of HCC

Sixty-nine people from the Beijing area of China were included in a discovery set (10 controls and 10 patients for each of the following groups: hepatitis B and C, stages 1–3, and nine patients for stage 4) of HCC staged using the EASL–EORTC Clinical Practice Guidelines for HCC (Table 1 and "Methods" section). To address the question of whether quantitative differences in DNA methylation states in PBMCs correlate with progression of HCC (see "Methods" section for staging criteria) and whether this DNA methylation signature differentiates between chronic hepatitis B and C and HCC, we performed a genome-wide measurement of DNA methylation states in ~ 480,000 CpGs using the Illumina Infinium Human Methylation 450K BeadChip Array platform as described in the "Methods" section. Following normalization and batch correction, we performed a Pearson correlation analysis with Bonferroni correction for multiple testing ($< 1 \times 10^{-7}$) between the quantitative distribution of DNA methylation in the batch-normalized CGs across the array and progression of HCC. The analysis revealed a broad signature of DNA methylation that correlates with progression of HCC. A genome-wide view of the intensifying change in DNA methylation of 3924 robust differentially methylated sites ($r > 0.8$; $r < -0.8$; delta beta > 0.2, > -0.2, $p < 10^{-7}$; Additional file 1: Table S1) during HCC progression is shown in Fig. 1a, b; notably, hypomethylation increases

Table 1 Clinical characteristics of the training set

Variable	Control ($n = 10$)	HepB ($n = 10$)	HepC ($n = 10$)	HCC1 ($n = 10$)	HCC2 ($n = 10$)	HCC3 ($n = 10$)	HCC4 ($n = 9$)	p value
Age (mean ± SD)	34.4 ± 8.5	36.9 ± 9.04	37.7 ± 12.08	52.9 ± 5.97	52.2 ± 12.2	56.1 ± 5.5	48 ± 15.06	5×10^{-6}
Sex								0.034
Male	3 (30%)	8 (80%)	7 (70%)	9 (90%)	8 (80%)	9 (90%)	7 (70%)	
Female	7	2	3	1	2	1	2	
Alcohol								N.S.
No	8 (80%)	6 (60%)	8 (80%)	4 (40%)	5 (50%)	4 (40%)	5 (50%)	
Infrequent	2	1	2	2	1	0	2	
Heavy	0	3	0	4	4	6	2	
Smoking								N.S.
No	8 (80%)	5 (50%)	7 (70%)	6 (60%)	5 (50%)	4 (40%)	6 (60%)	
Quit	0	0	0	0	3	2	1	
Low	0	0	0	0	1	0	0	
Heavy	2	5	3	4	1	4	2	
Cirrhosis	0	0	0	0	0	0	0	
Hepatitis B	0	10	0	10	9	10	8	2×10^{-16}
Hepatitis C	0	0	10	2	1	0		3×10^{-16}
AFP (> 500u/ng)	0	1 (10%)	0	0	1 (10%)	3 (30%)	5 (50%)	0.043

AFP alpha feto protein, *HBV* Hepatitis B virus, HCV Hepatitis C virus

with progression of HCC (Fig 1b). The heat map in Fig. 1c shows the increasing differences in DNA methylation as HCC progresses and the clustering of all individual HCC patients away from healthy controls and hepatitis B and C individuals (except CAN1-5 who is clustered at the boundary between HepC and HCC (Fig. 1c). Importantly, PBMC DNA methylation profiles differentiate individual HCC patients at the earliest stage from hepatitis B and C, a critical challenge in early diagnosis of HCC.

Unique and overlapping differentially methylated sites associate with different HCC stages and differentiate HCC from hepatitis B and C; cross validation across HCC samples

We delineated differentially methylated CGs between healthy controls and each of the HCC stages independently using the Bioconductor package Limma [44], [45] as implemented in ChAMP [46]. The number of differentially methylated CG sites ($p < 1 \times 10^{-7}$) between each stage of HCC and healthy controls increases with advance in stage; 14,375 for stage 1; 22,018 for stage 2; 30,709 for stage 3; and 54,580 for stage 4 (Additional file 2: Table S2, Additional file 3: Table S3, Additional file 4: Table S4 and Additional file 5: Table S5) with a notable increase in the fraction of hypomethylated sites (26% in stage 1 to 57% in stage 4) (Fig. 2a) as observed in the Pearson correlation analysis (Fig. 1b). We derived for each HCC stage a set of highly robust CG methylation markers ($p < 1 \times 10^{-7}$, delta beta $> \pm 0.3$; < -0.3 for HCC stage 1 and $p < 10^{-10}$ delta beta of ± 0.3; for stage 2–4, we used a

more stringent threshold for later stages to reduce the number of sites) (74 for stage 1, 14 for stage 2, 58 for stage 3, and 298 for stage 4). By combining the lists of markers derived independently for each stage and removing redundant CG sites between stages, we derived a combined non-redundant list of 350 CGs (Additional file 6: Table S6).

We used two methods of "cross-validation." In the first method, all samples from one stage were used for "training" comparing the HCC samples to healthy controls. Subsequently, we tested the ability of the obtained signature to classify other stages and differentiate them from chronic hepatitis B and C, which were not "trained" for the said "CGs." In the second method, samples of each group were randomized to two subsets, a "training set" and a "validation set," and the signature of 369 significant CG sites obtained for the training set was tested on the "validation" set (Additional file 7: Figure S2).

Using the first method, the differentially methylated sites for each of the stages were derived by comparing ten healthy control and ten stage-specific HCCs. HCC of other stages and hepatitis B and C samples were not "trained" for these differentially methylated CGs and could serve as "cross-validation sets" to determine whether markers "trained" on one stage of HCC cluster correctly other HCC samples and whether they also differentiate HCC from other hepatitis B and hepatitis C.

As seen in Fig. 2b, c (stage 1 and 4) and Additional file 7: Figure S1 (stage 2 and 3), each of the independently derived set of markers for specific stages of HCC was

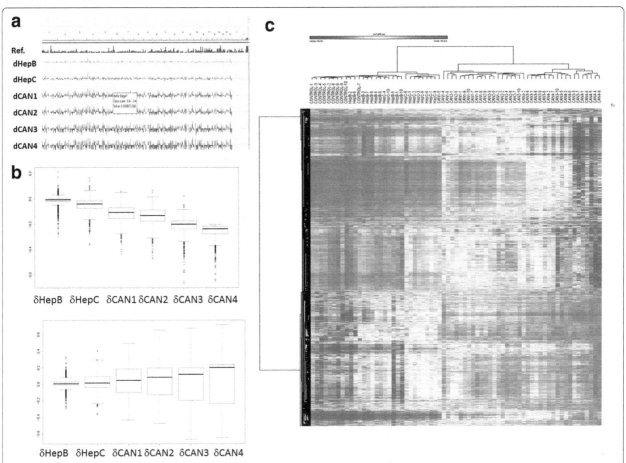

Fig. 1 Correlation between quantitative distribution of site-specific DNA methylation levels and progression of HCC. **a** A genome wide view (IGV genome browser) of the escalating differences in DNA methylation from healthy controls (delta beta) in 3924 CG sites whose quantitative levels of methylation correlate with HCC progression ($r > 0.8$, $r < -0.8$; delta beta > 0.2, < −0.2; $p < 10^{-7}$) in PBMC from HCC and hepatitis B and C patients. HepB-Hepatitis B; HepC-Hepatitis C; CAN1-stage 1 HCC; CAN2- Stage 2 HCC: CAN3- Stage 3 HCC; CAN4-Stage 4 HCC. **b** Box plot of DNA methylation delta beta values of the 3924 CG sites whose levels of methylation correlate significantly ($p < 10^{-7}$) with HCC progression. Sites that are hypomethylated relative to healthy control during progression of HCC (upper panel) and sites that are hypermethylated relative to healthy controls (bottom panel) are shown separately. **c** Heat map of hierarchical clustering using one minus Pearson correlation of 69 people by DNA methylation beta values of the 3924 CG sites

"cross-validated" by its ability to cluster with a sharp boundary, all other HCC stages separately from controls, and hepatitis B and C samples that were not "trained" for these CGs. Interestingly, these markers also cluster hepatitis C and B samples separately from each other.

The overlap between independently derived CG markers that differentiate each of the HCC stages from healthy controls (Fig. 2d) is highly significant for all possible overlaps between the stages (hypergeometric test, $p < 1.921718e^{-319}$) allowing for using of these differentially methylated CGs as peripheral markers of HCC.

We tested whether we could use the 350 CG list (described above) (Additional file 6: Table S6) to differentiate HCC stages from each other. Hierarchical clustering by one minus Pearson correlation of all samples using these 350 CGs correctly clustered the HCC samples by stage, and hepatitis B and C were clustered with healthy

controls even though they were not "trained" by these CGs (Fig. 3a).

Since the 350 CG signature that was used to classify HCC stages was obtained by combining the signatures obtained for each stage, the signature has already been "trained" with the data used for testing. We therefore used a second method to "train" and "validate" a DNA methylation profile that classifies HCC stages. First, we randomly split each group (CTRL, HepB and C, and the different HCC stages) to two sets, a "training set" and a "validation set." We then performed a correlation analysis between progression of HCC and levels of CG methylation. We selected the top 369 CGs (delta beta Can4-Can1 > 0.4, > −0.4, adjusted p value < 0.05) (Additional file 7: Figure S2a left panel; Additional file 6: Table S6). Hierarchical clustering by one minus Pearson correlation of the "validation set" using these 369 CGs

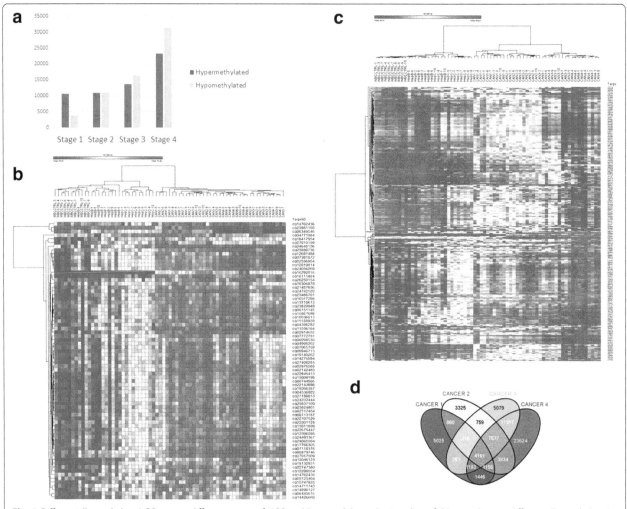

Fig. 2 Differentially methylated CG sites at different stages of HCC and "cross-validation." **a** Number of CG sites that are differentially methylated between different stages of HCC and healthy controls ($p < 10^{-7}$) green: hypomethylated, red: hypermethylated. **b** Heat map presentation of hierarchical clustering of 69 people by 74 differentially methylated CGs between HCC stage 1 and control. **c** Heat map of hierarchical clustering of 69 people by 298 differentially methylated CGs between HCC stage 4 and control. **d** Ven diagram of the overlap between differentially methylated CG sites at different HCC stages (1–4). Significance was determined using Fisher exact test for all overlaps and all overlaps were highly significant ($p < 1.7 \times 10^{-321}$)

(trained in the "training set") correctly clustered these other untrained HCC samples by stage while hepatitis B and C were clustered with healthy controls (Additional file 7: Figure S2a right panel). A randomized set of 369 CGs was unable to reveal the progressive alteration of the DNA methylation profile with advance of HCC stages (Additional file 7: Figure S2b).

To test whether we could delineate within the 350 CGs a shortlist of CG sites that differentiate early (stages 1 and 2) from late stages of HCC (stages 3 and 4), we performed a penalized regression on the "training set" that included randomized samples (five per group) from all HCC stages and all controls on the 350 CG list (Additional file 6: Table S6) using the R package "penalized" [47] which performs likelihood cross-validation and makes predictions on each left-out subject. The fitted

model identified seven CGs (Additional file 8: Table S7) whose combined coefficients predicted with 100% accuracy the likelihood of stage HCC 3 and 4 cases and 100% specificity in calling HCC stage 1 and 2 as well as all controls (healthy and hepatitis B and C) as false. The penalized model was then used on the "validation set" of samples of HCC cases and controls to predict likelihood of each case being late stage HCC (Fig. 3b). We included in the test in addition to the new PBMC samples ten samples of T cells from healthy controls and ten T cell samples from different stages of HCC (Fig. 3c). Importantly, neither the 350 CG sites "classifier" nor the "penalized" model was previously "trained" with the T cell data. The penalized model predicted all the late stage samples including three late-stage HCCs in the T cells samples with 100% sensitivity and 100% specificity.

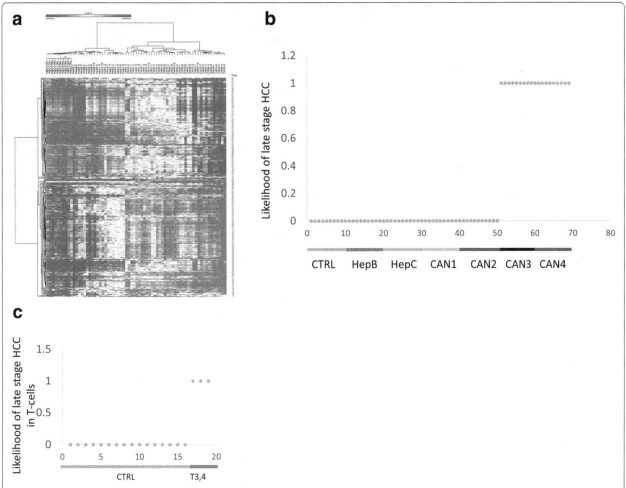

Fig. 3 Staging of HCC using differentially methylated CGs. **a** Heat map presentation of hierarchical clustering of 69 people by 350 non-redundant CGs that are differentially methylated between different HCC stages and healthy controls. **b** Prediction of late stage HCC in 69 patients using a penalized model trained on a randomized half of the HCC patients and controls ("training set") and tested on the other half ("validation set"). The plot shows all samples (the training and validation sets combined). The *y* axis indicates the predicted probability of late stage HCC for each person (from 0 to 1). **c** Prediction of late stages of HCC using the penalized model in T cell samples

However, since the 350 CG signature that was used to classify HCC stages was obtained by combining the signatures obtained for each stage and has already been "trained" with the data used for testing, we also used the list of 369 CGs obtained from a "training set" that included representative samples from all cases and controls. We then performed a penalized regression on this set to identify CG sites that differentiate early (stages 1, 2) from late HCC (stages 3, 4). The fitted model identified a different set of 15 CGs (Additional file 8: Table S7) whose combined coefficients predicted with 100% accuracy the likelihood of stage HCC 3 and 4 cases and 100% specificity in calling HCC stage 1 and 2 as well as all controls (healthy and hepatitis B and C) as false. The penalized model was then used on the "validation set" of other samples of HCC cases and controls that were not used in training of either the selection of the 369 sites or the penalized model, to predict

likelihood of each case being late stage HCC (Additional file 7: Figure S3). The penalized model predicted all the late stage samples with 100% sensitivity and 100% specificity. In summary, these data suggest that DNA methylation measurements could predict and differentiate HCC from controls and chronic hepatitis as well as early stage HCC.

The DNA methylation signature of HCC remains significant after correction for potential confounders: sex, age, alcohol, smoking, and cell count

HCC patients in our study and in clinical setting are a heterogeneous group with respect to alcohol, smoking [48–51], sex [52], and age [53], and each of these factors are known to affect DNA methylation. In addition, white blood cells are a heterogeneous mixture of cells and alterations in white cell distribution between individuals might affect DNA methylation as well. We first determined the cell count distribution for each case using the Houseman algorithm [54]. Two-way ANOVA followed by pairwise comparisons and correction for multiple testing found no significant difference in cell count between the groups. We then performed a multivariate linear regression on the normalized beta values of the 350 CG sites that differentiate HCC from all other groups using group (HCC versus non HCC), sex, alcohol, smoking, age, and cell-count as covariates. All CG sites remained highly significant for group covariate even after including the other covariates in the model. Following Bonferroni corrections for 350 measurements, 342 CG sites remained highly significant for the group (HCC versus non HCC) (Additional file 9: Table S8). We performed a multifactorial ANOVA analysis on the beta values of the 350 sites as dependent variables and group (HCC versus non-HCC), sex, and age as independent variables to determine whether there are possible interactions between either sex and group age and group and between sex + age and group on DNA methylation. While the group remained significant for all 350 CGs, no significant interactions with sex or age were found after Bonferroni corrections (Additional file 10: Table S9).

Differences in DNA methylation between HCC and healthy controls in T cells DNA overlap with differences in methylation in PBMC

Our multivariate analysis suggests that differences in PBMC DNA methylation between HCC and other groups (control and chronic Hepatitis) remain even when differences in cell count are taken into account. Furthermore, to determine whether differences in DNA methylation between cancer and control would disappear once the complexity of cell composition is reduced (although heterogeneity in cell subtypes remains), we analyzed the differences in DNA methylation profiles between T cells isolated from 10 of the 39 HCC patients included in the study (marked in Table 1) and all healthy controls ($n = 10$); the analysis revealed 24,863 differentially methylated sites at a threshold of $p < 1 \times 10^{-7}$ (Additional file 11: Table S10). Three hundred seventy robust sites ($p < 1 \times 10^{-7}$ and delta beta > 0.3, < -0.3) correctly cluster all individual samples into two groups: HCC and controls (Fig. 4a) as well as cluster correctly all PBMC samples ($n = 69$) (Fig. 4b). The clustering analysis presented in Fig. 4b shows that CG sites that are differentially methylated in T cell DNA cluster individual HCC, hepatitis, and healthy control DNA samples from white blood cells with 100% accuracy. Thus, the differentially methylated CGs discovered using T cell DNA were cross-validated on different samples (29 different patients with HCC and 20 with chronic hepatitis) of PBMC DNA that were not used in training these CGs. Conversely, the 350

CGs that were derived by analysis of PBMC DNA from cancer stages and controls clustered the T cell healthy controls and HCC samples correctly (Fig 4c). There is a highly significant (hypergeometric test $p = 0$) overlap between the significant CGs ($p < 1 \times 10^{-7}$) that differentiate healthy controls from HCC using T cell DNA and CGs that differentiate the different HCC stages and controls using PBMC DNA (Fig. 4d). These data support the hypothesis that the differences in DNA methylation between HCC and other samples remain even when the complexity of cell types is reduced by isolation of particular cell types and provides further "cross-validation" for the association of these CGs with HCC.

Differentially methylated genes in PBMC in HCC are enriched in immune-related canonical pathways

Progression of HCC has a broad footprint in the methylome (Fig. 1a). To gain insight into the functional footprint of the differentially methylated genes in PBMC and T cells from HCC patients, the gene lists generated from the differential methylation analyses were subjected to a gene set enrichment analysis using Ingenuity Pathway Analysis (IPA). We first subjected genes associated with CGs that showed linear correlation with stages of HCC in the Pearson correlation analysis (Fig. 1b) ($r > 0.8$; $r < -0.8$; delta beta > 0.2, < -0.2). Notably, the top upstream regulators of genes associated with these CGs are TGFbeta ($p < 1.09 \times 10^{-17}$), TNF ($p < 7.32 \times 10^{-15}$), dexamethasone ($p < 7.74 \times 10^{-12}$), and estradiol ($p < 4 \times 10^{-12}$) which are major immune, inflammation, and stress regulators of the immune system. Top diseases identified were cancer (p value 1×10^{-5} to 2×10^{-51}) and hepatic disease ($p < 1.24 \times 10^{-5}$ to 1.11×10^{-25}). A strong signal was noted for liver hyperplasia ($p < 6.19 \times 10^{-1}$ to 1.11×10^{-25}) and hepatocellular carcinoma ($p < 5.2 \times 10^{-1}$ to 3.76×10^{-25}). An inspection of the genes that are differentially methylated reveals a large representation of immune regulatory molecules such as IL2, IL4, IL5, IL16, IL7, Il10, IL18, Il24, Il1B and interleukin receptors such as IL12RB2, IL1B, IL1R1, IL1R2, IL2RA, IL4R, IL5RA; chemokines such as CCL1, CCL7, CCL18, CCL24, as well as chemokine receptors such CCR6, CCR7 and CCR9; cellular receptors such as CD2, CD6, CD14, CD38, CD44, CD80 and CD83; TGFbeta3 and TGFbetaI, NFKB, STAT1, STAT3 and TNFa. Notably, a CG site in the promoter of PD-1, a protein that triggers an immune checkpoint and is now recognized as promising clinical target for anti-immune-blockade cancer treatment [43], is gradually demethylated in PBMC as HCC progresses (Fig. 5a). Differential methylation of 29 out of 78 CG probes associated with PD1 in the 450K array strongly correlated with HCC progression ($R < -0.7$, $Q < 1 \times 10^{-8}$), a highly significant enrichment (hypergeometric test, $p = 4.3 \times 10^{-238}$) (Additional file 12: Table S11).

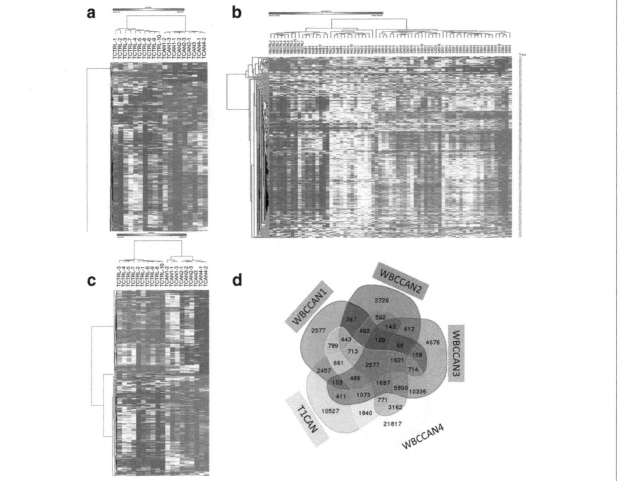

Fig. 4 Differences in DNA methylation between HCC and healthy controls in T cells DNA overlap with differences in methylation in PBMC. **a** Heat map presentation of hierarchical clustering of ten healthy controls and HCC samples from T cells by 370 significantly differentially methylated CGs. **b** Heat map presentation of hierarchical clustering of PBMC DNA methylation samples from 69 people by 370 CGs "trained" in T cells. **c** Heat map presentation of hierarchical clustering of T cell samples from ten healthy controls and ten HCC by 350 CGs "trained" on T cell DNA. **d** Overlap of differentially methylated CGs in HCC in T cells and differentially methylated CGs in PBMC at different stages of HCC

The average methylation of all significantly hypomethylated *PD-1* CGs was significantly correlated with HCC progression (Pearson cor $R = -0.9$, $p < 1.7 \times 10^{-321}$) (Fig. 5b).

A comparative IPA analysis between differentially methylated genes in PBMC and T cells revealed *NFKB*, *TNF*, *VEGF* and *IL4* and *NFAT* as common upstream regulators. Overall, the DNA methylation alterations in HCC PBMC and T cell show a strong signature in immune modulation functions and are consistent with the emerging role of the immune system in cancer. We have previously delineated differentially methylated promoters between HCC biopsies and noncancerous liver tissue [23]. We found a nonsignificant overlap ($n = 44$) (hypergeometric test; $p = 0.76$) between promoters that are differentially methylated in HCC in the cancer biopsies (1983) and PBMC (545) ($p < 1 \times 10-7$; delta beta < -0.2; > 0.2). These data support the hypothesis that changes in DNA methylation seen in PBMC reflect changes in the

immune system in HCC and are not a footprint of circulating DNA from tumors or tumor surrogates.

Since methylation of individual CGs across regions are never homogenous, we analyzed extended differentially methylated regions (DMRs) between all cancers and all controls including chronic hepatitis B and C using the champ.lasso function in Champ [55], shortlisted DMRs in 5′ regions of genes, and calculated their average methylation levels. We identified 4261 significant DMR (FDR adjusted $p < 0.05$) in promoters and 5′ upstream regions. We then tested whether the average methylation levels of these promoter DMRs correlate with cancer progression using Pearson correlation analysis. Five hundred thirty DMRs ranging from 19 to 8951 bases showed highly significant correlation with HCC progression ($r > 0.8$; $r < -0.8$, $p = 0$) (Additional file 13: Table S12) suggesting that differential methylation during HCC progression involves broad regulatory regions and is not

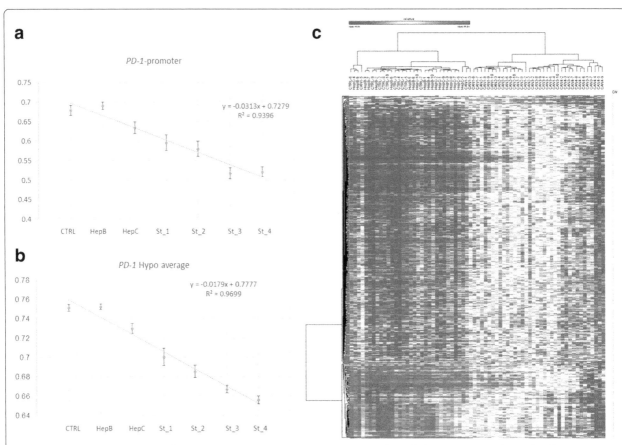

Fig. 5 Progressive hypomethylation of PD-1 gene during HCC progression. **a** Correlation (linear fit) between average beta values for cg14453145 positioned at the TSS region of *PD-1* and control (CTRL) (stage code 0), hepatitis B (HepB) (1), hepatitis C (HepC) (2), and the four stages of HCC (St_1 to St_4) (3–6) diagnoses (equation and R values are indicated). One way ANOVA showed a highly significant effect of diagnosis on DNA methylation ($p = 1 \times 10^{-13}$; $F = 20.77$). Bonferroni adjusted pairwise comparison revealed significant differences between HCC stage 1 and control ($p = 0.0058$) and hepatitis B ($p = 0.00079$); between stage 2 and control ($p = 0.0004$) and hepatitis B ($p = 4.9 \times 10^{-5}$); between stage 3 and control ($p = 4.8 \times 10^{-9}$), hepatitis B ($p = 4.9 \times 10^{-10}$), hepatitis C ($p = 1.8 \times 10^{-5}$) and stage 1($p = 0.00993$); between stage 4 and control ($p = 2.1 \times 10^{-8}$), hepatitis B ($p = 2.3 \times 10^{-9}$), hepatitis C ($p = 5.9 \times 10^{-5}$) and stage1 ($p = 0.00558$). **b** Correlation (linear fit) between average beta values of 24 CGs associated to the *PD-1* gene on the Illumina 450k arrays that were hypomethylated in HCC (average methylation score was calculated per person, the average of these scores were then calculated per group). There was a highly significant effect of diagnosis on DNA methylation as determined by one way ANOVA ($p = 2.2 \times 10^{-16}$, $F = 52.74$). Bonferroni adjusted pairwise comparison revealed significant differences between stage 1 HCC and control ($p = 1.2 \times 10^{-7}$), hepatitis B ($p = 7 \times 10^{-8}$), hepatitis C ($p = 0.00487$), stage 3 ($p = 0.00081$), and Stage 4 ($p = 6.4 \times 10^{-6}$); between stage 2 and control ($p = 4.7 \times 10^{-11}$), hepatitis B ($p = 2.8 \times 10^{-11}$), hepatitis C ($p = 3.8 \times 10^{-6}$), and stage 4 ($p = 0.00645$); between stage3 and control ($p = 3.4 \times 10^{-15}$), hepatitis B (2.1×10^{-15}), hepatitis C (2.1×10^{-10}), and stage 1 ($p = 0.00081$); and between stage 4 and control (2×10^{-16}), hepatitis B (2×10^{-16}), hepatitis C (1.8×10^{-12}), stage 1 (6.4×10^{-6}), and Stage 2 ($p = 0.00645$). **c** Heat map presentation of hierarchical clustering (city block) of 69 people by 5′DMR whose average methylation correlates with HCC progression

limited to scattered individual CG sites. These DMRs clearly cluster all HCC away from all controls including hepatitis B and C and nicely differentiate HCC from hepatitis (Fig. 5c).

Validation of differentially methylated CGs by pyrosequencing

We randomly selected CG sites that were significantly different between HCC and controls in T cells that were either hypermethylated *A Kinase (PRKA) Anchor protein 7 (AKAP7)* gene, the *Signal Transducing Adaptor Family 1 (STAP1)*, or hypomethylated the *Schlafen family member 14(SLFN14)* gene for validation using pyrosequencing (Fig. 6a). The *SLFN14* region that we validated contained three CGs which allowed us to calculate the average methylation of the region which is also significantly different between HCC and controls (Fig. 6a).

For our validation set, we used T cells DNA to reduce cell composition issues. The validation set included 80 people, ten healthy controls and ten individuals from each of the hepatitis B and C (total control $n = 30$) and 50 HCC (HCC stage1 $n = 8$; stage 2 $n = 12$, stage 3 $n = 11$ and stage 4 $n = 19$) (Table 2 and Additional file 14: Table S13) and examined the same genes as well as one

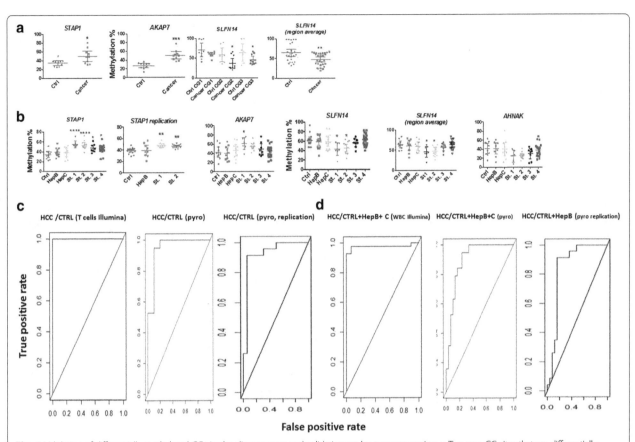

Fig. 6 Validation of differentially methylated CGs in the discovery set and validation set by pyrosequencing. **a** Top row, CG sites that are differentially methylated between HCC (*n* = 10) and healthy controls (*n* = 10) in T cells (significance was measured by student *t* test set at a threshold of < 0.05). The primers for pyrosequencing and conditions are listed in Additional file 18: Table S17. The scattered plot shows the mean and 95% confidence intervals (C.I.). The average methylation for three CG sites in the SLFN14 differentially methylated region is shown in the left panel. Summary of statistics including CI, SD, and SEM values are presented in Additional file 16: Table S15. **b** Validation by pyrosequencing of DNA extracted from T cells in the validation set. ANOVA was used to compare variance between the hepatitis B (HepB) control and other groups healthy (*n* = 10), hepatitis B (*n* = 10), hepatitis C (HepC) (*n* = 10) group and the HCC stages 1 (*n* = 8), 2 (*n* = 12), 3 (*n* = 8), and 4 (*n* = 22). *STAP1* replication presents pyrosequencing data from T cells DNA from the second replication cohort (Additional file 15: Table S14). **c** ROC curve measuring specificity (*Y* axis) and sensitivity (*X* axis) of *STAP1* methylation as a biomarker for discriminating HCC from healthy controls in T cells first cohort (Illumina 450 K data), in first validation set (pyrosequencing) and third validation set (pyrosequencing replication). **d**. ROC curve for *STAP1* methylation as a biomarker for distinguishing HCC from healthy persons and chronic hepatitis in PBMC (Illumina), first validation set (pyrosequencing), and third validation set (pyrosequencing, replication). Statistic code: * 0.05, ** 0.01, *** 0.001

additional hypomethylated gene: *Neuroblast differentiation-associated protein (AHNAK)* (cg14171514) (6b). Linear regression between all controls including chronic hepatitis B and C (healthy and hepatitis B and C) and HCC stages 1 and 2 (0+A) revealed significant association with early HCC stages (1, 2) for all four CGs after correction for multiple testing (*STAP1* $p = 4.04 \times 10^{-7}$; *AKAP7* $p = .0046$; *SLFNL14* $p = 0.012$; *AHNAK* $p = 0.003436$. Linear regression between all controls and all stages of HCC revealed significant association for *STAP1* ($p = 1.6 \times 10^{-6}$) and *AHNAK* ($p = 0.026$) with HCC after correction for multiple testing. One way ANOVA analysis was performed with methylation as the dependent variable, and the seven diagnosis groups (healthy, hepatitis B, hepatitis C, HCC stage 1 to 4) as independent variables revealed significant effect for

diagnosis ($F = 7.263$; $p < 7.49 \times 10^{-6}$) on *STAP1* methylation. Dunnett test, a multiple comparison procedure which compares each of the number of treatments with a single control, revealed significant differences between HCC stage 1 (BCLC 0) and HepB (*p < 0.01*) and stage 2 and HepB (*p < 0.01*) and no significant difference between chronic hepatitis B and healthy controls, hepatitis B and hepatitis C, and hepatitis B and late stages HCC. There was a significant effect of diagnosis on *AKAP 7* methylation ($F = 2.71155$; $p = 0.0198$). A multiple comparison test (Dunnett) between the different diagnosis groups and hepatitis B revealed significant differences between HCC stage 1 (BCLC 0) and hepatitis B (*p* < 0.05), but not between either stage 2 and hepatitis B, hepatitis B and healthy controls, hepatitis B and hepatitis C, and hepatitis B and late stages HCC. There was a

Table 2 Clinical characteristics of the validation set

Variable	Control ($n = 10$)	HepB ($n = 10$)	HepC ($n = 10$)	HCC1 ($n = 8$)	HCC2 ($n = 12$)	HCC3 ($n = 11$)	HCC4 ($n = 19$)	p value
Age (mean ± SD)	35.5 ± 7.5	43.9 ± 15.46	50.1 ± 14.7	61.75 ± 9.39	56 ± 7.45	48.09 ± 7.49	53.63 ± 9.77	5×10^{-5}
Sex								
Male	6 (60%)	7 (70%)	5 (50%)	8 (100%)	10 (83%)	11 (100%)	16 (84%)	
Female	4	3	5	0	2	0	3	
Alcohol								8×10^{-4}
No	NA	10 (100%)	9 (90%)	7 (87.5%)	8 (67%)	3 (27%)	9 (47%)	
Quit	NA	0	1	0	0	0	1	
Infrequent	NA	0	0	0	1	0	0	
Heavy	NA	0	0	1	3	8	9	
Smoking								0.035
No	NA	6 (60%)	8 80%)	4 (50%)	8 (67%)	9 (82%)	9 (47%)	
Quit	NA	1	0	0	1	0	0	
Low	NA	0	0	0	0	0	1	
Heavy	NA	3	2	4	3	2	9	
Cirrhosis	0	0	0	0	0	0	0	
Hepatitis B	0 (0%)	10 (100%)	1 (10%)	8 (100%)	12 (100%)	11 (100%)	19 (100%)	3×10^{-13}
Hepatitis C	0 (0%)	0 (0%)	10 (0%)	0 (0%)	0 (0%)	0 (0%)	0 (0%)	0
AFP	0 (0%)	0 (0%)	0 (0%)	1 (12.5%)	2 (16.6%)	4 (36%)	9 (47.3%)	0.34

significant effect of diagnosis on *SLFN14* methylation (Cg00974761) ($F = 3.877$; $p = 0.0018$). A multiple comparison test (Dunnett) between the different diagnosis groups and hepatitis B revealed significant differences between HCC stage 1 (BCLC 0) and hepatitis B ($p < 0.05$), but not between either stage 2 and hepatitis B, hepatitis B and healthy controls, hepatitis B and hepatitis C, and hepatitis B and late stage HCC. We also measured the average methylation of three CG sites in the *SLFN14* differentially methylated region. There was a significant effect of diagnosis on DNA methylation of this region ($F = 3.727$, $p = 0.0028$). A multiple comparison test (Dunnett) between the different diagnosis groups and hepatitis B revealed significant differences between early stages HCC stage 1+2 and controls when they were combined ($p < 0.05$), but no other significant differences were detected in the other pairwise comparisons. There was a significant effect of diagnosis on *AHNAK* methylation ($F = 2.461$; $p = 0.0323$). A multiple comparison test revealed significant differences between early stages HCC stage 1+2 and controls when they were combined ($p < 0.05$) but no other significant differences were detected in the other pairwise comparisons (Fig. 6b).

Since our validation test indicated that *STAP1* effectively discriminated between either healthy controls or chronic hepatitis and early stages of HCC, we further validated association of *STAP1* methylation with stages 1 and 2 HCC in T cell DNA derived from a third independent cohort ($n = 48$) of healthy controls ($n = 16$) HepB (9) and HCC stages 1 ($n = 11$) and 2 patients ($n = 12$)

(Additional file 15: Table S14). We included HepB samples as a control since all HCC samples were HepB positive. Linear regression between all healthy controls and stages 1 and 2 revealed significant association with early HCC stages (1, 2) ($p = 6.9 \times 10^{-7}$, $F = 35.62$), which remained significant even when sex and age were included in the model as covariates ($p = 1.9 \times 10^{-6}$). Linear regression between all controls including hepatitis B and HCC stages 1 and 2 revealed a significant association ($p = 3.3 \times 10^{-5}$, $F = 21.18$) (Fig. 6b). We performed a multifactorial ANOVA analysis on the methylation values of *STAP1* as a dependent variable and diagnosis (HCC versus non-HCC including hepatitis B), sex, and age as independent variables as well as interactions between sex, age, and diagnosis. Our analysis revealed a significant main effect of diagnosis ($p = 4.88 \times 10^{-5}$, $F = 20.48$) and no significant interaction between sex and diagnosis ($p = 0.96$, $F = 0.0024$) or age and diagnosis ($p = 0.829$, $F = 0.0473$). There was no significant effect of either age ($p = 0.1329$, $F = 2.349$) or sex ($p = 0.7529$, $F = 0.1004$) on DNA methylation. One way ANOVA analysis performed with methylation as the dependent variable and the seven diagnosis groups (healthy, HepB, HCC stage 1 to 2) as independent variables revealed significant effect for diagnosis on *STAP1* methylation ($F = 6.983$, $p = 0.0006$). A multiple comparison test (Dunnett) between the different diagnosis groups and HepB (since all HCC patients were positive for HepB) revealed significant differences between HCC stage 1 and HepB ($p < 0.01$) and stage 2 and HepB ($p < 0.05$), but no difference was detected between HepB and healthy controls (Fig. 6b) (summary of

ANOVA tests and descriptive statistics could be found in Additional file 16: Table S15).

Diagnostic value of differentially methylated CGs in peripheral T cells and PBMC

A measure of the diagnostic value of a biomarker is the receiver operating characteristic (ROC) which measures sensitivity as a function of specificity and determines a threshold value for a predictor which provides the highest accuracy as a biomarker for differentiating case from control [56] (Fig. 6c, d). We first determined ROC characteristics for the normalized Illumina 450K beta values for T cells from healthy controls and HCC (Fig. 6c). STAP1 (cg04398282) behaved as a potential biomarker (accuracy 100%, AUC 1 and both sensitivity and specificity 100%). The STAP1 biomarker was discovered by comparing T cell DNA methylation from HCC and healthy controls (Additional file 11: Table S10). We therefore cross-validated the biomarker properties of STAP1 cg04398282 by examining the ROC characteristics using normalized beta values from the PBMC DNA samples which included hepatitis B and hepatitis C patients as well as 29 additional HCC patients that were not included in the T cell DNA methylation analysis (Fig. 6d). The accuracy is 96% using a threshold beta value of 0.6729; AUC was 0.9741379 (sensitivity 0.975 and specificity 0.973). We then examined the ROC characteristics using pyrosequencing values of STAP1 in the validation set of T cell DNA (Fig. 6c). The accuracy of calling HCC 1,2 from all other controls (healthy and hepatitis B and C) is 85.7% using a threshold beta value of 0.50; AUC is 0.898 (89.5% sensitivity and 83% specificity) (Fig. 6d). In the third cohort of T cell DNA (Additional file 15: Table S14), the accuracy of calling HCC stages 1 and 2 from all other controls (healthy and hepatitis B) using a threshold of 44.5 is 87.5%; the AUC is 84.7% (91.3 sensitivity and 84% specificity). The accuracy of differentiating HCC stages 1 and 2 from healthy controls is 92.3%; the AUC is 0.924 (91.3% sensitivity and 93.75% specificity). We noted however a "batch effect" in pyrosequencing. While we can compare the groups within a study, overall methylation levels vary between experiments done at different times, which will require in the future a normalization procedure that will allow comparisons across different batches.

In summary, STAP1 provides proof of principle for potential DNA methylation biomarkers in HCC peripheral white blood cells and for discriminating Stage 1 from chronic hepatitis and healthy controls which is a critical hurdle in early diagnosis of liver cancer. The other three CGs that we have validated using pyrosequencing to be associated with stage 1 HCC in the validation set do not exhibit biomarker properties

in ROC curves. Further experiments are required to delineate and validate other high quality biomarkers from the list of associated DNA methylation sites that we have delineated in this paper.

Discussion

The focus in DNA methylation studies in cancer to date has been on the tumor, tumor microenvironment [27, 28], and circulating tumor DNA [24, 25], and major advances were made in this respect. In this study, we focused however on the host immune system since the idea that the qualities of the host immune system might define the clinical emergence and trajectory of cancer has been proposed almost a century ago [2] and there is an emerging line of evidence that is consistent with this hypothesis [6–19]. New approaches to cancer therapy are targeted at boosting the host immune system [43]. HCC is a very interesting example since it frequently progresses from preexisting chronic hepatitis and liver cirrhosis [42] and could provide a tractable clinical paradigm for addressing this question.

Our analysis revealed a large number of sites whose quantitative state of methylation strongly correlates ($r = \pm 0.8$–0.9) with progression of HCC which is consistent with the idea that DNA methylation alterations in the immune system are tightly linked with the development of HCC. Interestingly, the overall direction of the differences in DNA methylation changes as HCC advances, from hypermethylation to hypomethylation. Importantly, there is a sharp boundary between stage 1 HCC and chronic hepatitis B and C supporting the hypothesis that changes in DNA methylation are linked with the transition from chronic hepatitis B and C to HCC. It should be noted however that our study is a cross-sectional study and we cannot distinguish at this stage between progressive changes in the same individual from intrinsic differences between people who develop advanced cancer and those who have early stage HCC. Future longitudinal studies in the same patients will be required to address this important question.

Careful inspection of the DNA methylation profiles of chronic hepatitis and HCC in Fig. 1a suggests however that some of the CG sites that undergo large changes in methylation as HCC progresses are already slightly altered in chronic hepatitis. This is consistent with the fact that HCC often progresses from chronic hepatitis and suggests that the changes in DNA methylation in PBMC are seeded by chronic hepatitis and that they might be playing a role in the progression from chronic hepatitis to HCC. Nevertheless, the changes in DNA methylation between chronic hepatitis and early HCC are dramatic, and a clear boundary is seen in our heat maps between chronic hepatitis and HCC that might be utilized to diagnose early transition from chronic hepatitis to HCC.

We also used a "case-control" approach comparing each stage of HCC independently ($n = 10$) with healthy controls ($n = 10$). Cross validation revealed highly significant overlap between differentially methylated CGs in the four stages of HCC. DNA methylation data for chronic hepatitis B and C were not used for deriving the HCC-stage differentially methylated CGs; nevertheless, these CGs accurately clustered the hepatitis samples with the healthy controls separately from other HCC stages. Thus, these markers were "cross-validated" using a separate set of non-HCC samples and other HCC samples, further demonstrating clear boundary in DNA methylation profiles between all HCC samples and chronic hepatitis. Although differentially methylated CGs overlapped between different HCC stages, they nevertheless differentiated stages of cancer from each other because of the intensifying changes in DNA methylation at these sites with progression of HCC. Importantly, the differentially methylated sites remained significant even after taking into account in the regression model differences in sex, age, smoking, and alcohol abuse as well as differences in cell type distribution. Combined, these data support the conclusion that the emergence and progression of HCC is linked with robust and broad changes in DNA methylation in the peripheral immune system.

We further "cross-validated" this conclusion by examining DNA methylation profiles in isolated T cells from a subset of ten HCC patients and the ten healthy controls. If indeed changes in DNA methylation that were detected in PBMC were an artifact of differences in cellular distribution, they should "disappear" using an isolated subset of white blood cells. In addition, by using a different set of DNA samples analyzed independently, we could rule out technical and random effects. We show a robust signature of HCC in peripheral T cell DNA methylation that significantly overlaps with changes in DNA methylation in PBMC and correctly clusters other "untrained" HCC samples separately from chronic hepatitis using DNA methylation data from PBMC. Conversely, differentially methylated CG sites derived from comparing PBMC DNA methylation profiles accurately cluster HCC samples away from controls using T cell DNA methylation data.

Examination of the "gene set" that is differentially methylated in peripheral immune cells in HCC provides some cues as to the potential implications of these changes. It is important to note that in difference from circulating tumor DNA, differentially methylated CGs delineated in our study in PBMC and T cells do not overlap significantly with previously characterized differentially methylated genes in HCC tumors [23].

Analysis of the upstream regulators of the differentially methylated genes provides an overall picture of the functional pathways that are affected. The list includes members of nodal inflammatory and immune regulatory pathways such as TGFbeta, TNFalpha, and the glucocorticoid receptor. Notably, the upstream regulators TGFbeta and TNFalpha are differentially methylated themselves. Interestingly, the enrichment of stress responsive glucocorticoid-regulated pathway might indicate that a fraction of the changes in DNA methylation reflects activation of stress-related processes in the HCC patients. Differentially methylated genes include nodal transcription factors in the immune system and inflammatory response such as NFAT, STAT3, and NFKB; a rich representation of interleukins, chemokines, chemokine receptors and nodal cellular antigens that are involved in cellular fate and differentiation such as CD38 CD44 as well as PD-1 a protein that controls an immune response checkpoint and is now recognized as a promising clinical target for immunity-boosting anticancer treatment [43]. CD38 is an enzyme that synthesizes cyclic ADP-ribose and nicotinate-adenine dinucleotide phosphate, is expressed in leukocytes and functions in cell adhesion and calcium signaling. Interestingly, CD38 is associated with plasmablastic lymphoma and prolymphocytic leukemias. CD44 is a cell surface glycoprotein that is involved in multiple cellular functions relevant to the immune system including lymphocyte activation, recirculation and homing, hematopoiesis, and tumor metastasis. It is unclear how these broad and complex changes in methylation of genes in immune and inflammatory pathways in PBMC affect the overall output and function of the immune system and inflammatory reactions during progression of HCC. Our data provides however compelling data implicating an escalating epigenetic reprogramming of the immune and inflammatory systems during HCC advancement.

The relationship between changes in DNA methylation and steady state transcription is complex. Moreover, DNA methylation alterations might "program" genes in the immune system to respond to transient signals that are time and context dependent and are not captured by examining steady state mRNA levels. Extensive future experiments are required to understand how this "epigenetic" reprogramming of the immune system affects its function in promoting/suppressing HCC. One interesting question that remains to be answered is whether the changes described here for HCC occur in other cancers or whether these changes are unique to HCC, a disease that frequently emerges from chronic inflammation of the liver. However, the changes in DNA methylation in HCC are dramatically enhanced in comparison with the effects of chronic hepatitis on DNA methylation supporting the conclusion that these are cancer-related changes.

A cardinal question that our data is raising is whether these changes in DNA methylation are a cause or consequence of HCC. The fact that the changes in DNA methylation intensify with the advance of HCC is consistent with the idea that they respond to HCC progression. However, they might be still playing an important role in the escalation of the disease. This idea is consistent with recent approaches to cancer therapy that target the immune system to boost a cytotoxic T cell response to tumor cells such as current anti-PD-1 treatments [43]. It is tempting to speculate that targeting the epigenetic changes in the immune system in addition to the changes in the tumor itself might have therapeutic effects. Our data suggests that this possibility should at least be considered.

The observation that HCC has a DNA methylation signature in easily accessible PBMC and T cells points to the opportunity that these could serve as "noninvasive" biomarkers for detection of early transition from chronic hepatitis to HCC as well as HCC progression. We provide here several examples that present a "proof of principle" for using PBMC differentially methylated CGs as "biomarkers" in HCC. However, future extensive analyses of large samples of HCC and controls are required before it is possible to confirm that such "biomarkers" have sufficient accuracy to differentiate early stage HCC from controls. It is also expected that a single CG site will not have sufficient accuracy and that a combination of sites within a region will be required. Nevertheless, this study provides a "proof of principle" for further exploring this opportunity.

One limitation of our data is the relatively small number of samples. However, the effect sizes that we observe are large because of the low average variance in DNA methylation values across individuals. A power calculation using the pooled standard deviation of control and stage 1 HCC groups (0.022) and desired power of 0.8 shows that an extremely "small" (unreliable) sample size ($n = 4$) is required to detect a delta beta of 0.1 at genome-wide significance ($p = 1 \times 10^{-7}$). This large effect size might explain how sites that were discovered by comparing two groups with sample sizes of 10, cross-validated in other samples and were replicated. Power calculation suggests that increasing the sample size from 10 to 100 would not increase power as it is maximized with $n = 10$. However, it is clear that further replication is required to rule out random bias or stratification in our samples.

Conclusions

Our study shows that the host immune system has a distinct DNA methylation signature in cancer, that this signature intensifies as cancer progresses, and that this signature differentiates HCC from liver inflammatory diseases chronic hepatitis B and C. Importantly, distinct DNA methylation differences emerge at early stages and these might serve as noninvasive diagnostic markers of early stage HCC. The broad differences in DNA methylation that progress with HCC suggest a possible role for epigenetic modulation of the peripheral immune system in HCC and its progression that warrants further exploration.

Methods
Patient samples
HCC staging was diagnosed according to EASL–EORTC Clinical Practice Guidelines: Management of hepatocellular carcinoma. The patients were divided into four groups, including stage 0 (1), stage A (2), stage B (3), and stage C+D (4). For simplicity, we refer to stages 1–4 in the figures and manuscript. Chronic hepatitis B diagnosing was confirmed using AASLD practice guideline for chronic hepatitis B, and chronic hepatitis C diagnosing was according to AASLD recommendations for testing, managing, and treating Hepatitis C. A strict exclusion criterion was cirrhosis, any other known inflammatory disease (bacterial or viral infection with the exception of hepatitis B or C, diabetes, asthma, autoimmune disease, active thyroid disease) which could alter T cells and monocyte characteristics as well as presence of other cancers. Clinical characteristics of patients are provided in Tables 1 and 2 and additional information is found in Additional file 14: Table S13, Additional file 15: Table S14 and Additional file 17: Table S16. The participants in the study provided consent according to the regulations of the Capital Medical School. All methods were performed in accordance with the relevant guidelines and regulations. All the candidates were enrolled in the study since 2014 and all the patients prior to receiving the standard therapy according to the BCLC criteria. Whole-blood specimens were collected before the start of standard therapy for the second and third cohort. For the first cohort, blood was drawn either prior to initiation of therapy, prior to surgery, or on the day of surgery (see Additional file 17: Table S16 for details). Informed consent has been obtained from all participants and the study received ethical approval from The Capital Medical School in Beijing and McGill University (IRB Study Number A02-M34-13B).

Illumina Beadchip 450K analysis
DNA was extracted from T cells isolated using antiCD3 immuno-magnetic beads (Dynabeads Invitrogen), bisulfite converted, and subjected to Illumina Human-Methyaltion450k BeadChip analysis. Samples were randomized with respect to slide and position on arrays,

and all samples were hybridized and scanned concurrently to mitigate batch effects as recommended by McGill genome center using Illumina Infinum HD technology user guide. Illumina arrays were analyzed using the ChAMP Bioconductor package in R [46]. IDAT files were used as input in the champ.load function using minfi quality control and normalization options. Raw data were filtered for probes with a detection value of $P > 0.01$ in at least one sample. We filtered out probes on the X or Y chromosome to mitigate sex effects and probes with SNPs as identified in [57], as well as probes that align to multiple locations as identified in [57]. Batch effects were analyzed on the nonnormalized data using the function champ.svd. Five out of the first six principal components were associated with group and batch (slides). Intra-array normalization to adjust the data for bias introduced by the Infinium type 2 probe design was performed using beta-mixture quantile normalization (BMIQ) with function champ.norm (norm = "BMIQ") [46]. We corrected for batch effects after BMIQ normalization using champ.runcombat function. Cell count analysis for white blood cells distribution in our samples was performed according to the Houseman algorithm [54] using the function estimateCellCounts and FlowSorted.Blood.450k data as reference. We used the Beta values of the batch corrected normalized data for downstream statistical analyses. To compute linear correlation between HCC stages and quantitative distribution of DNA methylation at the 450K CG sites, we performed Pearson correlation between the normalized DNA methylation values and stages of HCC (with stage codes of 0 for control, 1 and 2 for hepatitis B and C, respectively, and 3–6 for the four stages of HCC) using the Pearson correlation function in R and correcting for multiple testing using the method "fdr" of Benjamini Hochberg (adjusted P value (Q) of <0.05 as well as the conservative Bonferroni correction ($Q < 1 \times 10^{-7}$). Differentially methylated CGs (MVP) were called using the Bioconductor package Limma [45] as implemented in ChAMP using either "fdr" for multiple testing correction (adjusted P value (Q) of < 0.05) or Bonferroni corrections. Multifactorial ANOVA with group, sex, and age as cofactors was performed for CGs that were shortlisted for association with HCC using loop_anova lmFit function with Bonferroni adjustment for multiple testing. Multivariate linear regression was performed on the shortlisted CG sites that were found to associate with HCC to test whether these associations will survive if we used cell counts, sex, age, and alcohol abuse as covariates in the linear regression model using the lmFit function in R. Comparison of differentially methylated (relative to control) gene lists in different groups was performed using Venny (Oliveros JC 2007; http://bioinfogp.cnb.csic.es/tools/venny/index.html).

Significance of overlap between two groups was determined using hypergeometric Fisher exact test in R. Hierarchical clustering was performed using one minus Pearson correlation, and heatmaps were generated in the Broad institute GeneE application (https://software.broadinstitute.org/GENE-E/).

Pyrosequencing

Pyrosequencing was performed using the Pyro Mark Q24 (Qiagen) machine, and results were analyzed with Pyro Mark Q24 Software 2.0 (Qiagen). All data were expressed as mean ± standard error of the mean (SEM). The statistical analysis was undertaken using Prism (GraphPad Software Inc., San Diego, California). Primers used for the analysis are listed in Additional file 18: Table S17. All data were analyzed using Student's t test. Significance was set at $P < 0.05$ for comparisons of two groups. When multiple groups were involved, ANOVA followed by Bonferroni corrections for multiple testing were used. We determined using multivariate linear regressions whether confounding clinical variables age, sex, smoking, drinking, or treatment were potential covariates. None of these confounding factors showed consistent correlation with CG methylation across the groups.

Additional files

Additional file 1: Table S1. CG sites whose quantitative level of DNA methylation correlates with the stage of HCC as determined by a Pearson correlation analysis ($P<1x10^{-7}$).

Additional file 2: Table S2. Differentially methylated sites between Stage 1 HCC and healthy controls.

Additional file 3: Table S3. Differentially methylated sites between Stage 2 HCC and healthy controls.

Additional file 4: Table S4. Differentially methylated sites between Stage 3 HCC and healthy controls.

Additional file 5: Table S5. Differentially methylated sites between Stage 4 HCC and healthy controls.

Additional file 6: Table S6. Annotated non-redundant list of 350CGs and 369 CGs that are differentially methylated between stages of HCC and healthy controls.

Additional file 7: Figure S1. Differentially Methylated CG Sites at different stages of HCC and "cross-validation". a. Heat map presentation of hierarchical clustering of 69 people by 14 differentially methylated CGs between HCC stage 2 and control. b. Heat map of hierarchical clustering of 69 people by 58 differentially methylated CGs between HCC stage 3 and control. **Figure S2.** Differentially Methylated CG Sites at different stages of HCC in a "training set" and "cross-validation" in a "validation set". a. Heat map presentation of hierarchical clustering of 35 people by a 369 CG signature that correlate with progression in a "training set" (right panel) classify HCC and controls in a "validation set" (left panel as well). b. Heat map of a randomized list of 350 CGs on all patients and controls. **Figure S3.** Prediction of late stage HCC using a penalized model using the 369 CG list which was trained on a randomized half of the HCC patients and controls ("training set") and tested on the other half ("validation set"). The plot shows the "validated" samples (The y axis indicates the predicted probability of late stage HCC for each person (from 0 to 1) (True if prediction >0.5 and False if prediction is <0.5). All late HCC stages in the "validation set" are TRUE and all other stages and controls are FALSE.

Additional file 8: Table S7. List of CG DNA methylation markers derived from penalized regression model on 350 CG and 369CG sites distinguishing early stages HCC from late stages.

Additional file 9: Table S8. Multivariate analysis of 350 CGs. Table provides p values on the right and adjusted values (350 measurements, Bonferroni) on the left.

Additional file 10: Table S9. Multifactorial ANOVA analysis of 350 CGs. No interaction detected between group (HCC) and sex and age as independent variables with CG methylation as a dependent variable.

Additional file 11: Table S10. Differentially methylated CG sites in T cell DNA between healthy controls and HCC.

Additional file 12: Table S11. Correlation of methylation of CG sites associated with the PD-1 gene and progression of HCC.

Additional file 13: Table S12. Differentially methylated that correlate with HCC progression in PBMC.

Additional file 14: Table S13. Clinical data of second cohort.

Additional file 15: Table S14. Clinical data of third cohort.

Additional file 16: Table S15. Descriptive statistics for. (XLS 87 kb)

Additional file 17: Table S16. Clinical data of first cohort.

Additional file 18: Table S17. Pyrosequencing primers.

Abbreviations

AHNAK: Neuroblast differentiation-associated protein; *AKAP7: A Kinase (PRKA) Anchor protein 7; CCL1*: Chemokine (C-C motif) ligand 1; *CCL18*: Chemokine (C-C motif) ligand 18; *CCL24*: Chemokine (C-C motif) ligand 24; *CCL7*: Chemokine (C-C motif) ligand 7; *CCR6*: Chemokine receptor 6; *CCR7*: Chemokine receptor 7; *CCR9*: Chemokine receptor 9; *CD14*: Cluster of differentiation 14; *CD2*: Cluster of differentiation 2; *CD38*: Cluster of differentiation 38; *CD44*: Cluster of differentiation 44; *CD6*: Cluster of differentiation 6; *CD80*: Cluster of differentiation 80; *CD83*: Cluster of differentiation 80; DMR: Differentially methylated region; EWAS: Epigenome wide association studies; HCC: Hepatocellular carcinoma; HepB: Hepatitis B; HepC: Hepatitis C; *Il10*: Interleukin 10; *IL12RB2*: Interleukin 1 receptor B2; *IL16*: Interleukin 16; *IL18*: Interleukin 18; *Il1B*: Interleukin 1 beta; *IL1R1*: Interleukin1 receptor 1; *IL1R2*: Interleukin 1 receptor 2; *IL2*: Interleukin 2; *Il24*: Interleukin 24; *IL2RA*: Interleukin 2 receptor A; *IL4*: Interleukin 4; *IL4R*: Interleukin 4 receptor; *IL5*: Interleukin5; *IL5RA*: Interleukin 5 receptor A; *IL7*: Interleukin 7; *NFAT*: Nuclear factor of activated T-cells; *NFKB*: Nuclear factor kappa B; PBMC: Peripheral blood mononuclear cells; PD-1: Programmed cell death 1; ROC: Receiver Operating Characteristic; *SLFN14*: Schlafen family member 14; *STAP1*: Signal Transducing Adaptor Family 1; *STAT1*: Signal transducer and activator of transcription 1; *STAT3*: Signal transducer and activator of transcription 3; *TGFbeta3*: Transforming growth factor beta 3; *TGFbeta1*: Transforming growth factor beta 1; *TNFa*: Tumor necrosis factor alpha; *VEGF*: Vascular endothelial growth factor

Acknowledgements
NA

Funding
This study was funded by a grant (PSR-SIIRI-635) from the ministère de l'Enseignement supérieur, de la Recherche, de la Science et de la Technologie (MESRST) of the government of Quebec, International cooperation project (2012DFA30850) and Beijing Municipal Science & Technology Commission (D131100005313004, D131100005313005) Canadian Institute of Health Research MOP-42411 and Canadian Institute of Health Research Post-Doctoral Fellowship (CIHR PDF) to SP. National Natural Science Foundation of China (81320108017), Beijing Municipal Science & Technology Commission (Z171100001017078), Beijing Key Laboratory (BZ0373).

Authors' contributions
MS, NL, and YHZ conceived and supervised the study; MS wrote the main manuscript text; NL and YHZ supervised collection of clinical material and clinical data; MS and SP analyzed the data; DC and RZ helped with data analysis; SP, LJ, KL, and SD performed validation experiments; and all authors contributed to the writing and reviewing of the text. All authors read and approved the final manuscript.

Consent for publication
Participants/patients have given their consent for their data to be published in the report.

Competing interests
The authors declare that MS, YZ, SP, and NL have applied for patent protection, MS has equity in HKG epitherapeutics and Montreal epiterapia, and DC has equity in Montreal epiterapia.

Author details
[1]Beijing Youan Hospital, Capital Medical School, Beijing, China. [2]Department of Pharmacology and Therapeutics, McGill University, 3655 Sir William Osler Promenade, Montreal, Quebec H3G 1Y6, Canada. [3]Deparment of Clinical Science, Karolinska Institutet, Alfred Nobels Allé 8, 141 52 Huddinge, Sweden. [4]Montreal EpiTerapia Inc., 4567 Cecile, H9K1N2, Montreal, QC, Canada.

References
1. Blair GE, Cook GP. Cancer and the immune system: an overview. Oncogene. 2008;27:5868.
2. Ehrlich P. Ueber den jetzigen Stand der Karzinomforschung. Ned Tijdschr Geneeskd. 1909;5:273–90.
3. Vesely MD, Kershaw MH, Schreiber RD, Smyth MJ. Natural innate and adaptive immunity to cancer. Annu Rev Immunol. 2011;29:235–71.
4. Dunn GP, Bruce AT, Ikeda H, Old LJ, Schreiber RD. Cancer immunoediting: from immunosurveillance to tumor escape. Nat Immunol. 2002;3:991–8.
5. Swann JB, Smyth MJ. Immune surveillance of tumors. J Clin Invest. 2007;117:1137–46.
6. Mackensen A, Ferradini L, Carcelain G, Triebel F, Faure F, Viel S, Hercend T. Evidence for in situ amplification of cytotoxic T-lymphocytes with antitumor activity in a human regressive melanoma. Cancer Res. 1993;53:3569–73.
7. Ferradini L, Mackensen A, Genevee C, Bosq J, Duvillard P, Avril MF, Hercend T. Analysis of T cell receptor variability in tumor-infiltrating lymphocytes from a human regressive melanoma. Evidence for in situ T cell clonal expansion. J Clin Invest. 1993;91:1183–90.
8. Zorn E, Hercend T. A natural cytotoxic T cell response in a spontaneously regressing human melanoma targets a neoantigen resulting from a somatic point mutation. Eur J Immunol. 1999;29:592–601.
9. Zorn E, Hercend T. A MAGE-6-encoded peptide is recognized by expanded lymphocytes infiltrating a spontaneously regressing human primary melanoma lesion. Eur J Immunol. 1999;29:602–7.
10. Carcelain G, Rouas-Freiss N, Zorn E, Chung-Scott V, Viel S, Faure F, Bosq J, Hercend T. In situ T-cell responses in a primary regressive melanoma and subsequent metastases: a comparative analysis. Int J Cancer. 1997; 72:241–7.
11. Knuth A, Danowski B, Oettgen HF, Old LJ. T-cell-mediated cytotoxicity against autologous malignant melanoma: analysis with interleukin 2-dependent T-cell cultures. Proc Natl Acad Sci U S A. 1984;81:3511–5.
12. Schumacher K, Haensch W, Roefzaad C, Schlag PM. Prognostic significance of activated CD8(+) T cell infiltrations within esophageal carcinomas. Cancer Res. 2001;61:3932–6.
13. Conejo-Garcia JR, Benencia F, Courreges MC, Gimotty PA, Khang E, Buckanovich RJ, Frauwirth KA, Zhang L, Katsaros D, Thompson CB, et al. Ovarian carcinoma expresses the NKG2D ligand Letal and promotes the survival and expansion of CD28- antitumor T cells. Cancer Res. 2004;64:2175–82.
14. Sato E, Olson SH, Ahn J, Bundy B, Nishikawa H, Qian F, Jungbluth AA, Frosina D, Gnjatic S, Ambrosone C, et al. Intraepithelial CD8+ tumor-infiltrating lymphocytes and a high CD8+/regulatory T cell ratio are associated with favorable prognosis in ovarian cancer. Proc Natl Acad Sci U S A. 2005;102:18538–43.
15. Naito Y, Saito K, Shiiba K, Ohuchi A, Saigenji K, Nagura H, Ohtani H. CD8+ T cells infiltrated within cancer cell nests as a prognostic factor in human colorectal cancer. Cancer Res. 1998;58:3491–4.
16. Galon J, Costes A, Sanchez-Cabo F, Kirilovsky A, Mlecnik B, Lagorce-Pages C, Tosolini M, Camus M, Berger A, Wind P, et al. Type, density, and location of immune cells within human colorectal tumors predict clinical outcome. Science. 2006;313:1960–4.

17. Pages F, Berger A, Camus M, Sanchez-Cabo F, Costes A, Molidor R, Mlecnik B, Kirilovsky A, Nilsson M, Damotte D, et al. Effector memory T cells, early metastasis, and survival in colorectal cancer. N Engl J Med. 2005;353:2654–66.

18. Aguirre-Ghiso JA. Models, mechanisms and clinical evidence for cancer dormancy. Nat Rev Cancer. 2007;7:834–46.

19. Teng MW, Vesely MD, Duret H, McLaughlin N, Towne JE, Schreiber RD, Smyth MJ. Opposing roles for IL-23 and IL-12 in maintaining occult cancer in an equilibrium state. Cancer Res. 2012;72:3987–96.

20. Baylin SB, Esteller M, Rountree MR, Bachman KE, Schuebel K, Herman JG. Aberrant patterns of DNA methylation, chromatin formation and gene expression in cancer. Hum Mol Genet. 2001;10:687–92.

21. Issa JP, Vertino PM, Wu J, Sazawal S, Celano P, Nelkin BD, Hamilton SR, Baylin SB. Increased cytosine DNA-methyltransferase activity during colon cancer progression. J Natl Cancer Inst. 1993;85:1235–40.

22. Ehrlich M. DNA methylation in cancer: too much, but also too little. Oncogene. 2002;21:5400–13.

23. Stefanska B, Huang J, Bhattacharyya B, Suderman M, Hallett M, Han ZG, Szyf M. Definition of the landscape of promoter DNA hypomethylation in liver cancer. Cancer Res. 2011;71:5891–903.

24. Jiao L, Zhu J, Hassan MM, Evans DB, Abbruzzese JL, Li D. K-ras mutation and p16 and preproenkephalin promoter hypermethylation in plasma DNA of pancreatic cancer patients: in relation to cigarette smoking. Pancreas. 2007;34:55–62.

25. Park JW, Baek IH, Kim YT. Preliminary study analyzing the methylated genes in the plasma of patients with pancreatic cancer. Scand J Surg. 2012;101:38–44.

26. Dirix L, Van Dam P, Vermeulen P. Genomics and circulating tumor cells: promising tools for choosing and monitoring adjuvant therapy in patients with early breast cancer? Curr Opin Oncol. 2005;17:551–8.

27. Finak G, Laferriere J, Hallett M, Park M. The tumor microenvironment: a new tool to predict breast cancer outcome. Med Sci (Paris). 2009;25:439–41.

28. Finak G, Sadekova S, Pepin F, Hallett M, Meterissian S, Halwani F, Khetani K, Souleimanova M, Zabolotny B, Omeroglu A, Park M. Gene expression signatures of morphologically normal breast tissue identify basal-like tumors. Breast Cancer Res. 2006;8:R58.

29. Finak G, Bertos N, Pepin F, Sadekova S, Souleimanova M, Zhao H, Chen H, Omeroglu G, Meterissian S, Omeroglu A, et al. Stromal gene expression predicts clinical outcome in breast cancer. Nat Med. 2008;14:518–27.

30. Sehouli J, Loddenkemper C, Cornu T, Schwachula T, Hoffmuller U, Grutzkau A, Lohneis P, Dickhaus T, Grone J, Kruschewski M, et al. Epigenetic quantification of tumor-infiltrating T-lymphocytes. Epigenetics. 2011;6:236–46.

31. Jeschke J, Collignon E, Fuks F. DNA methylome profiling beyond promoters: taking an epigenetic snapshot of the breast tumor microenvironment. FEBS J. 2014;

32. Kristensen VN, Vaske CJ, Ursini-Siegel J, Van Loo P, Nordgard SH, Sachidanandam R, Sorlie T, Warnberg F, Haakensen VD, Helland A, et al. Integrated molecular profiles of invasive breast tumors and ductal carcinoma in situ (DCIS) reveal differential vascular and interleukin signaling. Proc Natl Acad Sci U S A. 2011;

33. Teschendorff AE, Menon U, Gentry-Maharaj A, Ramus SJ, Gayther SA, Apostolidou S, Jones A, Lechner M, Beck S, Jacobs IJ, Widschwendter M. An epigenetic signature in peripheral blood predicts active ovarian cancer. PLoS One. 2009;4:e8274.

34. Widschwendter M, Apostolidou S, Raum E, Rothenbacher D, Fiegl H, Menon U, Stegmaier C, Jacobs IJ, Brenner H. Epigenotyping in peripheral blood cell DNA and breast cancer risk: a proof of principle study. PLoS One. 2008;3:e2656.

35. Xu Z, Bolick SC, DeRoo LA, Weinberg CR, Sandler DP, Taylor JA. Epigenome-wide association study of breast cancer using prospectively collected sister study samples. J Natl Cancer Inst. 2013;105:694–700.

36. Koestler DC, Marsit CJ, Christensen BC, Accomando W, Langevin SM, Houseman EA, Nelson HH, Karagas MR, Wiencke JK, Kelsey KT. Peripheral blood immune cell methylation profiles are associated with nonhematopoietic cancers. Cancer Epidemiol Biomark Prev. 2012;21:1293–302.

37. Langevin SM, Houseman EA, Accomando WP, Koestler DC, Christensen BC, Nelson HH, Karagas MR, Marsit CJ, Wiencke JK, Kelsey KT. Leukocyte-adjusted epigenome-wide association studies of blood from solid tumor patients. Epigenetics. 2014;9:884–95.

38. Fridley BL, Armasu SM, Cicek MS, Larson MC, Wang C, Winham SJ, Kalli KR, Koestler DC, Rider DN, Shridhar V, et al. Methylation of leukocyte DNA and ovarian cancer: relationships with disease status and outcome. BMC Med Genet. 2014;7:21.

39. Luo X, Huang R, Sun H, Liu Y, Bi H, Li J, Yu H, Sun J, Lin S, Cui B, Zhao Y. Methylation of a panel of genes in peripheral blood leukocytes is associated with colorectal cancer. Sci Rep. 2016;6:29922.

40. Kao WY, Yang SH, Liu WJ, Yeh MY, Lin CL, Liu CJ, Huang CJ, Lin SM, Lee SD, Chen PJ, MW Y. Genome-wide identification of blood DNA methylation patterns associated with early-onset hepatocellular carcinoma development in hepatitis B carriers. Mol Carcinog. 2016;

41. El-Serag HB. Hepatocellular carcinoma. N Engl J Med. 2011;365:1118–27.

42. Flores A, Marrero JA. Emerging trends in hepatocellular carcinoma: focus on diagnosis and therapeutics. Clin Med Insights Oncol. 2014;8:71–6.

43. Swaika A, Hammond WA, Joseph RW. Current state of anti-PD-L1 and anti-PD-1 agents in cancer therapy. Mol Immunol. 2015;67:4–17.

44. Smyth GK. Limma: linear models for microarray data. In: Gentleman VC R, Dudoit S, Irizarry R, Huber W, editors. Bioinformatics and computational biology solutions using R and bioconductor, vol. 1. New York: Springer; 2005. p. 397–420.

45. Smyth GK, Michaud J, Scott HS. Use of within-array replicate spots for assessing differential expression in microarray experiments. Bioinformatics. 2005;21:2067–75.

46. Morris TJ, Butcher LM, Feber A, Teschendorff AE, Chakravarthy AR, Wojdacz TK, Beck S. ChAMP: 450k Chip analysis methylation pipeline. Bioinformatics. 2014;30:428–30.

47. Goeman JJ. L1 penalized estimation in the Cox proportional hazards model. Biom J. 2010;52:70–84.

48. Wan ES, Qiu W, Carey VJ, Morrow J, Bacherman H, Foreman MG, Hokanson JE, Bowler RP, Crapo JD, DeMeo DL. Smoking associated site specific differential methylation in buccal mucosa in the COPDGene study. Am J Respir Cell Mol Biol. 2015;53:246–54.

49. Allione A, Marcon F, Fiorito G, Guarrera S, Siniscalchi E, Zijno A, Crebelli R, Matullo G. Novel epigenetic changes unveiled by monozygotic twins discordant for smoking habits. PLoS One. 2015;10:e0128265.

50. Cheng L, Liu J, Li B, Liu S, Li X, Tu H. Cigarette smoke-induced hypermethylation of the GCLC gene is associated with chronic obstructive pulmonary disease. Chest. 2016;149:474–82.

51. Li H, Hedmer M, Wojdacz T, Hossain MB, Lindh CH, Tinnerberg H, Albin M, Broberg K. Oxidative stress, telomere shortening, and DNA methylation in relation to low-to-moderate occupational exposure to welding fumes. Environ Mol Mutagen. 2015;

52. Liu J, Morgan M, Hutchison K, Calhoun VD. A study of the influence of sex on genome wide methylation. PLoS One. 2010;5:e10028.

53. Horvath S. DNA methylation age of human tissues and cell types. Genome Biol. 2013;14:R115.

54. Houseman EA, Accomando WP, Koestler DC, Christensen BC, Marsit CJ, Nelson HH, Wiencke JK, Kelsey KT. DNA methylation arrays as surrogate measures of cell mixture distribution. BMC Bioinformatics. 2012;13:86.

55. Butcher LM, Beck S. Probe Lasso: a novel method to rope in differentially methylated regions with 450K DNA methylation data. Methods. 2015;72:21–8.

56. Mandrekar JN. Receiver operating characteristic curve in diagnostic test assessment. J Thorac Oncol. 2010;5:1315–6.

57. Marzouka NA, Nordlund J, Backlin CL, Lonnerholm G, Syvanen AC, Carlsson Almlof J. CopyNumber450kCancer: baseline correction for accurate copy number calling from the 450k methylation array. Bioinformatics. 2016;32:1080–2.

9

A urine-based DNA methylation assay, ProCUrE, to identify clinically significant prostate cancer

Fang Zhao[1,2], Ekaterina Olkhov-Mitsel[1,2], Shivani Kamdar[1,2], Renu Jeyapala[1], Julia Garcia[1], Rachel Hurst[3], Marcelino Yazbek Hanna[4], Robert Mills[3], Alexandra V. Tuzova[5], Eve O'Reilly[5], Sarah Kelly[5], Colin Cooper[3], The Movember Urine Biomarker Consortium, Daniel Brewer[3,6], Antoinette S. Perry[5], Jeremy Clark[3], Neil Fleshner[7] and Bharati Bapat[1,2,7*]

Abstract

Background: Prevention of unnecessary biopsies and overtreatment of indolent disease remains a challenge in the management of prostate cancer. Novel non-invasive tests that can identify clinically significant (intermediate-risk and high-risk) diseases are needed to improve risk stratification and monitoring of prostate cancer patients. Here, we investigated a panel of six DNA methylation biomarkers in urine samples collected post-digital rectal exam from patients undergoing prostate biopsy, for their utility to guide decision making for diagnostic biopsy and early detection of aggressive prostate cancer.

Results: We recruited 408 patients in risk categories ranging from benign to low-, intermediate-, and high-risk prostate cancer from three international cohorts. Patients were separated into 2/3 training and 1/3 validation cohorts. Methylation biomarkers were analyzed in post-digital rectal exam urinary sediment DNA by quantitative MethyLight assay and investigated for their association with any or aggressive prostate cancers.

We developed a Prostate Cancer Urinary Epigenetic (ProCUrE) assay based on an optimal two-gene (*HOXD3* and *GSTP1*) LASSO model, derived from methylation values in the training cohort, and assessed ProCUrE's diagnostic and prognostic ability for prostate cancer in both the training and validation cohorts.

ProCUrE demonstrated improved prostate cancer diagnosis and identification of patients with clinically significant disease in both the training and validation cohorts. Using three different risk stratification criteria (Gleason score, D'Amico criteria, and CAPRA score), we found that the positive predictive value for ProCUrE was higher (59.4–78%) than prostate specific antigen (PSA) (38.2–72.1%) for all risk category comparisons. ProCUrE also demonstrated additive value to PSA in identifying GS $\geq$ 7 PCa compared to PSA alone (DeLong's test $p = 0.039$), as well as additive value to the PCPT risk calculator for identifying any PCa and GS $\geq$ 7 PCa (DeLong's test $p = 0.011$ and 0.022, respectively).

Conclusions: ProCUrE is a promising non-invasive urinary methylation assay for the early detection and prognostication of prostate cancer. ProCUrE has the potential to supplement PSA testing to identify patients with clinically significant prostate cancer.

Keywords: Prostate cancer, PSA, Urine, DNA methylation, Biomarker, Early detection, Overtreatment

* Correspondence: bapat@lunenfeld.ca
[1]Lunenfeld-Tanenbaum Research Institute, Sinai Health System, Toronto, Canada
[2]Department of Laboratory Medicine & Pathobiology, University of Toronto, Toronto, Canada
Full list of author information is available at the end of the article

Introduction

The introduction of circulating prostate specific antigen (PSA) test has increased the rate of diagnosis of prostate cancer (PCa) by as much as 50%. However, the majority of PCa patients diagnosed through PSA screening present with low-risk, localized, Gleason score (GS) 6 tumors. Although PSA has a high negative predictive value (NPV) for PCa, its lack of specificity, limited impact on reducing morbidity, and the harms of over-diagnosing indolent disease have raised concerns about PSA screening [1].

To reduce overtreatment and associated morbidity, the U.S. Preventive Services Task Force (USPSTF) recently recommended against PSA screening to prevent unnecessary biopsies of "clinically insignificant" PCa (CI-PCa) which included patients with benign and low-risk disease [1]. However, following these recommendations, there was a substantial decrease (42.9%) [2] in the detection of GS ≥ 7 disease, indicating the reduction in PSA screening could delay diagnosis of "clinically significant" PCa (CS-PCa) consisting of intermediate- and high-risk disease. The revised recommendations now include advising men under 70 about the potential benefits and limitations of PSA based screening. However, their impact on the diagnosis of CS-PCa is currently unknown.

Several nomograms have been developed to estimate PCa aggressiveness following biopsy, such as the well-established D'Amico criteria [3] which includes PSA, GS, and clinical T stage. Due to the limited number of variables, patients with the same D'Amico risk category may have vastly different outcomes. Alternatively, the recently developed UCSF-Cancer of the Prostate Risk Assessment (CAPRA) score [4] is more informative due to ease of calculation and inclusion of key clinical variables including age, PSA, percent of cores positive in biopsy (%core), clinical T stage, and Gleason patterns. There is also the Prostate Cancer Prevention Trial (PCPT) PCa risk calculator [5], which takes into account ethnicity, family history, PSA, age, and digital rectal exams (DREs) results to calculate the risk of finding any cancer or high-risk (GS ≥ 7) cancer upon biopsy. These nomograms are used to distinguish low-risk versus high-risk PCa patients for management decisions after biopsy.

Low-risk PCa patients may be recommended enrollment into an active surveillance (AS) protocol where they are monitored with DREs, PSA tests, multiparametric (mp) MRI where available, and periodic biopsies instead of definitive treatment [6]. Although AS is a preferable management option for patients with CI-PCa, many AS patients with indolent tumors still undergo additional unnecessary biopsies and suffer associated morbidities.

Consequently, there is an urgent need to develop non-invasive biomarkers to complement PSA screening for the early identification of aggressive PCa and to guide decision making for initial diagnostic prostate biopsy or repeat biopsies of low-risk patients on AS. To address this, the Movember foundation introduced the Global Action Plan (GAP) 1: Urine biomarker initiative, which brought together 12 research teams from seven different countries. Our study, as part of this initiative, investigated non-invasive DNA methylation biomarkers for improved prognostication of PCa.

Aberrant DNA methylation is a hallmark of PCa [7, 8]. Tumor-specific gene methylation alterations are ideal biomarkers due to their stability and ease of detection from patient samples with limited amounts of DNA such as urinary sediments. Detection of DNA methylation biomarkers in urine sediment is non-invasive and may be able to supplement PSA screening to identify CS-PCa patients.

We have previously discovered and/or characterized tumor-specific DNA methylation of six genes (*APC*, *GSTP1*, *HOXD3*, *KLK10*, *TBX15*, and *TGFβ2*) in radical prostatectomy tumor samples [9–11]. Increased methylation of these genes was found to be associated with higher GS and adverse clinical prognosis. We also examined these biomarkers in post-DRE urine samples from a Canadian AS PCa patient cohort [12]. In the current study, we investigated the utility of these urinary DNA methylation biomarkers for diagnosis and prognostication of CS-PCa in three international patient cohorts.

Results

Cohort characteristics

The clinicopathologic characteristics for patient cohorts are summarized in Table 1. To mitigate any inherent biases in patient recruitment, all patients were combined, randomized, and separated into training (2/3 of patients) and validation (1/3 of patients) cohorts [13, 14] (Table 2).

Age and PSA were significantly correlated with each other, as well as prostate volume, and %core. (Additional file 1: Table S1, Spearman's ρ $p < 0.01$).

Detection of urinary DNA methylation biomarkers and association with clinicopathologic variables

We assessed DNA methylation of our panel of biomarkers in the urinary sediment of patients recruited. Methylation frequencies (patients with percent methylated of reference (PMR) > 0) ranged from 39.5% (161/408 patients) for *GSTP1* to 92.6% (378/408 patients) for *HOXD3*. PMR distribution for individual markers among benign and PCa patients is shown in Fig. 1. Five of the six gene methylation showed significant increase in PCa

Table 1 Clinical characteristics from the University of East Anglia, UK (UEA), GU Biobank at UHN, Canada (UHN), Trinity College at Dublin, Ireland (Dublin)

Patient clinical characteristics	UEA	UHN	Dublin
n (%)	194 (48)	155 (38)	59 (14)
Benign	109 (56)	46 (30)	27 (46)
PCa	85 (44)	109 (70)	32 (54)
Gleason score			
6	17 (20)	64 (59)	15 (47)
7	42 (49)	32 (29)	9 (28)
8–10	26 (31)	13 (12)	8 (25)
Clinical T stage			
T1	38 (45)	91 (83)	20 (63)
T2	14 (16)	16 (15)	11 (34)
T3	19 (22)	2 (2)	1 (3)
T4	14 (16)	0	0
% Biopsy cores positive for PCa			
Median	57%	20%	21%
Range	7–100%	5–100%	6–100%
Interquartile range	33%–100%	9%–38%	13%–43%
N/A	9	1	0
Age at enrollment			
Median	67	64	65
Range	42–85	37–83	46–80
Interquartile range	62–73	57–69	58–71
PSA at presentation			
Median	8.4	5.8	5.9
Range	0.2–277.3	0.01–67.31	0.5–248
Interquartile range	5.8–12.2	3.97–9.07	3.85–8.64
Prostate volume			
Median	59.54	47	
Range	21.08–244.6	16.05–127.0	
Interquartile range	42.52–86.52	34–57	
N/A	92	18	59
Perineural invasion			
Yes	28	11	4
No	80	17	55
N/A	86	127	0
CAPRA risk			
CAPRA low	10 (12)	57 (52)	13 (41)
CAPRA intermediate	32 (38)	36 (33)	14 (44)
CAPRA high	43 (51)	16 (15)	5 (16)
D'Amico risk			
D'Amico low	8 (9)	52 (48)	9 (28)
D'Amico intermediate	29 (34)	39 (36)	13 (41)
D'Amico high	48 (56)	18 (17)	10 (31)

Table 2 Cohort characteristics of the training and validation cohorts

Patient clinical characteristics	Training	Validation
n (%)	268 (65.4)	140 (34.6)
Benign	123 (44)	59 (41)
PCa	145 (52)	81 (57)
Gleason score		
6	60 (41)	36 (44)
7	55 (38)	28 (35)
8–10	30 (21)	17 (21)
Clinical T stage		
T1	101 (70)	48 (59)
T2	25 (17)	16 (20)
T3	11 (8)	11 (14)
T4	8 (6)	6 (7)
% Biopsy cores positive for PCa		
Median	29%	33%
Range	5–100%	5–100%
Interquartile range	13%–55%	14%–63%
N/A	8	6
Age at enrollment		
Median	66	66
Range	42–85	37–85
Interquartile range	59–71	59–72
PSA at presentation		
Median	6.9	7.04
Range	0.01–248	0.04–377.00
Interquartile range	4.55–10.36	4.79–11
Prostate volume		
Median	7	49
Range	18.0–121.6	18.0–121.6
Interquartile range	39.15–68.68	37–65.1
N/A	114	58
Perineural invasion		
Yes	30	13
No	97	55
N/A	153	75
CAPRA risk		
CAPRA low	47 (32)	33 (41)
CAPRA intermediate	56 (39)	26 (32)
CAPRA high	42 (29)	22 (27)
D'Amico risk		
D'Amico low	41 (28)	28 (35)
D'Amico intermediate	57 (39)	24 (30)
D'Amico high	47 (32)	29 (36)

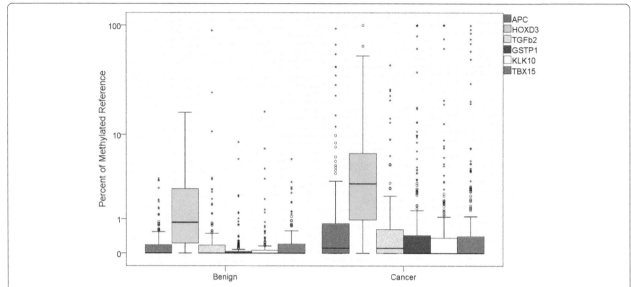

Fig. 1 Distribution of percent of methylated reference (PMR) values for individual biomarkers among benign and PCa (Cancer) patients. Number of patients = 408. *APC, HOXD3, TGFβ2, GSTP1,* and *KLK10* are able to significantly differentiate benign and PCa (Mann Whitney *U p* < 0.05). Circles indicate outliers within 1.5× IQR, stars indicate outliers > 1.5× IQR. Mann Whitney *U p* values can be found in Additional file 1: Table S2

compared to benign (Additional file 1: Table S2 Mann Whitney *U p* < 0.05).

Methylation levels of all six genes were significantly correlated with each other (Spearman's ρ *p* < 0.01). *APC, GSTP1, KLK10, TBX15,* and *TGFB2* showed significant association with age and %core (Additional file 1: Table S1; Spearman's ρ *p* < 0.05). Additionally, *GSTP1, KLK10,* and *TBX15* were associated with PSA (Spearman's ρ *p* < 0.05). *HOXD3* did not correlate with any clinical variables.

Building an optimal predictor gene model and ProCUrE assay

To investigate whether combinations of biomarkers were more informative compared to individual markers for detection of any PCa and/or aggressive PCa, we applied least absolute shrinkage and selection operator (LASSO) and constructed an optimal two-gene (*HOXD3* and *GSTP1*) classifier model (ProCUrE) in the training cohort comparing between benign vs CAPRA-HR patients. Receiver operating characteristic (ROC) curve analysis of ProCUrE showed an area under curve (AUC) of 0.795 (bootstrapped 1000 iterations) (Fig. 2), which was higher than any individual marker; thus, we did not analyze individual markers in the validation cohort. An optimal cut-off threshold for ProCUrE was established with the maximum combined sensitivity (57.1%) and specificity (97%). Patients with methylation levels above this threshold are considered positive for ProCUrE status (ProCUrE +ve).

Assessment of ProCUrE for improved PCa diagnosis

To determine ProCUrE's value for PCa diagnosis, we tested its association with PCa. ProCUrE +ve status was

significantly associated with PCa positive biopsies in both the training (Additional file 2: Figure S1A) and validation cohorts (Fig. 3a) (χ^2 *p* < 0.01) while age-adjusted PSA (see definition in the "Material and methods" section) [15, 16] was not (Fig. 3a; χ^2 *p* > 0.05). ProCUrE status identified 31.6% PCa patients with 11.9% false positive cases, while age-adjusted PSA detected 75.3% PCa patients but also had a high number (69.5%) of false positives. The positive predictive value (PPV) for ProCUrE was higher than for age-adjusted PSA (78.1% vs 59.8%) (Table 3A). These results demonstrate that ProCUrE +ve patients are more likely to harbor PCa.

Assessment of ProCUrE for early prognostication of PCa

To investigate ProCUrE's value for PCa prognostication, we assessed the ability of individual markers, ProCUrE, and clinical variables to differentiate CI-PCa and CS-PCa patients as determined by GS. Using univariable logistic regression analysis, ProCUrE, PSA, and age showed significant association with CS-PCa. Due to the difference in range of each variable, interquartile range odds ratios (IQR OR) were estimated. The IQR OR of ProCUrE in the validation cohort (OR = 1.58, 95% CI = 1.28–1.96) were of similar size to PSA (OR = 1.98, 95% CI = 1.46–2.68), and age (OR = 1.66, 95% CI = 1.13–2.45) for CS-PCa (Table 4A). Multivariable logistic regression of significant variables age, PSA, and ProCUrE showed that ProCUrE was an independently significant variable for CS-PCa in the validation cohort (Table 4B). These results show that ProCUrE is a robust prognosticator of CS-PCa.

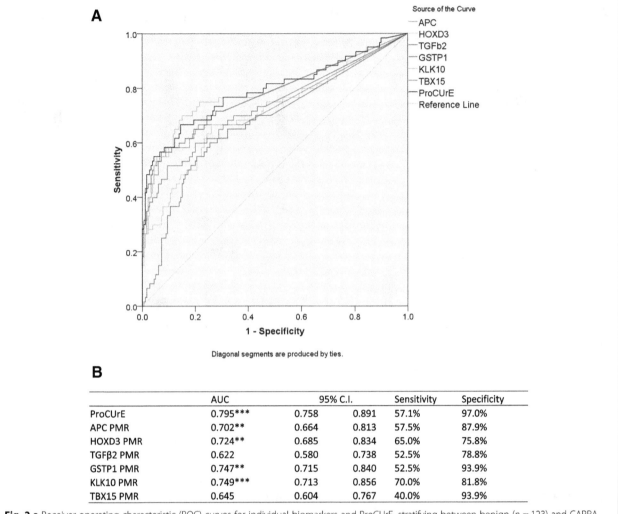

Fig. 2 a Receiver operating characteristic (ROC) curves for individual biomarkers and ProCUrE, stratifying between benign (*n* = 123) and CAPRA high-risk (*n* = 42) patients in the training cohort. PSA was not included in this figure since PSA is used to calculate CAPRA and will always have a very strong association with CAPRA high risk. **b** AUC (bootstrapped 1000 iterations), sensitivity, and specificity for each gene and ProCUrE. ROC ******$p < 0.01$; *******$p < 0.001$

	AUC	95% C.I.		Sensitivity	Specificity
ProCUrE	0.795***	0.758	0.891	57.1%	97.0%
APC PMR	0.702**	0.664	0.813	57.5%	87.9%
HOXD3 PMR	0.724**	0.685	0.834	65.0%	75.8%
TGFβ2 PMR	0.622	0.580	0.738	52.5%	78.8%
GSTP1 PMR	0.747**	0.715	0.840	52.5%	93.9%
KLK10 PMR	0.749***	0.713	0.856	70.0%	81.8%
TBX15 PMR	0.645	0.604	0.767	40.0%	93.9%

We examined ProCUrE among patients stratified into different risk categories based on GS, D'Amico criteria and CAPRA score. χ^2 analysis showed that both Pro-CUrE and age-adjusted PSA could differentiate between patients harboring no disease and/or CI-PCa versus CS-PCa in both the training cohort (Additional file 2: Figure S1B–D) and validation cohort (Fig. 3b–d) (χ^2 $p <$ 0.05). ProCUrE was able to differentiate low-risk vs intermediate- and high-risk PCa patients based on GS and CAPRA score, but not D'Amico criteria (Fig. 3b–d). Furthermore, ProCUrE's prognostic value was consistently more robust as demonstrated by a more stringent p value (χ^2 $p < 0.01$) compared to age-adjusted PSA (χ^2 $p < 0.05$) in the validation cohort.

ProCUrE exhibited higher PPV than age-adjusted PSA for several different prognostic assessments (PPV: 76% vs 70.5% for CAPRA, 76% vs 72.1% for D'Amico, and 76% vs 63.9% for GS risk, Table 3B–D) indicating its overall robust ability to identify CS-PCa. Additionally, we found that ProCUrE significantly differentiated high-grade PCa, including GS ≥ 7(4 + 3), and GS ≥ 8 from all other patients, with higher PPV (37.5% and 31.3%, respectively) compared to age-adjusted PSA (24.5% and 16.7%, respectively) (Fig. 3, Table 3B–D).

These results indicate that patients who are ProCUrE +ve have a higher likelihood of harboring CS-PCa and high-grade (GS ≥ 8) tumors.

Additional discriminative value of ProCUrE to PSA

To determine whether ProCUrE could add discriminatory value to PSA testing, we performed concordance statistics (c-statistic) analysis of PSA alone and ProCUrE with PSA combined in the training cohort using logistic regression. The c-statistic with PSA and ProCUrE

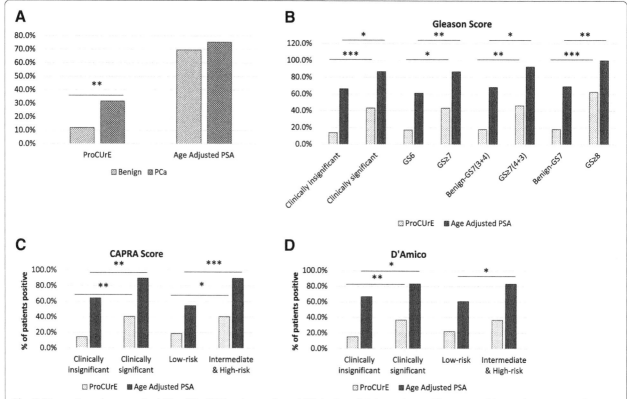

Fig. 3 Diagnostic and prognostic ability of ProCUrE and age-adjusted PSA in the validation cohort. **a** The percent false- and true-positive for ProCUrE or age-adjusted PSA separating benign and PCa patients. **b** The percent of patients positive for ProCUrE or age-adjusted PSA for clinically insignificant (benign and low-risk) vs clinically significant (intermediate- and high-risk) based on Gleason score. **c, d** The percent of patients positive for ProCUrE or age-adjusted PSA for clinically insignificant (benign and low-risk) vs clinically significant (intermediate- and high-risk) and low-risk vs intermediate- and high-risk as determined by CAPRA score and D'Amico. $N = 140$, χ^2*$p < 0.05$, **$p < 0.01$, ***$p < 0.001$

combined (0.775) is significantly improved over PSA alone (0.729) (DeLong test $p = 0.039$), indicating that ProCUrE has additional discriminatory value to PSA for detecting CS-PCa (Table 5). Only GS risk was analyzed since CAPRA score and D'Amico criteria is calculated using PSA.

Additional discriminative value of ProCUrE to PCPT

To determine whether ProCUrE could add additional value to current clinical nomograms, we used the PCPT risk calculator for risk assessment in a subset of 144 patients (out of 408 patients in total) that had family history, ethnicity, and DRE results available.

We assessed the diagnostic (detection of any PCa) and prognostic (detection of GS ≥ 7 PCa) value of PCPT (AUC = 0.741; 0.771, respectively) and ProCUrE (AUC = 0.746; 0.730, respectively) individually. Further, using c-statistic, we calculated the additive value of ProCUrE to PCPT for diagnosis of PCa using logistic regression, which increased from $c = 0.741$ for PCPT alone to $c = 0.817$ for PCPT with ProCUrE. Similarly, for the detection of CS-PCa (as determined by GS) addition of ProCUrE to PCPT increased from $c = 0.771$ to $c = 0.822$.

Both values represent a significant increase (DeLong's test $p = 0.011$ for diagnostic, $p = 0.022$ for prognostic value) and indicate that the information provided by ProCUrE could further improve current PCPT parameters for prognosticating PCa patients prior to biopsy.

Discussion

Our study developed a urinary DNA methylation biomarker-based actionable assay, ProCUrE, to identify CS-PCa that would warrant treatment. ProCUrE significantly improves risk stratification with a higher PPV compared to age-adjusted PSA. Patients who are positive for ProCUrE will be more likely to harbor aggressive tumors and thus ProCUrE has the potential to supplement PSA or other tests that focus on NPV. Importantly, ProCUrE has additive value to PSA assessment and to PCPT risk calculator for the detection of aggressive (GS ≥ 7) cancers.

PSA testing cannot reliably distinguish patients that have CS-PCa disease from those that do not require treatment. Therefore, invasive confirmation biopsy is necessary for PCa diagnosis and prognostication. A non-invasive adjunct test to PSA, such as ProCUrE, that

Table 3 Diagnosis (A) and prognostication (B–D) of PCa

	PPV	NPV
A		
Benign vs PCa		
ProCUrE	78.10%	49.10%
Age-adjusted PSA	59.80%	47.40%
B		
GS clinically insignificant vs clinically significant		
ProCUrE	59.40%	76.40%
Age-adjusted PSA	38.20%	84.20%
GS6 vs GS ≥ 7		
ProCUrE	76.00%	53.70%
Age-adjusted PSA	63.90%	70.00%
Benign, GS6, GS7(3 + 4) vs GS ≥ 7 (4 + 3)		
ProCUrE	37.5%	86.8%
Age-adjusted PSA	24.5%	94.7%
Benign, GS6, GS7 vs GS ≥ 8		
ProCUrE	31.3%	94.3%
Age-adjusted PSA	16.7%	100.0%
C		
CAPRA clinically insignificant vs clinically significant		
ProCUrE	59.40%	73.60%
Age-adjusted PSA	42.20%	86.80%
CAPRA low risk vs intermediate and high risk		
ProCUrE	76.00%	48.10%
Age-adjusted PSA	70.50%	75.00%
D		
D'Amico clinically insignificant vs clinically significant		
ProCUrE	59.40%	68.90%
Age-adjusted PSA	43.10%	76.30%
D'Amico low risk vs intermediate and high risk		
ProCUrE	76.00%	38.90%
Age-adjusted PSA	72.10%	55.00%

Positive (PPV) and negative (NPV) predictive values for ProCUrE and age-adjusted PSA in the validation cohort separating benign vs PCa (A); clinically insignificant (benign and low-risk) vs clinically significant (intermediate- and high-risk) and low-risk vs clinically significant (intermediate- and high-risk) as determined by GS, CAPRA score, D'Amico criteria. (x^2 p values for these comparisons could be found in Fig. 3)

can identify patients with CS-PCa would reduce overtreatment and prevent morbidity associated with unnecessary biopsies.

ProCUrE is comprised of the promoter methylation of *HOXD3* and *GSTP1* genes. *HOXD3* is a member of the homeobox gene family of transcription factors which play important roles in morphogenesis and cell adhesion [17, 18], while *GSTP1* is a member of the GST family of metabolic enzymes which function in regulation of cell cycle, DNA repair, and apoptosis [19]. Increased methylation levels of *HOXD3* and *GSTP1* are observed in prostate tumors and are correlated with aggressive PCa and/or adverse clinical outcomes [9, 18, 20, 21]. *GSTP1* methylation has been previously investigated in urine sediments and was found to be PCa specific when compared to benign patients [22].

In a recent study of urinary methylation biomarkers, *APC* and *GSTP1* methylation in conjunction with clinical variables demonstrated 100% NPV for distinguishing GS ≥ 7 PCa [23]. Although this study demonstrated that urine-based DNA methylation markers could be used to prognosticate PCa aggressiveness, their results showed a high (26%) false positive rate compared to only 13.8% false positive rate observed for ProCUrE. Thus, their combined panel of *APC* and *GSTP1* is less than favorable to address the current challenges for managing PCa, specifically, to minimize overtreatment of low-risk patients. Currently available non-invasive tests for PCa diagnosis and prognosis include the Prostate Health Index [24], SelectMDx [25], mpMRI [26], and PCA3. Similar to ProCUrE, SelectMDx is a post-DRE urine-based, two-gene (*HOXC6* and *DLX1*) expression assay that can detect CS-PCa (GS > 6) (AUC = 0.77). The Prostate Health Index (PHI) [27] is a FDA-approved blood test that measures total, free and -2proPSA with greater specificity than free and total PSA for CS-PCa [28]. MpMRI has a high NPV (95%) for GS ≥ 7 tumors [26]. However, high cost and limited availability remain a limitation for implementing mpMRI as a screening tool. The Progensa PCA3 test is the only FDA approved urine-based test for PCa diagnosis. With its high NPV for PCa (90%) [29], PCA3 can prevent unnecessary repeat biopsies. The Mi-Prostate score combines PCA3 and TMPRSS2:ERG fusion with the multivariable Prostate Cancer Prevention Trial risk calculator (PCPT) for prediction of PCa (AUC = 0.762) and high-risk PCa (AUC = 0.779) which is comparable to our ProCUrE assay in the training cohort (AUC = 0.795 for benign vs high-risk PCa, Fig. 2) [30] Additionally, we demonstrated that ProCUrE, when combined with PCPT, has even greater AUCs for diagnosis (0.817) and prognostication (0.822) of PCa than the Mi-Prostate score. However, it should be noted that this comparison was calculated on a subset of the total number of patients that had DRE, family history, and ethnicity information available.

All of the aforementioned tests are promising for PCa diagnosis or prognostication. However, all of these tests focus on NPV for PCa or high-risk PCa. Patients who are above the selected thresholds for these tests remain uncertain with respect to their disease status. Working in conjunction with the above tests or PSA, our ProCUrE assay fulfills a niche by focusing on PPV to offer a distinct advantage in identifying PCa patients with clinically significant tumors. Thus, patients who cannot be

Table 4 Prediction of CS-PCa (as determined by GS) by individual markers, clinical variables, and ProCUrE

A

Univariable	1st quartile	3rd quartile	Difference	OR	95% CI.		p value
ProCUrE	− 2.0006	− 1.0828	0.91789	1.58***	1.28	1.96	< 0.0001
APC	0	0.37652	0.37652	1.24***	1.10	1.39	0.0003
HOXD3	0.43965	5.2043	4.7646	1.54***	1.24	1.90	< 0.0001
GSTP1	0	0.1216	0.1216	1.06**	1.02	1.10	0.0013
KLK10	0	0.12665	0.12665	1.07**	1.02	1.11	0.0048
TGFβ2	0	0.309	0.309	1.03*	1.00	1.06	0.0481
TBX15	0	0.21295	0.21295	1.16***	1.08	1.24	< 0.0001
PSA	4.565	10.48	5.915	1.98***	1.46	2.68	< 0.0001
Age	59	71	12	1.66**	1.13	2.45	0.0105
PSA density	0.09	0.2	0.11	2.35***	1.51	3.68	0.0002
Prostate volume	40	70	30	0.70	0.47	1.06	0.0926

B

Multivariable	OR	95% CI.		p value
ProCUrE	1.358*	1.051	1.754	0.0194
PSA	0.816	0.373	1.785	0.6108
Age	2.718**	1.295	5.707	0.0082
PSA density	2.878**	1.455	5.694	0.0024

Using univariable and multivariable logistic regression, the ability of individual methylation markers, ProCUrE, and clinical variables to differentiate CI-PCa and CS-PCa as determined by GS was assessed in the training cohort. Since the scale of each variable is different, interquartile range odds ratios were estimated (logistic regression model *p < 0.05, **p < 0.01, ***p < 0.001)

ruled out as having indolent tumors could be tested with ProCUrE to assess whether they have aggressive disease. ProCUrE could also be combined with tests such as SelectMDx to build a more comprehensive multivariable urine test in the future. This will improve current clinical PCa patient management once validated in independent studies.

Our study has certain limitations, including the fact that patient cohorts recruited for our study had differences in size and composition (e.g., UEA cohort had patients with higher PSA, GS, T stage compared to other cohorts) and as such, they could not be analyzed as three independent cohorts, despite using consensus recruitment criteria. Histopathological-based cancer diagnosis of biopsies was performed at three different participating centers which may have contributed to

Table 5 C-statistic for distinguishing clinically significant disease based on GS

CI-PCa vs CS-PCa (GS)	C-statistic
PSA	0.729
ProCUrE	0.684
Combined	0.775*

C-statistics was used to determine any additive value of ProCUrE to PSA for discriminating CI-PCa vs CS-PCa as determined by GS in the training cohort. Only GS risk was analyzed since CAPRA score and D'Amico criteria is calculated using PSA. DeLong's test *p = 0.039

some variation in Gleason grading between cohorts. However, our strategy of combining patients from all three cohorts and subsequently randomizing into training and validation cohorts, overcomes these caveats as it ensured that the training and validation sets would include patients representing a broad spectrum of prostate status, from benign with low PSA to very high risk PCa.

Other potential caveats are that benign patients with abnormalities such as high PSA may have contributed to the lack of significance for age-adjusted PSA with PCa diagnosis. Patients who are false positive for ProCUrE +ve status may actually harbor occult tumors. In this regard, follow-up data collection is ongoing for future biopsies and/or MRIs, which will enable assessment of ProCUrE for prediction and confirmation of CS-PCa. Additionally, clinical stage information for the UEA and Dublin cohorts did not differentiate T2a, T2b, and T2c tumors. Therefore, D'Amico criteria was calculated with all T2 patients assigned as intermediate risk. This may have contributed to the lack of significance for ProCUrE to stratify patients based on D'Amico criteria. In previously published studies, we described a 4-gene methylation Classifier Panel (APC, GSTP1, CRIP3, HOXD8) in PCa patients monitored by active surveillance (AS) for the prediction of risk-reclassification [12, 31]. We were unable to screen two genes (CRIP3 and HOXD8) from this four-gene classifier panel in the current study due to limitations of DNA samples availability. Similar to our findings in the AS PCa patient cohort, we found that methylation frequencies of *APC* and *GSTP1* in urinary sediment were lower compared to those reported in tissue samples. Lastly, it is difficult to assess the additive value of ProCUrE to PSA for identifying patients based on the CAPRA score and D'Amico criteria, since both nomograms are calculated using PSA leading to strong association with PSA with these risk groups. Both CAPRA and D'Amico criteria are limited in that they require prostate biopsy to calculate risk. ProCUrE is advantageous in this regard since risk assessment can be performed prior to biopsy.

Conclusion

A non-invasive urine-based assay that can distinguish PCa patients with aggressive, clinically significant disease from those with benign and/or low risk disease would be valuable in reducing morbidity associated with over-diagnosis and preventing under-diagnosis of patients that would benefit from definitive treatment. Our ProCUrE assay could be used to supplement PSA screening and monitoring so those with aggressive disease would be identified early and those without will avoid unnecessary treatment.

Materials and methods
Patient cohorts

Participants were prospectively recruited between April 2012 and September 2015, from the University of East Anglia/Norfolk and Norwich University Hospital, UK (UEA cohort, $n = 194$), the University Health Network, Canada (UHN cohort, $n = 155$), and Trinity College, Ireland (Dublin cohort, $n = 59$), together as part of the Movember GAP1 Multi-Center Urine Biomarker (MoGAP-MUB) cohort. Patients underwent prostate TRUS biopsy due to increased PSA and/or abnormal DRE (PSA follow-up time 0–122 months). Benign patients with normal age-adjusted PSA were recruited due to symptoms of BPH or had microhematuria detectable on dipstick only (i.e., not gross hematuria). Less than 10% (39/408) patients had prior biopsies, all other patients were recruited at initial biopsy. Post-DRE first catch urine samples were either collected prior to biopsy or at least 1-month post-biopsy. Samples were mostly collected within 12 months from the date of biopsy. There were two patients that had > 12-month difference between biopsy and sample collection (range 14–146 months) and three patients with unknown biopsy dates. The patient with sample collected 146 months post-biopsy and all patients with unknown biopsy dates were benign patients. Informed consent was obtained following protocols approved by the research ethics boards of all centers and Sinai Health System, Toronto, Canada.

The cut-off for normal PSA (referred to as age-adjusted PSA) was determined following British Association of Urological Surgeon guidelines [15, 16]. Patients were classified based on the following criteria:

Benign indicates patients with negative biopsy.

GS, D'Amico criteria, and CAPRA score were utilized to stratify risk in PCa patients:

GS: low risk (GS ≤ 6), intermediate risk (GS7), and high risk (GS ≥ 8).

D'Amico criteria: low risk (GS ≤ 6 and T1–T2a and PSA < 10 ng/mL), intermediate risk (GS7 or T2b or PSA 10–20 ng/mL), and high risk (GS $\geq$ or > or PSA > 20 ng/mL) [3].

The CAPRA score is defined as the sum of the following variables: age at diagnosis (< 50 = 0, $\geq 50 = 1$), PSA at diagnosis(ng/mL) ($\leq 6 = 0$, 6.1–10 = 1, 10.1–20 = 2, 20.1–30 = 3, > 30 = 4), biopsy Gleason pattern (no pattern $\geq 4 = 0$, secondary pattern $\geq 4 = 1$, primary pattern $\geq 4 = 3$), clinical T stage (T1 or T2 = 0, $\geq$ T3 = 1; %core: < 34% = 0, $\geq 34\% = 1$) [4]. CAPRA risk categories are as follows: low risk (0–2 points), intermediate risk [3–5], and high risk (≥ 6).

Calculation of PCPT risk score

PCPT risk was calculated using the Cleveland Clinic Risk Calculator Library – PCPT Risk Calculator v2.0 [5].

Urine collection/processing

Up to 50 mL of first catch urine was collected from each patient following DRE and centrifuged at $1200 \times g$ for 5 min. Urine sediments were separated from supernatant and resuspended in 1 ml of PBS and stored at − 80 °C. Urinary sediment DNA was extracted using the AllPrep DNA/RNA mini-kit (Qiagen Inc.) Bisulfite conversion was as previously described [12].

MethyLight analysis

Multiplex MethyLight, a methylation-specific qPCR assay was used to determine the methylation levels of *APC*, *GSTP1*, *HOXD3*, *KLK10*, *TBX15*, and *TGFβ2* [32]. *ALU-C4* (*ALU*) was used as a methylation-independent, sodium bisulfite conversion-dependent internal input DNA control.

Primer/probe concentrations, cycling parameters, and data acquisition/analysis were as previously described, using Applied Biosystems 7500 (Life Technologies) [12].

Gene methylation was scored as percent methylated of reference (PMR) according to Eads et al. [33] CpGenome Universal Methylated DNA (EMD Millipore) was used as the positive control and to generate standard curves. Quality control criteria included genes of interest (GOIs) standard curve $R^2 > 0.95$, ALU $R^2 > 0.99$, and slope range from − 3.28 to − 4.86. Any sample with a higher cycling threshold (lower quantity) for *ALU* than the least concentrated standard curve point for which all GOI amplified was excluded from analysis. Samples were analyzed in duplicate and were reanalyzed if replicates had a difference in PMR of > 10%. Data development and analysis were carried out in accordance with the Minimum Information for Publication of Quantitative real-time PCR Experiments (MIQE) guideline [34].

Calculation for ProCUrE

Least absolute shrinkage and selection operator (LASSO) was applied to construct gene models using benign vs CAPRA high-risk (CAPRA-HR) patients in the training cohort. LASSO was used to eliminate genes that had insufficient contribution to the model. The remaining genes (*APC, GSTP1, HOXD3, KLK10, TGFβ2*) with non-zero coefficients as determined by LASSO were tested for every possible combination using the generalized linear model in the training cohort to determine their AUC, Akaike information criterion (AIC), and Bayesian information criterion (BIC). An optimal two-gene model consisting of *HOXD3* and *GSTP1*, which had the highest AUC with the lowest AIC and BIC, was selected for further analysis. We developed Prostate Cancer Urinary Epigenetic (ProCUrE) assay, based on the formula:

$$\text{Intercept} + \text{Coefficient(GSTP1)} * \text{PMR(GSTP1)} \\ + \text{Coefficient(HOXD3)} * \text{PMR(HOXD3)}$$

where the intercept is − 0.8395549, the coefficient for *HOXD3* is 0.1397128, and the coefficient for *GSTP1* is 0.8632709.

Additional comparisons (benign vs PCa, CI vs CS-PCa as determined by GS, CAPRA and D'Amico) were performed in the training cohort (Additional file 2: Figure S2). However, none of these comparisons yielded a model with as robust discriminative value (higher AUC) as observed with benign vs CAPRA-HR comparison. Therefore, we opted for the model constructed using benign vs CAPRA-HR for further analysis.

Statistical analysis

Spearman's ρ rank was used to compare PMR, age, %core, prostate volume (cc), and PSA at diagnosis. ROC curve analysis was used to determine ProCUrE's sensitivity and specificity at every cut-off value. The value with the highest sum of sensitivity and specificity was chosen as the optimized threshold. The same numerical values derived from the training cohort were used in the validation cohort: threshold derived from ROC analysis (threshold value = 0.574264899821094) and intercept and coefficients derived from generalized linear modeling. χ^2 tests were used to determine any significant association with overall cancer status or CS-PCa (≥ intermediate-risk cancer as determined by GS, CAPRA, or D'Amico criteria).

Univariable and multivariable logistic regression was performed to estimate odds ratios and corresponding 95% confidence intervals to assess the ability of individual markers, ProCUrE, and clinical variables to identify CS-PCa patients using the lrm function of the "rms" R package (5.1–2). C-statistic was calculated using ROC curves [35]. DeLong's test [36] was used to compare significance for c-statistic as part of the roc.test function of the pROC R package (v1.13.0).

LASSO analysis was carried out using the "glmnet" function of the "glmnet" R package (v2.0-13) [37] to determine the optimal value of the penalty coefficient lambda, with 10-fold cross-validation performed using the "cv.glmnet" function. Optimal lambda was chosen as the cross-validated lambda at the minimum binomial deviance. Model assessment was performed using the "ROCR" R package, AUC was determined via bootstrapping with 1000 iterations.

For all described methods, two-sided *p* values of < 0.05 were considered significant. All tests were conducted with IBM SPSS software (SPSS Inc. Released 2014. PASW Statistics for Windows, Version 22.0) or R version 3.4.0 [38]. Reporting recommendations for tumor marker prognostic studies (REMARK) guidelines were followed in analysis [39].

Additional files

Additional file 1: Table S1. Correlations. Spearman's rank correlations for the PMR values of each biomarker, ProCUrE, and clinical variables Spearman's ρ *p < 0.05; **p < 0.01. **Table S2.** Average PMR values of individual gene methylation for benign and PCa patients. All genes except *TBX15* was able to significantly differentiate between benign and PCa (Mann Whitney *U* *p* < 0.05). **Table S3.** Diagnosis (A) and prognostication (B-D) of PCa in the training cohort.

Additional file 2: Figure S1. Diagnostic and prognostic ability of ProCUrE and age-adjusted PSA in the training cohort. **Figure S2.** Receiver operating characteristic curve analysis of training cohort for (A) benign vs PCa, clinically insignificant vs clinically significant PCa as determined by (B) GS, (C) CAPRA, and (D) D'Amico.

Abbreviations

%core: Percent of cores positive in biopsy; AIC: Akaike information criterion; AS: Active surveillance; AUC: Area under curve; BIC: Bayesian information criterion; CAPRA: Cancer of the Prostate Risk Assessment; CI-PCa: Clinically insignificant prostate cancer; CS-PCa: Clinically significant prostate cancer; DRE: Digital rectal exam; GAP: Global Action Plan; GOI: Gene of interest; GS: Gleason score; IQR OR: Interquartile range odds ratios; LASSO: Least absolute shrinkage and selection operator; MIQE: Minimum Information for Publication of Quantitative real-time PCR Experiments; MoGAP-MUB: Movember GAP1 Multi-Center Urine Biomarker; MpMRI: Multiparametric magnetic resonance imaging; NPV: Negative predictive value; PCa: Prostate cancer; PCPT: Prostate Cancer Prevention Trial; PMR: Percent methylated of reference; PPV: Positive predictive value; ProCUrE: Prostate Cancer Urinary Epigenetic; PSA: Prostate specific antigen; ROC: Receiver operating characteristic; USPSTF: U.S. Preventive Services Task Force

Acknowledgements

We would like to acknowledge the Movember Urine Biomarker Consortium which brought together 12 research teams from seven different countries to collaborate on such an important project. A list of members is provided below.

Movember GAP1 Urine Biomarker Collaborative Members

1. Colin Cooper, Jeremy Clark, Schools of Medicine and Biological Sciences, University of East Anglia: Norwich, Norfolk. NR4 7TJ, UK.
2. Bharati Bapat, Lunenfeld-Tanenbaum Research Institute, Sinai Health Systems, University of Toronto, Toronto, Ontario, M5T 3L9, Canada.
3. Rob Bristow, Ontario Cancer Institute, Princess Margaret Hospital, Toronto, Ontario, Canada.
4. Chris Parker, Royal Marsden Hospital, Downs Road, Sutton, SM2 5PT, UK.
5. Ian Mills, Queen's University Belfast. School of Medicine, Dentistry and Biomedical Sciences. Institute for Health Sciences, Belfast, UK.
6. Hardev Pandha, University of Surrey: Guildford, Surrey, GU2 7XH. UK.
7. Hayley Whitaker, Molecular Diagnostics and Therapeutics Group, Cruciform Building, University College London, Gower Street. WC1E 6BT. UK.
8. David Neal, David Neal, Cancer Research UK Cambridge Research Institute, Robinson Way, Cambridge, CB2 0RE. UK
9. Mireia Olivan, Andreas Doll, Vall d'Hebron Research Institute and Hospital and Autonomous University of Barcelona: 08035 Barcelona, Spain.
10. Hing Leung, The Beatson Institute for Cancer Research, Joseph Black Building, University of Glasgow, Glasgow, G12 8QQ, UK.
11. Antoinette Perry, Cancer Biology and Therapeutics Laboratory, School of Biomolecular and Biomedical Science, Conway Institute, University College Dublin, Dublin 4, Ireland.
12. Martin Sanda, Department of Urology, Winship Cancer Institute, Emory University School of Medicine, Atlanta, GA, USA.
13. Jack Schalken, Nijmegen Medical Centre, Radboud University, The Netherlands. Nijmegen, The Netherlands

We gratefully acknowledge the patients from three international centers who participated in this study. We also acknowledge assistance of staff at The UHN Genito-Urinary BioBank (GUBioBank) (Toronto). The UHN GUBioBank is a REB approved investigator-initiated bio-banking program that collects and archives biological specimens and data obtained from consenting urologic oncology patients. The purpose of the program is to facilitate the discovery and validation of novel biomarkers, which ultimately will enable advances in the area of personalized medicine in Urology.

Funding

This work was supported by grant funding from Movember (GAP1 Urine Biomarker Award), Prostate Cancer Canada No. 2011-700, Movember PCC TAG No. 2014-01 (B. Bapat), The Ontario Student Opportunity Trust Funds Award (F. Zhao, S. Kamdar, R. Jeyapala), and Ontario Graduate Scholarships (F. Zhao, E. Olkhov-Mitsel, S. Kamdar).

Authors' contributions

BB, JC, and NF contributed to the conception and design. FZ, EO-M, JG, NF, JC, RH, MH, RM, AP, AT, EO'R, and SK took part in the acquisition of the data. FZ, EO-M, SK, RJ, DB, and BB were responsible for the analysis and interpretation of the data. FZ and BB drafted the manuscript. BB, EO-M, CC, DB, and JC did the critical revision of the manuscript for important intellectual content. FZ, EO-M, SK, and DB contributed to the statistical analysis. BB and CC obtained the funding. BB, DB, JC, NF, and AP gave administrative, technical, or material support. BB did the supervision. All authors read and approved the final manuscript.

Consent for publication

Informed consent was obtained from all participants at all centers.

Competing interests

The authors declare that they have no competing interests.

Author details

[1]Lunenfeld-Tanenbaum Research Institute, Sinai Health System, Toronto, Canada. [2]Department of Laboratory Medicine & Pathobiology, University of Toronto, Toronto, Canada. [3]Schools of Medicine and Biological Sciences, University of East Anglia, Norwich, Norfolk, UK. [4]Norfolk and Norwich University Hospital, Norwich, Norfolk, UK. [5]Cancer Biology and Therapeutics Laboratory, School of Biomolecular and Biomedical Science, Conway Institute, University College Dublin, Dublin 4, Ireland. [6]The Earlham Institute, Norwich, Norfolk, UK. [7]Division of Urology, University Health Network, University of Toronto, Toronto, Canada.

References

1. Force USPST. Final recommendation statement: prostate cancer. Screening. 2012; [updated November 2013. Available from: https://www.uspreventiveservicestaskforce.org/Page/Name/second-annual-report-to-congress-on-high-priority-evidence-gaps-for-clinical-preventive-services.
2. Bhindi B, Mamdani M, Kulkarni GS, Finelli A, Hamilton RJ, Trachtenberg J, et al. Impact of the U.S. Preventive Services Task Force recommendations against prostate specific antigen screening on prostate biopsy and cancer detection rates. J Urol. 2015;193(5):1519–24.
3. D'Amico AV, Whittington R, Malkowicz SB, Schultz D, Blank K, Broderick GA, et al. Biochemical outcome after radical prostatectomy, external beam radiation therapy, or interstitial radiation therapy for clinically localized prostate cancer. JAMA. 1998;280(11):969–74.
4. Cooperberg MR, Hilton JF, Carroll PR. The CAPRA-S score: a straightforward tool for improved prediction of outcomes after radical prostatectomy. Cancer. 2011;117(22):5039–46.
5. Thompson IM, Ankerst DP, Chi C, Goodman PJ, Tangen CM, Lucia MS, et al. Assessing prostate cancer risk: results from the Prostate Cancer Prevention Trial. J Natl Cancer Inst. 2006;98(8):529–34.
6. Klotz L, Vesprini D, Sethukavalan P, Jethava V, Zhang L, Jain S, et al. Long-term follow-up of a large active surveillance cohort of patients with prostate cancer. J Clin Oncol. 2015;33(3):272–7.
7. Ahmad AS, Vasiljevic N, Carter P, Berney DM, Moller H, Foster CS, et al. A novel DNA methylation score accurately predicts death from prostate cancer in men with low to intermediate clinical risk factors. Oncotarget. 2016;7(44):71833–40.
8. Vasiljevic N, Wu K, Brentnall AR, Kim DC, Thorat MA, Kudahetti SC, et al. Absolute quantitation of DNA methylation of 28 candidate genes in prostate cancer using pyrosequencing. Dis Markers. 2011;30(4):151–61.
9. Liu L, Kron KJ, Pethe VV, Demetrashvili N, Nesbitt ME, Trachtenberg J, et al. Association of tissue promoter methylation levels of APC, TGFβ2, HOXD3 and RASSF1A with prostate cancer progression. Int J Cancer. 2011;129(10): 2454–62.
10. Kron K, Liu L, Trudel D, Pethe V, Trachtenberg J, Fleshner N, et al. Correlation of ERG expression and DNA methylation biomarkers with adverse clinicopathologic features of prostate cancer. Clin Cancer Res. 2012; 18(10):2896–904.
11. Olkhov-Mitsel E, Van der Kwast T, Kron KJ, Ozcelik H, Briollais L, Massey C, et al. Quantitative DNA methylation analysis of genes coding for kallikrein-related peptidases 6 and 10 as biomarkers for prostate cancer. Epigenetics. 2012;7(9):1037–45.
12. Zhao F, Olkhov-Mitsel E, van der Kwast T, Sykes J, Zdravic D, Venkateswaran V, et al. Urinary DNA methylation biomarkers for noninvasive prediction of aggressive disease in patients with prostate cancer on active surveillance. J Urol. 2017;197(2):335–41.
13. Dobbin KK, Simon RM. Optimally splitting cases for training and testing high dimensional classifiers. BMC Med Genet. 2011;4:31.
14. Dupuy A, Simon RM. Critical review of published microarray studies for cancer outcome and guidelines on statistical analysis and reporting. J Natl Cancer Inst. 2007;99(2):147–57.
15. Surgeons TBAoU. PSA MEASUREMENTS Frequently-Asked Questions [Available from: http://www.tsft.nhs.uk/media/45224/PSA_levels.pdf. Accessed 15 Oct 2018.
16. DeAntoni EP, Crawford ED, Oesterling JE, Ross CA, Berger ER, McLeod DG, et al. Age- and race-specific reference ranges for prostate-specific antigen from a large community-based study. Urology. 1996;48(2): 234–9.

17. Hamada J, Omatsu T, Okada F, Furuuchi K, Okubo Y, Takahashi Y, et al. Overexpression of homeobox gene HOXD3 induces coordinate expression of metastasis-related genes in human lung cancer cells. Int J Cancer. 2001; 93(4):516–25.

18. Kron KJ, Liu L, Pethe VV, Demetrashvili N, Nesbitt ME, Trachtenberg J, et al. DNA methylation of HOXD3 as a marker of prostate cancer progression. Lab Investig. 2010;90(7):1060–7.

19. Tew KD, Manevich Y, Grek C, Xiong Y, Uys J, Townsend DM. The role of glutathione S-transferase P in signaling pathways and S-glutathionylation in cancer. Free Radic Biol Med. 2011;51(2):299–313.

20. Jeronimo C, Usadel H, Henrique R, Silva C, Oliveira J, Lopes C, et al. Quantitative GSTP1 hypermethylation in bodily fluids of patients with prostate cancer. Urology. 2002;60(6):1131–5.

21. Wu T, Giovannucci E, Welge J, Mallick P, Tang WY, Ho SM. Measurement of GSTP1 promoter methylation in body fluids may complement PSA screening: a meta-analysis. Br J Cancer. 2011;105(1):65–73.

22. Goessl C, Muller M, Heicappell R, Krause H, Straub B, Schrader M, et al. DNA-based detection of prostate cancer in urine after prostatic massage. Urology. 2001;58(3):335–8.

23. Jatkoe TA, Karnes RJ, Freedland SJ, Wang Y, Le A, Baden J. A urine-based methylation signature for risk stratification within low-risk prostate cancer. Br J Cancer. 2015;112(5):802–8.

24. Loeb S, Catalona WJ. The Prostate Health Index: a new test for the detection of prostate cancer. Ther Adv Urol. 2014;6(2):74–7.

25. Leyten GH, Hessels D, Smit FP, Jannink SA, de Jong H, Melchers WJ, et al. Identification of a candidate gene panel for the early diagnosis of prostate cancer. Clin Cancer Res. 2015;21(13):3061–70.

26. Ahmed HU, El-Shater Bosaily A, Brown LC, Gabe R, Kaplan R, Parmar MK, et al. Diagnostic accuracy of multi-parametric MRI and TRUS biopsy in prostate cancer (PROMIS): a paired validating confirmatory study. Lancet. 2017; 389(10071):815–22.

27. White J, Shenoy BV, Tutrone RF, Karsh LI, Saltzstein DR, Harmon WJ, et al. Clinical utility of the Prostate Health Index (phi) for biopsy decision management in a large group urology practice setting. Prostate Cancer Prostatic Dis. 2018;21(1):78-84.

28. Nordstrom T, Vickers A, Assel M, Lilja H, Gronberg H, Eklund M. Comparison between the four-kallikrein panel and prostate health index for predicting prostate cancer. Eur Urol. 2015;68(1):139–46.

29. Gittelman MC, Hertzman B, Bailen J, Williams T, Koziol I, Henderson RJ, et al. PCA3 molecular urine test as a predictor of repeat prostate biopsy outcome in men with previous negative biopsies: a prospective multicenter clinical study. J Urol. 2013;190(1):64–9.

30. Tomlins SA, Day JR, Lonigro RJ, Hovelson DH, Siddiqui J, Kunju LP, et al. Urine TMPRSS2:ERG plus PCA3 for individualized prostate cancer risk assessment. Eur Urol. 2016;70(1):45–53.

31. Bapat B. Re: urinary DNA methylation biomarkers for noninvasive prediction of aggressive disease in patients with prostate cancer on active surveillance: F. Zhao, E. Olkhov-Mitsel, T. van der Kwast, J. Sykes, D. Zdravic, V. Venkateswaran, A. R. Zlotta, A. Loblaw, N. E. Fleshner, L. Klotz, D. Vesprini and B. Bapat. J Urol 2017;197:335–341. J Urol. 2018. https://doi.org/10.1016/j.juro.2017.10.061.

32. Olkhov-Mitsel E, Zdravic D, Kron K, van der Kwast T, Fleshner N, Bapat B. Novel multiplex MethyLight protocol for detection of DNA methylation in patient tissues and bodily fluids. Sci Rep. 2014;4:4432.

33. Eads CA, Danenberg KD, Kawakami K, Saltz LB, Blake C, Shibata D, et al. MethyLight: a high-throughput assay to measure DNA methylation. Nucleic Acids Res. 2000;28(8):E32.

34. Bustin SA, Benes V, Garson JA, Hellemans J, Huggett J, Kubista M, et al. The MIQE guidelines: minimum information for publication of quantitative real-time PCR experiments. Clin Chem. 2009;55(4):611–22.

35. Steyerberg EW. Clinical prediction models a practical approach to development, validation, and updating. New York: Springer; 2009.

36. DeLong ER, DeLong DM, Clarke-Pearson DL. Comparing the areas under two or more correlated receiver operating characteristic curves: a nonparametric approach. Biometrics. 1988;44(3):837–45.

37. Friedman J, Hastie T, Tibshirani R. Regularization paths for generalized linear models via coordinate descent. J Stat Softw. 2010;33(1):1–22.

38. TR Core Team. R: A language and environment for statistical computing. R Foundation for Statistical Computing, Vienna, Austria. 2014. http://www.R-project.org/.

39. McShane LM, Altman DG, Sauerbrei W, Taube SE, Gion M, Clark GM, et al. Reporting recommendations for tumor marker prognostic studies. J Clin Oncol. 2005;23(36):9067–72.

10

A long-range interactive DNA methylation marker panel for the promoters of HOXA9 and HOXA10 predicts survival in breast cancer patients

Seong-Min Park[1,2†], Eun-Young Choi[1†], Mingyun Bae[3], Jung Kyoon Choi[3] and Youn-Jae Kim[1*]

Abstract

Background: Most DNA cancer methylation markers are based on the transcriptional regulation of the promoter-gene relationship. Recently, the importance of long-range interactions between distal CpGs and target genes has been revealed. Here, we attempted to identify methylation markers for breast cancer that interact with distant genes.

Results: We performed integrated analysis using chromatin interactome data, methylome data, transcriptome data, and clinical information for breast cancer from public databases. Using the chromatin interactome and methylome data, we defined CpG-distant target gene relationships. After determining the differences in methylation between tumor and paired normal samples, the survival association, and the correlation between CpG methylation and distant target gene expression, we selected CpG methylation marker candidates. Using Cox proportional hazards models, we combined the selected markers and evaluated the prognostic model. We identified six methylation markers in HOXA9 and HOXA10 promoter regions and their long-range target genes. We experimentally validated the chromatin interactions, methylation status, and transcriptional regulation. A prognostic model showed that the combination of six methylation markers was highly associated with poor survival in independent datasets. According to our multivariate analysis, the prognostic model showed significantly better prognostic ability than other histological and molecular markers.

Conclusions: The combination of long-range interacting HOXA9 and HOXA10 promoter CpGs predicted the survival of breast cancer patients, providing a comprehensive and novel approach for discovering new methylation markers.

Keywords: Biomarker, Prognosis, DNA methylation, Survival, Long-range interaction, Chromatin interaction, HOXA9, HOXA10

Background

Breast cancer is both the most common cancer and the most frequent cause of cancer-related deaths among women [1]. Based on the expression level of hormone receptors, such as the estrogen receptor (ER) and progesterone receptor (PR), or human epidermal growth factor receptor (Her2), breast cancers are divided into several subtypes, and small molecules or antibodies targeting ER, PR, and Her2 have been used in breast cancer therapies [2]. Breast cancer is conventionally diagnosed by mammography, but this method cannot be applied to

some cases, including women with premenopausal breast cancer [3]. Molecular markers and reference laboratory tests for breast cancer diagnosis and prognosis have been developed, but the methods are limited to specific subtypes, such as node-negative and ER-positive breast cancer [4, 5]. Thus, novel approaches for the diagnosis and prognosis of breast cancer are still needed.

DNA methylation is one of the most well-known aberrations in human cancers [6]. During tumor progression from normal tissue to invasive cancer, the total level of DNA methylation gradually decreases, but the frequency of hypermethylated CpG islands on promoters increases, causing the transcriptional silencing of tumor-suppressive genes [7, 8]. DNA methylation markers have advantages compared to other molecular markers. For example,

* Correspondence: yjkim@ncc.re.kr
†Equal contributors
[1]Translational Research Branch, Research Institute, National Cancer Center, Goyang, Gyeonggi 10408, Republic of Korea
Full list of author information is available at the end of the article

hypermethylation of promoter CpGs is a common and early event during the progression of various tumors [9, 10], and DNA methylation is more chemically and biologically stable than RNA or most proteins [6]. DNA methylation markers for cancer diagnosis and prognosis have been discovered, and some of them have been used in clinical trials [11, 12]. For breast cancer, researchers have also reported particular DNA methylation markers [13, 14], some of which need further development for clinical application.

The DNA methylation of promoters and CpG islands is known to inhibit target gene expression by regulating the binding of transcription modulators to the promoter [15, 16]. The long-range interaction between CpGs and target genes has been reported [17, 18]. A recent genomic study revealed that the correlation between DNA methylation at distal regulatory sites and long-range target gene expression is significantly stronger than the correlation with promoter methylation and that differences in DNA methylation between cancer and normal tissues at distal regulatory sites are significantly greater than differences in promoter methylation among various cancer types [19]. Nevertheless, most DNA methylation markers for cancer diagnosis and prognosis have been developed based on promoter-gene relationships because of the difficulty of defining the relationship between distal CpGs and target genes. The long-range action of distal CpG-target gene interaction and transcriptional regulation can be specified by chromatin interactome data, particularly data from RNA polymerase II (Pol II) chromatin interaction analysis by paired-end tag sequencing (ChIA-PET) [20]. Thus, novel approaches could be based on the long-range interaction between CpGs and their target genes.

In this study, we identified DNA methylation markers for breast cancer and the putative target genes that had long-range interactions using an integrated analysis incorporating the chromatin interactome, methylome, and transcriptome data for breast cancer from public databases. We tried to validate the chromatin interaction, methylation status, and transcriptional regulation. Selected marker candidates were combined to establish a prognostic model and evaluated as markers for breast cancer.

Methods
Public data analysis
The Cancer Genome Atlas (TCGA) methylome (Illumina Infinium Human Methylation 450k BeadChip microarray data, Infinium HM450k) and transcriptome (high-throughput RNA sequencing, RNA-seq) data containing clinical information were downloaded from the International Cancer Genome Consortium (ICGC) data portal (http://icgc.org/). The chromatin interactome

(chromatin interaction analysis by paired-end tag sequencing, ChIA-PET) data were downloaded from the Encyclopedia of DNA Elements (ENCODE) databases (https://genome.ucsc.edu/ENCODE/). Another Infinium HM450k methylome dataset for validation was downloaded from the NCBI GEO database (http://www.ncbi.nlm.nih.gov/geo/) (GSE39004). The expression microarray (Affymetrix Human Genome U133 Plus 2.0 microarray, affyU133P2) data for MCF7 breast cancer cells after 5-azacytidine (5-aza C) treatment and the untreated control data were downloaded from the NCBI GEO database (GSE22250). The methylome data were globally normalized using β values (methylation ratio). The RNA-seq data were normalized based on the RPKM (reads per kilobase per million mapped reads) values. The affyU133P2 data were globally normalized using the Robust Multi-array Average (RMA) method.

The genomic positions were defined by the human hg19 reference genome. Genomic loci from 2000 bp upstream to 500 bp downstream of the transcription start sites (TSS) were defined as promoters. ENCODE MCF7 Pol II ChIA-PET data deposited in the UCSC genome browser database (http://genome.ucsc.edu/) were used to define CpG-target gene relationships. Genes whose promoters were anchored by ChIA PET reads were defined as target genes, and CpGs that overlapped with opposite ends of promoter-anchored ends were defined as distal CpGs.

Statistical tests were performed using the R program (https://www.r-project.org/). Graphs and heatmaps were prepared using Excel (Microsoft) and R.

Cell culture and AZA treatment of the MCF7 breast cell line
The MCF7 cell line was purchased from the American Type Culture Collection (ATCC). MCF7 was maintained in complete Dulbecco's modified Eagle medium (DMEM, HyClone) at 37 °C in a humidified 5% CO_2 incubator. The complete medium was supplemented with 10% fetal bovine serum (HyClone), 100 U/ml penicillin/streptomycin (WelGENE), and 2 mM L-glutamine (HyClone).

The cells were treated with 1 µM 5-aza-2′-deoxycytidine (5-AZA C) (Sigma-Aldrich, A3656) dissolved in DMSO (Sigma-Aldrich, D2650), and the equivalent amount of DMSO was used as a control treatment. The cells were harvested after 72 h.

Chromosome conformation capture (3C)
For this process, 5.0×10^6 cells were cross-linked with 2% formaldehyde for 10 min at 25 °C. Five milliliters of NP-40 buffer (10 mM Tris-HCl, pH 7.5, 10 mM NaCl, 0.2% NP-40, and a protease inhibitor cocktail) was added to the cells, and the cells were incubated at 4 °C for 2 h.

The mixture was then centrifuged, and the pellet was resuspended in 0.5 ml of 1.2× DpnII restriction enzyme buffer (NEB). Fifteen microliters of 10% SDS was added to the sample, and the mixture was incubated at 37 °C for 1 h. Forty microliters of 25% Triton-X100 was added, and the sample was incubated at 37 °C for 1 h. The sample was digested with 400 units of DpnII restriction enzyme at 37 °C for approximately 18 h. To deactivate DpnII, 80 µl of 10% SDS was added to the sample, and the mixture was incubated at 65 °C for 20 min. Then, 6.125 ml of 1.15× ligation buffer and 300 µl of 25% Triton-X100 were added, and the sample was incubated at 37 °C for 1 h. The digested samples were ligated with 100 units of T4 DNA ligase (Promega) at 16 °C for 4 h and then at 25 °C for 30 min. Reverse cross-linking and proteinase K treatment were performed overnight at 65 °C. The chromatin was then treated with RNase A for 1 h at 37 °C. The DNA was purified using a phenol/chloroform extraction or with a QIAquick PCR Purification Kit.

3C–PCR assays were performed using amfiXpand PCR Master Mix. The data were normalized to "internal" primers for the GAPDH gene. At least three independent biological replicates were included for each 3C–PCR assay. The primer sequences are listed in Additional file 1: Table S1.

Reverse transcription PCR

The total RNA was extracted using the RNeasy Mini Kit (QIAGEN) according to the manufacturer's instructions. Reverse transcription was performed with 1 µg of total RNA as the template and M-MLV Reverse Transcriptase (Promega). RT-PCR assays were performed using AmfiXpand PCR Master Mix (GenDEPOT). The cDNA expression was normalized to the levels of GAPDH. The primers used for the PCR reactions were designed either manually or using the Primer3 program (http://biotool-s.umassmed.edu/bioapps/primer3_www.cgi). All primer sequences are listed in Additional file 1: Table S1.

Pyrosequencing

The total DNA was extracted using a QIAamp DNA Blood Mini Kit (QIAGEN) according to the manufacturer's protocol. In total, 0.5 µg of total DNA from each of the samples was used for bisulfite conversion using an EZ DNA Methylation Lightning kit (Zymo Research). The bisulfite-converted DNA was amplified using TOP-simple Premix (Enzynomics). Pyrosequencing was performed using the PyroMark Q96 ID (PSQ 96MA, QIAGEN) system according to the manufacturer's protocol. Pyrosequencing primers (forward, reverse, and sequencing) were designed using the PSQ Assay Design program (version 1.0.6). All primer sequences are listed in Additional file 1: Table S1.

Code accessibility

We provided our Python and R scripts in GitHub (https://github.com/lastmhc/long-range_interactive_DNA_methylation_marker).

Results

HOXA9 and HOXA10 promoter CpG selection from public data

To identify DNA methylation markers for breast cancer that physically interact with distant genes, we performed an integrated analysis and stepwise selection of the DNA methylation markers using publicly available chromatin interactome, methylome, transcriptome, and clinical information (Fig. 1a). The following data sets were used: the chromatin interaction analysis by paired-end tag sequencing (ChIA-PET) data for the MCF7 breast cancer cell line from the ENCODE database, Illumina Infinium Human Methylation 450k BeadChip (Infinium HM450k) microarray data, high-throughput RNA sequencing (RNA-seq) data and clinical information for breast cancer patients from TCGA database, and Affymetrix Human Genome U133 Plus 2.0 microarray (affyU133P2) data for MCF7 breast cancer cells after 5-aza C treatment from the NCBI GEO dataset (GSE22250). For the CpG sites probed by Infinium HM450k, CpGs that had missing values in the TCGA breast cancer dataset were removed. To select CpGs with long-range interactions, we used the MCF7 ChIA-PET data for RNA polymerase II (Pol II) and selected CpGs with genomic positions that overlapped with the ChIA-PET reads. Subsequent analyses were performed using the data from TCGA breast cancer patients who had both tumor and the paired solid normal tissue samples because we tried to select CpG methylation marker candidates that have both diagnostic and prognostic value. To determine the diagnostic value, the methylation difference of each CpG between the tumor and paired normal samples ($\Delta\beta$ value) was calculated, and hypermethylated CpGs (average $\Delta\beta > 0.2$) were selected. To determine the prognostic value, after calculating the association between CpG methylation and the survival of breast cancer patients, we selected significantly associated CpGs ($p < 0.05$, log-rank test). To identify putative target genes of the selected CpGs, we paired the hypermethylated CpGs and distant gene promoters using the MCF7 Pol II ChIA-PET data. Using this information, we calculated the Pearson correlation coefficient (R) between the methylation of the CpGs and the expression of paired genes. To identify the putative relationship between the hypermethylated CpGs and distant target genes, we selected CpG-target gene relationships that had a highly negative correlation ($R < -0.35$). To test the DNA methylation sensitivity of the putative target genes, we examined the expression change in the genes after 5-aza C treatment

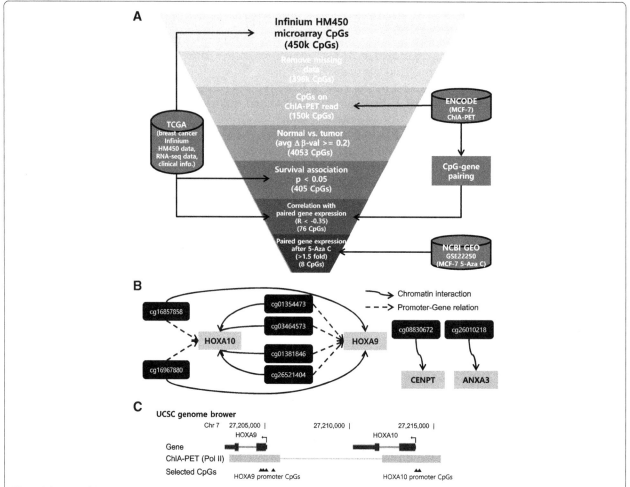

Fig. 1 Selection of long-range interacting CpG methylation markers for the diagnosis and prognosis of breast cancer patients. **a** Scheme for the marker selection using pubic data (from the TCGA breast cancer dataset: Infinium 450k array data (methylome), RNA-seq data (transcriptome), and clinical information; from ENCODE: ChIA-PET data (chromatin interactome) for MCF7 breast cancer cells; from NCBI GEO: expression microarray data (transcriptome, GSE22250) for MCF7 breast cancer cells after 5-Aza C treatment). **b** Selected CpG marker candidates and long-range interacting genes. **c** Gene structure, the selected CpG loci and chromatin status near the HOXA9 and HOXA10 locus (UCSC genome browser)

using the GSE22250 dataset. Differentially expressed (fold change >1.5) genes (DEGs) and the paired CpGs were selected. Thus, eight putative CpG-target gene relationships that were supposed to compose long-range regulation modules were selected (Fig. 1b). Interestingly, six existed on Homeobox gene 9 and 10 (HOXA9 and 10) promoter loci, and four CpGs on the HOXA9 promoter and two CpGs on the HOXA10 promoter physically interacted with one another in the single ChIA-PET read (Fig. 1c). Thus, we subsequently focused on the methylation status of four CpGs on the HOXA9 promoter and two CpGs on the HOXA10 promoter.

Long-range interplay of HOXA9 and HOXA10 promoter methylation markers

HOXA9 and HOXA10 have been reported to be tumor suppressor genes in breast cancer [21–23]. The methylation of the HOXA9 and HOXA10 promoters is associated

with cancer progression in various cancers [24, 25]. Previous analyses identified HOXA9 and HOXA10 promoter methylation marker candidates and their chromatin interactions. Thus, we further investigated the methylation status and the association with the survival of breast cancer patients using TCGA Infinium HM450k data and clinical information for paired tissues. First, we examined the DNA methylation differences between the normal and tumor samples. Comparing the DNA methylation percentage of each of the six CpGs on the HOXA9 and HOXA10 promoters in tumor tissues and their paired normal samples revealed that the methylation was significantly higher in the tumor samples than their paired normal samples ($p < 1.0 \times 10^{-14}$ for all six CpGs, $n = 90$, paired t test) (Fig. 2a). Using the survival data, we examined the association between the survival rate of breast cancer patients after surgery and the DNA methylation of each of the six CpGs on the HOXA9 and HOXA10

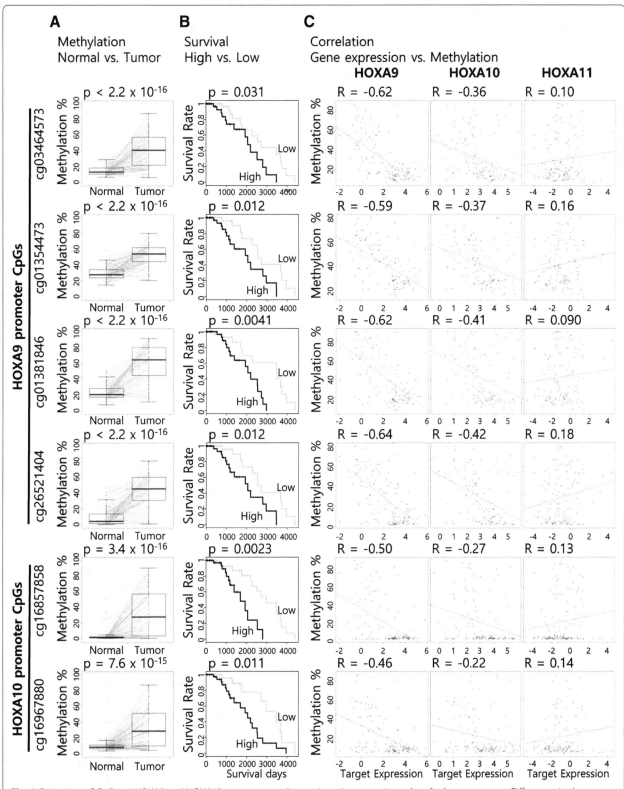

Fig. 2 Estimation of CpGs on HOXA9 and HOXA10 promoters as diagnostic and prognostic markers for breast cancer. **a** Differences in the methylation status of the CpGs between tumor and paired normal tissues (diagnostic value). **b** Association between survival and the CpG methylation status (prognostic value, *black*: high, *gray*: low). **c** Correlation between methylation level of the CpGs and expression level of the HOXA9, HOXA10, and HOXA11 genes

promoters. We divided the patients into two groups based on the median $\Delta\beta$ value. Survival was significantly lower in the patient group with higher methylation percentages than the patient group with lower methylation percentages ($p < 0.05$ for all six CpGs, $n = 82$, log-rank test). These results suggested that the six CpGs on the HOXA9 and HOXA10 promoters show possibility as diagnostic and prognostic markers for breast cancer.

After paring and integrating the TCGA Infinium HM450k and RNA-seq data, we investigated the target gene expression status, survival association, and correlation with neighbor genes. Comparing the expression level of HOXA9, HOXA10, and HOXA11 in tumors and their paired normal samples revealed that the expression of each HOXA9 and HOXA10 gene was significantly lower in the tumor samples than their paired normal samples ($p = 1.3 \times 10^{-14}$ for HOXA9, $p = 7.6 \times 10^{-9}$ for HOXA9, $n = 49$, paired t test) (Additional file 1: Figure S1A). Survival was not significantly associated with the expression of HOXA9, HOXA10, and HOXA11 (Additional file 1: Figure S1B). To define the CpG-target gene relationships, we examined the correlation among the six CpG methylations and the gene expression of HOXA9, HOXA10, and HOXA11 (Pearson correlation, $n = 142$). The methylation of all six CpGs was negatively correlated with the expression of HOXA9 (Fig. 2c). Interestingly, the methylation of the two HOXA10 promoter CpGs showed a higher correlation with HOXA9 expression than HOXA10 expression, whereas the four HOXA9 promoter CpGs highly correlated with HOXA9 expression (Fig. 2c). These results implied that the major target gene of the six CpGs is HOXA9 and that the two HOXA10 promoter CpGs regulate HOXA9 expression through a long-range interaction.

Validation of the long-range interplay between the promoters of HOXA9 and HOXA10

Considering previous analyses, we hypothesized that CpGs on the HOXA9 and HOXA10 promoters interplayed through a long-range interaction. Using a cell-based model, we tried to validate the chromatin interaction between the HOXA9 and HOXA10 promoters. By performing 3C PCR assays in MCF7 and MDA-MB-231 breast cancer cells and MCF10A normal breast cells, we confirmed that the HOXA9 and HOXA10 promoters physically interacted each other, while the HOXA11 promoter did not interact with the HOXA9 or HOXA10 promoter (Fig. 3a and Additional file 1: Figure S2). Next, we tried to investigate whether the gene expression of HOXA9, HOXA10, and HOXA11 could be influenced by methylation status. We treated MCF7 breast cancer cells with 5-aza C during culture. Using a pyrosequencing assay with 5-aza C-treated cells and the controls, we confirmed that 5-aza C treatment decreased the

methylation levels of four CpGs (Fig. 3b). Because we could not design good primers for two CpG sites on the HOXA9 promoter, the methylation of only two HOXA9 promoter CpG sites were validated. Consequently, we examined the gene expression change of HOXA9, HOXA10, and HOXA11 by performing a RT-PCR assay with 5-aza C-treated cells and the controls. We found that the expression of HOXA9 and HOXA10 increased with 5-aza C treatment, while the expression of HOXA11 did not (Fig. 3c). These results suggested that the CpGs on the HOXA9 and HOXA10 promoters regulated the target gene expression levels through long-range interactions.

Prognostic value of HOXA9 and HOXA10 methylation marker combinations

As previously shown, the methylation of each HOXA9 and HOXA10 promoter CpG showed potential as a prognostic marker (Fig. 2b). To increase the prognostic ability, we tried to combine the CpGs and evaluated the prognostic abilities of the combinations. By grouping neighboring CpGs based on the genomic loci, we made two combinations, the HOXA9 promoter CpG (H9) group and the HOXA10 promoter CpG (H10) group. After calculating the risk scores (RSs) of H9 and H10 based on the Cox proportional hazards model, we performed survival analyses with the previously used TCGA paired sample dataset ($n = 82$). We divided the patients into two groups based on the median of the $\Delta\beta$ value. The RS of the separate H9 and H10 combinations significantly predicted poor survival, but it did not better predict poor survival compared with the single CpG methylations shown in Fig. 2b (Fig. 4a top and middle). However, the RS of all CpG (H9 + H10) combinations highly significantly predicted and better predicted poor survival compared with the single CpG methylations shown in Fig. 2b ($p = 1.9 \times 10^{-4}$, $n = 82$, log-rank test) (Fig. 4a bottom).

The TCGA breast cancer dataset contains information about more than paired samples, but they do not have paired normal samples. However, this dataset could be used to evaluate the robustness of the combination of H9 and H10. In the case of this dataset, it is impossible to calculate the $\Delta\beta$ value because of the lack of paired normal samples. Thus, we divided the patient into two groups based on the median of the β value (methylation ratio) of the tumor samples. After the missing data were removed, the Infinium HM450k data and survival data of 781 patients were available (TCGA tumor). Using the dataset, we evaluated the prognostic abilities of the combinations of H9 and H10. Performing the same analysis as for Fig. 4a, we found that the RS of all CpG (H9 + H10) combinations showed better prognostic ability than H9 or H10 alone ($p = 7.1 \times 10^{-5}$, $n = 781$, log-

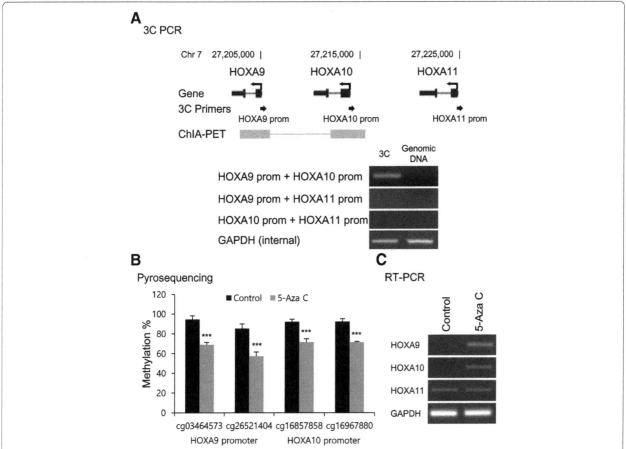

Fig. 3 Validation of the CpG-gene relationship of HOXA9 and HOXA10 promoters. **a** Validation of chromatin interaction between HOXA9 and HOXA10 promoters using 3C PCR assays. **b** Validation of the HOXA9 and HOXA10 promoter CpG methylation decrease by 5-Aza C treatment. **c** Validation of the HOXA9 and HOXA10 gene expression increase by 5-Aza C treatment

rank test) (Fig. 4b). Additionally, we performed survival analysis using another dataset from the NCBI GEO database (GSE39004). For survival analysis, we divided the patients into two groups based on the median of the β value of tumor samples. The RS of all CpG (H9 + H10) combination also significantly predicted poor survival in the GSE39004 dataset ($p = 2.1 \times 10^{-3}$, $n = 62$, log-rank test) (Fig. 4c). Thus, we hypothesize that the combination of HOXA9 and HOXA10 promoter CpG methylation markers are enhanced, robust prognostic markers.

Subtype independency of the combination of HOXA9 and HOXA10 methylation markers

Breast cancer patients are divided into several molecular subtypes based on the expression level of hormone receptors, such as ER positive, PR positive, Her2 positive, and triple negative [2]. Using Infinium HM450k data and clinical information, including molecular subtype markers in the TCGA tumor dataset, we divided the patients into molecular subtype groups. Breast cancer patients can also be divided into several molecular subtypes based on a well-known gene expression signature (PAM50)

[26]. We also divided the patients of the TCGA tumor dataset into PAM50 subtype groups. Then, we examined the association between subtypes and the combination of HOXA9 and HOXA10 promoter CpG methylations (RS). RS also separated the poor survival patients in each subtype except the Her2-positive type in both the TCGA tumor and GSE39004 datasets (Additional file 1: Figure S3). For the Her2-positive type, we observed a similar tendency as for the other subtypes, but this tendency was not significant because of the small number of patients ($n = 46$). To assess the value of the combination of the HOXA9 and HOXA10 promoter CpG methylation markers as a prognostic marker for breast cancer, we performed a multivariate Cox proportional hazards analysis with other subtype markers. For multivariate analysis, we selected patients from TCGA tumor dataset who had a distinct subtype annotation as positive or negative ($n = 249$). The RS better predicted poor survival in the TCGA tumor dataset than for any other marker (Table 1). For validation, we performed the same analysis using the GSE39004 dataset after removing samples with missing values ($n = 58$). The RS also better predicted poor survival in the GSE39004 than any other markers

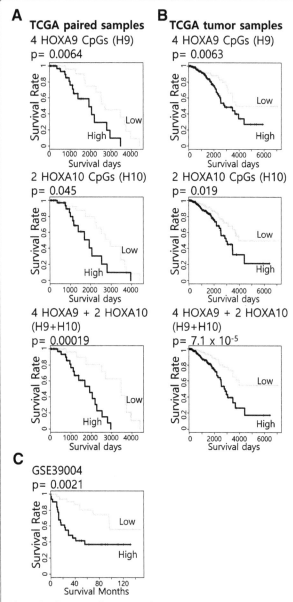 DNA Methylation: Current Research

Fig. 4 Combination of HOXA9 and HOXA10 promoter CpG methylation markers to enhance the prognostic ability. **a** Evaluation of the CpG combinations using the TCGA paired sample dataset. **b** Evaluation of the CpG combinations using the TCGA all sample dataset. **c** Evaluation of the combination of six HOXA9 and HOXA10 promoter CpGs using an independent dataset (NCBI GEO GSE39004) (*black*: high, *gray*: low)

Table 1 Multivariate Cox proportional hazards analysis for the prediction of breast cancer patient survival (TCGA tumor) (HR: hazard ratio, CI: confidence interval)

Variable	Survival	
	HR (95% CI)	*p* value
Risk score (high vs. low)	9.01 (1.59–50.9)	0.0128
HER2 (positive vs. negative)	8.41 (1.39–51.1)	0.0207
Ductal vs. lobular	11.7 (0.00867–0.842)	0.0351
Stage (I, II vs. III, IV)	3.86 (0.815–18.3)	0.0888
PR (positive vs. negative)	15.0 (0.656–344)	0.0898
Age (over vs. under 55)	3.01 (0.793–11.4)	0.105
ER (positive vs. negative)	0.277 (0.0105–7.32)	0.443
PAM50		
Luminal A	Reference	–
Basal	8.23 (0.383–177)	0.178
HER2	6.72 (0.353–128)	0.205
Luminal B	0.726 (0.116–4.54)	0.732

HR hazard ratio, *CI* confidence interval

inhibit the gene expression of the nearest target. Recently, the function of trans or long-range actions of CpG methylation has been revealed by genome-wide scale analyses [19], but distal CpGs are still excluded from methylation marker development because of the difficulty in defining CpG-target gene relationships. The trans or long-range action of the CpG-target gene interaction can be specified by chromatin interactome data, and Pol II ChIA-PET data can specify transcriptional regulation. To define the relationship between CpG and long-range target genes, we used Pol II ChIA-PET data. After testing the diagnostic and prognostic maker value, we selected several long-range CpG-gene interactions that could play an important role in breast cancer tumorigenesis and progression. The representative markers are the CpGs on the HOXA9 and HOXA10 promoters.

Homeobox (HOX) genes are highly conserved gene clusters that encode transcription factors mediating the development process [21–23]. In humans, HOX genes are divided into four clusters, HOXA through HOXD [27]. The HOX gene loci are reported to form three-dimensional

Table 2 Multivariate Cox proportional hazards analysis for the prediction of breast cancer patient survival (GSE39004)

Variable	Survival	
	HR (95% CI)	*p* value
Risk score (high vs. low)	3.23 (1.37–7.60)	0.00721
Age (over vs. under 55)	2.29 (0.938–5.57)	0.0688
ER (positive vs. negative)	0.413 (0.158–1.08)	0.0703
Stage (I, II vs. III, IV)	2.03 (0.840–4.93)	0.115
Triple negative or basal-like (yes vs. no)	1.05 (0.341–3.26)	0.927

HR hazard ratio, *CI* confidence interval

(Table 2). Thus, we suggest that the combination of HOXA9 and HOXA10 methylation markers is an independent prognostic marker for breast cancer.

Discussion

Many CpG methylations have been identified as diagnostic or prognostic markers for cancer [13, 14]. Some of them have been used in the clinical field [11, 12]. The methylation of CpGs on promoters has been known to

nuclear structures through long-range chromatin interactions [28–31]. An association has been reported between the methylation of several CpGs on HOXA loci and breast cancer progression [32, 33]. The expression of many HOX genes is associated with tumorigenesis and cancer progression in various cancers [34–37]. In breast cancer, HOXA9 and HOXA10 act as tumor suppressor genes [21–23]. The methylation of the HOXA9 and HOXA10 promoters is associated with cancer progression in various cancers [24, 25]. These associations imply that studies of HOXA loci will provide good models for the long-range interactions between distal CpG methylation markers and target genes. According to our analysis of the correlation between CpG methylation and target gene expression, HOXA10 promoter CpGs showed a different correlation pattern with HOXA9 promoter CpGs. HOXA9 promoter CpG methylation tended to correlate with HOXA9 gene expression, but HOXA10 promoter CpG methylation tended to correlate with long-range HOXA9 gene expression rather than the expression of the nearer HOXA10 gene. In other words, the HOXA10 promoter had an enhancer-like function, whereas the HOXA9 promoter had a promoter function. We suggest that the promoter-promoter interaction between HOXA9 and HOXA10 is important in breast cancer progression through the enhancer-like action of HOXA10 promoter CpGs.

To select both diagnostic and prognostic markers, we started marker selection from paired samples. In the case of the expression level of HOXA9 and HOXA10, there was the possibility of only a diagnostic marker but not a prognostic marker. This result could be caused by the instability of RNA markers. The HOXA9 and HOXA10 promoter methylation markers showed potential as both diagnostic and prognostic markers. The initial selection was performed based on the $\Delta\beta$ value, and the combination of the six HOXA9 and HOXA10 promoter methylation markers showed a highly significant prognostic value. Based on the tumor β value, the combination also showed a highly significant prognostic value in two independent datasets. Molecular markers that are clinically used can be applied to specific subtypes, such as node-negative and ER-positive breast cancer [4, 5]. The combination of the HOXA9 and HOXA10 promoter methylation markers showed a subtype-independent effect. Multivariate analysis indicated that the combination acted like an independent variable in the prediction of the prognosis of breast cancer. Thus, we suggest that the long-range interplay of HOXA9 and HOXA10 promoter CpGs is an efficient and robust methylation marker for breast cancer.

Molecular markers using single gene expression or CpG methylation have been identified [11], but more marker panels of multiple genes or CpGs have been developed due to advantages in efficiency and robustness [4, 5]. Many multiple marker panels have been identified by unsupervised methods without considering biological mechanisms [12–14]. The HOXA9 and HOXA10 promoter CpGs were identified by a supervised method based on long-range chromatin interactions, and the detailed action is specified in the HOXA9 and HOXA10 transcriptional regulation module. With respect to translational and biological relevance, this method has advantages for specifying therapeutic targets and strategies. We suggest that our method provides a comprehensive and novel approach for the development of molecular markers for personalized medicine and facilitating the precise determination of cancer prognosis.

Conclusions

Breast cancer is both the most common cancer and the most frequent cause of cancer-related deaths among women. Mammography and some molecular markers have been used for its diagnosis and prognosis, but these techniques have limitations in premenopausal breast cancer or specific subtypes. In this study, we show that a combination of HOXA9 and HOXA10 promoter methylation markers is significantly associated with the prognosis of breast cancer patients in independent datasets and compose a transcriptional regulation module through long-range chromatin interactions. In contrast to other clinically used methylation markers applied to specific subtypes, the combination of the HOXA9 and HOXA10 promoter methylation markers showed a subtype-independent manner. Therefore, we suggest that the prognostic model using the HOXA9 and HOXA10 promoter CpG combination has translational potential to facilitate determination of breast cancer prognosis and therapeutic strategies targeting a specific molecular regulation module.

Abbreviations

3C: Chromosome conformation capture; 5-aza C: 5-azacytidine; affyU133P2: Affymetrix Human Genome U133 Plus 2.0 microarray; ATCC: American Type Culture Collection; ChIA-PET: Chromatin interaction analysis by paired-end tag sequencing; CpG: A region of DNA where a cytosine nucleotide is followed by a guanine nucleotide; DMEM: Dulbecco's modified Eagle medium; DMSO: Dimethyl sulfoxide; ENCODE: Encyclopedia of DNA Elements; ER: Estrogen receptor; GAPDH: Glyceraldehyde 3-phosphate dehydrogenase gene; GEO: Gene Expression Omnibus; Her2: Human epidermal growth factor receptor 2; HOXA10: Homeobox gene A 10; HOXA11: Homeobox gene A 11; HOXA9: Homeobox gene A 9; ICGC: International Cancer Genome Consortium; Infinium HM450k: Illumina Infinium Human Methylation 450 k BeadChip microarray; NCBI: National Center for Biotechnology Information; Pol

II: RNase polymerase 2; PR: Progesterone receptor; RMA: Robust Multi-array Average; RNA-seq: High-throughput RNA sequencing; RPKM: Reads per kilobase per million mapped reads; RS: Risk score; TCGA: The Cancer Genome Atlas; TSS: Transcription start site

Acknowledgements
Not applicable.

Funding
This work was supported by grants from the National Cancer Center (NCC-1611800, NCC-1410300, NCC-1710260).

Authors' contributions
YJK conceived the study. SMP, EYC, and YJK designed the experiments. EYC performed the experiments. SMP, JKC, and MB analyzed the data, and SMP and YJK wrote the manuscript. All authors read and approved the final manuscript.

Consent for publication
Not applicable.

Competing interests
The authors declare that they have no competing interests.

Author details
[1]Translational Research Branch, Research Institute, National Cancer Center, Goyang, Gyeonggi 10408, Republic of Korea. [2]Personalized Genomic Medicine Research Center, KRIBB, Daejeon 34141, Republic of Korea. [3]Department of Bio and Brain Engineering, KAIST, Daejeon 34141, Republic of Korea.

References
1. Siegel RL, Miller KD, Jemal A. Cancer statistics, 2015. CA Cancer J Clin. 2015;65(1):5–29.
2. Fu Y, Zhuang Z, Dewing M, Apple S, Chang H. Predictors for contralateral prophylactic mastectomy in breast cancer patients. Int J Clin Exp Pathol. 2015;8(4):3748–64.
3. Elmore JG, Barton MB, Moceri VM, Polk S, Arena PJ, Fletcher SW. Ten-year risk of false positive screening mammograms and clinical breast examinations. N Engl J Med. 1998;338(16):1089–96.
4. Paik S, Tang G, Shak S, Kim C, Baker J, Kim W, Cronin M, Baehner FL, Watson D, Bryant J, et al. Gene expression and benefit of chemotherapy in women with node-negative, estrogen receptor-positive breast cancer. J Clin Oncol. 2006;24(23):3726–34.
5. Wittner BS, Sgroi DC, Ryan PD, Bruinsma TJ, Glas AM, Male A, Dahiya S, Habin K, Bernards R, Haber DA, et al. Analysis of the MammaPrint breast cancer assay in a predominantly postmenopausal cohort. Clin Cancer Res. 2008;14(10):2988–93.
6. Laird PW. The power and the promise of DNA methylation markers. Nat Rev Cancer. 2003;3(4):253–66.
7. Esteller M. Epigenetics in cancer. N Engl J Med. 2008;358(11):1148–59.
8. Jones PA, Laird PW. Cancer epigenetics comes of age. Nature Genet. 1999;21(2):163–7.
9. Lehmann U, Langer F, Feist H, Glockner S, Hasemeier B, Kreipe H. Quantitative assessment of promoter hypermethylation during breast cancer development. Am J Pathol. 2002;160(2):605–12.
10. Holst CR, Nuovo GJ, Esteller M, Chew K, Baylin SB, Herman JG, Tlsty TD. Methylation of p16(INK4a) promoters occurs in vivo in histologically normal human mammary epithelia. Cancer Res. 2003;63(7):1596–601.
11. Church TR, Wandell M, Lofton-Day C, Mongin SJ, Burger M, Payne SR, Castanos-Velez E, Blumenstein BA, Rosch T, Osborn N, et al. Prospective evaluation of methylated SEPT9 in plasma for detection of asymptomatic colorectal cancer. Gut. 2014;63(2):317–25.
12. Brock MV, Hooker CM, Ota-Machida E, Han Y, Guo M, Ames S, Glockner S, Piantadosi S, Gabrielson E, Pridham G, et al. DNA methylation markers and early recurrence in stage I lung cancer. N Engl J Med. 2008;358(11):1118–28.
13. Muller HM, Widschwendter A, Fiegl H, Ivarsson L, Goebel G, Perkmann E, Marth C, Widschwendter M. DNA methylation in serum of breast cancer patients: an independent prognostic marker. Cancer Res. 2003;63(22):7641–5.
14. Widschwendter M, Siegmund KD, Muller HM, Fiegl H, Marth C, Muller-Holzner E, Jones PA, Laird PW. Association of breast cancer DNA methylation profiles with hormone receptor status and response to tamoxifen. Cancer Res. 2004;64(11):3807–13.
15. Klose RJ, Bird AP. Genomic DNA methylation: the mark and its mediators. Trends Biochem Sci. 2006;31(2):89–97.
16. Cedar H, Bergman Y. Programming of DNA methylation patterns. Annu Rev Biochem. 2012;81:97–117.
17. Kurukuti S, Tiwari VK, Tavoosidana G, Pugacheva E, Murrell A, Zhao Z, Lobanenkov V, Reik W, Ohlsson R. CTCF binding at the H19 imprinting control region mediates maternally inherited higher-order chromatin conformation to restrict enhancer access to Igf2. Proc Natl Acad Sci U S A. 2006;103(28):10684–9.
18. Court F, Camprubi C, Garcia CV, Guillaumet-Adkins A, Sparago A, Seruggia D, Sandoval J, Esteller M, Martin-Trujillo A, Riccio A, et al. The PEG13-DMR and brain-specific enhancers dictate imprinted expression within the 8q24 intellectual disability risk locus. Epigenetics Chromatin. 2014;7(1):5.
19. Aran D, Sabato S, Hellman A. DNA methylation of distal regulatory sites characterizes dysregulation of cancer genes. Genome Biol. 2013;14(3):R21.
20. Kieffer-Kwon KR, Tang Z, Mathe E, Qian J, Sung MH, Li G, Resch W, Baek S, Pruett N, Grontved L, et al. Interactome maps of mouse gene regulatory domains reveal basic principles of transcriptional regulation. Cell. 2013; 155(7):1507–20.
21. Chu MC, Selam FB, Taylor HS. HOXA10 regulates p53 expression and matrigel invasion in human breast cancer cells. Cancer Biol Ther. 2004; 3(6):568–72.
22. Chen Y, Zhang J, Wang H, Zhao J, Xu C, Du Y, Luo X, Zheng F, Liu R, Zhang H, et al. miRNA-135a promotes breast cancer cell migration and invasion by targeting HOXA10. BMC Cancer. 2012;12:111.
23. Sun M, Song CX, Huang H, Frankenberger CA, Sankarasharma D, Gomes S, Chen P, Chen J, Chada KK, He C, et al. HMGA2/TET1/HOXA9 signaling pathway regulates breast cancer growth and metastasis. Proc Natl Acad Sci U S A. 2013;110(24):9920–5.
24. Yoshida H, Broaddus R, Cheng W, Xie S, Naora H. Deregulation of the HOXA10 homeobox gene in endometrial carcinoma: role in epithelial-mesenchymal transition. Cancer Res. 2006;66(2):889–97.
25. Hwang JA, Lee BB, Kim Y, Hong SH, Kim YH, Han J, Shim YM, Yoon CY, Lee YS, Kim DH. HOXA9 inhibits migration of lung cancer cells and its hypermethylation is associated with recurrence in non-small cell lung cancer. Mol Carcinog. 2015;54(Suppl 1):E72–80.
26. Parker JS, Mullins M, Cheang MC, Leung S, Voduc D, Vickery T, Davies S, Fauron C, He X, Hu Z, et al. Supervised risk predictor of breast cancer based on intrinsic subtypes. J Clin Oncol. 2009;27(8):1160–7.
27. Apiou F, Flagiello D, Cillo C, Malfoy B, Poupon MF, Dutrillaux B. Fine mapping of human HOX gene clusters. Cytogenet Cell Genet. 1996;73(1–2):114–5.
28. Lee JY, Min H, Wang X, Khan AA, Kim MH. Chromatin organization and transcriptional activation of Hox genes. Anat Cell Biol. 2010;43(1):78–85.
29. Acemel RD, Tena JJ, Irastorza-Azcarate I, Marletaz F, Gomez-Marin C, de la Calle-Mustienes E, Bertrand S, Diaz SG, Aldea D, Aury JM, et al. A single three-dimensional chromatin compartment in amphioxus indicates a stepwise evolution of vertebrate Hox bimodal regulation. Nat Genet. 2016;48(3):336–41.
30. Min H, Kong KA, Lee JY, Hong CP, Seo SH, Roh TY, Bae SS, Kim MH. CTCF-mediated chromatin loop for the posterior Hoxc gene expression in MEF cells. IUBMB Life. 2016;68(6):436–44.
31. Buxa MK, Slotman JA, van Royen ME, Paul MW, Houtsmuller AB, Renkawitz R. Insulator speckles associated with long-distance chromatin contacts. Biol Open. 2016;5(9):1266–74.
32. Pilato B, Pinto R, De Summa S, Lambo R, Paradiso A, Tommasi S. HOX gene methylation status analysis in patients with hereditary breast cancer. J Hum Genet. 2013;58(1):51–3.

33. Park SY, Kwon HJ, Lee HE, Ryu HS, Kim SW, Kim JH, Kim IA, Jung N, Cho NY, Kang GH. Promoter CpG island hypermethylation during breast cancer progression. Virchows Arch. 2011;458(1):73–84.
34. Henderson GS, van Diest PJ, Burger H, Russo J, Raman V. Expression pattern of a homeotic gene, HOXA5, in normal breast and in breast tumors. Cell Oncol. 2006;28(5–6):305–13.
35. Raman V, Martensen SA, Reisman D, Evron E, Odenwald WF, Jaffee E, Marks J, Sukumar S. Compromised HOXA5 function can limit p53 expression in human breast tumours. Nature. 2000;405(6789):974–8.
36. Svingen T, Tonissen KF. Altered HOX gene expression in human skin and breast cancer cells. Cancer Biol Ther. 2003;2(5):518–23.
37. Wu X, Chen H, Parker B, Rubin E, Zhu T, Lee JS, Argani P, Sukumar S. HOXB7, a homeodomain protein, is overexpressed in breast cancer and confers epithelial-mesenchymal transition. Cancer Res. 2006;66(19):9527–34.

11

Epigenome-wide association study of DNA methylation in panic disorder

Mihoko Shimada-Sugimoto[1], Takeshi Otowa[2*], Taku Miyagawa[1,3], Tadashi Umekage[4], Yoshiya Kawamura[5], Miki Bundo[6], Kazuya Iwamoto[6], Mamoru Tochigi[7], Kiyoto Kasai[8], Hisanobu Kaiya[9], Hisashi Tanii[10], Yuji Okazaki[11], Katsushi Tokunaga[1] and Tsukasa Sasaki[12]

Abstract

Background: Panic disorder (PD) is considered to be a multifactorial disorder emerging from interactions among multiple genetic and environmental factors. To date, although genetic studies reported several susceptibility genes with PD, few of them were replicated and the pathogenesis of PD remains to be clarified. Epigenetics is considered to play an important role in etiology of complex traits and diseases, and DNA methylation is one of the major forms of epigenetic modifications. In this study, we performed an epigenome-wide association study of PD using DNA methylation arrays so as to investigate the possibility that different levels of DNA methylation might be associated with PD.

Methods: The DNA methylation levels of CpG sites across the genome were examined with genomic DNA samples (PD, $N = 48$, control, $N = 48$) extracted from peripheral blood. Methylation arrays were used for the analysis. β values, which represent the levels of DNA methylation, were normalized via an appropriate pipeline. Then, β values were converted to M values via the logit transformation for epigenome-wide association study. The relationship between each DNA methylation site and PD was assessed by linear regression analysis with adjustments for the effects of leukocyte subsets.

Results: Forty CpG sites showed significant association with PD at 5% FDR correction, though the differences of the DNA methylation levels were relatively small. Most of the significant CpG sites (37/40 CpG sites) were located in or around CpG islands. Many of the significant CpG sites (27/40 CpG sites) were located upstream of genes, and all such CpG sites with the exception of two were hypomethylated in PD subjects. A pathway analysis on the genes annotated to the significant CpG sites identified several pathways, including "positive regulation of lymphocyte activation."

Conclusions: Although future studies with larger number of samples are necessary to confirm the small DNA methylation abnormalities associated with PD, there is a possibility that several CpG sites might be associated, together as a group, with PD.

Keywords: DNA methylation, Panic disorder, Epigenome-wide association study, Epigenetics, Psychiatric disorder

* Correspondence: totowa-psy@umin.org
[2]Graduate School of Clinical Psychology, Teikyo Heisei University Major of Professional Clinical Psychology, 2-51-4 Higashiikebukuro, Toshima Ward, Tokyo 171-0014, Japan
Full list of author information is available at the end of the article

Background

Panic disorder (PD) is a major anxiety disorder characterized by recurrent unexpected panic attacks and anticipatory anxiety. According to previous twin and family studies [1–3], PD is considered to be a multifactorial disorder emerging from the interactions between multiple genetic and environmental factors. Recently, genome-wide association studies (GWASs), whole-exome sequencing and meta- analyses were performed [4–8] and identified transmembrane protein 132D (*TMEM132D*) and catechol-O-methyltransferase (*COMT*) as PD susceptibility genes [4, 5]. However, there would be other genetic factors associated with PD.

Epigenetics is one of the biological fields that is considered to play an important role in the etiology of complex diseases [9]. The term "epigenetics" is now generally understood to refer to potentially heritable and functionally relevant to gene expression and chromatin structure with no changes to genetic sequences [9, 10]. DNA methylation is one of the major forms of epigenetic modifications that was found to play important roles in the context of gene regulation [11]. Moreover, a part of the DNA methylation is reported to be involved in the pathogenesis of psychiatric disorders, including anxiety disorders [12, 13].

Previous studies on DNA methylation in anxiety disorders have focused mainly on candidate genes that were reported to be involved in the stress response, neurotransmission, and neuroplasticity [14]. One recent study conducted in social anxiety disorder (SAD) patients reported an association between SAD and oxytocin receptor (*OXTR*) gene hypomethylation [15]. Hypomethylation of the promoter and intron 2 region in another candidate gene, glutamate-decarboxylase 1 (*GAD1*), was also reported in PD patients [16]. Another preliminary study showed that CpG sites in monoamine oxidase A (*MAOA*) were significantly less methylated in PD patients than in healthy controls [17, 18] and negative life events were associated with this lower level of DNA methylation [18]. In another study, solute carrier family 6, member 4 (*SLC6A4*) and serotonin transporter (*SERT*) were examined in children with anxiety disorders before and after cognitive behavior therapy; a DNA methylation change in *SLC6A4* was related to response to the psychological therapy, as responders had increased *SLC6A4* methylation [19]. Methylation of another neurotransmitter transporter, noradrenaline transporter (*NET*), also known as solute carrier family 6, member 2 (*SLC6A2*), was also studied in subjects with PD and hypertension; results showed that DNA hypermethylation in the promoter region of *NET* caused *NET* gene silencing through the binding of methyl-CpG binding protein 2 (MeCP2), a methylation-related inhibitory transcription factor [20, 21]. However, results

from another study did not support the finding of significant changes in *SLC6A2* promoter methylation in the patients with PD or major depressive disorder [22]. Overall, the results from such previous studies suggest the importance of DNA methylation abnormalities in the pathogenesis of PD, although the number of studies and the sample sizes have been limited and most of the findings have not been confirmed in replication studies.

Recently, a genome-wide approach has enabled the examination of DNA methylation patterns without any prior information. A methylation array (Infinium® Human Methylation 450 K BeadChip, Illumina Inc., San Diego, CA, USA) can simultaneously detect the DNA methylation status of more than 480,000 cytosine residues across the genome. This array has been used to successfully identify DNA methylation marks related to aging [23, 24], leukocyte subsets [25], smoking [26], and disease outcomes [27, 28]. As far as we know, there has been no report of the examination of genome-wide DNA methylation patterns in PD.

In this study, we performed an epigenome-wide association study (EWAS) of PD using the array technology. DNA samples extracted from peripheral blood were utilized. Although an EWAS using brain tissue would be more appropriate for identifying disease-associated differentially methylated positions (DMPs), peripheral blood is more accessible and might enable the development of diagnostic biomarkers. Here, we examined the genome-wide DNA methylation profiles of 48 PD subjects and 48 age- and sex-matched control subjects to investigate aberrant differences in DNA methylation that are related to PD.

Results

Quality check of the DNA methylation array data

In the DNA methylation array analysis, each probe signal for a sample had a detection P value calculated as the probability that a target signal is distinguishable from the negative controls to show the overall probe performance. We confirmed that more than 99% of all probes in all samples had a detection P value ≤0.05, showing that the overall performance of the assay was high. Principal component analysis using probes on the X chromosome was performed to predict the gender of samples in this study. The result showed that all samples were correctly labeled in the gender groups (Additional file 1: Figure S1). Density plots of the β values were prepared from the raw data of each sample for a visual inspection. All plots showed a standard bimodal distribution of the β values (Additional file 1: Figure S2) with the same characteristics of the distribution described in a previous study [29]. The distribution of DNA methylation was bimodal with a minority of probes showing intermediate DNA methylation levels. The DNAm age was estimated using the results of approximately 350 probes, and the Pearson's correlation coefficients between the estimated DNAm age and chronological age were calculated to be 0.77 and 0.90 in PD and control groups, respectively, which is considered to support the data quality of this method (Additional file 1: Figure S3).

Prediction of the distribution of leukocyte subsets

The proportions of leukocyte subsets (natural killer cells, B cells, CD4$^+$ T cells, CD8$^+$ T cells, monocytes, and granulocytes) were estimated from the DNA methylation array data using a published algorithm [30]. Wilcoxon's rank-sum tests using the estimated proportion of each leukocyte subset were performed to examine whether the compositions of leukocyte cells differed between the PD and control subjects. The proportion of CD4$^+$ T cells was significantly higher in the PD subjects than in the control subjects ($P = 0.0034$) (Fig. 1). To further interpret the result of the prediction of leukocyte subsets, the abundance measures of plasmablasts, CD8$^+$CD28$^-$ CD45RA$^-$ T cells, naive CD8$^+$T cells, and naive CD4$^+$ T cells were estimated and compared between PD and control subjects, resulting that no significant difference was observed (Additional file 1: Figure S4).

EWAS of PD and control subjects

With the DNA methylation array experiment, the methylation status of a total of 485,512 cytosine residues were examined. We filtered out the low quality

probes and those on sex chromosomes, and finally, 376,602 probes remained (Additional file 1: Figure S5). The data were normalized via the pipeline, Lumi: QN + BMIQ + ComBat. We then performed an EWAS of PD. The Q-Q plot was showed in Additional file 1: Figure S6. After excluding three possible cross-reactive probes, 40 probes showed significant association with PD when the false discovery rate (FDR) was set to 5% (Table 1, Fig. 2). The most significant probe, cg25270498, was located upstream of the meteorin, glial cell differentiation regulator-like (*METRNL*) gene and was significantly hypomethylated in PD patients (FDR q value = 1.19×10^{-4}), followed by cg05910615 (*HSPB6*; *C19orf55*, FDR q value = 1.19×10^{-4}) and cg20340149 (*CLASP1*, FDR q value = 8.64×10^{-4}). At many of the CpG sites with significantly different levels of DNA methylation, the cytosine residues were less methylated in the PD subjects than in the control subjects (Fig. 2). Only eight CpG sites were found to be significantly more methylated in the PD subjects than in the control subjects. Most of the significant CpG sites (37/40 CpG sites) were located in CpG island or CpG island shore and shelf that span up to 2 kb and 2–4 kb from CpG islands, respectively (Table 1). Many of them (27/40 CpG sites) were located upstream (within 1500 bp from the transcription start site, the 5′ untranslated region and the first exon) of genes, and these 27 CpG sites other than two were all hypomethylated in PD subjects (Table 1).

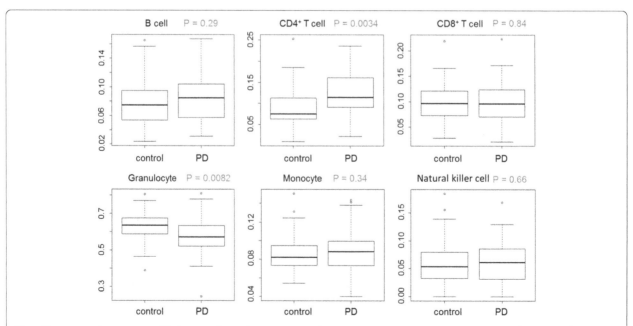

Fig. 1 The estimated proportions of leukocyte subsets. The proportions of leukocyte subsets (natural killer cell, B cell, CD4$^+$ T cell, CD8$^+$ T cell, monocyte, and granulocyte) were estimated using the results of the DNA methylation array. Wilcoxon rank-sum tests using estimated proportions of leukocyte subsets were performed between the PD and control subjects. *P* values are indicated in *blue characters*. Significance level after the Bonferroni correction was set as $\alpha = 0.005$

Table 1 Probes with significant differential DNA methylation status between PD and control

CHR	Position (hg19)	Target ID	Mean β value		Mean Adjusted M value				Genes	Location with respect to genes	Relation to CpG Islad[a]
			PD	Control	PD	Control	P value	q value			
17	81037414	cg25270498	0.257	0.283	-1.71	-1.53	5.67×10^{-10}	1.19×10^{-4}	METRNL	TSS200	Island
19	36248877	cg05910615	0.222	0.251	-2.05	-1.81	6.30×10^{-10}	1.19×10^{-4}	HSPB6;C19orf55	TSS1500;TSS200	N_Shore
2	122407145	cg20340149	0.152	0.175	-2.73	-2.49	6.88×10^{-9}	8.64×10^{-4}	CLASP1	TSS200	Island
13	80055594	cg14777817	0.134	0.153	-2.90	-2.69	5.17×10^{-8}	4.87×10^{-3}	NDFIP2	1stExon	Island
10	135088451	cg25526061	0.147	0.164	-2.79	-2.56	1.30×10^{-7}	7.64×10^{-3}	ADAM8	Body	N_Shore
17	27224823	cg04266864	0.133	0.151	-2.97	-2.73	1.42×10^{-7}	7.64×10^{-3}	FLOT2;DHRS13	TSS200;3'UTR	Island
16	12142335	cg10475689	0.606	0.576	0.82	0.63	2.09×10^{-7}	9.38×10^{-3}	RUNDC2A	Body	
16	23568708	cg05742564	0.192	0.215	-2.30	-2.08	2.48×10^{-7}	9.38×10^{-3}	UBFD1;EARS2	TSS200;TSS200	Island
1	228604037	cg02931001	0.263	0.280	-1.62	-1.49	2.49×10^{-7}	9.38×10^{-3}	TRIM17	5'UTR	Island
13	28024472	cg08209163	0.147	0.165	-2.73	-2.54	3.45×10^{-7}	0.0118	MTIF3	TSS200;5'UTR	Island
12	50017361	cg10727759	0.147	0.165	-2.74	-2.54	4.51×10^{-7}	0.0139	PRPF40B	TSS200	Island
22	24236284	cg12738349	0.290	0.304	-1.40	-1.30	4.80×10^{-7}	0.0139	MIF	TSS1500	Island
3	197409980	cg08942682	0.830	0.848	2.06	2.28	6.05×10^{-7}	0.0163	KIAA0226	Body	
6	170597377	cg05228964	0.602	0.580	0.71	0.59	7.23×10^{-7}	0.0172	DLL1	Body	Island
12	4381997	cg08553284	0.316	0.330	-1.20	-1.11	7.57×10^{-7}	0.0172	CCND2	TSS1500	Island
1	155164676	cg03425468	0.215	0.240	-2.10	-1.88	7.78×10^{-7}	0.0172	MIR92B	TSS1500	Island
22	38202626	cg11029475	0.202	0.219	-2.23	-2.02	8.43×10^{-7}	0.0176	GCAT;H1F0;H1F0	TSS1500;1stExon;3'UTR	N_Shore
16	2732724	cg02205746	0.295	0.311	-1.36	-1.26	9.40×10^{-7}	0.0182	KCTD5	1stExon	Island
17	44270511	cg10256219	0.116	0.103	-2.71	-2.93	9.64×10^{-7}	0.0182			Island
3	50375496	cg09386807	0.236	0.256	-1.86	-1.70	1.14×10^{-6}	0.0204	RASSF1	TSS1500;Body;5'UTR	Island
8	28243934	cg13411962	0.168	0.181	-2.46	-2.31	1.25×10^{-6}	0.0211	ZNF395	5'UTR;1stExon	Island
4	4861398	cg01959412	0.325	0.339	-1.17	-1.07	1.29×10^{-6}	0.0211	MSX1	5'UTR;1stExon	Island
11	61197477	cg03342113	0.264	0.278	-1.58	-1.48	1.53×10^{-6}	0.0230	CPSF7;SDHAF2	TSS200;TSS200	Island
17	42293627	cg24247482	0.556	0.557	0.50	0.39	1.53×10^{-6}	0.0230	UBTF	Body	N_Shelf
17	80189962	cg17932802	0.358	0.366	-0.92	-0.85	1.65×10^{-6}	0.0239	SLC16A3	TSS200;5'UTR	Island
1	204159498	cg13065121	0.233	0.250	-1.89	-1.73	1.84×10^{-6}	0.0253	KISS1	3'UTR	N_Shore
6	32055370	cg26997880	0.225	0.247	-1.98	-1.79	1.95×10^{-6}	0.0253	TNXB	Body	Island
12	121148158	cg19464320	0.123	0.107	-2.58	-2.83	2.26×10^{-6}	0.0284	UNC119B	1stExon;5'UTR	N_Shore
9	87284706	cg13965062	0.230	0.243	-1.85	-1.75	2.39×10^{-6}	0.0290	NTRK2	5'UTR;1stExon	Island
1	245316477	cg07124903	0.102	0.088	-2.88	-3.14	2.47×10^{-6}	0.0290			N_Shore
19	11074303	cg08315613	0.574	0.570	0.59	0.48	2.54×10^{-6}	0.0290	SMARCA4	5'UTR	S_Shore
2	242254519	cg13009927	0.196	0.214	-2.22	-2.05	3.28×10^{-10}	0.0353	SEPT2;HDLBP	TSS1500;5'UTR	Island
2	217559020	cg03222971	0.233	0.251	-1.90	-1.73	3.42×10^{-6}	0.0357	IGFBP5	Body	N_Shore
1	206223719	cg26795730	0.196	0.215	-2.23	-2.05	3.53×10^{-6}	0.0357	AVPR1B	TSS1500	Island
2	73144353	cg15921587	0.302	0.318	-1.33	-1.21	3.69×10^{-6}	0.0357	EMX1	TSS1500	Island
13	25621328	cg18098400	0.132	0.150	-2.96	-2.74	3.83×10^{-6}	0.0357			Island
15	66993412	cg25048202	0.290	0.306	-1.41	-1.30	3.84×10^{-6}	0.0357	SMAD6	TSS1500	N_Shore
2	27718181	cg04015759	0.250	0.267	-1.71	-1.58	3.89×10^{-6}	0.0357	FNDC4	TSS200	Island
15	101690195	cg24378951	0.753	0.728	1.80	1.61	4.53×10^{-6}	0.0406			
2	44059266	cg15889012	0.194	0.213	-2.25	-2.07	4.74×10^{-6}	0.0415	ABCG5	Body	S_Shore

Abbreviation: CHR chromosome
TSS, transcription start site
UTR, untranslated region
[a] Each category of "Relation to CpG island" column defines the following regions: Island, CpG island; N_Shore, 0-2 kb upstream of CpG island; S_Shore, 0-2 kb downstream of CpG island; N_Shelf, 2–4 kb upstream of CpG island

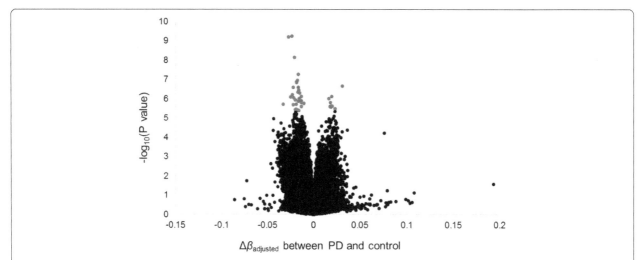

Fig. 2 Results of the EWAS comparing between the PD and control subjects. Log-transformed *P* values of all the probes were plotted. The *horizontal axis* represents average adjusted β value differences ($\Delta\beta_{adjusted}$ = average $\beta_{adjusted\ (PD)}$ − average $\beta_{adjusted\ (control)}$) between PD and control subjects. Significant probes at 5% FDR correction are shown in *red dots*

For confirmation, we examined the possibility that the significant associations were influenced by smoking status. We checked the distributions of adjusted *M* values of the significant CpG sites between smokers and non-smokers among the PD subjects and found that there was no effect of smoking for these sites in PD subjects (Additional file 1: Figure S6).

Pathway analysis

To assess the overall influence of the significant differences in DNA methylation between the PD and control subjects, a pathway analysis was performed. Annotation information on the 40 significantly associated CpG sites (Table 1) was used for the analysis; in total, 42 genes were annotated to the CpG sites. Three gene sets showed significant associations at a FDR of 5% (Table 2) after we excluded pathways that showed no association in the GOseq pathway analysis in which gene length was taken into consideration. The identified pathways included the "positive regulation of lymphocyte activation" gene set.

Discussion

In this study, we performed an EWAS of PD using a DNA methylation array and examined the genome-wide

DNA methylation profiles of PD for the first time, as far as we know, although replication is necessary in future studies. This array technology can target 99% of genes and 95% of CpG island regions [31] and enables us the analysis of DNA methylation status in a genome-wide manner [32]. Recently, a number of studies have employed this platform to identify differentially DNA methylation sites according to phenotypes. In particular, in the cancer field, this platform has been used to identify a number of DMPs accompanying large β value differences in cancer cells ($\geq$0.2) [28]. DNA methylation is considered to change according to environmental factors [33, 34]; as such, in psychiatric disorders, it was predicted that the DNA methylation levels at specific sites would differ between patients and healthy subjects [12]. However, in the present study, no DMP showed a large β value difference ($\geq$0.2) between the PD and control subjects.

In the psychiatric field, epigenetics has been considered to play a role in disease pathogenesis and several recent studies have examined the relationships between DNA methylation and psychiatric disorders in a genome-wide manner. For example, an EWAS of major depressive disorder identified more than 350 CpG sites that were associated with the disease and all of these CpG sites were hypomethylated in the major depressive disorder patients8

Table 2 Result of the pathway analysis

Gene sets	Number of genes in pathways	P value	FDR	Number of genes in the data	Associated genes in the data
Epidermis development	426	1.5×10^{-4}	0.017	5	*DLL1, FLOT2, IGFBP5, MIF, SMARCA4*
Positive regulation of cell cycle	493	3.3×10^{-4}	0.019	5	*CCND2, AVPR1B, MIF, RASSF1, MSX1*
Positive regulation of lymphocyte activation	469	1.9×10^{-3}	0.037	5	*MIF, ADAM8, IGFBP5, AVPR1B, FLOT2*

FDR false discovery rate

[35] Other EWASs of suicidal behavior or early life stress-associated depression found that the DNA methylation status differed globally between the patients and control subjects [36–38]. In addition, EWASs of schizophrenia identified numerous DMPs associated with the disease; a part of these DMPs were annotated in gene regions previously reported as candidate genes of schizophrenia or psychiatric diseases [39–42]. The results of these previous studies suggest the possibility that in psychiatric disorders, multiple DMPs with small effects (β value difference ≤ 0.1) might be associated with diseases together. This would be one explanation for an inflation of P values of the regression analysis observed in this study (Additional file 1: Figure S6), although it cannot be denied that several factors other than age, sex ,and proportions of leukocyte subsets potentially confounded the result.

In this study, 40 CpG sites were found to be significantly associated with PD. Among these, 27 CpG sites were located upstream of genes and with the exception of two CpG sites, they were all hypomethylated in PD subjects when compared with control subjects. According to previous studies, DNA hypomethylation of upstream gene regions is often associated with a higher level of gene expression [11, 43, 44]. Therefore, such DNA methylation differences upstream of genes may be related to a higher expression level of the annotated genes. We further performed pathway analyses to evaluate the overall influence of the DNA methylation differences. Among the detected gene sets, "positive regulation of lymphocyte activation," which reflected the significant probes annotated to MIF or ADAM8, IGFBP5, AVPR1B, and FLOT2, was particularly intriguing. We previously reported the associations of the immune pathways and the specific HLA allele, HLA-DRB1*13:02, with PD [45]. Furthermore, the most significant probe, cg25270498, located upstream of METRNL, which was reported to possibly act as a cytokine and to exert effects on immune process [46].

In the current study, we also examined the DNA methylation around candidate genes of which DNA methylation statuses were previously reported to be associated with anxiety disorders. The gene regions of OXTR, GAD1, SLC6A4, SLC6A2, and MAOA were individually examined. As a result, with our sample set, no significant differences between PD and healthy subjects were observed in these candidate gene regions (Additional file 2: Tables S1–S5). None of the genes annotated to the significant CpG sites in this study have been identified in previous studies of PD, including GWASs [8, 47]. This might have been partly due to the relatively small sizes of the samples analyzed, or because the CpGs examined using the array were not sufficient in density. Therefore, future detailed studies with larger samples are necessary to further investigate the relationship between the PD candidate genes and DNA methylation.

The proportions of leukocyte subset estimated with the DNA methylation array data were compared between the PD and healthy subjects in this study. As a result, a higher proportion of CD4+ T cells ($P = 0.0034$) was found in the PD subjects. A higher tendency of the abundance measure of naive CD4+ T cell was also observed in PD subjects. As major histocompatibility complex class II molecules, including HLA-DR, interact mainly with CD4+ T cells, an increase in CD4+ T cells might be a part of an immune abnormality in PD patients. However, results of previous studies on surface immune phenotypes of lymphocytes did not report consistent results on the up- and down-regulation of leukocyte subsets [48–51]. Additionally, lymphocytes can be affected by environmental factors and/or infection status. Further, the comparisons for leukocyte subsets in this study were not based on cell count or abundance and they were not independent; therefore, further analysis of surface markers by flow cytometry with larger samples is needed to validate the association of CD4+ T cells with PD.

There are several limitations to this study: lack of validation and replication studies, use of blood, potential confounding of other factors, and small sample size. First, the current study lacks replication analysis using other DNA methylation measurement such as pyrosequencing. Although replication with pyrosequencing makes the results more reliable, almost all the significant CpG sites in this study are located in or around CpG islands. Consequently, it is difficult to design the primers for pyrosequencing. Second, we used DNA extracted from peripheral blood rather than brain samples. Several studies have reported that disease-associated DNA methylation abnormalities can be detected across tissues [39, 42, 52, 53], but there are clear tissue-specific differences in DNA methylation profiles [54, 55]. We checked the blood-brain correlations of the significant CpG sites of this study, using Blood Brain DNA Methylation Comparison Tool [55]. We found that nine out of 40 significant CpG sites showed the blood-brain correlation >0.3, in at least one region of the four examined brain regions (Additional file 3: Table S6). However, we consider that there is a possibility that biological processes in blood such as immune abnormalities are associated with PD [45]. A previous study of multiple sclerosis, an autoimmune disease of the central nervous system, has reported an association of DNA methylation status of a CpG site in blood with the disease [56], supporting our hypothesis. Nevertheless, detailed studies using brain samples are needed to find additional and/or tissue-specific DNA methylation differences associated with PD. Moreover, DNA methylation was found to be influenced by SNP genotypes [55, 56].

The DNA methylation status of a CpG site in multiple sclerosis, which we have mentioned above, is also affected by SNP genotypes [56]. As for the 40 significant probes, two probes, cg07124903 and cg20340149, have previously been reported to be methylation QTLs (meQTLs) in developing brain and are influenced by genetic variations (Additional file 3: Table S7) [57]. We also checked meQTLs identified with blood samples and found that only one probe cg07124903 was reported to be meQTLs (Additional file 3: Table S7) [58]. Additional studies combining the GWAS SNP data and EWAS data might provide new knowledge on the relationships among DNA methylation, SNPs, and PD. Finally, we were unable to take into account the effects of medication for PD and smoking. A previous study reported that long-term medication for schizophrenia decreased DNA methylation of the *GAD1* promoter, which was hypermethylated in a mouse model of schizophrenia [59]. Most of the PD patients in the present study were prescribed psychotropic medications. As such there is a possibility that the drugs affected the DNA methylation differences between the PD and healthy subjects. In addition, some DNA methylation sites have been reported to be influenced by smoking [60]. In this study, we could not adjust for such smoking effects as we did not have data on the smoking status of the healthy control subjects. However, when we examined the distributions of M values of the significant CpG sites between smokers and non-smokers among the PD subjects, we found that there was no effect of smoking for these sites (Additional file 1: Figure S8).

In conclusion, there might not be any CpG sites with DNA methylation differences that have a large effect on PD. However, we obtained some intriguing results: the hypomethylated CpG sites annotated to genes associated with the leukocyte activation pathway and the higher proportion of CD4+ T cells in PD. There is a possibility that several CpG sites with small effects, especially those that are related to immunity, are associated, together as a group, with PD. Further replication studies with larger number of samples are necessary to confirm the findings of this study.

Conclusions

We performed the EWAS of PD and identified 40 CpG sites of which the levels of DNA methylation were significantly different between PD and healthy control subjects. Some of these CpG sites have the possibility to be related the "positive regulation of lymphocyte activation" pathway. Such CpG sites with small effects might be associated, together as a group, with PD.

Methods
Subjects
DNA samples for the EWAS were obtained from our PD and healthy control sample set: patients with PD ($N = 48$) and age- and sex-matched healthy control subjects ($N = 48$) were recruited from among Japanese individuals living in Tokyo and Nagoya, located in the center of mainland Japan (Additional file 4: Table S8. These samples other than one discordant monozygotic twins were from unrelated PD patients and healthy control subjects. Each PD patient was diagnosed according to the Diagnostic and Statistical Manual of Mental Disorders, 4th Edition (DSM-IV) criteria [61] based on responses to the Mini International Neuropsychiatric Interview (MINI) [62] and clinical records. Healthy control subjects were interviewed by psychiatrists and were asked to fill out a questionnaire, MINI, in order to exclude those with a history of a major psychiatric illness, including PD.

Epigenome-wide DNA methylation analysis
Genomic DNA (PD, $N = 48$; control, $N = 48$) was extracted from leukocytes in whole blood by the standard phenol chloroform method (Wizard genomic DNA purification kit, Promega Corporation, WI, USA). DNA samples were first bisulfite-converted using a kit for the bisulfite conversion of DNA (EZ DNA Methylation™ Kit, Zymo Research, Irvine, CA, USA). For all samples, the DNA methylation levels of cytosine residues across the genome were examined with a DNA methylation array (Infinium® Human Methylation 450K BeadChip, Illumina Inc.) according to the manufacturer's protocol. Briefly, the bisulfite-converted DNA samples underwent whole-genome amplification and were fragmented and hybridized on BeadChip. After hybridization of the fragmented DNA with their complementary probe sequences, the DNA methylation status was determined through a single-base extension step. The arrays were imaged with a high-precision scanner (iScan system, Illumina Inc.), and the signal intensities were extracted using a software package (GenomeStudio Software, Illumina Inc.). The DNA methylation status of each cytosine residue was evaluated with the β value, which is the ratio of the signal from the methylated probe divided by the total signal intensity. The β value ranges from 0 (unmethylated) to 1 (completely methylated).

Data filtering and normalization
Data filtering and processing were performed for quality control of the calculated β values. β values with a detection P value <0.01 were treated as missing values. We then calculated the ratio of the detected β values to all of the examined β values ($N = 96$) for each probe; this was defined as the probe call rate. Probes that met the following conditions were used in the subsequent

analyses: (1) probe call rate >95%; (2) probe not on a sex chromosome; (3) probe not including a single-nucleotide polymorphism (SNP) with a minor allele frequency ≥0.05; and (4) probe not reported to be cross-reactive [63] (Additional file 1: Figure S5). As mentioned in the last criterion, we excluded cross-reactive probes that were reported to co-hybridize to alternate sequences that are highly homologous (<4 base mismatches among 50 bases) to the intended targets [63]. Furthermore, we created a list of possible cross-reactive probes that have unintended target sequences identical to the 20-base sequence from the 5′ end of each intended target (Additional file 5: Table S9). The 20 bases from the 5′ end of each target were mapped against the reference sequence (Genome Reference Consortium Human Reference 37 (GCA_000001405.1)) using BLAST (https://blast.ncbi.nlm.nih.gov/Blast.cgi). In examining significant probes, we excluded the probes that were on the list.

After the filtering, data normalization was performed, with the following pipeline, Lumi: quantile normalization (QN; correction for the distributions of the pooled probes) + beta-mixture quantile (BMIQ) normalization (correction for probe design bias) + correction for the batch effect (ComBat). First, the distributions of the pooled methylated and unmethylated probes were quantile normalized using the Lumi package under the assumption that they were similar between different samples [64]. A beta-mixture QN method was used to correct probe design bias with BMIQ normalization [65]. Finally, an empirical Bayes batch-correction method, ComBat [66], was employed to control for batch effects among arrays. In order to detect PD-associated DMPs, the original β values were converted to M values via the logit transformation [67] and used for performing the case control analysis. As for probes on X chromosome, the data filtering and normalization were performed in the same way separately only with the female samples (PD, $N = 31$, control, $N = 31$) (Additional file 1: Figure S8).

Prediction of the DNA methylation age and the distributions of leukocyte subsets

DNA methylation age (DNAm age) was defined as the age estimated from the DNA methylation status data of several CpG sites. We predicted the DNAm age to evaluate the reliability of the assay in both PD and healthy control subjects. DNAm age was estimated using the EWAS data following the algorithm reported in a previous study [24]. As for the analysis of the estimated DNAm age, the data were normalized according to the previous report [24], because the probes used in this analysis were a portion of the total probes and they did not include any type II probe. The estimated DNAm age

was compared with chronological age by calculating Pearson's correlation coefficients.

Furthermore, DNA methylation data were used to predict the cell mixture distributions of leukocyte subsets [30] to examine the possibility that cell mixture distributions differ between PD and healthy subjects. The proportions of leukocyte subsets (natural killer cells, B cells, CD4+ T cells, CD8+ T cells, monocytes, and granulocytes) were estimated using a published algorithm [30] with an R package, Minfi. Briefly, the β values of CpG sites, which correspond to putative differentially methylated sites among leukocyte subsets and that enable them to be distinguished, were selected. The selected β values were applied to the analysis, which resembled a regression calibration, as it can be considered a surrogate measure of the distribution of leukocyte cell mixtures [30]. The estimated proportions of leukocyte subsets were compared between the PD and control subjects. Additionally, we estimated abundance measures of plasmablasts, CD8+CD28−CD45RA− T cells, naive CD8+ T cells, and naive CD4+ T cells using the epigenetic clock software [24].

Pathway analysis

A pathway analysis was performed using the MetaCore™ platform (version 6.24 build 67895, Thomson Reuters, New York, NY, USA). Genes annotated to significant CpG sites were examined to determine whether they had any enrichment of gene sets for biological processes and molecular functions in the GO database (http://geneontology.org/) [68]. To be more precise, if the significant CpG sites were located in regions within 1500 bp from a transcription start site, 5′ UTR, body, and 3′ UTR of genes, the genes were annotated to the CpG sites and included in the pathway analysis. Since gene sets with large numbers of genes have a tendency to represent broader categories and have no useful biological meaning, gene sets with more than 500 genes were disregarded [69]. Gene sets with less than five registered genes or five consequent genes annotated from a list of examined genes were also disregarded because such gene sets are worthy of little attention in a pathway-based approach. Furthermore, we also used another method of GO-based pathway analysis, GOseq [70], in which bias caused by the different numbers of probes associated with each gene can be corrected [70, 71]. Pathways identified using the MetaCore™ platform, but not replicated with GOseq analysis, were excluded from the list of significant pathways.

Statistical analysis

The Wilcoxon rank sum test was employed to compare the proportions of the leukocyte subsets between the PD and control subjects. The Bonferroni correction was applied to adjust for multiple comparisons.

For the EWAS, significant associations were assessed by linear regression analysis with adjustments for the effects of the predicted proportions of leukocyte subsets using M values at a false discovery rate (FDR) of 5%. To check the effect of smoking on the significant sites, adjusted M values were compared between smokers and non-smokers in PD group using t test.

All analyses were performed using R software.

Additional files

Additional file 1: Figures S1–S8. Figures which were used for the quality check of the DNA methylation array data, filtering procedure, and additional leukocyte subset analysis and check for smoking effect are indicated.

Additional file 2: Tables S1–S5. Results of probes in the gene regions of which DNA methylation were previously reported to be associated with anxiety disorders.

Additional file 3: Table S6 and S7. Blood-brain correlations of the significant CpG sites and meQTL sites found in the significant CpG sites.

Additional file 4: Table S8. Demographic characteristics of samples.

Additional file 5: Table S9. Probe IDs of possible cross-reactive probes.

Acknowledgements
We would like to express our thanks to Associate Professor Dr. Akihiro Fujimoto, Department of Drug Discovery Medicine, Graduate School of Medicine, Kyoto University, for the technical assistance.

Funding
This study is funded by JSPS (JSPS KAKENHI Grant Numbers 15J04964, 26461712, and 25461723) and the Takeda Science Foundation in Japan.

Authors' contributions
MS-S, TO, TM, KT, and TS designed the study. MS-S, TO, TM, KT, and TS wrote the manuscript. MS-S contributed to the data analyses. TU, YK, MT, KK, HK, HT, and YO collected the samples. TO, TU, KT, and TS contributed to the reagents, materials, and analysis tools. MB and KI provided technical assistance. All authors read and approved the final manuscript.

Competing interests
The authors declare that they have no competing interests.

Author details
[1]Department of Human Genetics, Graduate School of Medicine, The University of Tokyo, 7-3-1 Hongo, Bunkyo Ward, Tokyo 113-0033, Japan. [2]Graduate School of Clinical Psychology, Teikyo Heisei University Major of Professional Clinical Psychology, 2-51-4 Higashiikebukuro, Toshima Ward, Tokyo 171-0014, Japan. [3]Department of Psychiatry and Behavioral Sciences, Tokyo Metropolitan Institute of Medical Science, 2-1-6 Kamikitazawa, Setagaya Ward, Tokyo 156-8506, Japan. [4]Division for Environment, Health and Safety, The University of Tokyo, 7-3-1 Hongo, Bunkyo Ward, Tokyo 113-0033, Japan. [5]Department of Psychiatry, Shonan Kamakura General Hospital, 1370-1 Okamoto, Kamakura City, Kanagawa 247-8533, Japan. [6]Department of Molecular Brain Science, Graduate School of Medical Sciences, Kumamoto University, 1-1-1 Honjo, Chuo Ward, Kumamoto City, Kumamoto 860-8556, Japan. [7]Department of Neuropsychiatry, Teikyo University School of Medicine, 2-11-1 Kaga, Itabashi Ward, Tokyo 173-0003, Japan. [8]Department of Neuropsychiatry, Graduate School of Medicine, The University of Tokyo, 7-3-1 Hongo, Bunkyo Ward, Tokyo 113-0033, Japan. [9]Panic Disorder Research Center, Warakukai Med Corp, 3-9-18 Akasaka, Minato Ward, Tokyo 107-0052, Japan. [10]Department of Psychiatry, Institute of Medical Life Science, Graduate School of Medicine, Mie University, 2-174 Edobashi, Tsu City, Mie 514-8502, Japan. [11]Department of Psychiatry, Koseikai Michinoo Hospital, 1-1 Nijigaokamachi, Nagasaki City, Nagasaki 852-8055, Japan. [12]Department of Physical and Health Education, Graduate School of Education, The University of Tokyo, 7-3-1 Hongo, Bunkyo Ward, Tokyo 113-0033, Japan.

References
1. Hettema JM, Neale MC, Kendler KS. A review and meta-analysis of the genetic epidemiology of anxiety disorders. Am J Psychiatry. 2001;158(10):1568–78.
2. Crowe RR, Noyes R, Pauls DL, Slymen D. A family study of panic disorder. Arch Gen Psychiatry. 1983;40(10):1065–9.
3. Goldstein RB, Wickramaratne PJ, Horwath E, Weissman MM. Familial aggregation and phenomenology of 'early'-onset (at or before age 20 years) panic disorder. Arch Gen Psychiatry. 1997;54(3):271–8.
4. Erhardt A, Czibere L, Roeske D, Lucae S, Unschuld PG, Ripke S, Specht M, Kohli MA, Kloiber S, Ising M, et al. TMEM132D, a new candidate for anxiety phenotypes: evidence from human and mouse studies. Mol Psychiatry. 2011;16(6):647–63.
5. Erhardt A, Akula N, Schumacher J, Czamara D, Karbalai N, Müller-Myhsok B, Mors O, Borglum A, Kristensen AS, Woldbye DP, et al. Replication and meta-analysis of TMEM132D gene variants in panic disorder. Transl Psychiatry. 2012;2, e156.
6. Otowa T, Yoshida E, Sugaya N, Yasuda S, Nishimura Y, Inoue K, Tochigi M, Umekage T, Miyagawa T, Nishida N, et al. Genome-wide association study of panic disorder in the Japanese population. J Hum Genet. 2009;54(2):122–6.
7. Otowa T, Kawamura Y, Nishida N, Sugaya N, Koike A, Yoshida E, Inoue K, Yasuda S, Nishimura Y, Liu X, et al. Meta-analysis of genome-wide association studies for panic disorder in the Japanese population. Transl Psychiatry. 2012;2, e186.
8. Gregersen NO, Lescai F, Liang J, Li Q, Als T, Buttenschøn HN, Hedemand A, Biskopstø M, Wang J, Wang AG, et al. Whole-exome sequencing implicates DGKH as a risk gene for panic disorder in the Faroese population. Am J Med Genet B Neuropsychiatr Genet. 2016.
9. Portela A, Esteller M. Epigenetic modifications and human disease. Nat Biotechnol. 2010;28(10):1057–68.
10. Hodes GE. Sex, stress, and epigenetics: regulation of behavior in animal models of mood disorders. Biol Sex Differ. 2013;4(1):1.
11. Lou S, Lee HM, Qin H, Li JW, Gao Z, Liu X, Chan LL, Kl Lam V, So WY, Wang Y, et al. Whole-genome bisulfite sequencing of multiple individuals reveals complementary roles of promoter and gene body methylation in transcriptional regulation. Genome Biol. 2014;15(7):408.
12. Murphy TM, O'Donovan A, Mullins N, O'Farrelly C, McCann A, Malone K. Anxiety is associated with higher levels of global DNA methylation and altered expression of epigenetic and interleukin-6 genes. Psychiatr Genet. 2014.
13. Gelernter J. Genetics of complex traits in psychiatry. Biol Psychiatry. 2015;77(1):36–42.
14. Shimada-Sugimoto M, Otowa T, Hettema JM. Genetics of anxiety disorders: genetic epidemiological and molecular studies in humans. Psychiatry Clin Neurosci. 2015;69(7):388–401.
15. Ziegler C, Dannlowski U, Bräuer D, Stevens S, Laeger I, Wittmann H, Kugel H, Dobel C, Hurlemann R, Reif A, et al. Oxytocin receptor gene methylation—converging multi-level evidence for a role in social anxiety. Neuropsychopharmacology. 2015.
16. Domschke K, Tidow N, Schrempf M, Schwarte K, Klauke B, Reif A, Kersting A, Arolt V, Zwanzger P, Deckert J. Epigenetic signature of panic disorder: a role of glutamate decarboxylase 1 (GAD1) DNA hypomethylation? Prog Neuropsychopharmacol Biol Psychiatry. 2013;46:189–96.
17. Ziegler C, Richter J, Mahr M, Gajewska A, Schiele MA, Gehrmann A, Schmidt B, Lesch KP, Lang T, Helbig-Lang S, et al. MAOA gene hypomethylation in panic disorder-reversibility of an epigenetic risk pattern by psychotherapy. Transl Psychiatry. 2016;6, e773.
18. Domschke K, Tidow N, Kuithan H, Schwarte K, Klauke B, Ambrée O, Reif A, Schmidt H, Arolt V, Kersting A, et al. Monoamine oxidase A gene DNA

hypomethylation—a risk factor for panic disorder? Int J Neuropsychopharmacol. 2012;15(9):1217–28.

19. Roberts S, Lester KJ, Hudson JL, Rapee RM, Creswell C, Cooper PJ, Thirlwall KJ, Coleman JR, Breen G, Wong CC, et al. Serotonin tranporter methylation and response to cognitive behaviour therapy in children with anxiety disorders. Transl Psychiatry. 2014;4, e444.

20. Esler M, Alvarenga M, Pier C, Richards J, El-Osta A, Barton D, Haikerwal D, Kaye D, Schlaich M, Guo L, et al. The neuronal noradrenaline transporter, anxiety and cardiovascular disease. J Psychopharmacol. 2006;20(4 Suppl):60–6.

21. Esler M, Eikelis N, Schlaich M, Lambert G, Alvarenga M, Kaye D, El-Osta A, Guo L, Barton D, Pier C, et al. Human sympathetic nerve biology: parallel influences of stress and epigenetics in essential hypertension and panic disorder. Ann N Y Acad Sci. 2008;1148:338–48.

22. Bayles R, Baker EK, Jowett JB, Barton D, Esler M, El-Osta A, Lambert G. Methylation of the SLC6a2 gene promoter in major depression and panic disorder. PLoS One. 2013;8(12), e83223.

23. Heyn H, Li N, Ferreira HJ, Moran S, Pisano DG, Gomez A, Diez J, Sanchez-Mut JV, Setien F, Carmona FJ, et al. Distinct DNA methylomes of newborns and centenarians. Proc Natl Acad Sci U S A. 2012;109(26):10522–7.

24. Horvath S. DNA methylation age of human tissues and cell types. Genome Biol. 2013;14(10):R115.

25. Houseman EA, Accomando WP, Koestler DC, Christensen BC, Marsit CJ, Nelson HH, Wiencke JK, Kelsey KT. DNA methylation arrays as surrogate measures of cell mixture distribution. BMC Bioinformatics. 2012;13:86.

26. Gao X, Jia M, Zhang Y, Breitling LP, Brenner H. DNA methylation changes of whole blood cells in response to active smoking exposure in adults: a systematic review of DNA methylation studies. Clin Epigenetics. 2015;7:113.

27. Toperoff G, Aran D, Kark JD, Rosenberg M, Dubnikov T, Nissan B, Wainstein J, Friedlander Y, Levy-Lahad E, Glaser B, et al. Genome-wide survey reveals predisposing diabetes type 2-related DNA methylation variations in human peripheral blood. Hum Mol Genet. 2012;21(2):371–83.

28. Kaneda A, Matsusaka K, Sakai E, Funata S. DNA methylation accumulation and its predetermination of future cancer phenotypes. J Biochem. 2014; 156(2):63–72.

29. Slieker RC, Bos SD, Goeman JJ, Bovée JV, Talens RP, van der Breggen R, Suchiman HE, Lameijer EW, Putter H, van den Akker EB, et al. Identification and systematic annotation of tissue-specific differentially methylated regions using the Illumina 450 k array. Epigenetics Chromatin. 2013;6(1):26.

30. Jaffe AE, Irizarry RA. Accounting for cellular heterogeneity is critical in epigenome-wide association studies. Genome Biol. 2014;15(2):R31.

31. Bibikova M, Barnes B, Tsan C, Ho V, Klotzle B, Le JM, Delano D, Zhang L, Schroth GP, Gunderson KL, et al. High density DNA methylation array with single CpG site resolution. Genomics. 2011;98(4):288–95.

32. Wang T, Guan W, Lin J, Boutaoui N, Canino G, Luo J, Celedón JC, Chen W. A systematic study of normalization methods for Infinium 450 K methylation data using whole-genome bisulfite sequencing data. Epigenetics. 2015;10(7):662–9.

33. Szyf M. Nongenetic inheritance and transgenerational epigenetics. Trends Mol Med. 2014.

34. Feil R, Fraga MF. Epigenetics and the environment: emerging patterns and implications. Nat Rev Genet. 2011;13(2):97–109.

35. Numata S, Ishii K, Tajima A, Iga J, Kinoshita M, Watanabe S, Umehara H, Fuchikami M, Okada S, Boku S, et al. Blood diagnostic biomarkers for major depressive disorder using multiplex DNA methylation profiles: discovery and validation. Epigenetics. 2015;10(2):135–41.

36. Dempster EL, Wong CC, Lester KJ, Burrage J, Gregory AM, Mill J, Eley TC. Genome-wide methylomic analysis of monozygotic twins discordant for adolescent depression. Biol Psychiatry. 2014;76(12):977–83.

37. Wang D, Liu X, Zhou Y, Xie H, Hong X, Tsai HJ, Wang G, Liu R, Wang X. Individual variation and longitudinal pattern of genome-wide DNA methylation from birth to the first two years of life. Epigenetics. 2012;7(6):594–605.

38. Córdova-Palomera A, Fatjó-Vilas M, Gastó C, Navarro V, Krebs MO, Fañanás L. Genome-wide methylation study on depression: differential methylation and variable methylation in monozygotic twins. Transl Psychiatry. 2015;5, e557.

39. Wockner LF, Noble EP, Lawford BR, Young RM, Morris CP, Whitehall VL, Voisey J. Genome-wide DNA methylation analysis of human brain tissue from schizophrenia patients. Transl Psychiatry. 2014;4, e339.

40. Wockner LF, Morris CP, Noble EP, Lawford BR, Whitehall VL, Young RM, Voisey J. Brain-specific epigenetic markers of schizophrenia. Transl Psychiatry. 2015;5, e680.

41. Kinoshita M, Numata S, Tajima A, Ohi K, Hashimoto R, Shimodera S, Imoto I, Takeda M, Ohmori T. Aberrant DNA methylation of blood in schizophrenia

by adjusting for estimated cellular proportions. Neuromolecular Med. 2014; 16(4):697–703.

42. Dempster EL, Pidsley R, Schalkwyk LC, Owens S, Georgiades A, Kane F, Kalidindi S, Picchioni M, Kravariti E, Toulopoulou T, et al. Disease-associated epigenetic changes in monozygotic twins discordant for schizophrenia and bipolar disorder. Hum Mol Genet. 2011;20(24):4786–96.

43. Ball MP, Li JB, Gao Y, Lee JH, LeProust EM, Park IH, Xie B, Daley GQ, Church GM. Targeted and genome-scale strategies reveal gene-body methylation signatures in human cells. Nat Biotechnol. 2009;27(4):361–8.

44. Rauch TA, Wu X, Zhong X, Riggs AD, Pfeifer GP. A human B cell methylome at 100-base pair resolution. Proc Natl Acad Sci U S A. 2009;106(3):671–8.

45. Shimada-Sugimoto M, Otowa T, Miyagawa T, Khor SS, Kashiwase K, Sugaya N, Kawamura Y, Umekage T, Kojima H, Saji H, et al. Immune-related pathways including HLA-DRB1(*)13:02 are associated with panic disorder. Brain Behav Immun. 2015;46:96–103.

46. Zheng SL, Li ZY, Song J, Liu JM, Miao CY. Metrnl: a secreted protein with new emerging functions. Acta Pharmacol Sin. 2016;37(5):571–9.

47. Howe AS, Buttenschøn HN, Bani-Fatemi A, Maron E, Otowa T, Erhardt A, Binder EB, Gregersen NO, Mors O, Woldbye DP, et al. Candidate genes in panic disorder: meta-analyses of 23 common variants in major anxiogenic pathways. Mol Psychiatry. 2015.

48. Marazziti D, Ambrogi F, Vanacore R, Mignani V, Savino M, Palego L, Cassano GB, Akiskal HS: Immune cell imbalance in major depressive and panic disorders. Neuropsychobiology. 1992;26(1-2):23-26

49. Schleifer SJ, Keller SE, Bartlett JA. Panic disorder and immunity: few effects on circulating lymphocytes, mitogen response, and NK cell activity. Brain Behav Immun. 2002;16(6):698–705.

50. Manfro GG, Pollack MH, Otto MW, Worthington JJ, Rosenbaum JF, Scott EL, Kradin RL. Cell-surface expression of L-selectin (CD62L) by blood lymphocytes: correlates with affective parameters and severity of panic disorder. Depress Anxiety. 2000;11(1):31–7.

51. Park JE, Kim SW, Park Q, Jeong DU, Yu BH. Lymphocyte subsets and mood states in panic disorder patients. J Korean Med Sci. 2005;20(2):215–9.

52. Sapienza C, Lee J, Powell J, Erinle O, Yafai F, Reichert J, Siraj ES, Madaio M. DNA methylation profiling identifies epigenetic differences between diabetes patients with ESRD and diabetes patients without nephropathy. Epigenetics. 2011;6(1):20–8.

53. Kaminsky Z, Tochigi M, Jia P, Pal M, Mill J, Kwan A, Ioshikhes I, Vincent JB, Kennedy JL, Strauss J, et al. A multi-tissue analysis identifies HLA complex group 9 gene methylation differences in bipolar disorder. Mol Psychiatry. 2012;17(7):728–40.

54. Davies MN, Volta M, Pidsley R, Lunnon K, Dixit A, Lovestone S, Coarfa C, Harris RA, Milosavljevic A, Troakes C, et al. Functional annotation of the human brain methylome identifies tissue-specific epigenetic variation across brain and blood. Genome Biol. 2012;13(6):R43.

55. Hannon E, Lunnon K, Schalkwyk L, Mill J. Interindividual methylomic variation across blood, cortex, and cerebellum: implications for epigenetic studies of neurological and neuropsychiatric phenotypes. Epigenetics. 2015;10(11):1024–32.

56. Andlauer TF, Buck D, Antony G, Bayas A, Bechmann L, Berthele A, Chan A, Gasperi C, Gold R, Graetz C, et al. Novel multiple sclerosis susceptibility loci implicated in epigenetic regulation. Sci Adv. 2016;2(6), e1501678.

57. Hannon E, Spiers H, Viana J, Pidsley R, Burrage J, Murphy TM, Troakes C, Turecki G, O'Donovan MC, Schalkwyk LC, et al. Methylation QTLs in the developing brain and their enrichment in schizophrenia risk loci. Nat Neurosci. 2016;19(1):48–54.

58. Lemire M, Zaidi SH, Ban M, et al. Long-range epigenetic regulation is conferred by genetic variation located at thousands of independent loci. Nat Commun. 2015;6:6326.

59. Dong E, Nelson M, Grayson DR, Costa E, Guidotti A. Clozapine and sulpiride but not haloperidol or olanzapine activate brain DNA demethylation. Proc Natl Acad Sci U S A. 2008;105(36):13614–9.

60. Joehanes R, Just AC, Marioni RE, Pilling LC, Reynolds LM, Mandaviya PR, Guan W, Xu T, Elks CE, Aslibekyan S, et al. Epigenetic signatures of cigarette smoking. Circ Cardiovasc Genet. 2016;9(5):436–47.

61. American Psychiatric Association. Diagnostic and Statistical Manual of Mental Disorders, Fourth Edition (DSM-4). Washington, DC: American psychiatric press; 1994.

62. Sheehan DV, Lecrubier Y, Sheehan KH, Amorim P, Janavs J, Weiller E, Hergueta T, Baker R, Dunbar GC. The Mini-International Neuropsychiatric Interview (M.I.N.I.): the development and validation of a structured

diagnostic psychiatric interview for DSM-IV and ICD-10. J Clin Psychiatry. 1998;59 Suppl 20:22–33. quiz 34–57.

63. Chen YA, Lemire M, Choufani S, Butcher DT, Grafodatskaya D, Zanke BW, Gallinger S, Hudson TJ, Weksberg R. Discovery of cross-reactive probes and polymorphic CpGs in the Illumina Infinium HumanMethylation450 microarray. Epigenetics. 2013;8(2):203–9.

64. Du P, Kibbe WA. Lin SM: lumi: a pipeline for processing Illumina microarray. Bioinformatics. 2008;24(13):1547–8.

65. Teschendorff AE, Marabita F, Lechner M, Bartlett T, Tegner J, Gomez-Cabrero D, Beck S. A beta-mixture quantile normalization method for correcting probe design bias in Illumina Infinium 450 k DNA methylation data. Bioinformatics. 2013;29(2):189–96.

66. Johnson WE, Li C, Rabinovic A. Adjusting batch effects in microarray expression data using empirical Bayes methods. Biostatistics. 2007;8(1):118–27.

67. Du P, Zhang X, Huang CC, Jafari N, Kibbe WA, Hou L, Lin SM. Comparison of beta-value and M-value methods for quantifying methylation levels by microarray analysis. BMC Bioinformatics. 2010;11:587.

68. Ashburner M, Ball CA, Blake JA, Botstein D, Butler H, Cherry JM, Davis AP, Dolinski K, Dwight SS, Eppig JT, et al. Gene ontology: tool for the unification of biology. The Gene Ontology Consortium. Nat Genet. 2000;25(1):25–9.

69. Pan W, Kwak IY, Wei P. A powerful pathway-based adaptive test for genetic association with common or rare variants. Am J Hum Genet. 2015;97(1):86–98.

70. Young MD, Wakefield MJ, Smyth GK, Oshlack A. Gene ontology analysis for RNA-seq: accounting for selection bias. Genome Biol. 2010;11(2):R14.

71. Geeleher P, Hartnett L, Egan LJ, Golden A, Raja Ali RA, Seoighe C. Gene-set analysis is severely biased when applied to genome-wide methylation data. Bioinformatics. 2013;29(15):1851–7.

12

Genome-wide DNA methylation profiling shows a distinct epigenetic signature associated with lung macrophages in cystic fibrosis

Youdinghuan Chen[1,2†], David A. Armstrong[3*†] (iD), Lucas A. Salas[1,2], Haley F. Hazlett[4], Amanda B. Nymon[5], John A. Dessaint[3], Daniel S. Aridgides[3], Diane L. Mellinger[3], Xiaoying Liu[6], Brock C. Christensen[1,2,7] and Alix Ashare[3,4,5]

Abstract

Background: Lung macrophages are major participants in the pulmonary innate immune response. In the cystic fibrosis (CF) lung, the inability of lung macrophages to successfully regulate the exaggerated inflammatory response suggests dysfunctional innate immune cell function. In this study, we aim to gain insight into innate immune cell dysfunction in CF by investigating alterations in DNA methylation in bronchoalveolar lavage (BAL) cells, composed primarily of lung macrophages of CF subjects compared with healthy controls. All analyses were performed using primary alveolar macrophages from human subjects collected via bronchoalveolar lavage. Epigenome-wide DNA methylation was examined via Illumina MethylationEPIC (850 K) array. Targeted next-generation bisulfite sequencing was used to validate selected differentially methylated CpGs. Methylation-based sample classification was performed using the recursively partitioned mixture model (RPMM) and was tested against sample case-control status. Differentially methylated loci were identified by fitting linear models with adjustment of age, sex, estimated cell type proportions, and repeat measurement.

Results: RPMM class membership was significantly associated with the CF disease status ($P = 0.026$). One hundred nine CpG loci were differentially methylated in CF BAL cells (all FDR ≤ 0.1). The majority of differentially methylated loci in CF were hypo-methylated and found within non-promoter CpG islands as well as in putative enhancer regions and DNase hyper-sensitive regions.

Conclusions: These results support a hypothesis that epigenetic changes, specifically DNA methylation at a multitude of gene loci in lung macrophages, may participate, at least in part, in driving dysfunctional innate immune cells in the CF lung.

Keywords: Lung macrophages, Bronchoalveolar lavage, DNA methylation, Epigenetics

* Correspondence: David.A.Armstrong@hitchcock.org
†Youdinghuan Chen and David A. Armstrong contributed equally to this work.
³Department of Medicine, Dartmouth-Hitchcock Medical Center, Lebanon, NH, USA
Full list of author information is available at the end of the article

Background

The role of the innate immune system in the pathogenesis of cystic fibrosis (CF) lung disease has been an emerging research focus [1–3]. The inability to regulate both the chronic infections and an excessive inflammatory response suggests that the innate immune system is dysfunctional in CF [1]. Studies have revealed numerous physiologic defects associated with CF macrophages including dysregulation of phagocytic/signaling receptors [4], hyper-responsiveness to microbial stimuli [5–7], and impairment in removal of apoptotic cells [8, 9]. Additionally, increased levels of inflammatory mediators, typically secreted from macrophages, in sputum and bronchoalveolar lavage (BAL) fluid have been described in CF patients [10–12].

The multitude of biological processes seemingly affected in CF macrophages begs the following question: is there a broader epigenetic mechanism influencing a myriad of biological functions in CF macrophages? Epigenetics is the study of heritable changes in gene function caused by mechanisms other than changes in the underlying DNA sequence [13]. The most widely studied of the epigenetic modifications is DNA methylation [14]. Epigenetic mechanisms have emerged as modulators of host defenses that can lead to a more prominent immune response and shape the course of inflammation in the host, both driving the production of specific inflammatory mediators and controlling the magnitude of the host response [15].

To examine possible underlying epigenetic mechanisms related to dysregulation of innate immunity in CF, we initiated an epigenome-wide DNA methylation profiling study from primarily lung macrophages isolated from BAL fluid in subjects with and without CF.

Results

Subject characteristics

BAL samples were sequentially taken from the right upper lobe (RUL) and right lower lobe (RLL) of heathy ($n = 4$) and CF subjects ($n = 4$). Subject characteristics are listed in Table 1.

Gender and age distribution were as follows: two male and two female CF subjects with a mean age of 22.0 ± 4.90 years; healthy subjects (three females/one male) had a mean age of 26.0 ± 5.35 years. Three CF individuals were genotype *F508del/F508del*, and one participant had genotype *F508del/Y1092X*. Forced expiration volume (FEV_1) measurement range was 77–96% across the CF study group. The lung microbiology based on standard BAL culture was recorded for each CF patient. Three subjects cultured positive for *Staphylococcus aureus*, and two cultured positive for *Achromobacter xylosoxidans*. Three CF subjects were on antibiotic and/or modulator therapy including one or more of the following: inhaled aztreonam, inhaled colistimethate, oral doxycycline, inhaled tobramycin, or ivacaftor/lumacaftor.

DNA methylation landscape between CF and healthy individuals

We determined methylation subclasses with an unsupervised, model-based method, recursively partitioned mixture model (RPMM, Fig. 1). RPMM clustering of the 10,000 CpG sites with the highest variance in DNA methylation revealed greater adjacency of CF samples, as well as the clustering of healthy samples. In addition, CF subjects which showed pronounced heterogeneity (Fig. 1a). Subsequently, we formally tested BAL sample disease status against DNA methylation cluster membership. All BAL samples predicted to be in RPMM cluster *L* were from healthy subjects, and the majority of BAL samples in RPMM cluster *R* were from CF subjects (Fig. 1b, $P = 0.026$, two-tailed Fisher's exact test).

Cellular composition and heterogeneity profiling

To assess the potential contribution of BAL sample cell type heterogeneity on DNA methylation, we utilized an approach to cell type deconvolution that does not require a reference library of differentially methylated loci for specific cell types [16]. Across all samples, the reference-free cell type deconvolution identified two

Table 1 Subject characteristics

Patient number	Status	Age	Sex	Genotype	FEV_1 (%)	Microbiology	Antibiotic and/or modulator therapy
1	CF	18	Male	F508del/Y1092X	77	*S. aureus* many mixed	None
2	CF	24	Male	F508del/F508del	96	*S. aureus*	Aztreonam, colistimethate, doxycycline, ivacaftor/lumacaftor
3	CF	28	Female	F508del/F508del	91	*A. xylosoxidans* moderate mixed	Colistimethate, tobramycin
4	CF	18	Female	F508del/F508del	88	*S. aureus* many mixed *A. xylosoxidans*, moderate mixed	Colistimethate, tobramycin, ivacaftor /lumacaftor
5	Healthy	26	Female				
6	Healthy	33	Male				
7	Healthy	25	Female				
8	Healthy	20	Female				

S. aureus Staphylococcus aureus, A. xylosoxidans Achromobacter xylosoxidans

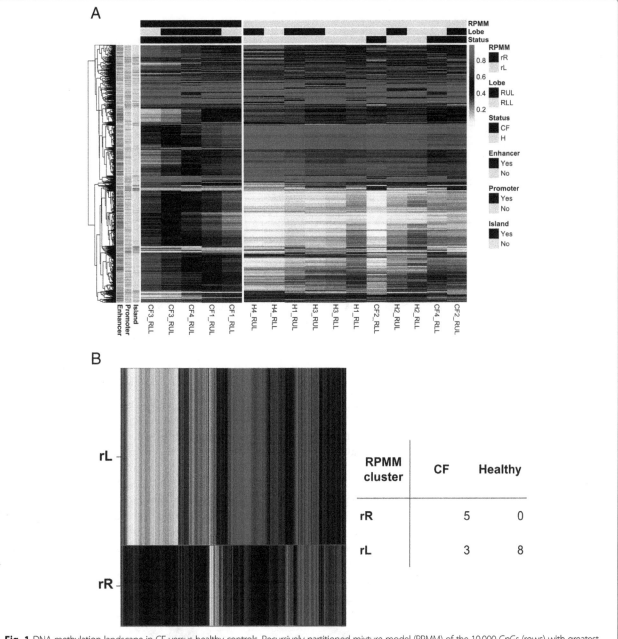

Fig. 1 DNA methylation landscape in CF versus healthy controls. Recursively partitioned mixture model (RPMM) of the 10,000 CpGs (rows) with greatest sample variance across subjects. Individual samples 1–16 are shown in columns with sample status bar at the top: black (CF) and gray (healthy). Blue color represents increased sample methylation (**a**). RPMM determined the similarity of methylation among subjects resulting in two methylation classes L and R (**b**). Methylation class membership was associated with class status; inset contingency table depicts subject distribution in each class via a two-tailed Fisher's exact test (P = 0.026)

putative cell types (Fig. 2a). Putative cell type 1 proportions were higher in healthy subjects compared to CF subjects (Fig. 2b, $P < 0.05$), and putative cell type 2 proportions were lower in healthy controls compared to CF subjects ($P < 0.05$). To confirm cell-type heterogeneity in CF subjects, cytospins were performed on RUL BAL cells isolated from a second group of healthy subjects and a second group of CF subjects ($n = 3$) genotypes

(*E60X/A455E*, *R1162X/W1282G*, and *F508del/F508del*). BAL cells from healthy subjects have characteristic lung macrophage "fried egg" or monocytoid appearance with reniform nuclei and ample cytosol (Fig. 3a, open arrow), with this cell phenotype comprising > 95% of the total cell population. The majority of BAL cells from CF subjects were similar to the macrophages seen in healthy subjects. However, the CF subjects had subpopulations

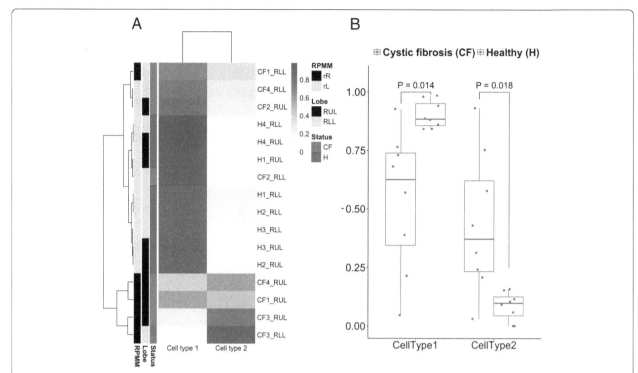

Fig. 2 Cellular composition and heterogeneity profiling. Heat map illustrates reference-free deconvolution (RefFreeEWAS) of putative cell type and size proportions across the *n* = 16 samples included in this study (**a**). Healthy subjects had higher proportions of putative cell type/size 1 (*P* = 0.014), and CF subjects had a higher proportion of cell type/size 2 (*P* = 0.018) (**b**)

of macrophages that were phenotypically diverse in size and appearance (Fig. 3b), as previously demonstrated in other respiratory disease states [17].

Epigenome-wide association analysis reveals differentially methylated loci

Next, we investigated the relationship between DNA methylation and CF disease status in BAL samples.

Because we identified differential proportions of putative cell types in CF subjects compared with controls, to identify differential methylation independent of cell type, we fit linear mixed effects models comparing methylation of the 26,733 most variable CpG sites with CF disease status, adjusted for subject age and sex, and included a term to account for repeat measurements from the same subject (RUL, RLL). We identified 109

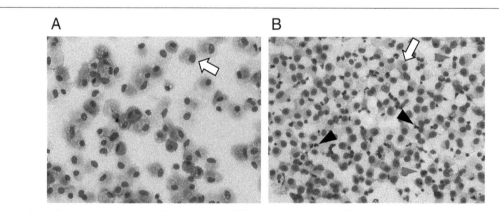

Fig. 3 Bronchoalveolar lavage cell cytospins. Bronchoalveolar lavage (BAL) samples were obtained from tertiary airways in the right upper lobe of subjects. BAL cells were isolated and prepared as cytospins as described in the "Methods" section. BAL cells from healthy subjects are a mostly homogeneous population of cells (lung macrophages (LM)) with oval to reniform nuclei and abundant cytosol (open arrow) (**a**). Cytospins of BAL cells from CF subjects (CD15-depleted) (**b**) show a majority population of LMs (open arrow) as well as smaller roundish cells with darker staining nuclei and less cytosol (blue arrowheads) and cells containing variably shaped and stained nuclei (black arrowheads). Images shown are representative of multiple subjects

differentially methylated CpGs, of which 51 are hyper-methylated and 58 are hypo-methylated in CF cases compared with controls (FDR-adjusted $P < 0.1$, $|\log_2 FC_{M\ value}| \geq 3.50$, Fig. 4a). There is a 31.6% median increase and 27.2% median reduction in the proportion of methylated alleles (beta-values) for differentially hyper- and hypo-methylated loci, respectively. The top five hypo- and hyper-methylated CpGs associated with known genes are shown in Additional file 1: Table S1.

We utilized targeted next-generation bisulfite sequencing (tNGS) for the validation of the EPIC methylation array. Genes that showed some of the greatest Δ-beta values, including *CD6*, *HOOK2*, *LSP1*, *RGS12*, *SH3PXD2A*, and *UPP1*, were selected to compare CpG methylation in CF vs. healthy subjects. Δ-Beta methylation values were consistent between the array and the sequencing approaches to measure DNA methylation (Table 2). Additionally, an example of tNGS assay design is shown for *LSP1* in Additional file 2:

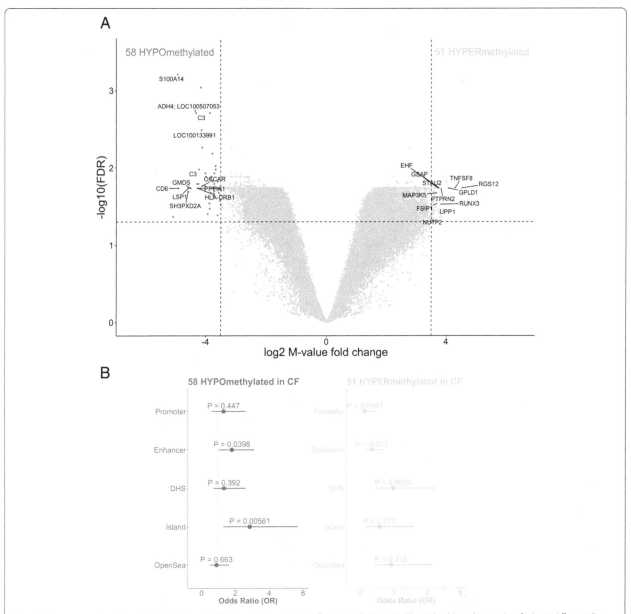

Fig. 4 Epigenome-wide differential methylation in CF. Comparative analysis of DNA methylation in CF and healthy subjects identified 109 differentially methylated CpGs (FDR $P < 0.1$, **a**). CpG hyper-methylated in CF compared to controls (green) and hypo-methylated CpGs (blue) are plotted as $\log_2$ fold increase or decrease in methylation M value (x-axis) versus $\log_{10}$ FDR-adjusted P value (y-axis). Statistically significant CpGs associated with specific genes are labeled, and unlabeled points represent CpGs associated with no known gene at that location. Enrichment of differentially methylated CpGs to genomic and transcriptional context is shown in forest plots (**b**), illustrating that hypo-methylated CpGs in CF BAL are enriched for enhancer regions and CpG islands and that hyper-methylated CpGs in CF BAL are under-represented for gene promoter regions

Table 2 Comparison of % methylation changes in CF subjects at gene-specific CpGs: EPIC Δ-beta vs targeted NGS Δ-beta value

Gene	Location	EPIC Δ-beta value	tNGS Δ-beta value
CD6	cg26427109	(−) 0.42	(−) 0.344
HOOK2	cg11738485	(−) 0.32	(−) 0.32
LSP1	cg18723409	(−) 0.44	(−) 0.292
RGS12	cg03132824	(+) 0.40	(+) 0.291
SH3PXD2A	cg06888746	(−) 0.28	(−) 0.184
UPP1	cg10317717	(+) 0.24	(+) 0.263

Figure S1, and detailed information about the tNGS assays performed including chromosomal location, number of CpG sites, average sample reads, and Δ-beta values across the entire assay is shown in Additional file 3: Table S2.

Next, we investigated the distribution of differentially methylated loci in CF subjects versus controls by genomic and transcriptional context. Among the hypo-methylated CpGs, enhancer regions and CpG islands were enriched (all OR > 1 and $P < 0.05$), and promoter-associated loci were significantly depleted in the hyper-methylated loci (OR = 0.25, $P < 0.05$) (Fig. 4b and Additional file 4: Table S3).

Unique genes associated with the top 5% (1337/26,377) most significant hyper-methylated and hypo-methylated CpGs were used as separate inputs for Kyoto Encyclopedia of Genes and Genomes (KEGG) pathway analysis against the 8515-gene universe associated with the 26,733 background loci (Additional file 5: Table S4).

In addition, we noted that 34.8% (38/109) differentially methylated loci do not track to any known genes. Compared to all CpGs tested for differential methylation, these "gene-less" CpG sites were over twice as likely as to co-occur with known single nucleotide polymorphisms (SNPs) (OR = 2.60, 95% CI = 1.27–5.18, $P = 5.09E-3$, Fisher's exact test). We have provided the complete list of DMPs with FDR < 0.05 as Additional file 6: Table S5.

Discussion

The primary objective of this study was to examine DNA methylation changes in CF BAL cells, primarily lung macrophages to gain mechanistic insight into innate immune regulation in cystic fibrosis. Our study is the first, which we are aware of, to report DNA methylation differences from lung macrophages isolated from CF subjects compared to healthy controls. One study has previously shown altered DNA methylation at a select number of lung modifier genes in nasal epithelial cells and whole blood of CF subjects [18]. Other studies have focused only on either histone-based macrophage epigenome profiling associated with macrophage phenotype [19–21] or the mapping of the lung proteome in

cystic fibrosis [22]. However, no investigation to date has used epigenome-wide DNA methylation profiling to study cystic fibrosis BAL cells. Our cell collection methodology (flexible bronchoscopy) is a rare approach used in CF research studies and affords us a unique opportunity to analyze innate immune cells isolated directly from the airway and alveolar space.

Our initial strategy was to collect BAL cells of both CF and healthy subjects to analyze epigenome-wide DNA methylation patterns. DNA methylation profiling revealed a distinct clustering pattern associated with CF subjects as compared to healthy subjects. Additionally, we identified 109 differentially methylated CpGs, of which 51 are hyper-methylated and 58 are hypo-methylated. Hyper-methylated CpGs included those associated with genes such as *TNFSF8* and *RUNX3*. *TNFSF8*, also known as *CD30L*, has been suggested to contribute to pro-inflammatory immune response in a cross-talk role between innate and adaptive immune cells [23] and has noted to be expressed at high levels in alveolar macrophages from sarcoid subjects [24]. The transcription factor RUNX3 has been shown to regulate chemokines CCL5, CCL19, and CXCL11, chemotactic molecules with the potential to recruit various leukocytes into inflammatory sites [25, 26]. Hypo-methylated CpGs in CF patients compared to controls included those associated with genes such as *S100A14*, *LSP1*, and *OSCAR*. S100A14 is a member of a S100 family of proteins, a family of calcium-binding cytosolic proteins composed of 25 known members that have a broad range of intracellular and extracellular functions including regulating calcium balance, cell apoptosis, migration, and proliferation [27, 28]. Studies demonstrate that when released into extracellular space, S100 proteins have crucial activities in the regulation of immune homeostasis, post-traumatic injury, and inflammation [28]. Another hypo-methylated CpG is associated with the gene for leukocyte-specific protein 1 (*LSP1*). LSP1 is an actin-associated protein expressed in macrophages, neutrophils, and endothelial cells and has been localized to nascent phagocytic cups during Fcγ receptor-mediated phagocytosis, where it displays the same spatial and temporal distribution as actin filaments. Downregulation of LSP1 severely reduces phagocytic activity of macrophages, clearly indicating a crucial role for this protein in Fcγ receptor-mediated phagocytosis [29]. OSCAR, an immuno-receptor for surfactant protein D (SP-D), has been found in alveolar macrophages and together with SP-D contributes to lung homeostasis and innate mucosal defense [30]. In humans, OSCAR has been reported to be expressed on monocytes, macrophages, neutrophils, and dendritic cells [31] and shown to enhance the pro-inflammatory response of monocytes [31, 32].

Kyoto Encyclopedia of Genes and Genomes (KEGG) analysis of the top 10 pathways for 803 unique genes associated with the top 5% CpGs based on P value revealed pathways such as biosynthesis of unsaturated fatty acids, glycerolipid metabolism, Fc gamma R-mediated phagocytosis, and Fc epsilon RI signaling as the top CpG-gene-associated pathways likely affected in CF subjects based on changes in DNA methylation pattern.

We were able to identify enrichment of hyper- or hypo-methylated loci to specific genomic contexts suggesting their importance for gene regulation. Hypomethylated CpGs in CF subjects were enriched for putative enhancer regions and in CpG islands, regions of high frequency of CpG sites [33]. Interestingly, enhancer regions are classically defined as *cis*-acting DNA sequences that can increase the transcription of genes. They generally function independently of orientation and at various distances from their target promoter (or promoters). Furthermore, they do not necessarily act on the respective closest promoter but can bypass neighboring genes to regulate genes located more distantly along a chromosome [34]. In some cases, individual enhancers have been found to regulate multiple genes [35].

Myeloid regulatory cells (MRC) have come into focus recently in lung disease including cystic fibrosis [36, 37] and in bacterial infections [38, 39]. Our observation of multiple cell sizes/types (CD15 (−)) in our CF population suggests that subpopulations of macrophages or even MRCs could be contributing to the observed differences in putative cell types in BAL between CF subjects and controls. To control for this, we performed a reference-free deconvolution of putative cell type proportions from the EPIC DNA methylation data set. Semi-supervised reference-free methods allow estimating proportions of cell types in the absence of a known or reliable reference of purified cell types [40]. Specifically, RefFreeEWAS first regresses out the effect of the phenotype of interest on the data and then uses a singular value decomposition on an augmented matrix based on the estimated regression and residual variation matrix [16]. Unlike other reference-free methods, RefFreeEWAS does not assume that the top components of data variation are associated with cell-type composition. Instead, it assumes that the top components in the regression and residual variation space are the cell types present in the bio-specimen. While it is possible that the putative cell types identified using reference-free deconvolution may represent distinct terminally differentiated cell types, it is also possible that cells with different activation states or in different stages of differentiation are captured by one or more putative cell types. With the apparent identification of more than one cell subtype in the DNA methylation data, we conducted follow-up studies on additional CF subjects for confirmation. We performed cytospins to visualize BAL cells before and after CD15 negative selection on CF and healthy subject

BAL fluid and noticed a heterogeneous cell population isolated from CF subjects. This observation is consistent with our previous work suggesting a size range of CD206(+) CF lung macrophages as identified by forward scatter height using flow cytometry [41]. Additionally, a recent study in COPD identifies "small" and "large" cells from alveolar spaces and lung interstitium in lung resection tissue of COPD subjects [17]. Although our study suggests the presence of these macrophage subpopulations in the CF lung, it is beyond the scope of this work to specifically identify these subpopulations and will be the focus of future studies.

Although our study has a limited sample size, this is the first study to report differential DNA methylation associated with cystic fibrosis using a whole-genome approach. In CF BAL cells, we identified a substantial number of differentially methylated CpGs after adjusting for potential confounders. Further, the extent of observed differential methylation was quite high and highlights their biological relevance. Taken together, the differential methylation status of these genes might indicate differential gene expression and imply a unique biology of the CF lung microenvironment.

Conclusions

Through the use of epigenome-wide DNA methylation profiling, we have identified 109 differentially methylated CpG loci in CF BAL cells. In addition, unsupervised DNA methylation cluster membership was significantly associated with the CF disease status. These observations support a hypothesis that epigenetic changes, specifically DNA methylation in lung macrophages, may participate, at least in part, in driving dysfunctional innate immune cells in the CF lung.

Methods
Study population
This study was approved by the Committee for the Protection of Human Subjects at the Geisel School of Medicine at Dartmouth (#22781). All subjects provided written informed consent and were clinically stable and in their baseline state of health. CF subjects had not had a pulmonary exacerbation within the preceding 4 weeks. All subjects were non-smokers.

Bronchoalveolar lavage and macrophage isolation
Subjects underwent flexible bronchoscopy following local anesthesia with lidocaine to the posterior pharynx and intravenous sedation. A bronchoscope was inserted transorally and advanced through the vocal cords. BAL fluid was obtained from tertiary airways in the right upper and lower lobes (RUL and RLL, respectively). BAL was performed sequentially in the RUL and RLL with 20 ml of sterile saline followed by 10 ml of air, and this was repeated for a total of

five times per airway. Lung macrophages were isolated as previously described [41, 42]. Briefly, BAL fluid was filtered through a two-layer gauze, centrifuged, and washed twice in 0.9% NaCl. Cells were counted with a T10-automated cell counter (Bio-Rad, Hercules, CA).

CF BAL cells were incubated with CD15 microbeads and run over an LD column on the QuadroMACS magnet (Miltenyi Biotec, Auburn, CA), according to the manufacturer's instructions, to deplete neutrophils. Our previous work [41] and the current routine cytospin cell monitoring indicate a final neutrophil population post-negative selection of less than 2% of total BAL cells. Cytospins were performed using a Shandon Cytospin 3 centrifuge (ThermoFisher Scientific, Waltham, MA). Briefly, 75,000 cells resuspended in 200 μl of 0.9% NaCl were loaded into a cytology funnel (Fisher Scientific, Pittsburgh, PA) and centrifuged for 10 min. Cells were allowed to air dry and processed for viewing via Hema 3 Stat pack (Fisher). Imaging was done on an Olympus (Waltham, MA) BX41 microscope with DP2-BSW software (version 2007).

DNA methylation array

Epigenome-wide DNA methylation profiling was performed via the Infinium Methylation EPIC Bead Chips (Illumina Inc., San Diego, CA) for the determination of methylation levels of more than 850,000 CpG sites as previously described [43]. Briefly, DNA was extracted from bronchoalveolar lavage-derived cells via Qiagen (Germantown, MD) DNeasy Blood and Tissue Kit. DNA was quantitated on a Qubit 3.0 Fluorometer (Life Technologies, Carlsbad, CA). Bisulfite conversion of DNA was carried out with the Zymo EZ DNA methylation kit (Zymo Research, Irvine, CA), and EPIC array hybridization and scanning were performed at the University of Southern California Molecular Genomics Core.

DNA methylation array data processing

Raw intensity data files (IDATs) from the MethylationE-PIC BeadChips were processed by the *minfi* R/Bioconductor analysis pipeline (version 1.21) [44] with annotation file version *ilm10b3.hg19*. Probes associated with known SNPs, non-CpGs and sex chromosomes, as well as those failing to meet a detection *P* value of 0.05 in ≥ 20% samples, were excluded. This pre-processing procedure left 813,096 CpGs with high-quality methylation data in the final data set.

Targeted, next-generation bisulfite sequencing and data analysis

tNGBS was performed by EpigenDx Inc. (Hopkinton, MA) on the same eight BAL cell specimens as in the EPIC DNA methylation array. Briefly, DNA bisulfite modification was done using EZ-96 DNA methylation

kit (Zymo, Irvine, CA) followed by multiplex PCR with Qiagen (Gaithersburg, MD) HotStar Taq and products purified with QIAquick PCR purification kit. Libraries were prepared using the KAPA Library Preparation Kit for Ion Torrent platforms and Ion XpressTM Barcode Adapters (ThermoFisher, Waltham, MA). Library products were purified using Agencourt AMPure XP beads (Beckman Coulter, Indianapolis, IN) and quantified using the Qiagen QIAxcel Advanced System. Barcoded samples were then pooled in an equimolar fashion before template preparation and enrichment were performed on the Ion ChefTM system (ThermoFisher) using Ion 520TM and Ion 530TM Chef reagents. Enriched, template-positive library products were sequenced on the Ion S5TM sequencer using Ion 530TM sequencing chips (ThermoFisher). FASTQ files from the Ion Torrent S5 server were aligned to the local reference database using open-source Bismark Bisulfite Read Mapper with the Bowtie2 alignment algorithm. Methylation levels were calculated in Bismark by dividing the number of methylated reads by the total number of reads.

Statistical analysis

Unsupervised hierarchical clustering with the Euclidean distance and complete linkage (default) was performed on CpG loci with greatest sample variances. At different variance thresholds, clustering structure appeared to be stable. The recursively partitioned mixture model (RPMM) [45] assuming two terminal clusters was applied to 10,000 most variable CpGs. RPMM was implemented in the *RPMM* R package (version 1.25). A two-sided Fisher's exact test was used to test the relation of supervised RPMM cluster membership with case-control status. A *P* value of 0.05 was used as the threshold for statistical significance.

The distribution of methylation beta-value variances was examined prior to statistical analysis: 26,733 CpGs with beta-value variance exceeding 0.01 were selected for further investigation. To account for the repeat measurements from a single subject, the correlation coefficient (= 0.762) for the 26,733 CpG beta-values among eight unique subjects (including four cases and four controls) was first calculated by the *duplicateCorrelation* function in the *limma* R/Bioconductor package (v.3.34.9). Differential methylation analysis was carried out by passing logit-transformed beta-values (i.e., *M* values), matched pairs, and associated correlation coefficient into the *lmFit* and *eBayes* functions in *limma*, with adjustment of subject age and sex as fixed effects and subject as a random effect in the model, such that $Y = \beta_0 + \beta_{CF} X_{CF} + \beta_{age} X_{age} + \beta_{sex} X_{sex}$ + RandomEffect(Subject), where Y is the methylation beta-values, β_0 is the intercept, X is a given covariate, and β is the respective model coefficient.

To assess the contribution of cell type heterogeneity to differential methylation, we reconstructed a linear model adjusting for the presence of putative cell types. Briefly, the *RefFreeEWAS* algorithm (R package version 2.1) [16] was applied to 10,000 most variable CpGs across all samples. The proportion of putative cell types were calculated iteratively for the number of such cell types K from 2 to 10. The optimal number of putative cell types $K = 2$ was selected as it minimized the variance of the bootstrapped deviance. The difference between cell type proportions were determined by a linear mixed effects model adjusting for age, sex, and repeat measurement, implemented in R packages *lme4* (version 1.1.17) and *lmerTest* (version 2.0.36).

Genomic contexts (Open Sea and Island) were provided in the Illumina EPIC annotation file. The "promoter" transcriptional context was defined as having either a "TSS200" or "TSS1500" annotation, or both in the column *UCSC_RefGene_Group* (TSS, transcription start site). Likewise, the "gene body" transcriptional context was defined as having a "Body" annotation. The "enhancer" context was defined as having a FANTOM4/5 enhancer record or Illumina array enhancer annotation. For each genomic or transcriptional context, odds ratios (OR) for the significant loci relative to the input loci were determined by a two-sided Fisher's exact test. A P value ≤ 0.05 was the threshold for statistical significance. To demonstrate outputs from two different linear models that were similar, a test for enrichment using the Kyoto Encyclopedia of Genes and Genomes (KEGG) was performed using the WebGestalt tool [46]. The differentially methylated CpG loci were used as the input and compared to genes associated with the universe set of CpGs. A pathway was considered significant if the pathway has a Benjamini-Hochberg FDR < 0.05, at least five genes up to a maximum of 2000 genes.

Additional files

Additional file 1: Table S1. Top hypo- and hyper-methylated CpGs in CF that are associated with known genes.

Additional file 2: Figure S1. Human leukocyte-specific protein-1 (LSP1) gene targeted next-generation sequencing (tNGS) assay region. A tNGS assay was designed for LSP1 surrounding EPIC cg18723409 located in intron 11 of Ensembl Gene ID: ENSG00000130592. A total of 13 CpGs were interrogated in this tNGS assay.

Additional file 3: Table S2. Targeted next-generation bisulfite sequencing of EPIC identified genes.

Additional file 4: Table S3. Summary of genomic context related with differentially methylated loci in CF. DHS, DNase I hyper-sensitivity sites.

Additional file 5: Table S4. Top 10 pathways for 803 unique genes associated with top 5% CpGs based on P value in the model without cell type adjustment.

Additional file 6: Complete list of DMPs with FDR < 0.05

Abbreviations
AM: Alveolar macrophage; BAL: Bronchoalveolar lavage; CF: Cystic fibrosis; FDR: False discovery rate; FEV1: Forced expiration volume; KEGG: Kyoto Encyclopedia of Genes and Genomes; RPMM: Recursively partitioned mixture model; tNGS: Targeted next-generation bisulfite sequencing

Funding
This work is supported by the NIH R01HL122372 (to A.A.) and NIH R01CA216265 (to B.C.C.). Subject enrollment and clinical sample collection were achieved with the assistance of the CF Translational Research Core at Dartmouth, which is jointly funded by the NIH (P30GM106394 to Bruce Stanton) and CFF (STANTO11R0 to Bruce Stanton). Additionally, we would like to acknowledge and thank the Burroughs-Wellcome Big Data in Life Sciences training grant.

Disclosures
No conflicts of interest, financial or otherwise, are declared by the authors.

Authors' contributions
The conception and design of the study were contributed by AA, DAA, and BCC. The acquisition of the data was done by DAA, YC, LAS, ABN, JAD, HFH, DSA, and DLM. The analysis and interpretation were carried out by DAA, YC, XL, BCC, and AA. The drafting of the manuscript for important intellectual content was done by DAA, YC, BCC, and AA. All authors contributed toward the critical review and approval of the final manuscript and are accountable for all aspects of the work published.

Competing interests
The authors declare that they have no competing interests.

Author details
[1]Department of Epidemiology, Geisel School of Medicine at Dartmouth, Lebanon, NH, USA. [2]Department of Molecular and Systems Biology, Geisel School of Medicine at Dartmouth, Lebanon, NH, USA. [3]Department of Medicine, Dartmouth-Hitchcock Medical Center, Lebanon, NH, USA. [4]Program in Experimental and Molecular Medicine, Geisel School of Medicine at Dartmouth, Hanover, NH, USA. [5]Department of Microbiology and Immunology, Geisel School of Medicine at Dartmouth, Dartmouth-Hitchcock Medical Center, Lebanon, NH, USA. [6]Department of Pathology and Laboratory Medicine, Dartmouth-Hitchcock Medical Center, Lebanon, NH, USA. [7]Department of Community and Family Medicine, Geisel School of Medicine at Dartmouth, Lebanon, NH, USA.

References
1. Bonfield T, Chmiel JF. Impaired innate immune cells in cystic fibrosis: is it really a surprise? J Cyst Fibros. 2017;16(4):433–5.
2. Paemka L, McCullagh BN, Abou Alaiwa MH, Stoltz DA, Dong Q, Randak CO, et al. Monocyte derived macrophages from CF pigs exhibit increased inflammatory responses at birth. J Cyst Fibros. 2017;16(4):471–4.
3. Tarique AA, Sly PD, Holt PG, Bosco A, Ware RS, Logan J, et al. CFTR-dependent defect in alternatively-activated macrophages in cystic fibrosis. J Cyst Fibros. 2017;16(4):475–82.
4. Simonin-Le Jeune K, Le Jeune A, Jouneau S, Belleguic C, Roux PF, Jaguin M, et al. Impaired functions of macrophage from cystic fibrosis patients: CD11b, TLR-5 decrease and sCD14, inflammatory cytokines increase. PLoS One. 2013;8(9): e75667.
5. Bruscia EM, Zhang PX, Satoh A, Caputo C, Medzhitov R, Shenoy A, et al. Abnormal trafficking and degradation of TLR4 underlie the elevated inflammatory response in cystic fibrosis. J Immunol. 2011;186(12):6990–8.
6. Kopp BT, Abdulrahman BA, Khweek AA, Kumar SB, Akhter A, Montione R, et al. Exaggerated inflammatory responses mediated by Burkholderia cenocepacia in human macrophages derived from cystic fibrosis patients. Biochem Biophys Res Commun. 2012;424(2):221–7.
7. Pfeffer KD, Huecksteadt TP, Hoidal JR. Expression and regulation of tumor necrosis factor in macrophages from cystic fibrosis patients. Am J Respir Cell Mol Biol. 1993;9(5):511–9.
8. Vandivier RW, Fadok VA, Hoffmann PR, Bratton DL, Penvari C, Brown KK, et al. Elastase-mediated phosphatidylserine receptor cleavage impairs apoptotic cell clearance in cystic fibrosis and bronchiectasis. J Clin Invest. 2002;109(5):661–70.

9. Vandivier RW, Fadok VA, Ogden CA, Hoffmann PR, Brain JD, Accurso FJ, et al. Impaired clearance of apoptotic cells from cystic fibrosis airways. Chest. 2002;121(3 Suppl):89S.

10. Bonfield TL, Panuska JR, Konstan MW, Hilliard KA, Hilliard JB, Ghnaim H, et al. Inflammatory cytokines in cystic fibrosis lungs. Am J Respir Crit Care Med. 1995;152(6 Pt 1):2111–8.

11. Osika E, Cavaillon JM, Chadelat K, Boule M, Fitting C, Tournier G, et al. Distinct sputum cytokine profiles in cystic fibrosis and other chronic inflammatory airway disease. Eur Respir J. 1999;14(2):339–46.

12. Sagel SD, Chmiel JF, Konstan MW. Sputum biomarkers of inflammation in cystic fibrosis lung disease. Proc Am Thorac Soc. 2007;4(4):406–17.

13. Wu C, Morris JR. Genes, genetics, and epigenetics: a correspondence. Science. 2001;293(5532):1103–5.

14. Marsit CJ, Brummel SS, Kacanek D, Seage GR 3rd, Spector SA, Armstrong DA, et al. Infant peripheral blood repetitive element hypomethylation associated with antiretroviral therapy in utero. Epigenetics. 2015;10(8):708–16.

15. Morandini AC, Santos CF, Yilmaz O. Role of epigenetics in modulation of immune response at the junction of host-pathogen interaction and danger molecule signaling. Pathog Dis. 2016;74(7):ftw082.

16. Houseman EA, Kile ML, Christiani DC, Ince TA, Kelsey KT, Marsit CJ. Reference-free deconvolution of DNA methylation data and mediation by cell composition effects. BMC Bioinformatics. 2016;17:259.

17. Dewhurst JA, Lea S, Hardaker E, Dungwa JV, Ravi AK, Singh D. Characterisation of lung macrophage subpopulations in COPD patients and controls. Sci Rep. 2017;7(1):7143.

18. Magalhaes M, Rivals I, Claustres M, Varilh J, Thomasset M, Bergougnoux A, et al. DNA methylation at modifier genes of lung disease severity is altered in cystic fibrosis. Clin Epigenetics. 2017;9:19.

19. Logie C, Stunnenberg HG. Epigenetic memory: a macrophage perspective. Semin Immunol. 2016;28(4):359–67.

20. Cabanel M, Brand C, Oliveira-Nunes MC, Cabral-Piccin MP, Lopes MF, Brito JM, et al. Epigenetic control of macrophage shape transition towards an atypical elongated phenotype by histone deacetylase activity. PLoS One. 2015;10(7):e0132984.

21. Kittan NA, Allen RM, Dhaliwal A, Cavassani KA, Schaller M, Gallagher KA, et al. Cytokine induced phenotypic and epigenetic signatures are key to establishing specific macrophage phenotypes. PLoS One. 2013;8(10):e78045.

22. Gharib SA, Vaisar T, Aitken ML, Park DR, Heinecke JW, Fu X. Mapping the lung proteome in cystic fibrosis. J Proteome Res. 2009;8(6):3020–8.

23. Simhadri VL, Hansen HP, Simhadri VR, Reiners KS, Bessler M, Engert A, et al. A novel role for reciprocal CD30-CD30L signaling in the cross-talk between natural killer and dendritic cells. Biol Chem. 2012;393(1–2):101–6.

24. Nicod LP, Isler P. Alveolar macrophages in sarcoidosis coexpress high levels of CD86 (B7.2), CD40, and CD30L. Am J Respir Cell Mol Biol. 1997;17(1):91–6.

25. Kim HJ, Park J, Lee SK, Kim KR, Park KK, Chung WY. Loss of RUNX3 expression promotes cancer-associated bone destruction by regulating CCL5, CCL19 and CXCL11 in non-small cell lung cancer. J Pathol. 2015; 237(4):520–31.

26. Aldinucci D, Colombatti A. The inflammatory chemokine CCL5 and cancer progression. Mediat Inflamm. 2014;2014:292376.

27. Donato R. Intracellular and extracellular roles of S100 proteins. Microsc Res Tech. 2003;60(6):540–51.

28. Xia C, Braunstein Z, Toomey AC, Zhong J, Rao X. S100 proteins as an important regulator of macrophage inflammation. Front Immunol. 2017;8: 1908.

29. Maxeiner S, Shi N, Schalla C, Aydin G, Hoss M, Vogel S, et al. Crucial role for the LSP1-myosin1e bimolecular complex in the regulation of Fcgamma receptor-driven phagocytosis. Mol Biol Cell. 2015;26(9):1652–64.

30. Barrow AD, Palarasah Y, Bugatti M, Holehouse AS, Byers DE, Holtzman MJ, et al. OSCAR is a receptor for surfactant protein D that activates TNF-alpha release from human CCR2+ inflammatory monocytes. J Immunol. 2015; 194(7):3317–26.

31. Merck E, Gaillard C, Sciuller M, Scapini P, Cassatella MA, Trinchieri G, et al. Ligation of the FcR gamma chain-associated human osteoclast-associated receptor enhances the proinflammatory responses of human monocytes and neutrophils. J Immunol. 2006;176(5):3149–56.

32. Auffray C, Sieweke MH, Geissmann F. Blood monocytes: development, heterogeneity, and relationship with dendritic cells. Annu Rev Immunol. 2009;27:669–92.

33. Gardiner-Garden M, Frommer M. CpG islands in vertebrate genomes. J Mol Biol. 1987;196(2):261–82.

34. Pennacchio LA, Bickmore W, Dean A, Nobrega MA, Bejerano G. Enhancers: five essential questions. Nat Rev Genet. 2013;14(4):288–95.

35. Mohrs M, Blankespoor CM, Wang ZE, Loots GG, Afzal V, Hadeiba H, et al. Deletion of a coordinate regulator of type 2 cytokine expression in mice. Nat Immunol. 2001;2(9):842–7.

36. Kolahian S, Oz HH, Zhou B, Griessinger CM, Rieber N, Hartl D. The emerging role of myeloid-derived suppressor cells in lung diseases. Eur Respir J. 2016; 47(3):967–77.

37. Rieber N, Brand A, Hector A, Graepler-Mainka U, Ost M, Schafer I, et al. Flagellin induces myeloid-derived suppressor cells: implications for Pseudomonas aeruginosa infection in cystic fibrosis lung disease. J Immunol. 2013;190(3):1276–84.

38. Ost M, Singh A, Peschel A, Mehling R, Rieber N, Hartl D. Myeloid-derived suppressor cells in bacterial infections. Front Cell Infect Microbiol. 2016;6:37.

39. Oz HH, Zhou B, Voss P, Carevic M, Schroth C, Frey N, et al. Pseudomonas aeruginosa airway infection recruits and modulates neutrophilic myeloid-derived suppressor cells. Front Cell Infect Microbiol. 2016;6:167.

40. Zheng SC, Beck S, Jaffe AE, Koestler DC, Hansen KD, Houseman AE, et al. Correcting for cell-type heterogeneity in epigenome-wide association studies: revisiting previous analyses. Nat Methods. 2017;14(3):216–7.

41. Bessich JL, Nymon AB, Moulton LA, Dorman D, Ashare A. Low levels of insulin-like growth factor-1 contribute to alveolar macrophage dysfunction in cystic fibrosis. J Immunol. 2013;191(1):378–85.

42. Monick MM, Powers LS, Barrett CW, Hinde S, Ashare A, Groskreutz DJ, et al. Constitutive ERK MAPK activity regulates macrophage ATP production and mitochondrial integrity. J Immunol. 2008;180(11):7485–96.

43. Kling T, Wenger A, Beck S, Caren H. Validation of the MethylationEPIC BeadChip for fresh-frozen and formalin-fixed paraffin-embedded tumours. Clin Epigenetics. 2017;9:33.

44. Aryee MJ, Jaffe AE, Corrada-Bravo H, Ladd-Acosta C, Feinberg AP, Hansen KD, et al. Minfi: a flexible and comprehensive Bioconductor package for the analysis of Infinium DNA methylation microarrays. Bioinformatics. 2014;30(10):1363–9.

45. Houseman EA, Christensen BC, Yeh RF, Marsit CJ, Karagas MR, Wrensch M, et al. Model-based clustering of DNA methylation array data: a recursive-partitioning algorithm for high-dimensional data arising as a mixture of beta distributions. BMC Bioinformatics. 2008;9:365.

46. Wang J, Vasaikar S, Shi Z, Greer M, Zhang B. WebGestalt 2017: a more comprehensive, powerful, flexible and interactive gene set enrichment analysis toolkit. Nucleic Acids Res. 2017;45(W1):W130–W7.

13

The impact of methylation quantitative trait loci (mQTLs) on active smoking-related DNA methylation changes

Xu Gao[1†] ⓘ, Hauke Thomsen[2†], Yan Zhang[1], Lutz Philipp Breitling[1] and Hermann Brenner[1,3,4*]

Abstract

Background: Methylation quantitative trait loci (mQTLs) are the genetic variants that may affect the DNA methylation patterns of CpG sites. However, their roles in influencing the disturbances of smoking-related epigenetic changes have not been well established. This study was conducted to address whether mQTLs exist in the vicinity of smoking-related CpG sites ($\pm$ 50 kb) and to examine their associations with smoking exposure and all-cause mortality in older adults.

Results: We obtained DNA methylation profiles in whole blood samples by Illumina Infinium Human Methylation 450 BeadChip array of two independent subsamples of the ESTHER study (discovery set, $n = 581$; validation set, $n = 368$) and their corresponding genotyping data using the Illumina Infinium OncoArray BeadChip. After correction for multiple testing (FDR), we successfully identified that 70 out of 151 previously reported smoking-related CpG sites were significantly associated with 192 SNPs within the 50 kb search window of each locus. The 192 mQTLs significantly influenced the active smoking-related DNA methylation changes, with percentage changes ranging from 0.01 to 18.96%, especially for the weakly/moderately smoking-related CpG sites. However, these identified mQTLs were not directly associated with active smoking exposure or all-cause mortality.

Conclusions: Our findings clearly demonstrated that if not dealt with properly, the mQTLs might impair the power of epigenetic-based models of smoking exposure to a certain extent. In addition, such genetic variants could be the key factor to distinguish between the heritable and smoking-induced impact on epigenome disparities. These mQTLs are of special importance when DNA methylation markers measured by Illumina Infinium assay are used for any comparative population studies related to smoking-related cancers and chronic diseases.

Keywords: DNA methylation, Active smoking, Methylation quantitative trait loci, Epigenetic epidemiology

Background

Active smoking has been recognized as a critical lifestyle factor for cardiovascular, respiratory, and neoplastic diseases and contributes to the leading causes of preventable morbidity and mortality [1, 2]. DNA methylation, one of the main forms of epigenetic modification, is involved in the pathways of smoking and smoking-induced diseases [3, 4]. Previous epigenome-wide association studies (EWASs) based on whole blood samples have successfully discovered an increasing number of tobacco smoking-related CpG sites in various genes, such as *AHRR* and *F2RL3* [5–7]. These DNA methylation patterns have been shown to be useful as quantitative biomarkers to reflect both current and lifetime smoking exposure and to enhance the prediction of smoking-related risks [8–11].

DNA methylation of particular genomic loci might be influenced by neighboring genetic sequence variants [12]. The single nucleotide polymorphisms (SNPs) that are associated with methylation levels of CpG sites are known as methylation quantitative trait loci (mQTLs) [13]. This genetic effect has been determined across different tissues [13–16] and has been highlighted in several diseases, including neurological disorders, arthritis,

* Correspondence: h.brenner@dkfz-heidelberg.de
†Equal contributors
[1]Division of Clinical Epidemiology and Aging Research, German Cancer Research Center (DKFZ), Im Neuenheimer Feld 581, 69120 Heidelberg, Germany
[3]Division of Preventive Oncology, German Cancer Research Center (DKFZ) and National Center for Tumor Diseases (NCT), Im Neuenheimer Feld 460, 69120 Heidelberg, Germany
Full list of author information is available at the end of the article

and cancer [17–21]. Recently, the mQTLs have been further reported to play a modifying role in the associations between DNA methylation levels at specific CpG sites and environmental exposures. For instance, Zhang et al. identified 238 mQTLs that were associated with 65 alcohol dependence-related CpG sites in African Americans and 305 mQTLs for 44 unique CpG sites in European Americans [22]. In 2016, Gonseth et al. found out that three of the strongest maternal smoking-related CpG sites in newborns were significantly associated with SNPs located in the vicinity of each gene [23]. Thus, these hereditary traits provide a possible mechanism by which methylation patterns could be different under environmental exposures, if the distribution of risk alleles differs between the exposed and the unexposed. In addition, the linkages of epigenetic signatures to genotypes might also further provide more mechanistic evidence on the genetic and environmental risk factors for various forms of diseases [24].

However, such genetic influences have not been well addressed or even overlooked by previous EWASs of active smoking exposure; to our knowledge, no study has so far investigated their contributions to the methylation intensities of active smoking-related CpG sites and smoking-related health outcomes in the general population. Therefore, we conducted a comprehensive analysis in a large population-based study of older adults in Germany with the aim of exploring the hitherto unknown association between active smoking-related DNA methylation and individual genetic variations. In particular, we aimed to identify the mQTLs within ± 50 kb from each of 151 previously reported active smoking-related CpG sites in whole blood samples [25] and to assess their relationships with active smoking exposure and all-cause mortality.

Results

Participant characteristics

Characteristics of the study population in the discovery and the validation panel with respect to smoking behaviors and lifestyle factors are summarized in Table 1. The average age of the participants of both subsets at the baseline was about 61 years. About half of the participants in each subset were ever smokers (current or former smokers), and around 18% still smoked at the time of recruitment. Female participants included a larger proportion of never smokers than males (discovery set, 67.9 vs. 28.6%; validation set, 63.3 vs. 21.4%). Average cumulative smoking exposure (pack-years) of current smokers was considerably higher than that of former smokers in both subsets (discovery set, 34.6 vs. 22.0; validation set, 33.1 vs. 19.4). Average time after smoking cessation (years) of former smokers in both subsets was also similar, approximately 17 years. The majority of participants in both subsets of the study

Table 1 Study population characteristics in discovery and validation panels (mean values (SD) for continuous variables and n (%) for categorical variables)

Characteristics	Discovery panel	Validation panel	p value
N	581	368	
Age (years)	61.0 (6.3)	61.1 (6.4)	0.809
Sex (male)	241 (41.5%)	117 (31.8%)	< 0.001
Smoking status			0.864
Current smoker	108 (18.6%)	65 (17.7%)	
Former smoker	173 (29.8%)	119 (32.3%)	
Never smoker	300 (51.6%)	184 (50.0%)	
Pack-years of smoking[a]			
Current smokers	34.6 (18.2)	33.1 (18.2)	0.250
Former smokers	22.0 (17.5)	19.4 (15.5)	0.033
Smoking cessation time (years)[b]	16.5 (11.3)	17.2 (10.2)	0.742
Body mass index[c]			0.248
Underweight or normal weight (< 25.0)	143 (24.7%)	116 (31.5%)	
Overweight (25.0–< 30.0)	290 (50.2%)	151 (41.0%)	
vObese (≥ 30.0)	145 (25.1%)	101 (27.5%)	
Alcohol consumption[d]			0.509
Abstainer	194 (36.3%)	128 (38.0%)	
Low	301 (56.4%)	188 (55.8%)	
Intermediate	30 (5.6%)	17 (5.0%)	
High	9 (1.7%)	4 (1.2%)	
Physical activity[e]			0.058
Inactive	109 (18.8%)	82 (22.3%)	
Low	245 (42.2%)	176 (47.8%)	
Medium or high	227 (39.0%)	110 (29.9%)	
Prevalence of CVD at baseline[f]			0.621
Prevalent	86 (14.8%)	58 (15.8%)	
Prevalence of diabetes at baseline[g]			0.617
Prevalent	86 (14.9%)	60 (16.6%)	
Prevalence of cancer at baseline			
Prevalent	33 (5.7%)	22 (6.0%)	0.744

[a]For subgroups of former and current smokers; data missing for 38 and 24 participants, respectively, in discovery and validation panels; a pack-year was defined as having smoked 20 cigarettes per day for 1 year
[b]Former smokers only, data missing for 5 and 2 participants, respectively, in discovery and validation panels; cessation time equals age at recruitment minus age at cessation
[c]Data missing for 3 participants in discovery panel
[d]Data missing for 47 and 31 participants, respectively, in discovery and validation panels. Categories defined as follows: abstainer, low [women, 0–< 20 g/d; men, 0–< 40 g/d], intermediate [20–< 40 g/d and 40–< 60 g/d, respectively], high [≥ 40 g/d and ≥ 60 g/d, respectively]
[e]Categories defined as follows: inactive [< 1 h of physical activity/week], medium or high [≥ 2 h of vigorous or ≥ 2 h of light physical activity/week], low (other)
[f]CVD cardiovascular disease. Data missing for 1 participant in discovery panel
[g]Data missing for 5 and 7 participants, respectively, in discovery and validation panels

population were overweight or obese, reported no or only low physical activity, and no or low amounts of alcohol drinking. During a median follow-up time of about 12 years (discovery set, 12.6 years; validation set, 12.2 years), 94 participants died in the discovery set (CVD = 30, cancer = 46, other diseases = 18) and 49 died in the validation set (CVD = 17, cancer = 21, other diseases = 11).

Identification of mQTLs for smoking-related CpG sites

For the 1396 SNP-CpG pairs consisting of 150 smoking-related CpG sites and 909 corresponding SNPs (Fig. 1), 380 pairs were significant at a FDR < 0.05 in the discovery panel even after controlling for covariates (Additional file 1: Table S2; Additional file 2: Figure S1). These 380 pairs were then replicated in the validation panel by applying the fully adjusted mixed linear regression model. A subset of 246 pairs formed of 70 CpG sites and 192 SNPs reached the statistical significance level after FDR correction (FDR < 0.05; Table 2, Additional file 1: Table S3, Additional file 1: Table S4; Fig. 2). Eventually, 192 SNPs were designated as the mQTLs of 70 CpG sites. The pair cg23576855/rs75509302 showed the strongest inter-relationship (FDR-corrected p value = 8.86 e – 103). Among the 70 CpG sites, five were highly smoking-related loci (reported ≥ 6 times; Table 3), 14 were moderately smoking-related (4 or 5 times), and 51 were weakly smoking-related (2 or 3 times). These CpG sites with mQTLs were mainly located in the gene body (37/70), ten were located in transcription start sites (TSS1500) and 23 were in untranslated regions (UTR) or intergenetic regions (Additional file 1: Table S3). The largest number of

mQTLs (n = 8) was found for locus cg06126421 within *6p21.33* (Table 3; Fig. 3). The coefficients of mQTLs ranged from –0.54 to 0.15.

The 192 mQTLs were mainly mapped on chromosomes 1 (16%), 6 (10%), and 7 (25%). Three SNPs were the most frequently identified mQTLs, rs75509392 (MAF = 0.144) for eight CpG sites within *AHRR*, rs79050605 (MAF = 0.202), and rs34835481 (MAF = 0.210) for five and six CpG sites located in *GFI1*, respectively (Table 4). We assessed the effects of the mQTLs on the DNA methylation changes as the absolute values of the coefficient changes of smoking status between the models without and with adjusting for corresponding mQTLs (carrier/non-carrier) among the 246 SNP-CpG pairs (Additional file 1: Table S5). Part of the mQTLs had opposite effects on different CpG sites (Additional file 1: Table S5, Additional file 3: Figure S2). For example, the minor allele of the SNP rs75509302 attenuated the association of smoking exposure with the methylation of cg11902777 by 5.2% (Additional file 1: Table S5, Additional file 3: Figure S2). In contrast, this variant strengthened the demethylation of cg17287155 in response to different smoking behaviors by 2.44%.

Genetic contributions of mQTLs to the DNA methylation changes

As shown in Additional file 1: Table S5, the associations between smoking exposure and DNA methylation were changed by between 0.01 and 18.96% by the mQTLs and were categorized by the distances between genetic variants and CpG sites and the reported frequencies of CpG sites. We observed that the closest SNPs (distance < 10 kb) had a slightly lower impact on DNA methylation

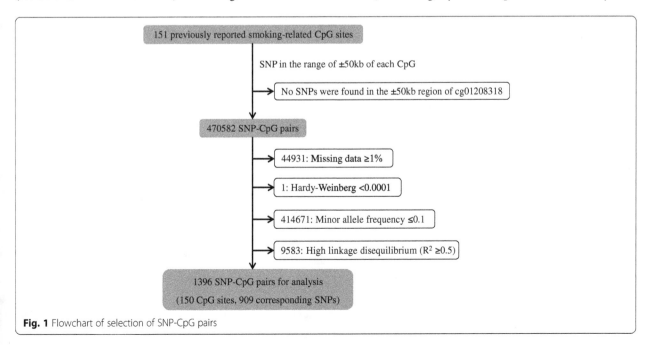

Fig. 1 Flowchart of selection of SNP-CpG pairs

Table 2 List of 246 significant SNP-CpG pairs (chromosomal and CpG sites positions were based on GRCh37/hg19)

Chromosome	Gene	CpG site	Position	Number of SNP candidates	Number of mQTLs
1	AVPR1B	cg09069072	15,482,754	13	5
	GFI1	cg09662411	92,946,132	10	2
		cg09935388	92,947,588	10	2
		cg10399789	92,945,668	9	6
		cg12876356	92,946,825	10	2
		cg18146737	92,946,701	10	2
		cg18316974	92,947,035	10	2
	GNG12	cg25189904	68,299,493	5	2
	NOS1AP	cg11231349	162,050,657	4	2
	TMEM51	cg21913886	15,485,346	14	9
	unassigned	cg03547355	227,003,061	8	2
	unassigned	cg12547807	9,473,751	9	1
	unassigned	cg21393163	12,217,630	8	2
	unassigned	cg26764244	68,299,511	3	3
2	2q37.1	cg05951221	233,284,402	5	1
		cg03329539	233,283,329	5	2
	ALPP	cg23667432	233,244,439	5	2
	NFE2L2	cg26271591	178,125,956	8	6
	SNED1	cg26718213	241,976,081	9	4
	unassigned	cg27241845	233,250,371	5	2
3	GPX1	cg18642234	49,394,623	10	5
5	AHRR	cg03604011	400,201	15	5
		cg03991871	368,448	9	1
		cg11902777	368,843	9	4
		cg12806681	368,395	9	2
		cg14817490	392,920	15	4
		cg17287155	393,347	15	1
		cg23576855	373,300	9	7
		cg23916896	368,805	9	4
6	6p21.33	cg06126421	30,720,081	16	8
	CDKN1A	cg15474579	36,645,813	22	8
	IER3	cg15342087	30,720,210	16	2
		cg24859433	30,720,204	16	3
	TIAM2	cg00931843	155,442,993	6	1
	VARS	cg17619755	31,760,629	16	8
	unassigned	cg14753356	30,720,109	8	2
7	C7orf40	cg03440944	45,023,330	5	1
	CNTNAP2	cg11207515	146,904,206	14	7
		cg25949550	145,814,306	11	8
	GNA12	cg19717773	2,847,554	22	22
	HOXA7	cg08396193	27,193,709	7	1
	LRRN3	cg11556164	110,738,316	5	5
	MYO1G	cg12803068	45,002,919	10	1
		cg22132788	45,002,487	10	1

Table 2 List of 246 significant SNP-CpG pairs (chromosomal and CpG sites positions were based on GRCh37/hg19) *(Continued)*

	TNRC18	cg09022230	5,457,226	24	3
8	ZC3H3	cg26361535	144,576,604	8	1
	unassigned	cg19589396	103,937,374	15	2
9	unassigned	cg01692968	108,005,349	2	2
10	ARID5B	cg25953130	63,753,550	6	3
11	KCNQ1	cg07123182	2,722,391	13	2
		cg26963277	2,722,408	13	3
		cg01744331	2,722,358	13	3
	KCNQ1OT1	cg16556677	2,722,402	13	2
	PRSS23	cg11660018	86,510,915	9	2
		cg23771366	86,510,999	9	2
	unassigned	cg16611234	58,870,075	10	10
14	C14orf43	cg01731783	74,211,789	6	1
		cg10919522	74,227,441	5	1
15	ANPEP	cg23161492	90,357,203	19	6
	SEMA7A	cg00310412	74,724,919	13	9
16	ITGAL	cg09099830	30,485,486	3	2
	XYLT1	cg16794579	17,562,419	3	1
17	LOC100130933	cg07251887	73,641,810	6	2
	STXBP4	cg07465627	53,167,407	8	4
19	CPAMD8	cg15159987	17,003,890	15	4
	CRTC1	cg23973524	18,873,223	12	1
	F2RL3	cg03636183	17,000,586	17	1
	PPP1R15A	cg03707168	49,379,127	10	4
21	ETS2	cg23110422	40,182,073	6	3
22	NCF4	cg02532700	37,257,404	8	2
Total				590	246

levels than the mQTLs located ≥ 10 kb (Fig. 4a). Compared with the highly smoking-related CpG sites, the mQTLs affect the methylation levels of weakly smoking-related CpG sites the most, and the changes of moderately smoking-related CpG sites stayed in the intermediate position (Fig. 4b, $F = 4.91$, p value = 0.008). Additionally, potential gene-environment interactions of the 192 mQTLs with smoking behaviors (current/never smoking) were assessed. Only for rs75509302, a significant interaction with smoking was observed regarding the methylation levels of cg23576855 (Tables 5, Additional file 1: Table S6).

Associations of mQTLs with active smoking exposure and all-cause mortality

Finally, we tested the associations of the 192 mQTLs (carrier/non-carrier) with different measurements of smoking exposure [ever smoking (current and former smoking) vs. never smoking, current smoking vs. never smoking, current smoking vs. former smoking, cumulative smoking (pack-years), durations of smoking (years), and the age of smoking initiation]. None of them was significantly associated with any smoking indicators after the correction of FDR (Additional file 1: Table S7). Similarly, the 192 SNPs were not significantly associated with all-cause mortality (death from CVD, cancer, and other chronic diseases) based on the results of the COX model (Additional file 1: Table S8).

Discussion

We conducted the first association study of 150 active smoking-related CpG sites and their corresponding SNPs located in the ± 50 kb region utilizing the genomic and epigenomic data of 949 participants from the ESTHER study. We found the DNA methylation levels of 70 CpG sites to be influenced by 192 proximal SNPs. These 192 mQTLs modified the DNA methylation changes in response to active smoking exposure, especially for the weakly/moderately active smoking-related CpG sites, but we did not observe any direct associations with active smoking exposure or all-cause mortality.

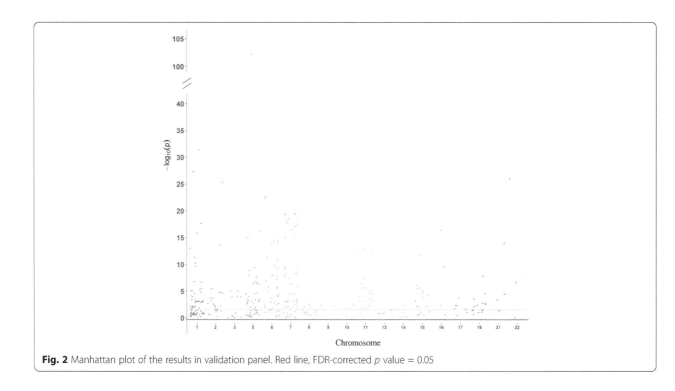

Fig. 2 Manhattan plot of the results in validation panel. Red line, FDR-corrected *p* value = 0.05

Table 3 Five frequently reported (≥ 6) CpG sites and corresponding mQTLs

CpG site	Frequency[a]	Gene	Chr	SNP	SNP position[b]	Minor allele	Distance (bp)[c]	FDR[d]	MAF[e]
cg03636183	12	*F2RL3*	19	rs2227357	17,003,553	A	2967	0.048	0.125
cg05951221	8	*2q37.1*	2	rs790051	30,718,035	A	− 1866	6.2 e − 4	0.226
cg06126421	7	*6p21.33*	6	rs2535324	30,727,983	C	− 2046	1.8 e − 9	0.3
				rs3095339	30,728,290	G	7902	2.6 e − 4	0.252
				rs3131036	30,728,360	A	8209	2.6 e − 4	0.253
				rs3094122	30,737,552	G	8279	3.2 e − 3	0.206
				rs13217914	30,739,657	A	17,471	2.4 e − 21	0.157
				rs6911571	30,753,639	T	19,576	0.007	0.16
				rs4713361	30,756,066	A	33,558	1.3 e − 21	0.159
				rs13201769	30,718,035	A	35,985	6.9 e − 7	0.326
cg03329539	6	*2q37.1*	2	rs790051	233,282,536	A	− 793	0.031	0.226
				rs34547337	233,300,755	T	17,426	1.5 e − 5	0.314
cg14817490	6	*AHRR*	5	rs75509302	365,653	C	− 27,267	0.002	0.144
				rs11746079	410,980	C	18,060	1.5 e − 3	0.154
				rs72717419	431,996	T	39,076	0.021	0.207
				rs2672725	434,981	G	42,061	0.042	0.117

[a]The reported times of CpG in previous studies (based on systematic review [25])
[b]Positions of CpG sites and SNPs were based on GRCh37/hg19
[c]The distance between SNP and CpG (SNP position–CpG position)
[d]The FDR-corrected *p* values of SNPs in fully adjusted mixed linear regression models, which controlled for age (years), sex, smoking status, random batch effect of methylation measurement, leukocyte distribution (Houseman algorithm), alcohol consumption (abstainer/low/intermediate/high), body mass index (BMI, underweight or normal weight/overweight/obese), physical activity (inactive/low/medium or high), prevalence of cardiovascular diseases (yes/no), prevalence of diabetes (yes/no), and prevalence of cancer (yes/no)
eMAF minor allele frequency

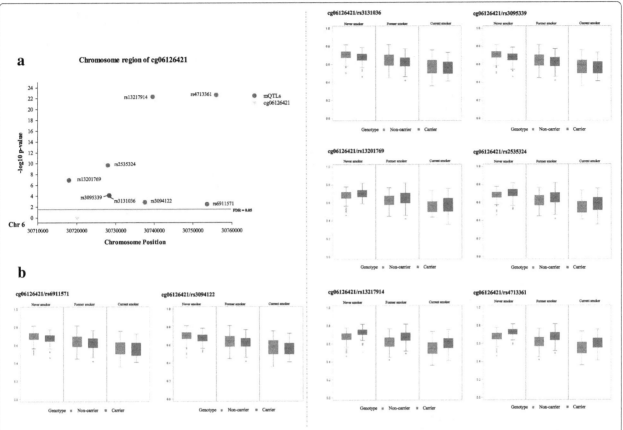

Fig. 3 Locations of cg06126421 and eight mQTLs (carrier/non-carrier) (**a**) and distributions of methylation levels based on smoking status (**b**) in validation panel. Red line, FDR-corrected *p* value = 0.05; red dot, mQTLs; yellow triangle, cg06126421; blue bar, non-carriers of minor allele; red bar, carriers of minor allele

The mQTLs are presented in the vicinity of active smoking-related CpG sites. Locus cg06126421 (*6p21.33*) is one of the pronounced smoking-related CpG sites with eight mQTLs. Four of the SNPs could impair the hypomethylation of cg06126421, while others could accelerate this process. All of these mQTLs were located in a genomic region relating to inflammation and/or immune-related (malignant) diseases, such as allergies [26], multiple myeloma [27], diffuse large B cell lymphoma [28], and lung cancer [29]. We also identified a mQTL rs2227357 which could slightly promote the demethylation of cg03636183 (*F2RL3*), but no mQTLs were discovered near other well-established smoking-related loci, like cg05575921 (*AHRR*) or cg19859270 (*GPR15*). However, eight CpG sites within the *AHRR* region were found to be modified by mQTLs. Among them, cg23576855 manifested the strongest connection with rs75509302 and was also the only hit which could be affected by the interaction between smoking status and corresponding mQTL. This phenomenon might be highly contributed by the genetic feature of this locus, which is also a CG → CA SNP annotated as rs6869832 [5, 30], and shares a very high LD with rs75509302 in

this study (R² = 0.94). Nevertheless, to our knowledge, the biological functions for most of the identified mQTLs in our study are not fully understood yet and need to be explored in further research.

The mQTLs may help to distinguish between the genetic and environmental effects on epigenome disparities. Researchers usually observed several outliers out of the predictive range of epigenetic signatures in EWASs [6, 7, 9, 31]. One of the most plausible explanations is measurement bias that may result from recall bias or intentional underreporting [32]. Our finding provides another possibility that the deviations of DNA methylation levels might be caused by neighboring genetic variants. For highly smoking-related CpG sites, active smoking is still the main driver of DNA methylation changes. For instance, the SNP rs2227357 only contributed to about 0.01% of the changes of the methylation level of cg03636183 (*F2RL3*), and the SNP rs790051 altered only 0.37% of the methylation level of cg05951221 (*2q37.1*). However, the mQTLs affected the methylation levels of less robustly smoking-related loci much more. For instance, the SNPs rs78131 and rs2741302 explained nearly 19% of the changes of cg26963277 (*KCNQ1*) and

Table 4 Three most frequently identified mQTLs and corresponding CpG sites

SNP	Chr	SNP position[a]	Minor allele	MAF[b]	CpG	Distance (bp)[c]	FDR[d]
rs75509302	5	365,653	C	0.144	cg23576855	− 7647	3.4 e − 100
					cg11902777	− 3190	1.4 e − 7
					cg17287155	− 27,694	7.8 e − 5
					cg03991871	− 2795	1.0 e − 4
					cg12806681	− 2742	1.2 e − 4
					cg23916896	− 3152	9.5 e − 4
					cg03604011	− 34,548	1.1 e − 3
					cg14817490	− 27,267	2.1 e − 3
rs34835481	1	92,991,624	T	0.210	cg10399789	45,956	2.2 e − 5
					cg12876356	44,799	1.3 e − 3
					cg09662411	45,492	1.9 e − 3
					cg18146737	44,923	2.0 e − 3
					cg18316974	44,589	3.0 e − 3
					cg09935388	44,036	0.016
rs79050605	1	92,925,962	G	0.202	cg12876356	− 20,863	3.7 e − 4
					cg18146737	− 20,739	1.1 e − 3
					cg18316974	− 21,073	1.7 e − 3
					cg09662411	− 20,170	1.9 e − 3
					cg09935388	− 21,626	2.2 e − 3

[a]SNPs positions were based on GRCh37/hg19
[b]MAF minor allele frequency
[c]The distance between SNP and CpG (SNP position–CpG position)
[d]The FDR-corrected p values of SNPs in fully adjusted mixed linear regression models, which controlled for age (years), sex, smoking status, random batch effect of methylation measurement, leukocyte distribution (Houseman algorithm), alcohol consumption (abstainer/low/intermediate/high), body mass index (BMI, underweight or normal weight/overweight/obese), physical activity (inactive/low/medium or high), prevalence of cardiovascular diseases (yes/no), prevalence of diabetes (yes/no), and prevalence of cancer (yes/no)

cg27241845, respectively. While additional external validation studies certainly are needed, we speculate that this strong diversity of results might be a result of undetermined biological interactions between the SNP and CpG sites.

Smoking-related CpG sites have been recognized as informative signatures of smoking exposure and smoking-related health outcomes [3, 4]. Part of these 70 CpG sites with mQTLs have been reported to be highly associated with long-term smoking exposure [cg03636183 (*F2RL3*) and cg06126421 (*6p21.33*)] [9, 10], aging-related health outcomes, such as telomere length (cg21393163) [33] and the development of frailty [cg14753356, cg19589396, cg23667432 (*ALPP*) and cg25189904 (*GNG12*)] [34], and were even employed to construct a comprehensive index to predict smoking impact in buccal cells [35]. Together with previous studies [8, 23], we suggest that future investigations utilizing smoking-related CpG sites might need to take the genotypes, especially the mQTLs of less robustly smoking-related loci, into consideration to account for their potential impact on DNA methylation levels.

Beyond the SNP-CpG associations, null associations of 192 mQTLs with active smoking and all-cause mortality

additionally imply that these novel genetic variants might be independently associated with the DNA methylation changes and might not be involved in the pathophysiological development of smoking-related health outcomes. Therefore, these mQTLs might have the potential to be used in the causal inference tests between the CpG sites and smoking-related health outcomes as instrumental variables (Mendelian Randomization, MR) [36]. Recently, researchers have suggested a two-stage MR test to establish the causal role of epigenetic processes in pathways of diseases [37]. Larger population-based investigations with longitudinal design and repeated measurements of smoking exposure and epigenome data are warranted to evaluate these potential instrumental variables and obtain further insights into the plausibility of suggested causal effects of DNA methylation in the development of smoking-related diseases.

Major strengths of the present study include comprehensive information on a broad range of covariates in a population-based cohort and validation in an independent subgroup. Some limitations still have to be acknowledged in the interpretation of study results. First, smoking-related shifts in leukocyte distribution might affect the

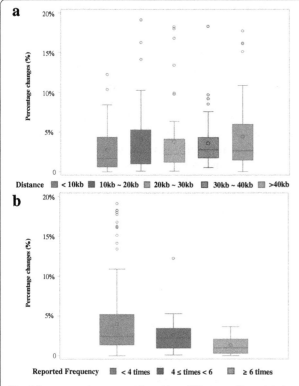

Fig. 4 Percentage changes contributed by mQTLs to smoking-related DNA methylation changes based on SNP-CpG distance (**a**) and reported frequencies of CpG sites (**b**)

(± 50 kb) in whole blood DNA due to the consideration of controlling for the pleiotropic effect or reverse causations from unknown genetic or epigenetic factors and the limited coverage of OncoArray [40], more mQTLs (*cis-* or *trans-*) for smoking-related CpG sites need to be established by expanding the search window. Since DNA methylation is highly tissue-specific, larger cohorts with various human tissues are also needed for more comprehensive evaluation of the whole landscape of genetic impact on the epigenome.

Conclusions

In conclusion, this study identified 192 mQTLs for 70 smoking-related CpG sites. These variants might theoretically reflect inherited differences in epigenetic states of people and their susceptibilities to smoking-related health outcomes. Incorporation of mQTLs might enhance the epigenetic-based assessments of smoking or smoking-related health outcomes by accounting for potential confounding from genetic background. Our results need to be further validated and confirmed in additional studies with larger number of participants and more detailed assessment of genomic and epigenomic data, including the CpG sites that have not previously been replicated. Along with previous investigations on the epigenetic changes related to other environmental exposures or lifestyle factors, our study adds evidence for the complex interplays among genetic traits, epigenetic signatures, and environmental factors.

Methods

Study design and population

Study subjects were selected from the ESTHER study, an ongoing statewide population-based cohort study conducted in Saarland, a state located in southwestern Germany. Details of the study design have been reported previously [41]. Briefly, 9949 older adults (aged 50–75 years) were enrolled by their general practitioners during a routine health check-up between July 2000 and December 2002 and followed up thereafter. The current cross-sectional analysis is based on data and biospecimen collected at baseline. Two independent subgroups

associations of DNA methylation in whole blood samples with active smoking [38]. Hence, we adjusted for leukocyte distribution by the Houseman algorithm to restrict potential confounding from differential blood counts to the greatest possible extent [39]. Further studies are also needed to evaluate to what extent our results can be generalized to middle-aged individuals or non-Caucasians, as the ESTHER study was conducted in the older (aged 50–75 years), almost exclusively Caucasian population in southern Germany during a routine screening program. In addition, our study had limited power for detecting direct associations of mQTLs with smoking exposure and all-cause mortality due to limited numbers of cases. Finally, we only measured a relatively small window of genetic regions

Table 5 Impact of rs75509302-smoking interaction on the methylation level of cg23576855[a]

Gene	CpG site	SNP	SNP-smoking interaction[b]				Smoking status[d]		
			Genotype[c]	Coefficient	SE	p value	Coefficient	SE	p value
AHRR	cg23576855	rs75509302	TT	Ref			−0.182	7.1 e − 3	3.8 e − 95
			CT	0.128	0.013	1.1 e − 20			
			CC	0.268	0.033	2.5 e − 15			

[a]Model is fully adjusted for age, sex, BMI, smoking status (current and never smoking only), alcohol consumption, physical activity, prevalence of CVD, diabetes and cancer at baseline. The methylation levels of CpG sites were responses, the SNPs and SNP-smoking interactions were predictors;
[b]The never smoking * genotype groups and current smoking * TT group were used as references;
[c]The group of interaction between current smoking and listed genotype;
[d]Never smoking was used as reference

were selected as the discovery and the validation panel for DNA methylation analyses as previously described [33]. Briefly, the discovery panel included 581 participants recruited consecutively at the start of the ESTHER study between July and October 2000. The validation panel included 368 participants randomly selected from the participants recruited between October 2000 and March 2001. The study was approved by the ethics committees of the University of Heidelberg and the state medical board of Saarland, Germany. Written informed consent was obtained from all participants.

Data collection

Information on socio-demographic characteristics, lifestyle factors, and health status at baseline was obtained by standardized self-administered questionnaires. In particular, detailed information on lifetime smoking history was obtained, including current smoking status and intensity, age at initiation, and smoking intensities at various ages, as well as the age of quitting smoking for former smokers [42]. Additional information on body mass index (BMI) was extracted from a standardized form filled by the general practitioners during the health check-ups. Prevalent cardiovascular disease (CVD) at baseline was defined by either physician-reported coronary heart disease or a self-reported history of a major cardiovascular event, such as myocardial infarction, stroke, pulmonary embolism, or revascularization of coronary arteries. Prevalent diabetes was defined by physician diagnosis or the use of glucose-lowering drugs. Prevalent cancer [ICD-10 C00-C99 except non-melanoma skin cancer (C44)] was determined by self-report or record linkage with data from the Saarland Cancer Registry (http://www.krebsregister.saarland.de/ziele/ziel1.html; in German). Deaths during the follow-up (between 2000 and end of 2014) were identified by record linkage with population registries in Saarland. Participants migrated out of Saarland were censored at the date last known to be alive. Information about the major cause of death was obtained from death certificates provided by the local public health offices and was coded with ICD-10 codes.

DNA methylation data

Blood samples were taken during the health check-up and stored at −80 °C until further processing. DNA from whole blood samples was collected using a salting out procedure [43]. DNA methylation profiles were extracted by the Illumina Human Methylation 450K BeadChip (Illumina, San Diego, CA, USA). As previously described [44], samples were analyzed following the manufacturer's instruction at the Genomics and Proteomics Core Facility of the German Cancer Research Center, Heidelberg, Germany. Illumina's GenomeStudio® (version 2011.1; Illumina, Inc.) was employed to extract DNA methylation

signals from the scanned arrays (Module version 1.9.0; Illumina, Inc.). The methylation level of a specific CpG site was quantified as a β value ranging from 0 (no methylation) to 1 (full methylation). According to the manufacturer's protocol, no background correction was done and data were normalized to internal controls provided by the manufacturer. All controls were checked for inconsistencies in each measured plate. Probes with a detection p value > 0.05 were excluded from analysis. We utilized the Illumina normalization and preprocessing method implemented in Illumina's GenomeStudio®. We selected the profiles of 151 smoking-related loci which had been identified ≥ 2 times in previous smoking EWASs for the present analysis [25].

Genotyping data

Extracted DNA from blood cells was genotyped using the Illumina Infinium OncoArray BeadChip (Illumina, San Diego, CA, USA). General genotyping quality control assessment was as previously described [45]. Genotypes for common variants across the genome were imputed using data from 1000 Genomes Project (phase 3, Oct. 2014) with IMPUTE2 v2.3.2 after pre-phasing with SHAPEIT software v2.12. We set thresholds for imputation quality to retain both potential common and rare variants for validation. Specifically, poorly imputed SNPs defined by an information metric $I < 0.70$ were excluded. All genomic locations are given in NCBI Build 37/UCSC hg19 coordinates. All SNPs having a MAF < 1% were excluded. After imputation, the SNP set consisted of 9,198,808 genotyped and imputed SNPs. PLINK v1.90 was then used to extract SNPs for the required regions of interest [46]. As shown in Fig. 1, we first identified SNPs within 50 kb upstream and downstream from each of the 151 smoking-related CpG sites (470,582 SNP-CpG pairs), a window in which most SNPs with significant cis associations with CpG sites are located [13]. The locus cg01208318 was excluded without any corresponding SNPs in this restricted region. For each of the remaining 150 CpG sites, we excluded any SNPs with ≥ 1% missing values ($n = 44,931$), deviating from the Hardy-Weinberg equilibrium (HWE exact test's p value < 0.0001, $n = 1$), with a minor allele frequency ≤ 0.1 ($n = 414,671$) or with high linkage disequilibrium (LD, $R^2 ≥ 0.5$) (Additional file 1: Table S1). After the final quality control, 1396 SNP-CpG pairs with strongest SNPs remained for analysis, which were constituted of 150 CpG sites and 909 corresponding SNPs (Additional file 1: Table S2).

Statistical analyses

First, major socio-demographic characteristics, lifestyle factors, smoking behavior, and prevalence of major chronic diseases in both the discovery and the validation panel were summarized by descriptive statistics.

We then evaluated the associations between the methylation intensities of the 150 CpG sites and corresponding SNPs to identify mQTLs as follows. For all SNP-CpG pairs, we used a mixed linear regression model with methylation batch as a random effect in which the methylation level of CpG site was the outcome and each regional SNP was the predictor (categorical variable, coded into 0, 1, and 2 based on the numbers of the minor allele). The model was fully adjusted for the following covariates that have been shown to be associated with DNA methylation changes [47–54]: age (years), sex (male/female), smoking status (current/former/never smoking), alcohol consumption (abstainer, low [women, 0 to < 20 g/d; men, 0 to < 40 g/d], intermediate [20 to < 40 g/d and 40 to < 60 g/d, respectively], high [≥ 40 g/d and ≥ 60 g/d, respectively]), body mass index (BMI, kg/m^2, underweight [< 18.5, < 1% of the study population] or normal weight [18.5 to < 25], overweight [25 to < 30], obese [≥ 30]), physical activity (inactive [< 1 h of physical activity/week], medium or high [≥ 2 h of vigorous or ≥ 2 h of light physical activity/week], low [other]), the leukocyte distribution estimated by the Houseman algorithm [39], the prevalence of CVD (yes/no), diabetes (yes/no), and cancer (yes/no) at the baseline. After correction for multiple testing by false discovery rate (FDR, Benjamini-Hochberg method [55]), SNP-CpG pairs with a FDR < 0.05 were selected and then analyzed in the validation panel. SNPs of the pairs with a FDR < 0.05 in the validation panel were eventually identified as the mQTL for the corresponding CpG site.

Furthermore, we tested the contributions of the identified mQTLs to the DNA methylation levels of corresponding CpG sites. Due to the limited number of individuals in the subgroup of minor homozygotes, we recoded the SNPs in order to use the dominant model, in which the heterozygote and minor homozygotes were combined as the carrier of minor allele and the major homozygotes were considered as non-carrier of the minor allele. We compared the coefficients of active smoking exposure (current vs. never smoking) in the fully adjusted model without the mQTLs (β_1) with the fully adjusted model including the mQTLs (β_2). The changes of coefficients were calculated as 100% * ($\beta_1-\beta_2$)/β_1, and their absolute values were determined as the percentage change contributed by mQTLs. The percentage changes were categorized by the absolute distances (bp) between CpG sites and corresponding mQTLs and the reported frequencies of CpG sites. To explore whether the gene-environment interactions could modify the DNA methylation changes of smoking-related CpG sites, we also tested whether interactions between the identified mQTLs and active smoking exposure (current vs. never smoking) could affect the impact of smoking on the methylation levels of corresponding

CpG sites. The mQTLs, smoking status, and their interaction (mQTLs*smoking status) were added in the model as predictors, and the methylation levels of CpG sites were outcomes. After controlling for all the potential covariates, the interactions with a FDR < 0.05 were considered as methylation-related interactions for corresponding CpG sites.

Finally, we examined whether the identified mQTLs (carrier/non-carrier) were associated with six active smoking indicators: ever smoking (current and former smoking) vs. never smoking, current smoking vs. never smoking, current smoking vs. former smoking, cumulative smoking (pack-years), durations of smoking (years), and the age of smoking initiation. The mixed linear models were fully adjusted for age (years), sex, smoking status, alcohol consumption, BMI, physical activity, the prevalence of CVD, diabetes, and cancer as described above. The mQTLs with a FDR < 0.05 were identified as smoking-related SNPs. We also assessed the associations of the significant mQTLs (carrier/non-carrier) with all-cause mortality in ESTHER study. Due to the limited number of deaths, we combined both subsets and performed the analysis using a multiple COX regression model. The model was adjusted for the above potential covariates and SNPs with a FDR < 0.05 were considered as all-cause mortality related variants.

Data cleaning and all aforementioned statistical analyses were performed by SAS version 9.4 (SAS Institute Inc., Cary, NC, USA). Manhattan plots for both panels were plotted by R package "ggplot2."

Additional files

Additional file 1: Table S1. Basic information of genomic data for 151 smoking-related CpG sites. **Table S2** List of 1396 significant SNP-CpG pairs in discovery panel. **Table S3** List of 246 significant SNP-CpG pairs in validation panel. **Table S4** Results of regression analyses for 246 SNP-CpG pairs in validation panel. **Table S5** Estimate changes of smoking status (never/current smoking) between models with and without adjusting for mQTLs. **Table S6** Impacts of mQTL-smoking interaction on the methylation levels of corresponding CpG sites in 949 participants (never/current smoking). **Table S7** Associations of 192 mQTLs with active smoking indicators. **Table S8.** Associations of 192 mQTLs with all-cause mortality. (XLSX 263 kb)

Additional file 2: Figure S1. Manhattan plot of the results in discovery panel. (PDF 85 kb)

Additional file 3: Figure S2. Locations and distributions of methylation levels of 19 smoking-related CpG sites based on the three most frequently identified mQTLs (carrier/non-carrier) and smoking status in validation panel.

Abbreviations
CpG sites: Cytosine-phosphate-guanine sites; CVD: Cardiovascular disease; EWASs: Epigenome-wide association studies; FDR: False discovery rate; mQTLs: Methylation quantitative trait loci; SNPs: Single nucleotide polymorphisms

Acknowledgements
The authors gratefully acknowledge contributions of DKFZ Genomics and Proteomics Core Facility, especially Dr. Melanie Bewerunge-Hudler and Dr. Matthias Schick, in the processing of DNA samples and performing the

laboratory work, Dr. Jonathan Heiss for providing the estimation of leukocyte distribution and Ms. Chen Chen for the language assistance.

Funding
The ESTHER study was supported in part by the Baden-Württemberg state Ministry of Science, Research and Arts (Stuttgart, Germany), and by the German Federal Ministry of Education and Research (Berlin, Germany). Xu Gao is supported by the grant from the China Scholarship Council (CSC).

Authors' contributions
XG conceived the study, carried out the main data analyses, interpreted the data, and drafted the manuscript. HT conducted genotyping and imputation and contributed to the interpretation of the study. YZ and LPB contributed to the design of the study. HB conducted the ESTHER study and contributed to all aspects of this work. All authors contributed to revision of the manuscript and approved the final version for submission.

Competing interests
The authors declare that they have no competing interests.

Author details
[1]Division of Clinical Epidemiology and Aging Research, German Cancer Research Center (DKFZ), Im Neuenheimer Feld 581, 69120 Heidelberg, Germany. [2]Division of Molecular Genetic Epidemiology, German Cancer Research Center (DKFZ), Im Neuenheimer Feld 580, 69120 Heidelberg, Germany. [3]Division of Preventive Oncology, German Cancer Research Center (DKFZ) and National Center for Tumor Diseases (NCT), Im Neuenheimer Feld 460, 69120 Heidelberg, Germany. [4]German Cancer Consortium (DKTK), German Cancer Research Center (DKFZ), Im Neuenheimer Feld 280, 69120 Heidelberg, Germany.

References
1. Babizhayev MA, Yegorov YE. Smoking and health: association between telomere length and factors impacting on human disease, quality of life and life span in a large population-based cohort under the effect of smoking duration. Fundam Clin Pharmacol. 2011;25:425–42.
2. Mathers CD, Loncar D. Projections of global mortality and burden of disease from 2002 to 2030. PLoS Med. 2006;3:e442.
3. Philibert RA, Beach SR, Brody GH. The DNA methylation signature of smoking: an archetype for the identification of biomarkers for behavioral illness. Neb Symp Motiv. 2014;61:109–27.
4. Lee KW, Pausova Z. Cigarette smoking and DNA methylation. Front Genet. 2013;4:132.
5. Philibert RA, Beach SR, Lei MK, Brody GH. Changes in DNA methylation at the aryl hydrocarbon receptor repressor may be a new biomarker for smoking. Clin Epigenetics. 2013;5:19.
6. Zeilinger S, Kuhnel B, Klopp N, Baurecht H, Kleinschmidt A, Gieger C, Weidinger S, Lattka E, Adamski J, Peters A, et al. Tobacco smoking leads to extensive genome-wide changes in DNA methylation. PLoS One. 2013;8:e63812.
7. Breitling LP, Yang R, Korn B, Burwinkel B, Brenner H. Tobacco-smoking-related differential DNA methylation: 27K discovery and replication. Am J Hum Genet. 2011;88:450–7.
8. Qiu W, Wan E, Morrow J, Cho MH, Crapo JD, Silverman EK, DeMeo DL. The impact of genetic variation and cigarette smoke on DNA methylation in current and former smokers from the COPDGene study. Epigenetics. 2015;10:1064–73.
9. Shenker NS, Ueland PM, Polidoro S, van Veldhoven K, Ricceri F, Brown R, Flanagan JM, Vineis P. DNA methylation as a long-term biomarker of exposure to tobacco smoke. Epidemiology. 2013;24:712–6.
10. Zhang Y, Yang R, Burwinkel B, Breitling LP, Brenner H. F2RL3 methylation as a biomarker of current and lifetime smoking exposures. Environ Health Perspect. 2014;122:131–7.
11. Tsaprouni LG, Yang TP, Bell J, Dick KJ, Kanoni S, Nisbet J, Vinuela A, Grundberg E, Nelson CP, Meduri E, et al. Cigarette smoking reduces DNA methylation levels at multiple genomic loci but the effect is partially reversible upon cessation. Epigenetics. 2014;9:1382–96.
12. Gutierrez-Arcelus M, Lappalainen T, Montgomery SB, Buil A, Ongen H, Yurovsky A, Bryois J, Giger T, Romano L, Planchon A, et al. Passive and active DNA methylation and the interplay with genetic variation in gene regulation. elife. 2013;2:e00523.
13. Bell JT, Pai AA, Pickrell JK, Gaffney DJ, Pique-Regi R, Degner JF, Gilad Y, Pritchard JK. DNA methylation patterns associate with genetic and gene expression variation in HapMap cell lines. Genome Biol. 2011;12:R10.
14. Gibbs JR, van der Brug MP, Hernandez DG, Traynor BJ, Nalls MA, Lai SL, Arepalli S, Dillman A, Rafferty IP, Troncoso J, et al. Abundant quantitative trait loci exist for DNA methylation and gene expression in human brain. PLoS Genet. 2010;6:e1000952.
15. Schalkwyk LC, Meaburn EL, Smith R, Dempster EL, Jeffries AR, Davies MN, Plomin R, Mill J. Allelic skewing of DNA methylation is widespread across the genome. Am J Hum Genet. 2010;86:196–212.
16. Kerkel K, Spadola A, Yuan E, Kosek J, Jiang L, Hod E, Li K, Murty VV, Schupf N, Vilain E, et al. Genomic surveys by methylation-sensitive SNP analysis identify sequence-dependent allele-specific DNA methylation. Nat Genet. 2008;40:904–8.
17. Gamazon ER, Badner JA, Cheng L, Zhang C, Zhang D, Cox NJ, Gershon ES, Kelsoe JR, Greenwood TA, Nievergelt CM, et al. Enrichment of cis-regulatory gene expression SNPs and methylation quantitative trait loci among bipolar disorder susceptibility variants. Mol Psychiatry. 2013;18:340–6.
18. Rushton MD, Reynard LN, Young DA, Shepherd C, Aubourg G, Gee F, Darlay R, Deehan D, Cordell HJ, Loughlin J. Methylation quantitative trait locus analysis of osteoarthritis links epigenetics with genetic risk. Hum Mol Genet. 2015;24:7432–44.
19. Heyn H, Sayols S, Moutinho C, Vidal E, Sanchez-Mut JV, Stefansson OA, Nadal E, Moran S, Eyfjord JE, Gonzalez-Suarez E, et al. Linkage of DNA methylation quantitative trait loci to human cancer risk. Cell Rep. 2014;7:331–8.
20. Liu Y, Aryee MJ, Padyukov L, Fallin MD, Hesselberg E, Runarsson A, Reinius L, Acevedo N, Taub M, Ronninger M, et al. Epigenome-wide association data implicate DNA methylation as an intermediary of genetic risk in rheumatoid arthritis. Nat Biotechnol. 2013;31:142–7.
21. Li Q, Seo JH, Stranger B, McKenna A, Pe'er I, Laframboise T, Brown M, Tyekucheva S, Freedman ML. Integrative eQTL-based analyses reveal the biology of breast cancer risk loci. Cell. 2013;152:633–41.
22. Zhang H, Wang F, Kranzler HR, Yang C, Xu H, Wang Z, Zhao H, Gelernter J. Identification of methylation quantitative trait loci (mQTLs) influencing promoter DNA methylation of alcohol dependence risk genes. Hum Genet. 2014;133:1093–104.
23. Gonseth S, de Smith AJ, Roy R, Zhou M, Lee ST, Shao X, Ohja J, Wrensch MR, Walsh KM, Metayer C, Wiemels JL. Genetic contribution to variation in DNA methylation at maternal smoking sensitive loci in exposed neonates. Epigenetics. 2016;0:.
24. Ladd-Acosta C, Fallin MD. The role of epigenetics in genetic and environmental epidemiology. Epigenomics. 2016;8:271–83.
25. Gao X, Jia M, Zhang Y, Breitling LP, Brenner H. DNA methylation changes of whole blood cells in response to active smoking exposure in adults: a systematic review of DNA methylation studies. Clin Epigenetics. 2015;7:113.
26. Hinds DA, McMahon G, Kiefer AK, Do CB, Eriksson N, Evans DM, St Pourcain B, Ring SM, Mountain JL, Francke U, et al. A genome-wide association meta-analysis of self-reported allergy identifies shared and allergy-specific susceptibility loci. Nat Genet. 2013;45:907–11.
27. Chubb D, Weinhold N, Broderick P, Chen B, Johnson DC, Forsti A, Vijayakrishnan J, Migliorini G, Dobbins SE, Holroyd A, et al. Common variation at 3q26.2, 6p21.33, 17p11.2 and 22q13.1 influences multiple myeloma risk. Nat Genet. 2013;45:1221–5.
28. Cerhan JR, Berndt SI, Vijai J, Ghesquieres H, McKay J, Wang SS, Wang Z, Yeager M, Conde L, de Bakker PI, et al. Genome-wide association study identifies multiple susceptibility loci for diffuse large B cell lymphoma. Nat Genet. 2014;46:1233–8.
29. Jin G, Zhu M, Yin R, Shen W, Liu J, Sun J, Wang C, Dai J, Ma H, Wu C, et al. Low-frequency coding variants at 6p21.33 and 20q11.21 are associated with lung cancer risk in Chinese populations. Am J Hum Genet. 2015;96:832–40.
30. Shenker NS, Polidoro S, van Veldhoven K, Sacerdote C, Ricceri F, Birrell MA, Belvisi MG, Brown R, Vineis P, Flanagan JM. Epigenome-wide association study in the European Prospective Investigation into Cancer and Nutrition (EPIC-Turin) identifies novel genetic loci associated with smoking. Hum Mol Genet. 2013;22:843–51.
31. Zhang Y, Florath I, Saum KU, Brenner H. Self-reported smoking, serum cotinine, and blood DNA methylation. Environ Res. 2016;146:395–403.
32. Connor Gorber S, Schofield-Hurwitz S, Hardt J, Levasseur G, Tremblay M. The accuracy of self-reported smoking: a systematic review of the relationship between self-reported and cotinine-assessed smoking status. Nicotine Tob Res. 2009;11:12–24.
33. Gao X, Mons U, Zhang Y, Breitling LP, Brenner H. DNA methylation changes

in response to active smoking exposure are associated with leukocyte telomere length among older adults. Eur J Epidemiol. 2016;31:1231–41.

34. Gao X, Zhang Y, Saum KU, Schottker B, Breitling LP, Brenner H. Tobacco smoking and smoking-related DNA methylation are associated with the development of frailty among older adults. Epigenetics. 2017;12:149–56.

35. Teschendorff AE, Yang Z, Wong A, Pipinikas CP, Jiao Y, Jones A, Anjum S, Hardy R, Salvesen HB, Thirlwell C, et al. Correlation of smoking-associated DNA methylation changes in buccal cells with DNA methylation changes in epithelial cancer. JAMA Oncol. 2015;1:476–85.

36. Davey Smith G, Hemani G. Mendelian randomization: genetic anchors for causal inference in epidemiological studies. Hum Mol Genet. 2014;23:R89–98.

37. Relton CL, Davey Smith G. Two-step epigenetic Mendelian randomization: a strategy for establishing the causal role of epigenetic processes in pathways to disease. Int J Epidemiol. 2012;41:161–76.

38. Schwartz J, Weiss ST. Cigarette smoking and peripheral blood leukocyte differentials. Ann Epidemiol. 1994;4:236–42.

39. Houseman EA, Accomando WP, Koestler DC, Christensen BC, Marsit CJ, Nelson HH, Wiencke JK, Kelsey KT. DNA methylation arrays as surrogate measures of cell mixture distribution. BMC Bioinformatics. 2012;13:86.

40. Lande R. The genetic covariance between characters maintained by pleiotropic mutations. Genetics. 1980;94:203–15.

41. Schöttker B, Haug U, Schomburg L, Kohrle J, Perna L, Muller H, Holleczek B, Brenner H. Strong associations of 25-hydroxyvitamin D concentrations with all-cause, cardiovascular, cancer, and respiratory disease mortality in a large cohort study. Am J Clin Nutr. 2013;97:782–93.

42. Gao X, Gao X, Zhang Y, Breitling LP, Schottker B, Brenner H. Associations of self-reported smoking, cotinine levels and epigenetic smoking indicators with oxidative stress among older adults: a population-based study. Eur J Epidemiol. 2017;32:443–56.

43. Miller SA, Dykes DD, Polesky HF. A simple salting out procedure for extracting DNA from human nucleated cells. Nucleic Acids Res. 1988;16:1215.

44. Florath I, Butterbach K, Heiss J, Bewerunge-Hudler M, Zhang Y, Schöttker B, Brenner H. Type 2 diabetes and leucocyte DNA methylation: an epigenome-wide association study in over 1,500 older adults. Diabetologia. 2015;59:130–8.

45. Anderson CA, Pettersson FH, Clarke GM, Cardon LR, Morris AP, Zondervan KT. Data quality control in genetic case-control association studies. Nat Protoc. 2010;5:1564–73.

46. Chang CC, Chow CC, Tellier LC, Vattikuti S, Purcell SM, Lee JJ. Second-generation PLINK: rising to the challenge of larger and richer datasets. Gigascience. 2015;4:7.

47. Shen-Orr SS, Tibshirani R, Khatri P, Bodian DL, Staedtler F, Perry NM, Hastie T, Sarwal MM, Davis MM, Butte AJ. Cell type-specific gene expression differences in complex tissues. Nat Methods. 2010;7:287–9.

48. Philibert RA, Plume JM, Gibbons FX, Brody GH, Beach SR. The impact of recent alcohol use on genome wide DNA methylation signatures. Front Genet. 2012;3:54.

49. Jones MJ, Goodman SJ, Kobor MS. DNA methylation and healthy human aging. Aging Cell. 2015;14:924–32.

50. Dick KJ, Nelson CP, Tsaprouni L, Sandling JK, Aissi D, Wahl S, Meduri E, Morange PE, Gagnon F, Grallert H, et al. DNA methylation and body-mass index: a genome-wide analysis. Lancet. 2014;383:1990–8.

51. Zhang FF, Cardarelli R, Carroll J, Zhang S, Fulda KG, Gonzalez K, Vishwanatha JK, Morabia A, Santella RM. Physical activity and global genomic DNA methylation in a cancer-free population. Epigenetics. 2011;6:293–9.

52. Nilsson E, Jansson PA, Perfilyev A, Volkov P, Pedersen M, Svensson MK, Poulsen P, Ribel-Madsen R, Pedersen NL, Almgren P, et al. Altered DNA methylation and differential expression of genes influencing metabolism and inflammation in adipose tissue from subjects with type 2 diabetes. Diabetes. 2014;63:2962–76.

53. Breitling LP. Current genetics and epigenetics of smoking/tobacco-related cardiovascular disease. Atertio Thromb Vasc Biol. 2013;33:1468–72.

54. Shen H, Laird PW. Interplay between the cancer genome and epigenome. Cell. 2013;153:38–55.

55. Benjamini Y, Hochberg Y. Controlling the false discovery rate—a practical and powerful approach to multiple testing. J R Stat Soc Series B Stat Methodol. 1995;57:289–300.

14

DNA methylation signatures for 2016 WHO classification subtypes of diffuse gliomas

Yashna Paul[†], Baisakhi Mondal[†], Vikas Patil and Kumaravel Somasundaram[*]

Abstract

Background: Glioma is the most common of all primary brain tumors with poor prognosis and high mortality. The 2016 World Health Organization classification of the tumors of central nervous system uses molecular parameters in addition to histology to redefine many tumor entities. The new classification scheme divides diffuse gliomas into low-grade glioma (LGG) and glioblastoma (GBM) as per histology. LGGs are further divided into isocitrate dehydrogenase (IDH) wild type or mutant, which is further classified into either oligodendroglioma that harbors 1p/19q codeletion or diffuse astrocytoma that has an intact 1p/19q loci but enriched for ATRX loss and TP53 mutation. GBMs are divided into IDH wild type that corresponds to primary or de novo GBMs and IDH mutant that corresponds to secondary or progressive GBMs. To make the 2016 WHO subtypes of diffuse gliomas more robust, we carried out Prediction Analysis of Microarrays (PAM) to develop DNA methylation signatures for these subtypes.

Results: In this study, we applied PAM on a training set of diffuse gliomas derived from The Cancer Genome Atlas (TCGA) and identified DNA methylation signatures to classify LGG IDH wild type from LGG IDH mutant, LGG IDH mutant with 1p/19q codeletion from LGG IDH mutant with intact 1p/19q loci and GBM IDH wild type from GBM IDH mutant with an accuracy of 99–100%. The signatures were validated using the test set of diffuse glioma samples derived from TCGA with an accuracy of 96 to 99%. In addition, we also carried out additional validation of all three signatures using independent LGG and GBM cohorts. Further, the methylation signatures identified a fraction of samples as discordant, which were found to have molecular and clinical features typical of the subtype as identified by methylation signatures.

Conclusions: Thus, we identified methylation signatures that classified different subtypes of diffuse glioma accurately and propose that these signatures could complement 2016 WHO classification scheme of diffuse glioma.

Keywords: Glioma, DNA methylation classification signature, IDH1/IDH2 mutation, 2016 WHO, PAM, PCA

Background

The neoplasia of non-neuronal glial cells in the brain is referred to as glioma and is the most common type of primary central nervous system (CNS) tumors [1]. The different histological subtypes of glioma are as follows: astrocytoma being the most common, accounting for 70% of all cases, while oligodendroglioma comprises 9% which includes classic oligodendrogliomas as well as mixed oligoastrocytomas and ependymoma comprises 6% [2].

Over the past decades, classification of brain tumors was based on the histopathological and microscopic features in hematoxylin- and eosin-stained sections, like cell type, level of differentiation, identifying necrotic lesions, and presence of lineage-specific markers. According to the WHO 2007-based classification, grade II/ diffused astrocytoma (DA) was described as low grade while high-grade glioma comprised of grade III/anaplastic astrocytoma (AA) and grade IV/glioblastoma (GBM) [3]. The vast majority of GBM develop de novo in elderly patients with no prior clinical or histological evidence and are referred to as primary GBM. Secondary GBM progresses through low-grade diffuse astrocytoma or anaplastic astrocytoma and is manifested in younger patients. Several studies have shown that glioma is highly heterogeneous which indicates that tumors of same grade have diverse genetic and epigenetic molecular aberrations [4–9]. With the invent of new technologies, many high-throughput studies have reported

* Correspondence: ksomasundaram1@gmail.com; skumar@mcbl.iisc.ernet.in
†Equal contributors
Department of Microbiology and Cell Biology, Indian Institute of Science, Bangalore 560012, India

different molecular signatures based on glioma CpG island methylator phenotype (GCIMP), expression-based studies for mRNA, miRNA, and lncRNA in GBM [10–13]. One of the most exciting and clinically relevant observations was the discovery that a high percentage of grade II/III and grade IV secondary glioblastoma harbor mutations in the genes isocitrate dehydrogenase 1 and 2 [2]. Growing data indicate that these mutations play a causal role in gliomagenesis, have a major impact on tumor biology, and also have clinical and prognostic importance [2].

Nearly 12% of GBM patients have been identified to have point mutation in codon 132 (R132H) of the isocitrate dehydrogenase 1 (IDH1) gene located in the chromosome locus 2q33 [14]. IDH1 codes for a cytosolic protein that controls oxidative cellular damage [14, 15]. Several studies showed that the IDH1 mutation is inversely associated with grade in diffuse glial tumors, affecting 71% of grade II, 64% of grade III, and 6% of primary glioblastomas [14]. Interestingly, IDH mutation is found to be present in the secondary glioblastoma (76%) probably because these tumors have been derived from the lower grade gliomas [16]. IDH1 is an enzyme and it catalyzes the oxidative decarboxylation of isocitrate to produce α-ketoglutarate (α-KG) [17].

IDH mutation has been shown to be associated with alterations in the methylome thus being sufficient to establish glioma hypermethylator phenotype [18]. At present, 2016 WHO CNS tumor classification has included both molecular markers along with histological features to identify and classify different subtypes of diffuse glioma which includes the WHO grade II and grade III astrocytic tumors, the grade II and III oligodendrogliomas, and the grade IV glioblastomas. The low-grade gliomas (LGGs), which include the WHO grade II and grade III astrocytic tumors and the grade II and III oligodendrogliomas, are classified based on IDH mutation status. The LGG IDH mutant subtype is further classified based on the codeletion of 1p/19q where LGG IDH mutant patients harboring 1p/19q codeletion is termed as oligodendrogliomas (ODG) while LGG IDH mutant patients having intact 1p/19q loci are termed as diffuse astrocytoma which may be enriched in TP53 mutation/ATRX loss. The other axis is the glioblastoma (GBM) which, similar to LGG, is further classified into IDH WT and mutant. The deficiency in this classification is that factors like intra-tumoral heterogeneity and insufficient molecular information could result in our ability to classify certain samples to any specific categories. In such cases, signatures based on whole tumor studies to classify the glioma subtypes might further complement 2016 WHO classification.

In the present study, we investigated the altered methylation pattern among the different subtypes of diffuse gliomas as per 2016 WHO CNS tumor classification [19] and derived methylation-based classification signature for distinguishing different subtypes. Our study sets up the premise of using methylation signature in combination to the 2016 WHO classification system with a higher precision of classification of the diffuse glioma patients, thereby helping better diagnosis and appropriate treatment therapy.

Result

The overall work flow of methylation-based signatures to distinguish diffuse glioma subtypes of 2016 WHO classification

To develop methylation-based signatures to distinguish diffuse glioma subtypes as per 2016 WHO CNS tumor classification (Fig. 1), we subjected the 450K DNA methylation data of The Cancer Genome Atlas (TCGA) diffuse glioma samples (https://cancergenome.nih.gov/) to various statistical tools and validation steps (Fig. 2). The methylation signatures were developed to distinguish LGG IDH mutant from LGG IDH WT, LGG IDH mutant with 1p/19q codeletion (oligodendroglioma) from LGG IDH mutant with intact 1p/19q loci (diffuse astrocytoma) and GBM IDH mutant (progressive GBM) from GBM IDH WT (de novo GBM). The TCGA samples were classified into these groups as per 2016 WHO classification scheme (Fig. 1). For methylation signature development, to begin with, we performed a Wilcoxon-rank sum test between different diffuse glioma subtypes to identify a list of significantly differentially methylated CpG probes, which were further subjected to a differential β value ($\Delta\beta$) of 0.4 between groups. The TCGA samples were then divided randomly into two equal groups as training and test sets (Additional file 1: Table S1). The training set was subjected to Prediction Analysis of Microarrays (PAM) [20] to identify the methylation signatures containing minimum number of CpGs with least error. The robustness of the identified signatures was internally cross validated within training set using Support Vector Machine (SVM) [21] and subset validation. The signatures were further applied on the test set for the additional validation. Further, the signatures were subjected to external validation by using independent cohorts. We also used principal component analysis (PCA) to test the ability of methylation signatures to separate the two compared groups into two distinct clusters. Additionally, 10-fold cross-validation by PAM was carried out to identify the discordant samples, which were then subjected to further analysis to find out the true nature of these samples.

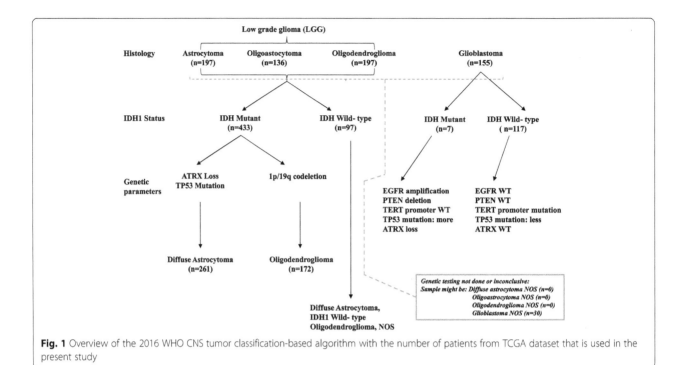

Fig. 1 Overview of the 2016 WHO CNS tumor classification-based algorithm with the number of patients from TCGA dataset that is used in the present study

14 CpG methylation signatures to distinguish LGG IDH mutant from LGG IDH wild type (WT): identification and validation

PAM analysis of differentially methylated CpGs (Additional file 1: Table S2) in the training (TCAG) set (Additional file 1: Table S1) identified a set of 14 CpGs to distinguish IDH mutant from IDH WT in LGG at a threshold value of 18.9 with least error (Fig. 3a, Additional file 2: Figure S1A). The robustness of this probe set was tested by internal cross-validation using SVM, which gave a classification accuracy of 100% and subset validation with an accuracy of 100% (Additional file 2: Figure S2A and B respectively; see the Methods section for more details). The CpG probes of the signature were found to be hypermethylated in IDH mutant LGGs compared to IDH WT LGGs (Fig. 3b and Table 1). Further, upon subjecting the 14 CpG probes to PCA, the two principal components were able to form two distinct clusters for IDH mutant and IDH WT LGGs (Fig. 3c). Prediction accuracy estimation by 10-fold cross-validation using PAM showed that the 14 CpG probe methylation signatures predicted all LGG IDH mutant samples accurately with no error (Fig. 3d). Similarly, all LGG IDH WT samples were rightly predicted to be LGG with WT IDH samples based on the 14 CpG probe methylation signatures (Fig. 3d). Thus, the 14 CpG DNA methylation signatures were able to discriminate LGG IDH mutant from LGG IDH WT with an overall classification accuracy of 100%. The sensitivity and specificity of the signature for IDH mutant and WT in LGG are 100% (Table 2).

Next, we validated the strength of 14 CpG methylation signatures using the test set (Additional file 1: Table S1). The 14 discriminatory probes were observed to be differentially methylated between LGG IDH mutant and LGG IDH WT in the test set also (Additional file 2: Figure S3A and Additional file 1: Table S3A). The PCA demonstrated that the probes were able to distinguish IDH mutant from the WT group as two distinct clusters (Additional file 2: Figure S3B). Prediction accuracy estimation by 10-fold cross-validation using PAM showed that the 14 CpG probe methylation signatures predicted all IDH mutant LGG samples accurately except one with an error rate of 0.004 (Additional file 2: Figure S3C). Among IDH WT LGG samples, all of them were accurately predicted by the signature (Additional file 2: Figure S3C). Thus, the 14 CpG methylation signatures were able to discriminate between IDH mutant and WT LGG samples with an overall diagnostic accuracy of 99.62% in the test set. The sensitivity of the signature for IDH mutant LGG is 99.53% while for IDH WT LGG is 100%, and the specificity for IDH mutant is 100% whereas for those of the IDH WT, it is 99.53% (Table 2). The 14 CpG methylation signatures, as identified in the training set and validated in the test set, were also used to classify the entire set of TCGA LGG. We found that the 14 discriminatory probes distinguished two groups (Additional file 2: Figure S4A, B, and C) with an overall accuracy of 99.81% (Table 2).

Next, we have also carried out additional validation of 14 CpG methylation signatures using two independent external LGG cohorts (GSE58218 [22] and GSE48462 [23]). In GSE58218, the 14 CpG methylation signatures

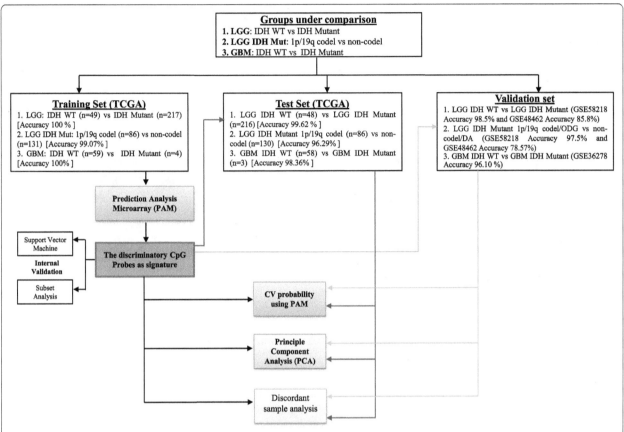

Fig. 2 The schematic representation of the work flow of statistical analysis. PAM identified 14 discriminatory CpG probes of DNA methylation between (*1*) IDH Mut (LGG IDH Mut) and WT (LGG IDH WT) which was further validated by principal component analysis (PCA). Fourteen CpG probe methylation signatures were then validated in test set. Here, TCGA dataset (450K methylation) was randomly divided into equal halves to form the training and test set. Similar protocol was performed for (*2*) LGG IDH Mut 1p/19q intact (diffuse astrocytoma/DA) versus LGG IDH Mut 1p/19q codel (oligodendroglioma/ODG) and (*3*) GBM IDH Mut versus WT. All the derived methylation signatures are validated in independent validation datasets with high accuracy

were able to discriminate IDH mutant from WT LGG samples with an overall diagnostic accuracy of 98.5% (Tables 1 and 2; Fig. 4a–c). Similarly, the 14 CpG methylation signatures were able to discriminate IDH mutant from WT LGG samples with an overall diagnostic accuracy of 85.8% in GSE48462 (Table 2; Additional file 1: Table S3A; Additional file 2: Figure S5A, B, and C). Thus, from these experiments, we conclude that the 14 CpG methylation signatures developed as above distinguished LGG IDH mutant from WT samples with high accuracy.

14 CpG probe methylation signatures to classify oligodendrogliomas (ODG) and diffuse astrocytoma (DA): identification and validation

PAM analysis of differentially methylated CpGs (Additional file 1: Table S4) on the training (TCGA) set (Additional file 1: Table S1) identified a set of 14 CpGs to distinguish IDH mutant with 1p/19q codeletion (designated as oligodendroglioma) from LGG IDH mutant with

intact 1p/19q loci (designated as diffuse astrocytoma) at a threshold value of 9.491 with minimal error (Fig. 5a, Additional file 2: Figure S1B). The robustness of this probe set was tested by internal cross-validation using SVM, which gave a classification accuracy of 97.67 to 100% and subset validation with an accuracy of 99 to 100% (Additional file 2: Figure S2C and D, respectively; see the Methods section for more detail). The CpG probes that correspond to this signature were found to be hypermethylated in oligodendroglioma compared to diffuse astrocytoma (Fig. 5b and Table 3). Further, upon subjecting the 14 CpG probes to PCA, the two principal components were able to separate these two groups into two distinct clusters (Fig. 5c). Prediction accuracy estimation by 10-fold cross-validation using PAM showed that the 14 CpG probe methylation signatures predicted all oligodendroglioma samples accurately with no error (Fig. 5d). With respect to diffuse astrocytoma, all samples except two were accurately predicted

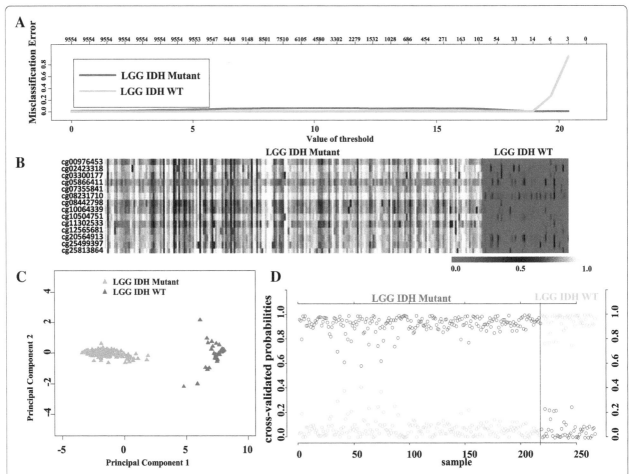

Fig. 3 Identification of 14 CpG probe methylation signatures of LGG IDH mutant versus WT in training set (TCGA). **a** Plot demonstrating classification error for 9554 CpG probes from PAM analysis in training set. The threshold value 18.9 corresponded to 14 discriminatory CpG probes which classified IDH mutant (n = 217) and WT (n = 49) LGG samples with classification error of 0%. **b** Heat map of the 14 CpG discriminatory probes identified from the PAM analysis between LGG IDH Mut and WT patient samples in the training set (TCGA). A *dual color code* was used where *yellow* indicates more methylation (hypermethylation) and *blue* indicates less methylation (hypomethylation). **c** PCA was performed using beta (methylation) values of 14 PAM-identified CpG probes between IDH mutant (n = 217) and WT (n = 49) LGG samples in training set. A scatter plot is generated using the first two principal components for each sample. The *color code* of the samples is as indicated. **d** The detailed cross-validation probabilities of 10-fold cross-validation for the samples of training set based on the beta values of 14 CpG probes are shown. For each sample, its probability as LGG IDH Mut (*red color*) and WT (*green color*) is shown and it was predicted by the PAM program as either IDH Mut or WT in LGG samples based on which grade's probability is higher. The original histological grade of the samples is shown on the *top*

to be diffuse astrocytoma based on the 14 CpG probe methylation signatures with an error rate of 0.0153 (Fig. 5d). Thus, the 14 CpG DNA methylation signatures were able to discriminate oligodendroglioma from diffuse astrocytoma with an overall diagnostic accuracy of 99.07%. The sensitivity of the signature for oligodendroglioma is 100% while for diffuse astrocytoma is 98.47%, and the specificity for oligodendroglioma is 98.47% whereas for those of the diffuse astrocytomas is 100% (Table 2).

Next, we validated the strength of 14 CpG methylation signatures using the test (TCGA) set (Additional file 1: Table S1). The 14 discriminatory probes were observed to be differentially methylated between oligodendrogliomas and diffused astrocytoma similar to as seen in the training set (Additional file 2: Figure S6A and Additional file 1:

Table S3B). The PCA demonstrated that the probes were able to distinguish oligodendrogliomas from diffused astrocytoma as two distinct clusters (Additional file 2: Figure S6B). Prediction accuracy estimation by 10-fold cross-validation using PAM showed that the 14 CpG probe methylation signatures predicted all oligodendroglioma samples except one accurately with an error rate of 0.0117 (Additional file 2: Figure S6C). Among diffused astrocytoma, except seven, all samples were accurately predicted by the signature with an error rate of 0.0539 (Additional file 2: Figure S6C). Thus, the 14 CpG methylation signatures were able to discriminate between oligodendroglioma and diffused astrocytoma samples with an overall diagnostic accuracy of 96.29% in the test set. The sensitivity of the signature for oligodendrogliomas is

Table 1 List of the 14 CpG methylation signatures for LGG IDH mutant versus IDH WT in the training set and validation set (GSE58218)

No.	CpG ID	Gene name	Training set (TCGA cohort)					Validation set (GSE58218 cohort)				
			Average β in mutant	Average β in WT	$\Delta\beta$ = (avg β in mutant–avg β in WT)	p value	FDR	Average β in mutant	Average β in WT	$\Delta\beta$ = (avg β in mutant–avg β in WT)	p value	FDR
1	cg00976453	KCNB1	0.795	0.037	0.758	1.31E–27	1.67E–27	0.806	0.162	0.644	1.18E–19	1.38E–19
2	cg02423318	NA	0.860	0.096	0.764	9.56E–28	1.67E–27	0.876	0.167	0.709	7.35E–21	2.87E–20
3	cg03300177	GNAO1	0.841	0.063	0.777	9.78E–28	1.67E–27	0.839	0.171	0.667	1.47E–20	2.95E–20
4	cg05866411	FGFRL1	0.784	0.102	0.682	8.35E–28	1.67E–27	0.681	0.208	0.473	3.96E–20	5.54E–20
5	cg07355841	TPPP3	0.835	0.055	0.781	1.25E–27	1.67E–27	0.819	0.184	0.635	1.93E–20	3.38E–20
6	cg08231710	MMP23A	0.874	0.127	0.747	9.35E–28	1.67E–27	0.809	0.293	0.516	1.47E–18	1.58E–18
7	cg08442798	NA	0.772	0.023	0.749	8.35E–28	1.67E–27	0.824	0.090	0.734	5.43E–21	2.87E–20
8	cg10064339	UCP2	0.779	0.042	0.737	8.35E–28	1.67E–27	0.805	0.115	0.690	8.82E–21	2.87E–20
9	cg10504751	GNAO1	0.846	0.067	0.779	1.23E–27	1.67E–27	0.846	0.173	0.673	2.61E–20	4.06E–20
10	cg11302533	NA	0.784	0.037	0.747	1.8E–27	1.94E–27	0.834	0.102	0.732	1.19E–20	2.87E–20
11	cg12565681	RHBDF2	0.834	0.053	0.781	1.44E–27	1.68E–27	0.822	0.295	0.527	5.87E–18	5.87E–18
12	cg20564913	FGFRL1	0.837	0.108	0.729	8.94E–28	1.67E–27	0.832	0.222	0.610	1.23E–20	2.87E–20
13	cg25499397	GPR62	0.822	0.076	0.746	8.35E–28	1.67E–27	0.798	0.272	0.527	1.12E–19	1.38E–19
14	cg25813864	RAPGEFL1	0.851	0.064	0.787	2.64E–27	2.64E–27	0.855	0.160	0.695	9.37E–21	2.87E–20

NA not associated with any gene

Table 2 For the methylation-based signatures: overall diagnostic accuracy, sensitivity, and specificity

1. Low-grade glioma IDH WT versus mutant: for 14 CpG methylation signatures

Cohort	Dataset	Overall accuracy (%)[a]	Sensitivity (%)[b]		Specificity (%)[c]		Overall error (%)	IDH mutant error (%)	IDH WT error (%)
			IDH mutant	IDH WT	IDH mutant	IDH WT			
TCGA	Training set	100 (266/266)	100 (217/217)	100 (49/49)	100 (49/49)	100 (217/217)	0	0	0
TCGA	Test set	99.62 (263/264)	99.53 (215/216)	100 (48/48)	100 (48/48)	99.53 (215/216)	0.38	0.47	0
TCGA	Combined set	99.81 (529/530)	99.76 (432/433)	100 (97/97)	100 (97/97)	99.76 (432/433)	0.19	0.24	0
GSE58218	Validation dataset	98.5 (192/195)	99.36 (156/157)	94.7 (36/38)	94.7 (36/38)	99.36 (156/157)	1.5	0.64	5.3
GSE48462	Validation dataset	85.8 (48/56)	96.55 (28/29)	74.07 (20/27)	74.07 (20/27)	96.55 (28/29)	14.2	3.4	25.9

2. Diffuse astrocytoma (IDH mutant and non-codeletion of 1p/19q; DA) versus oligodendroglioma (IDH mutant and 1p/19q codeletion; ODG): for 14 CpG methylation signatures

Cohort	Dataset	Overall accuracy (%)[a]	Sensitivity (%)[b]		Specificity (%)[c]		Overall error (%)	DA error (%)	ODG error (%)
			DA	ODG	DA	ODG			
TCGA	Training set	99.07 (215/217)	98.47 (129/131)	100 (86/86)	100 (86/86)	98.47 (129/131)	0.93	1.53	0
TCGA	Test set	96.29 (208/216)	94.61 (123/130)	98.83 (85/86)	98.83 (85/86)	94.61 (123/130)	3.71	5.39	1.17
TCGA	Combined set	97.69 (423/433)	96.55 (252/261)	99.41 (171/172)	99.41 (171/172)	96.55 (252/261)	2.31	3.45	0.59
GSE58218	Validation dataset	97.5 (153/157)	96.25 (77/80)	98.70 (77/78)	98.70 (77/78)	96.25 (77/80)	2.5	3.75	1.29
GSE48462	Validation dataset	78.57 (22/28)	71.42 (10/14)	85.71 (12/14)	85.71 (12/14)	71.42 (10/14)	21.43	28.58	14.29

3. For GBM IDH WT versus mutant: for 13 CpG methylation signatures

Cohort	Dataset	Overall accuracy (%)[a]	Sensitivity (%)[b]		Specificity (%)[c]		Overall error (%)	GBM IDH mutant error (%)	GBM IDH WT error (%)
			GBM IDH mutant	GBM IDH WT	GBM IDH mutant	GBM IDH WT			
TCGA	Training set	100 (63/63)	100 (4/4)	100 (59/59)	100 (59/59)	100 (4/4)	0	0	0
TCGA	Test set	98.36 (60/61)	100 (3/3)	98.27 (57/58)	98.27 (57/58)	100 (3/3)	1.64	0	1.73
TCGA	Combined set	99.19 (123/124)	100 (7/7)	99.14 (116/117)	99.14 (116/117)	100 (7/7)	0.81	0	0.86
GSE36278	Validation dataset	96.10 (74/77)	87.5 (14/16)	98.36 (60/61)	98.36 (60/61)	87.5 (14/16)	3.9	12.5	1.64

1. low-grade glioma IDH WT versus mutant, 2. diffuse astrocytoma (DA) versus oligodendroglioma (ODG), 3. GBM IDH WT versus mutant
[a](the number of samples predicted correctly)/(total number of samples analyzed)×100
[b](the number of positive samples predicted)/(the number of true positives)×100
[c](the number of negative samples predicted)/(the number of true negatives)×100

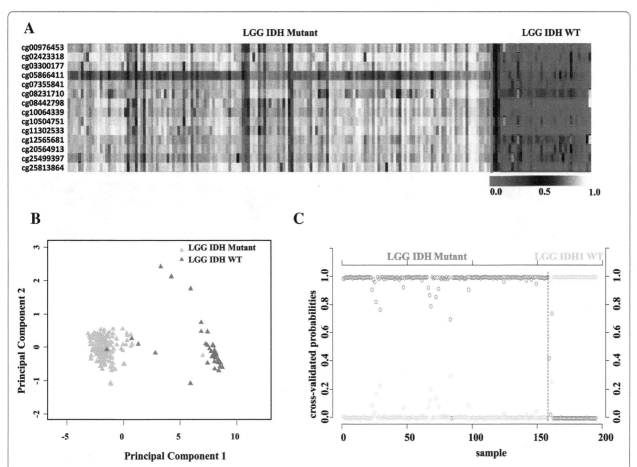

Fig. 4 Validation of the 14 CpG methylation signatures of LGG IDH mutant versus WT in an independent validation dataset GSE58218. **a** Heat map of the 14 CpG discriminatory probes identified in PAM analysis in IDH mutant (*n* = 157) and WT (*n* = 38) LGG patient samples in the entire TCGA dataset. A *dual color code* was used where *yellow* indicates more methylation (hypermethylation) and *blue* indicates less methylation (hypomethylation). **b** PCA was performed using *β* (methylation) values of 14 PAM-identified CpG probes between IDH mutant (*n* = 157) and WT (*n* = 38) LGG patient samples in the entire TCGA dataset. A scatter plot is generated using the first two principal components for each sample. The *color code* of the samples is as indicated. **c** The detailed probabilities of 10-fold cross-validation for the samples of training set based on the *β* values of 14 CpG probes are shown. For each sample, its probability as IDH mutant (*red color*) and WT (*green color*) of LGG patient samples is shown and it was predicted by the PAM program as either LGG IDH mutant or WT based on which grade's probability is higher. The original histological grade of the samples is shown on the *top*

98.83% while for diffused astrocytoma, it is 94.61%, and the specificity for oligodendrogliomas is 94.61% whereas for diffused astrocytoma, it is 98.83% (Table 2). The 14 CpG methylation signatures, as identified in the training set and validated in the test set, were also used to classify the entire TCGA LGG IDH mutant samples into oligodendroglioma and diffuse astrocytoma samples. We found that the 14 discriminatory probes behaved similar in the classification (Additional file 2: Figure S7A, B and C) with an overall accuracy of 97.69% (Table 2).

In addition, we have also carried out additional validation of 14 CpG methylation signatures to distinguish oligodenroglioma from diffuse astrocytoma using two independent external LGG cohorts (GSE58218 and GSE48462). In GSE58218, the 14 CpG methylation signatures were able to discriminate oligodenroglioma from

diffuse astrocytoma samples with an overall diagnostic accuracy of 97.5% (Tables 2 and 3; Fig. 6a–c). Similarly, the 14 CpG methylation signatures were also able to discriminate oligodenroglioma from diffuse astrocytoma samples with an overall diagnostic accuracy of 78.57% in GSE48462 (Table 2; Additional file 1: Table S3B; Additional file 2: Figure S8A, B and C). Thus, from these experiments, we conclude that the 14 CpG methylation signatures developed as above distinguished oligodenroglioma from diffuse astrocytoma samples with high accuracy.

13 CpG probe methylation signatures to classify IDH mutant from wild type (WT) in glioblastoma (GBM): identification and validation

PAM analysis of differentially methylated CpGs (Additional file 1: Table S5) in the training (TCGA)

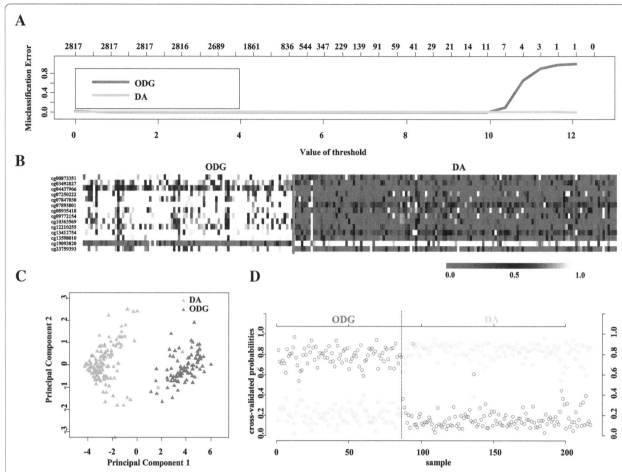

Fig. 5 Identification of 14 CpG probe methylation signatures in training set (TCGA) for diffuse astrocytoma (DA) and oligodendroglioma (ODG). **a** Plot demonstrating classification error for 2817 CpG probes from PAM analysis in training set. The threshold value of 9.491 corresponded to 14 discriminatory CpG probes which classified DA (LGG IDH Mut with intact 1p/19q; *n* = 131) and ODG (LGG IDH Mut with 1p/19q codel; *n* = 86) LGG samples with classification error of 0.93%. **b** Heat map of the 14 CpG discriminatory probes identified from the PAM analysis between DA and ODG patient samples in the training set (TCGA). A *dual color code* was used where *yellow* indicates more methylation (hypermethylation) and *blue* indicates less methylation (hypomethylation). **c** PCA was performed using beta (methylation) values of 14 PAM-identified CpG probes between DA (*n* = 131) and WT (*n* = 86) LGG samples in training set. A scatter plot is generated using the first two principal components for each sample. The *color code* of the samples is as indicated. **d** The detailed cross-validation probabilities of 10-fold cross-validation for the samples of training set based on the beta values of 14 CpG probes are shown. For each sample, its probability as ODG (*red color*) and DA (*green color*) is shown and it was predicted by the PAM program as either ODG or DA in LGG samples based on which grade's probability is higher. The original histological grade of the samples is shown on the *top*

set (Additional file 1: Table S1) identified a set of 13 CpGs to distinguish GBM IDH mutant from IDH WT samples at a threshold value of 2.694 with no error (Fig. 7a, Additional file 2: Figure S1C). The robustness of this probe set was tested by internal cross-validation using SVM, which gave a classification accuracy of 100% and subset validation with an accuracy of 100% (Additional file 2: Figure S2E and F, respectively; see the Methods section for more details). The CpG probes of the signature were found to be hypermethylated in IDH mutant GBMs compared to IDH WT GBMs (Fig. 7b and Table 4). Further, upon subjecting the 13 CpG probes to PCA, the two principal components were able to form two distinct clusters for IDH mutant and IDH WT GBMs (Fig. 7c). Prediction

accuracy estimation by 10-fold cross-validation using PAM showed that the 13 CpG probe methylation signatures predicted all the samples accurately with no error (Fig. 7d). Similarly, among GBM IDH wild-type samples, all were rightly predicted by the 13 CpG methylation signatures (Fig. 7d). Thus, the 13 CpG DNA methylation signatures were able to discriminate GBM IDH mutant from GBM IDH WT with an overall classification accuracy of 100%. The sensitivity and specificity of the signature for IDH mutant and WT in GBM are 100% (Table 2).

Next, we validated the strength of 13 CpG methylation signatures using the test set (Additional file 1: Table S1). The 13 discriminatory probes were observed to be differentially methylated between GBM IDH mutant and

Table 3 List of the 14 CpG methylation signatures for oligodendroglioma (ODG) versus diffuse astrocytoma (DA) in the training set and validation set (GSE58218)

No.	CpG ID	Gene name	Training set (TCGA cohort)					Validation set (GSE58218 cohort)				
			Average β in ODG	Average β in DA	Δβ = (avg β in ODG–avg β in DA)	p value	FDR	Average β in ODG	Average β in DA	Δβ = (avg β in ODG–avg β in DA)	p value	FDR
1	cg00873351	CD300LB	0.755	0.227	0.528	1.67E-32	3.9E-32	0.576	0.225	0.351	1.92E-22	2.99E-22
2	cg03492827	NA	0.647	0.192	0.455	8.18E-34	5.73E-33	0.701	0.286	0.415	3.01E-26	2.11E-25
3	cg04437966	FLJ37543	0.500	0.088	0.412	3.99E-33	1.39E-32	0.459	0.102	0.357	2.33E-25	8.16E-25
4	cg07250222	FGFR2	0.753	0.186	0.567	8.68E-30	8.68E-30	0.795	0.301	0.494	5.03E-21	7.04E-21
5	cg07847030	TCF7L1	0.730	0.163	0.567	6.78E-31	1.06E-30	0.798	0.368	0.429	2.62E-19	2.62E-19
6	cg07893801	PLCG1	0.807	0.312	0.495	1.28E-32	3.59E-32	0.737	0.378	0.359	1.61E-20	2.04E-20
7	cg08935418	PTPRN2	0.735	0.231	0.505	6.25E-30	6.73E-30	0.771	0.313	0.458	4.01E-23	7.53E-23
8	cg09772154	FGFR2	0.733	0.194	0.540	5.94E-30	6.73E-30	0.779	0.308	0.472	6.35E-20	6.84E-20
9	cg10363569	PRKAG2	0.675	0.162	0.513	5.1E-30	6.5E-30	0.574	0.196	0.377	5.22E-20	6.09E-20
10	cg12210255	NA	0.662	0.139	0.523	2.53E-33	1.18E-32	0.699	0.237	0.462	1.8E-24	5.05E-24
11	cg13412754	MAPKAP1	0.782	0.311	0.471	4.67E-32	9.34E-32	0.736	0.340	0.396	1.74E-25	8.1E-25
12	cg13598010	NA	0.778	0.175	0.603	1.89E-34	2.65E-33	0.790	0.254	0.536	5.09E-27	7.13E-26
13	cg19093820	GPR156	0.210	0.716	−0.506	3.56E-31	6.23E-31	0.211	0.655	−0.444	4.3E-23	7.53E-23
14	cg23759393	PTPRN2	0.722	0.203	0.518	1.08E-30	1.51E-30	0.790	0.300	0.490	8.76E-24	2.04E-23

NA not associated with any gene

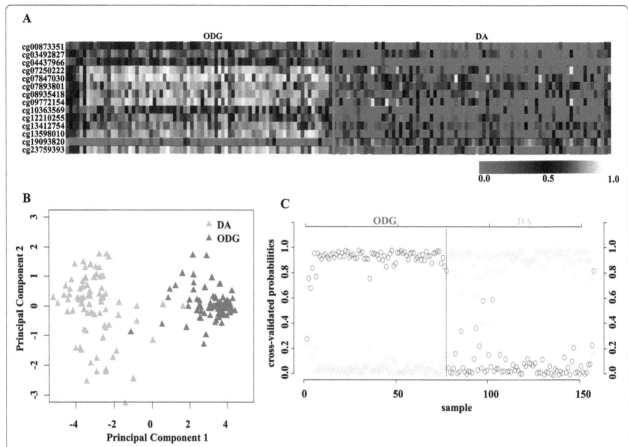

Fig. 6 Validation of the 14 CpG methylation signatures of oligodendroglioma (ODG) versus diffuse astrocytoma (DA) in an independent validation dataset GSE58218. **a** Heat map of the 14 CpG discriminatory probes identified in PAM analysis in ODG (*n* = 77) and DA (*n* = 80) LGG patient samples in the entire TCGA dataset. A *dual color code* was used where *yellow* indicates more methylation (hypermethylation) and *blue* indicates less methylation (hypomethylation). **b** PCA was performed using β (methylation) values of 14 PAM-identified CpG probes between ODG (*n* = 77) and DA (*n* = 80) LGG patient samples in the entire TCGA dataset. A scatter plot is generated using the first two principal components for each sample. The *color code* of the samples is as indicated. **c** The detailed probabilities of 10-fold cross-validation for the samples of training set based on the β values of 14 CpG probes are shown. For each sample, its probability as ODG (*red color*) and DA (*green color*) of LGG patient samples is shown and it was predicted by the PAM program as either LGG DA or ODG based on which grade's probability is higher. The original histological grade of the samples is shown on the *top*

GBM IDH WT in the test set also (Additional file 2: Figure S9A and Additional file 1: Table S3C). The PCA demonstrated that the probes were able to distinguish IDH mutant from the WT group as two distinct clusters (Additional file 2: Figure S9B). Prediction accuracy estimation by 10-fold cross-validation using PAM showed that the 13 CpG methylation signatures predicted all IDH mutant GBM samples accurately with no error rate (Additional file 2: Figure S9C). Among IDH WT GBM samples, all samples except one were accurately predicted by the signature with an error rate of 0.0173 (Additional file 2: Figure S9C). Thus, the 13 CpG methylation signatures were able to discriminate IDH mutant from WT GBM samples with an overall diagnostic accuracy of 98.36% in the test set. The sensitivity of the signature for IDH mutant GBM is 100% while for IDH WT GBM is 98.27%, and the specificity for IDH mutant

is 98.27% whereas for those of the IDH WT, it is 100% (Table 2). The 13 CpG methylation signatures, as identified in the training set and validated in the test set, were also used to classify the entire set of TCGA GBM set (117 IDH WT samples and 7 IDH mutant samples). We found that the 13 discriminatory probes distinguished two groups (Additional file 2: Figure S10A, B, and C) with an overall accuracy of 99.19% (Table 2). Further, we have also carried out additional validation of 13 CpG methylation signatures to distinguish GBM IDH mutant from WT samples using an independent external GBM cohort (GSE36278 [24]). Analysis revealed that the 13 CpG methylation signatures were able to discriminate GBM IDH mutant from WT samples with an overall diagnostic accuracy of 96.10% (Tables 2 and 4; Fig. 8a–c). Thus, from these experiments, we conclude that the 13 CpG methylation signatures developed as above

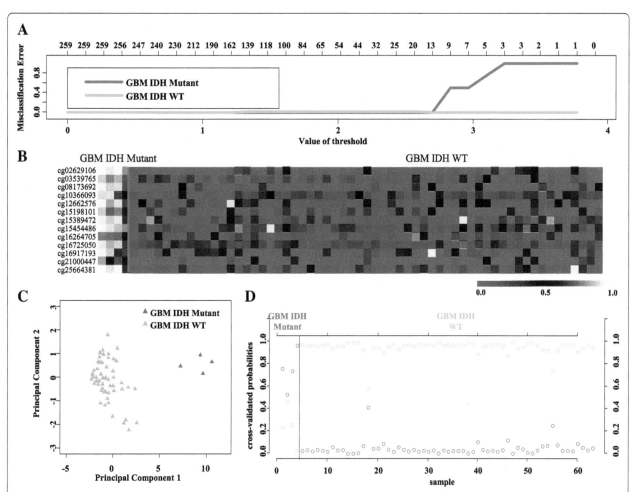

Fig. 7 Identification of 13 CpG probe methylation signatures in training set (TCGA) for IDH Mut and WT in GBM. **a** Plot demonstrating classification error for 259 CpG probes from PAM analysis in training set. The threshold value of 2.694 corresponded to 13 discriminatory CpG probes which classified IDH Mut (n = 4) and WT (n = 59) GBM samples with classification error of 0%. **b** Heat map of the 13 CpG discriminatory probes identified from the PAM analysis between IDH Mut and WT GBM patient samples in the training set (TCGA). A *dual color code* was used where *yellow* indicates more methylation (hypermethylation) and *blue* indicates less methylation (hypomethylation). **c** PCA was performed using beta (methylation) values of 13 PAM-identified CpG probes between IDH Mut (n = 4) and WT (n = 59) GBM samples in training set. A scatter plot is generated using the first two principal components for each sample. The *color code* of the samples is as indicated. **d** The detailed cross-validation probabilities of 10-fold cross-validation for the samples of training set based on the beta values of 14 CpG probes are shown. For each sample, its probability as IDH Mut (*red color*) and WT (*green color*) GBM samples is shown and it was predicted by the PAM program as either IDH Mut or WT in GBM samples based on which grade's probability is higher. The original histological grade of the samples is shown on the *top*

distinguished GBM IDH mutant from WT samples with high accuracy.

Molecular analysis of discordant samples

While the DNA methylation signatures were able to distinguish different diffuse glioma subtypes, it also identified a fraction of samples as discordant. It is of our interest to find out the accurate molecular nature of these samples in order to assess the true nature of them. While we could use TCGA cohort for this purpose as it had all relevant histological and molecular markers, external validation cohorts could not be subjected to molecular discordant analysis as they do not have these features. In the classification of LGG IDH mutant from

IDH WT, the 14 CpG signatures identified one IDH mutant LGG sample in the test set as discordant. We carried out a careful assessment of the molecular markers of this sample using c-Bioportal (http://www.cbioportal.org/) from the TCGA dataset. For this purpose, we analyzed TP53 mutation, ATRX loss, and 1p/19q codeletion status of all the samples (Additional file 1: Table S6, Table S7 A, B, and C, and Table S8). As per 2016 WHO CNS tumor classification, all LGG IDH mutant samples that have 1p/19q codeletion are designated as oligodendroglioma and those with intact 1p/19q loci and enriched for TP53 mutation/ATRX loss are designated as diffuse astrocytoma. The LGG IDH mutant discordant sample had intact 1p/19q, WT TP53, and ATRX genes indicating that this

Table 4 List of the 13 CpG methylation signatures for GBM IDH mutant versus IDH WT in the training set and validation set (GSE36278)

No.	CpG ID	Gene name	Training set (TCGA cohort)					Validation set (GSE36278 cohort)				
			Average β in mutant	Average β in WT	Δβ = (avg β in mutant–avg β in WT)	p value	FDR	Average β in mutant	Average β in WT	Δβ = (avg β in mutant–avg β in WT)	p value	FDR
1	cg02629106	PCDP1	0.845	0.113	0.732	0.00093	0.00163	0.76	0.18	0.58	4E-08	2.6E-07
2	cg03539765	LOC144571	0.746	0.127	0.618	0.00093	0.00163	0.63	0.18	0.45	2.7E-07	5.9E-07
3	cg08173692	PRR18	0.791	0.091	0.699	0.00093	0.00163	0.79	0.24	0.54	9.1E-06	1.1E-05
4	cg10366093	YPEL4	0.710	0.189	0.520	0.00102	0.00163	0.67	0.25	0.42	1.2E-06	2.3E-06
5	cg12662576	NA	0.780	0.172	0.608	0.00125	0.00163	0.76	0.29	0.48	2.7E-07	5.9E-07
6	cg15198101	SRRM3	0.695	0.109	0.586	0.00102	0.00163	0.79	0.15	0.64	1.7E-08	2.2E-07
7	cg15389472	GLUL	0.786	0.162	0.624	0.00184	0.00184	0.66	0.24	0.42	8.6E-06	1.1E-05
8	cg15454486	OBFC2A	0.823	0.231	0.592	0.00125	0.00163	0.71	0.32	0.39	3.1E-06	4.5E-06
9	cg16264705	ATP5G2	0.665	0.132	0.533	0.00167	0.00181	0.74	0.23	0.51	2.4E-07	5.9E-07
10	cg16725050	TUBA4B	0.769	0.214	0.556	0.00125	0.00163	0.78	0.32	0.46	2.7E-07	5.9E-07
11	cg16917193	NA	0.736	0.150	0.586	0.00138	0.00163	0.72	0.23	0.49	1.1E-05	1.2E-05
12	cg21000447	CHADL	0.626	0.077	0.549	0.00125	0.00163	0.70	0.19	0.51	3.1E-06	4.5E-06
13	cg25664381	NA	0.808	0.166	0.642	0.00138	0.00163	0.66	0.22	0.44	0.00016	0.00016

NA not associated with any gene

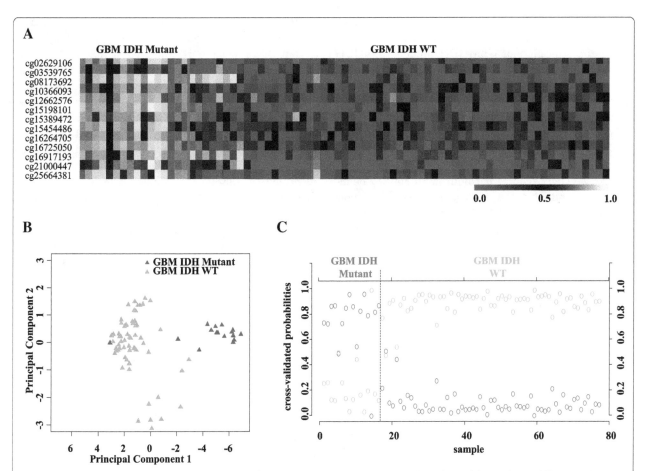

Fig. 8 Validation of the 13 CpG methylation signatures of GBM IDH mutant versus WT in an independent validation dataset GSE36278. **a** Heat map of the 13 CpG discriminatory probes identified in PAM analysis in IDH mutant (n = 16) and WT (n = 61) GBM patient samples in the entire TCGA dataset. A *dual color code* was used where *yellow* indicates more methylation (hypermethylation) and *blue* indicates less methylation (hypomethylation). **b** PCA was performed using β (methylation) values of 13 PAM-identified CpG probes between IDH mutant (n = 16) and WT (n = 61) GBM patient samples in the entire TCGA dataset. A scatter plot is generated using the first two principal components for each sample. The *color code* of the samples is as indicated. **c** The detailed probabilities of 10-fold cross-validation for the samples of training set based on the β values of 13 CpG probes are shown. For each sample, its probability as IDH mutant (*red color*) and WT (*green color*) of GBM patient samples is shown and it was predicted by the PAM program as either GBM IDH mutant or WT based on which grade's probability is higher. The original histological grade of the samples is shown on the *top*

sample is not an oligodendroglioma. The presence of WT TP53 and ATRX genes raises the possibility of it not being a diffuse astrocytoma. Interestingly, additional analysis revealed that the discordant sample is indeed carrying WT IDH as per DNA sequencing even though IDH antibody-based scoring classified it as IDH mutant. Therefore, it appears that IDH mutation scoring by IHC could be an error as evidenced by DNA sequencing and that the 14 CpG methylation signatures are able classify the LGGs more accurately.

In the classification of LGG oligodendroglioma from LGG diffuse astrocytoma, 14 CpG probe methylation signatures identified ten samples as discordant which did not match the WHO 2016 tumor grading. In order to understand the true status of the discordant samples, we analyzed the clinical information and molecular

markers using c-Bioportal (http://www.cbioportal.org/) from the TCGA dataset. For this purpose, we analyzed TP53 mutation, ATRX mutation, and 1p/19q codeletion status in DA, ODG, and discordant samples of LGG (Additional file 1: Table S6, Table S7 A, B, and C, and Table S8). Based on the WHO 2016 CNS tumor classification, IDH mutant LGGs having intact 1p/19q with an enrichment of TP53 mutation and ATRX loss are classified as diffuse astrocytoma. IDH mutant LGG samples with 1p/19q codeletion are classified as oligodendroglioma. The analysis of discordant samples for the molecular markers and histological features revealed some interesting findings. While the single ODG discordant sample had 1p/19q codeletion and WT TP53/ATRX genes, this sample was identified as oligoastrocytoma as per histology. Among nine DA discordant samples, while all of them had intact 1p/19q loci, a majority of them were found to have WT TP53/ATRX genes.

In the classification of GBM IDH mutant from IDH WT, the 13 CpG probe methylation signatures identified one GBM IDH WT sample as discordant. In order to understand the true nature of the discordant sample, we analyzed the clinical information and molecular markers using c-Bioportal (http://www.cbioportal.org/) from the TCGA dataset (Additional file 1: Table S6, Table S8, and Table S9 A and B). The discordant GBM IDH WT sample had WT IDH gene as per both immunohistochemical staining and DNA sequencing. However, this sample had no amplification of EGFR locus with an intact PTEN gene, unlike what is expected for a IDH WT GBM sample.

Discussion

Glioma is the most common and highly malignant primary brain tumor. The 2007 WHO classification of the glioma tumors was majorly based on microscopic appearance of cell type and histopathological markers largely segregating into three subtypes such as astrocytoma, oligodendroglioma, and oligoastrocytoma (mixed) [3]. With the advent of the high-throughput technologies, comprehensive understanding of the heterogeneous genetic and epigenetic landscape of both glioblastoma and the low grades became vibrant [25, 26]. The histopathological grading of glioma tumors could be subjected to inter-observer variation which would lead to misclassification with a potential possibility of not providing the right kind of treatment [27]. To combat this shortcoming, several groups including work from our laboratory carried out extensive studies and have identified several prognostic markers and molecular signatures based on mRNA, miRNA, and DNA methylation that would aid in better classification and identifying best choice of therapy [10–13, 15, 28–31].

The meeting by the International Society of Neuropathology held in Haarlem, Netherland, established guidelines for how to incorporate molecular findings into brain tumor diagnosis thereby setting the platform for a major revision of the 2007 CNS WHO classification [32]. The current updated version is summarized in the 2016 CNS WHO classifications [19]. In this study, using TCGA 450K DNA methylation data, we developed methylation signatures that could distinguish different classes of diffuse glioma with high accuracy. The signatures developed in this study using TCGA data are also validated extensively using TCGA data as well as independent datasets.

Infinium HumanMethylation450K BeadChip array data for astrocytoma (grade II, III, and IV/GBM), oligodendroglioma, and oligoastrocytoma tumor samples from TCGA dataset was used in this study. By using

PAM, we have successfully developed and validated DNA methylation signatures to distinguish LGG IDH mutant from LGG IDH wild-type samples, LGG IDH mutant samples into diffuse astrocytoma and IDH mutant GBM from the IDH WT GBMs. The signatures classified these groups with very high accuracy and also validated successfully in multiple independent datasets. We also used PCA to test the ability of signatures to divide the two groups in comparison into two distinct classes. Further, the 10-fold cross-validation using PAM identified the discordant samples, which upon further analysis revealed that majority of misclassified samples were indeed due to inadequacies of the current methods used for classification.

Thus, the present study enabled us to identify DNA methylation fingerprint for each of the groups in comparison (LGG IDH1 WT versus mutant, ODG versus DA, and GBM IDH mutant versus WT). The 2016 WHO classification system fails to classify some samples accurately in occasions like absence of certain molecular markers, errors due to antibody-based scoring, and intra-tumoral heterogeneity. We believe that DNA methylation signatures based on whole tumor developed in this study could complement the 2016 WHO classification of diffuse glioma subtypes.

Conclusions

In conclusion, we were able to classify diffuse glioma subtypes with high accuracy. The discordant samples identified by the methylation signature were found to be either due to technical errors or mixed histological types. More importantly, we believe that the high levels of intra-tumoral heterogeneity reported in glioma could also be a reason for their misclassification [7, 27]. Collectively, our study indicates that the methylation-based molecular profiles in combination with the revised 2016 WHO CNS tumor classification guidelines might be able to classify the samples more precisely.

Methods

Tumor samples and clinical details

Glioma TCGA dataset was used for this study. Methylation data for histologically defined WHO classification glioma types, which include astrocytoma ($n = 197$), oligoastrocytoma ($n = 136$), oligodendroglioma ($n = 197$), and glioblastoma ($n = 124$) samples, was used. Samples were then segregated according to the WHO 2016 CNS tumor IHC-based grading classification into three distinct groups, namely 1. lower grade glioma IDH wild-type and mutant (LGG IDH WT and mutant), 2. lower grade glioma IDH mutant with intact 1p/19q termed as diffuse astrocytoma and with 1p/19q codeletion termed as oligodendroglioma (DA and ODG), and 3. glioblastoma IDH mutant and wild type (GBM IDH WT and mutant). The clinical information for the same was also procured from TCGA.

With an aim to identify methylation differences between the diffuse glioma subtypes (based on IDH mutation and 1p/19q codeletion status) of each group, a supervised machine learning approach through PAM (Prediction Analysis of Microarrays) [20] was used. For this purpose, the first step was to identify significantly differentially methylated CpG probes between lower grade glioma IDH WT and mutant, between DA and ODG, and between GBM IDH mutant and WT which are described in details below.

Identification of differentially methylated CpGs

In this study, three different comparisons were carried out—1. LGG: IDH mutant versus WT, 2. LGG IDH mutant: 1p/19q codel (ODG) versus non-codel (DA), and 3. GBM: IDH mutant versus WT. For the first comparison between LGG IDH mutant and WT, we have performed a Wilcoxon-rank sum test between IDH mutant and WT which yielded 269,442 CpG probes significantly (FDR ≤0.0001) differentially methylated in mutant versus WT. Next, a stringent cutoff of 0.4 absolute $\Delta\beta$ value was applied that showed 9,554 significantly differentially methylated (26 CpGs were hypomethylated and 9528 CpGs were hypermethylated in IDH mutant LGG; Additional file 1: Table S2) CpG probes in mutant as compared to WT IDH LGG patients. Firstly, the TCGA 450K human methylation dataset for LGG patients with IDH mutation ($n = 433$) and LGG patients with WT IDH ($n = 97$) was randomized and 50% of each of the two classes formed the training set, and the remaining 50% was used as the test set. We randomized TCGA dataset ten times to obtain ten different training sets and their corresponding test sets. After performing PAM on each of the ten training sets, the training set that gave least error with minimum number of CpGs was selected for further studies. This process gave a set of 14 discriminatory CpG probes which were further tested through SVM and subset analysis before testing on the test set and external validation sets (Fig. 2; Table 1).

Similarly, analysis was carried out for LGG IDH mutant cohort with and without 1p/19q codeletion (ODG and DA, respectively) patients (Fig. 2). For this comparison, between LGG IDH mutant 1p/19q codel (ODG) and non-codel (DA), we have performed a Wilcoxon-rank sum test which yielded 160,288 CpG probes significantly differentially methylated in ODG versus DA. Next, a stringent cutoff of 0.2 absolute $\Delta\beta$ value was applied that showed 2817 significantly differentially methylated (627 CpGs were hypomethylated and 2190 CpGs were hypermethylated in ODG; Additional file 1: Table S4) CpG probes in mutant as compared to WT IDH LGG patients. The TCGA 450K human methylation

dataset for LGG patients with 1p/19q codel ($n = 172$) and non-codel ($n = 261$) was randomized and 50% of each of the two classes formed the training set, and the remaining 50% was used as the test set. We randomized TCGA dataset ten times to obtain ten different training sets and their corresponding test sets. After performing PAM on each of the ten training sets, the training set that gave least error with minimum number of CpGs was selected for further studies. This process gave a set of 14 discriminatory CpG probes which were further tested through SVM and subset analysis before testing on the test set and external validation set (Fig. 2; Table 3).

Likewise, the same work flow was followed to identify a methylation-based signature that could distinguish the GBM IDH WT from mutant samples (Fig. 2). In this comparison, between GBM IDH mutant and WT patient samples, we have performed a Wilcoxon-rank sum test which yielded 69,669 CpG probes significantly differentially methylated in mutant versus WT. Next, a stringent cutoff of 0.2 absolute $\Delta\beta$ value was applied that showed 259 significantly differentially methylated (33 CpGs were hypomethylated and 226 CpGs were hypermethylated in mutant; Additional file 1: Table S5) CpG probes in mutant as compared to WT IDH GBM patients. The TCGA 450K human methylation dataset for GBM patients with IDH mutation ($n = 7$) and WT ($n = 117$) was randomized and 50% of each of the two classes formed the training set, and the remaining 50% was used as the test set. We randomized TCGA dataset ten times to obtain ten different training sets and their corresponding test sets. After performing PAM on each of the ten training sets, the training set that gave least error with minimum number of CpGs was selected for further studies. This process gave a set of 13 discriminatory CpG probes which were further tested through SVM and subset analysis before testing on the test set and external validation set (Fig. 2; Table 4).

Prediction Analysis of Microarray (PAM)

To identify a list of a minimal set of signatory probes from the significantly differentially methylated CpGs between each compared groups, Prediction Analysis of Microarrays (PAM) using the package pamr available in R software (version 3.1.0) were applied. PAM uses nearest shrunken centroid method for classifying samples. This method "shrinks" each of the class centroids towards the overall centroid by the threshold. In case of selecting a signature, it is ideal to choose a threshold value that would achieve a set of minimum number of genes with maximum accuracy thereby least error. For preparing input files for PAM analysis, the list of significantly methylated probes between each compared groups across all the tumor samples was randomized and 50% of each of the two classes formed the training set, and the remaining 50% was used as the test set. This randomization was performed ten times which resulted into ten different compositions of training set and their corresponding test set. Thereafter, each of these ten training sets was subjected to PAM analysis that uses 10-fold cross-validation to identify a predictive signature. Ten different training sets that were used to construct the PAM classifier resulted in ten non-identical predictive signatures, one for each iteration. The most promising signature which had the maximum training and test set accuracies was chosen. We also performed an internal cross-validation on the training set of the most promising signature as predicted by PAM.

Internal cross-validation using Support Vector Machine (SVM) and random subset sampling

For internal cross-validation, we have used Support Vector Machine (SVM) [21]. Many prediction methods use SVM for classification of dataset into two or more classes. For a given set of binary classes training examples, SVM can map the input space into higher dimensional space and seek a hyperplane to separate the positive data examples from the negative ones with the largest margin. SVM-based internal cross-validation is used for the training sets of 1. LGG IDH mutant versus WT, 2. diffuse astrocytoma versus oligodendroglioma, and 3. GBM IDH mutant versus WT. For each of the abovementioned cases, the samples were divided randomly into five subgroups containing equal number of the respective samples. These five subgroups of each cases, example LGG IDH mutant and WT, were made into five groups where each group contained one subgroup of LGG IDH mutant and one subgroup of LGG IDH WT samples. Consequently, one group of LGG IDH WT plus LGG IDH mutant was considered as a test set while the rest four groups were considered as training set and this is referred to as a "fold." In this way, SVM models were built five times to give fivefolds, wherein every group was considered as a test set and the remaining groups as training set. The accuracy for each fold was checked by this method.

The predictive accuracy of the three signatures was also analyzed in a subset of the following cases: 1. LGG IDH mutant (217) versus WT ($n = 49$), 2. diffuse astrocytoma ($n = 131$) versus oligodendroglioma ($n = 86$), and 3. GBM IDH mutant ($n = 4$) versus WT ($n = 59$) by random subset sampling. PAM was used to predict the respective accuracies in the random subset sampling.

Principal component analysis

Principal component analysis (PCA) uses orthogonal transformation to convert a set of variables into a set of values of linearly uncorrelated variables that are called principal components. The number of principal components can be less than or equal to the number of original variables. The first two principal components account for the largest possible variation in the dataset. PCA was performed using R package (version 3.1.0), on the training and test sets to know how well the identified methylation signature classifies LGG IDH mutant and WT.

This process was repeated for identifying a methylation signature between IDH mutant DA and ODG and between GBM IDH mutant and WT (a cutoff of 0.2 absolute $\Delta\beta$ was used here to identify significantly differently methylated probes between the two classes).

Additional files

Additional file 1: Table S1. Sample size and diffuse glioma subsets of various cohorts used in this study. **Table S2.** List of differentially methylated CpGs between LGG IDH mutant and WT used as PAM input in the training (TCGA) set. **Table S3A.** List of the 14 CpG methylation signatures for LGG IDH mutant versus IDH WT in the test set (TCGA) and validation set (GSE48462). **Table S3B.** List of the 14 CpG methylation signatures for oligodendroglioma (ODG) versus diffuse astrocytoma (DA) in the test set (TCGA) and validation set (GSE48462). **Table S4.** List of differentially methylated CpGs between oligodendroglioma and diffuse astrocytoma used as PAM input in the training (TCGA) set. **Table S5.** List of 259 differentially methylated CpG probes between GBM IDH mutant and WT used as PAM input in the training (TCGA) set. **Table S6.** Molecular analysis of discordant samples identified by CpG methylation signatures. **Table S7A.** Molecular status for IDH, TP53, ATRX, and 1p/19q in LGG samples from TCGA used in this study. **Table S7B.** Molecular status for IDH, TP53, ATRX, and 1p/19q in LGG samples from GSE58218 used in this study. **Table S7C.** Molecular status for IDH, TP53, ATRX, and 1p/19q in LGG samples from GSE48462 used in this study. **Table S8.** Patient IDs of the discordant samples derived from all datasets used in this study. **Table S9A.** Molecular status of IDH, TP53, ATRX, EGFR, and PTEN in GBM samples from TCGA dataset used in this study. **Table S9B.** Molecular status of IDH, TP53, ATRX, EGFR, and PTEN in GBM samples from GSE38278 dataset used in this study. (ZIP 1601 kb)

Additional file 2: Has ten additional figures and their corresponding figure legends.

Abbreviations
AA: Anaplastic astrocytoma; DA: Diffuse astrocytoma; GBM: Glioblastoma; IDH: Isocitrate dehydrogenase; ODG: Oligodendroglioma; PAM: Prediction Analysis of Microarray; PCA: Principal component analysis; SVM: Support Vector Machine; TCGA: The Cancer Genome Atlas; WHO: World Health Organization

Acknowledgements
The results published here are in whole or part based upon data generated by The Cancer Genome Atlas (TCGA) pilot project established by the NCI and NHGRI. Information about TCGA and the investigators and institutions which constitute the TCGA research network can be found at http://cancergenome.nih.gov. We also acknowledge the use of GSE58218, GSE48462, and GSE36278 in this study. Infrastructure support by funding from DST-FIST, DBT grant-in-aid, and UGC (Centre for Advanced Studies in Molecular Microbiology) to MCB is acknowledged. KS thanks DBT, Government of India for financial support. KS is a JC Bose Fellow of the Department of Science and Technology.

Funding
Infrastructure support by funding from DST-FIST, DBT grant-in-aid, and UGC (Centre for Advanced Studies in Molecular Microbiology) to MCB is acknowledged. KS thanks DBT, Government of India for financial support. KS is a JC Bose Fellow of the Department of Science and Technology.

Authors' contributions
KS coordinated the study. KS and BM conceived and wrote the paper. KS and BM designed while YP performed the analysis of the TCGA dataset for all the experiments. VP performed all the data analysis related to the SVM, subset analysis, statistical analysis, and preparation of the revised manuscript. All authors read and approved the final manuscript.

Competing interests
The authors declare that they have no competing interests.

References
1. Ostrom QT, Gittleman H, Fulop J, Liu M, Blanda R, Kromer C, Wolinsky Y, Kruchko C, Barnholtz-Sloan JS. CBTRUS statistical report: primary brain and central nervous system tumors diagnosed in the United States in 2008–2012. Neuro Oncol. 2015;17 Suppl 4:iv1–iv62.
2. Cohen AL, Holmen SL, Colman H. IDH1 and IDH2 mutations in gliomas. Curr Neurol Neurosci Rep. 2013;13(5):345.
3. Louis DN, Ohgaki H, Wiestler OD, Cavenee WK, Burger PC, Jouvet A, Scheithauer BW, Kleihues P. The 2007 WHO classification of tumours of the central nervous system. Acta Neuropathol. 2007;114(2):97–109.
4. Dunn GP, Rinne ML, Wykosky J, Genovese G, Quayle SN, Dunn IF, Agarwalla PK, Chheda MG, Campos B, Wang A, et al. Emerging insights into the molecular and cellular basis of glioblastoma. Genes Dev. 2012;26(8):756–84.
5. Holland EC. Glioblastoma multiforme: the terminator. Proc Natl Acad Sci U S A. 2000;97(12):6242–4.
6. Meyer M, Reimand J, Lan X, Head R, Zhu X, Kushida M, Bayani J, Pressey JC, Lionel AC, Clarke ID, et al. Single cell-derived clonal analysis of human glioblastoma links functional and genomic heterogeneity. Proc Natl Acad Sci U S A. 2015;112(3):851–6.
7. Patel AP, Tirosh I, Trombetta JJ, Shalek AK, Gillespie SM, Wakimoto H, Cahill DP, Nahed BV, Curry WT, Martuza RL, et al. Single-cell RNA-seq highlights intratumoral heterogeneity in primary glioblastoma. Science. 2014;344(6190):1396–401.
8. Scherer HJ. A critical review: the pathology of cerebral gliomas. J Neurol Psychiatry. 1940;3(2):147–77.
9. Stupp R, Reni M, Gatta G, Mazza E, Vecht C. Anaplastic astrocytoma in adults. Crit Rev Oncol Hematol. 2007;63(1):72–80.
10. Noushmehr H, Weisenberger DJ, Diefes K, Phillips HS, Pujara K, Berman BP, Pan F, Pelloski CE, Sulman EP, Bhat KP, et al. Identification of a CpG island methylator phenotype that defines a distinct subgroup of glioma. Cancer Cell. 2010;17(5):510–22.
11. Shukla S, Pia Patric IR, Thinagararjan S, Srinivasan S, Mondal B, Hegde AS, Chandramouli BA, Santosh V, Arivazhagan A, Somasundaram K. A DNA methylation prognostic signature of glioblastoma: identification of NPTX2-PTEN-NF-kappaB nexus. Cancer Res. 2013;73(22):6563–73.
12. Srinivasan S, Patric IR, Somasundaram K. A ten-microRNA expression signature predicts survival in glioblastoma. PLoS One. 2011;6(3):e17438.
13. Verhaak RG, Hoadley KA, Purdom E, Wang V, Qi Y, Wilkerson MD, Miller CR, Ding L, Golub T, Mesirov JP, et al. Integrated genomic analysis identifies clinically relevant subtypes of glioblastoma characterized by abnormalities in PDGFRA, IDH1, EGFR, and NF1. Cancer Cell. 2010;17(1):98–110.
14. Cairns RA, Mak TW. Oncogenic isocitrate dehydrogenase mutations: mechanisms, models, and clinical opportunities. Cancer Discov. 2013;3(7):730–41.
15. Parsons DW, Jones S, Zhang X, Lin JC, Leary RJ, Angenendt P, Mankoo P, Carter H, Siu IM, Gallia GL, et al. An integrated genomic analysis of human glioblastoma multiforme. Science. 2008;321(5897):1807–12.
16. Yan H, Parsons DW, Jin G, McLendon R, Rasheed BA, Yuan W, Kos I, Batinic-Haberle I, Jones S, Riggins GJ, et al. IDH1 and IDH2 mutations in gliomas. N Engl J Med. 2009;360(8):765–73.
17. Zhao S, Lin Y, Xu W, Jiang W, Zha Z, Wang P, Yu W, Li Z, Gong L, Peng Y, et al. Glioma-derived mutations in IDH1 dominantly inhibit IDH1 catalytic activity and induce HIF-1alpha. Science. 2009;324(5924):261–5.
18. Turcan S, Rohle D, Goenka A, Walsh LA, Fang F, Yilmaz E, Campos C, Fabius AW, Lu C, Ward PS, et al. IDH1 mutation is sufficient to establish the glioma hypermethylator phenotype. Nature. 2012;483(7390):479–83.
19. Louis DN, Perry A, Reifenberger G, von Deimling A, Figarella-Branger D, Cavenee WK, Ohgaki H, Wiestler OD, Kleihues P, Ellison DW. The 2016 World Health Organization classification of tumors of the central nervous system: a summary. Acta Neuropathol. 2016;131(6):803–20.
20. Tibshirani R, Hastie T, Narasimhan B, Chu G. Diagnosis of multiple cancer types by shrunken centroids of gene expression. Proc Natl Acad Sci U S A. 2002;99(10):6567–72.
21. Cortes C, Vapnik V. Support-vector networks. Mach Learn. 1995;20(3):273–97.
22. Wiestler B, Capper D, Sill M, Jones DT, Hovestadt V, Sturm D, Koelsche C, Bertoni A, Schweizer L, Korshunov A, et al. Integrated DNA methylation and copy-number profiling identify three clinically and biologically relevant groups of anaplastic glioma. Acta Neuropathol. 2014;128(4):561–71.
23. van den Bent MJ, Erdem-Eraslan L, Idbaih A, de Rooi J, Eilers PH, Spliet WG, den Dunnen WF, Tijssen C, Wesseling P, Sillevis Smitt PA, et al. MGMT-STP27 methylation status as predictive marker for response to PCV in anaplastic oligodendrogliomas and oligoastrocytomas. A report from EORTC study 26951. Clin Cancer Res. 2013;19(19):5513–22.
24. Sturm D, Witt H, Hovestadt V, Khuong-Quang DA, Jones DT, Konermann C, Pfaff E, Tonjes M, Sill M, Bender S, et al. Hotspot mutations in H3F3A and IDH1 define distinct epigenetic and biological subgroups of glioblastoma. Cancer Cell. 2012;22(4):425–37.
25. Brennan CW, Verhaak RG, McKenna A, Campos B, Noushmehr H, Salama SR, Zheng S, Chakravarty D, Sanborn JZ, Berman SH, et al. The somatic genomic landscape of glioblastoma. Cell. 2013;155(2):462–77.
26. Frattini V, Trifonov V, Chan JM, Castano A, Lia M, Abate F, Keir ST, Ji AX, Zoppoli P, Niola F, et al. The integrated landscape of driver genomic alterations in glioblastoma. Nat Genet. 2013;45(10):1141–9.
27. Coons SW, Johnson PC, Scheithauer BW, Yates AJ, Pearl DK. Improving diagnostic accuracy and interobserver concordance in the classification and grading of primary gliomas. Cancer. 1997;79(7):1381–93.
28. Nijaguna MB, Patil V, Hegde AS, Chandramouli BA, Arivazhagan A, Santosh V, Somasundaram K. An eighteen serum cytokine signature for discriminating glioma from normal healthy individuals. PLoS One. 2015;10(9):e0137524.
29. Rao SA, Srinivasan S, Patric IR, Hegde AS, Chandramouli BA, Arimappamagan A, Santosh V, Kondaiah P, Rao MR, Somasundaram K. A 16-gene signature distinguishes anaplastic astrocytoma from glioblastoma. PLoS One. 2014;9(1):e85200.
30. Hegi ME, Diserens AC, Gorlia T, Hamou MF, de Tribolet N, Weller M, Kros JM, Hainfellner JA, Mason W, Mariani L, et al. MGMT gene silencing and benefit from temozolomide in glioblastoma. N Engl J Med. 2005;352(10):997–1003.
31. Colman H, Zhang L, Sulman EP, McDonald JM, Shooshtari NL, Rivera A, Popoff S, Nutt CL, Louis DN, Cairncross JG, et al. A multigene predictor of outcome in glioblastoma. Neuro Oncol. 2010;12(1):49–57.
32. Louis DN, Perry A, Burger P, Ellison DW, Reifenberger G, von Deimling A, Aldape K, Brat D, Collins VP, Eberhart C, et al. International Society Of Neuropathology—Haarlem consensus guidelines for nervous system tumor classification and grading. Brain Pathol. 2014;24(5):429–35.

15

Causal effect of smoking on DNA methylation in peripheral blood: a twin and family study

Shuai Li[1], Ee Ming Wong[2,3], Minh Bui[1], Tuong L. Nguyen[1], Ji-Hoon Eric Joo[2,3], Jennifer Stone[4], Gillian S. Dite[1], Graham G. Giles[1,5], Richard Saffery[6,7], Melissa C. Southey[2,3] and John L. Hopper[1*]

Abstract

Background: Smoking has been reported to be associated with peripheral blood DNA methylation, but the causal aspects of the association have rarely been investigated. We aimed to investigate the association and underlying causation between smoking and blood methylation.

Methods: The methylation profile of DNA from the peripheral blood, collected as dried blood spots stored on Guthrie cards, was measured for 479 Australian women including 66 monozygotic twin pairs, 66 dizygotic twin pairs, and 215 sisters of twins from 130 twin families using the Infinium HumanMethylation450K BeadChip array. Linear regression was used to estimate associations between methylation at ~ 410,000 cytosine-guanine dinucleotides (CpGs) and smoking status. A regression-based methodology for twins, Inference about Causation through Examination of Familial Confounding (ICE FALCON), was used to assess putative causation.

Results: At a 5% false discovery rate, 39 CpGs located at 27 loci, including previously reported *AHRR*, *F2RL3*, *2q37.1* and *6p21.33*, were found to be differentially methylated across never, former and current smokers. For all 39 CpG sites, current smokers had the lowest methylation level. Our study provides the first replication for two previously reported CpG sites, cg06226150 (*SLC2A4RG*) and cg21733098 (*12q24.32*). From the ICE FALCON analysis with smoking status as the predictor and methylation score as the outcome, a woman's methylation score was associated with her co-twin's smoking status, and the association attenuated towards the null conditioning on her own smoking status, consistent with smoking status causing changes in methylation. To the contrary, using methylation score as the predictor and smoking status as the outcome, a woman's smoking status was not associated with her co-twin's methylation score, consistent with changes in methylation not causing smoking status.

Conclusions: For middle-aged women, peripheral blood DNA methylation at several genomic locations is associated with smoking. Our study suggests that smoking has a causal effect on peripheral blood DNA methylation, but not vice versa.

Keywords: DNA methylation, Smoking, Epigenome-wide association study, Causal inference, Family study

* Correspondence: j.hopper@unimelb.edu.au
[1]Centre for Epidemiology and Biostatistics, Melbourne School of Population and Global Health, University of Melbourne, Parkville, Victoria, Australia
Full list of author information is available at the end of the article

Background

Epigenetics is a mechanism modifying gene expression without changing underlying DNA sequence. DNA methylation, a phenomenon that typically a methyl group (-CH3) is added to a cytosine-guanine dinucleotide (CpG) at which the cytosine is converted to a 5-methylcytosine, has been proposed to play a role in the aetiology of complex traits and diseases [1, 2].

At least 21 epigenome-wide association studies (EWASs) have reported that methylation in the blood of adults at a great many CpGs is associated with smoking status [3–23]. A recent, and the largest meta-analysis so far, reported 18,760 CpGs annotated to 7201 genes, which account for approximately one third of the known human genes, were differentially methylated between 2433 current smokers and 6956 never smokers [11]. Associations for several loci, such as *AHRR, F2RL3, GPR15, GFI1, 2q37.1* and *6p21.33*, have been consistently reported, and a systematic review published in 2015 found that associations for 62 CpGs had been reported at least three times [24]. Apart from smoking status, other smoking exposures such as cumulative smoking [3, 4, 8–12, 16–18, 20, 22] and years since quitting [4, 9–12, 15, 16, 19, 20, 22] have also been found to be associated with blood DNA methylation.

Most of the reported associations are from cross-sectional designs; thus, the causal nature of the association, i.e. whether DNA methylation has a causal effect on smoking or vice versa, is unknown. There is also a possibility that cross-sectional epigenetic associations are due to familial confounding [25]. Studies have suggested that smoking-related blood DNA methylation mediates the effects of smoking on lung cancer [26, 27], death [28], leukocyte telomere length [29], and subclinical atherosclerosis [30]. These studies assume that smoking has a causal effect on methylation without evidence of causality. To the best of our knowledge, the only causal evidence comes from a study using a two-step Mendelian randomisation (MR) approach to investigate the mediating role of methylation between smoking and inflammation [31]. This study found that smoking had a causal effect on methylation at CpGs located at *F2RL3* and *GPR15* genes.

In this study, we aimed to investigate association between smoking and blood DNA methylation, to replicate associations previously reported and to investigate putative causal nature of the association using regression methods for related individuals.

Methods

Study sample

The sample comprised women from the Australian Mammographic Density Twins and Sisters Study [32]. A total of 479 women including 66 monozygotic twin pairs, 66 dizygotic twin pairs and 215 sisters from 130 families were selected [33].

Smoking data collection

A telephone-administered questionnaire was used to collect participants' self-reported information on smoking. Participants were asked the question 'Have you ever smoked at least one cigarette per day for 3 months or longer?' Participants who answered 'No' were classified as never smokers, and the rest ever smokers. Ever smokers were further questioned for age at starting smoking, the average number of cigarettes smoked per day, and age at stopping smoking, if any. Ever smokers who had stopped smoking before the interview were classified as former smokers, and the rest current smokers.

DNA methylation data

DNA was extracted from dried blood spots stored on Guthrie cards using a method previously described [34]. Methylation was measured using the Infinium Human-Methylation450K BeadChip array. Raw intensity data were processed by Bioconductor *minfi* package [35], which included normalisation of data using Illumina's reference factor-based normalisation methods (*preprocessIllumina*) and subset-quantile within array normalisation (*preprocessSWAN*) [36] for type I and II probe bias correction. An empirical Bayes batch-effects removal method *ComBat* [37] was applied to minimise technical variation across batches. Probes with missing values (detection P value> 0.01) in one or more samples, with documented SNPs at the target CpG, with beadcount < 3 in more than 5% samples, binding to multiple locations [38] or binding to X chromosome, and the 65 control probes were excluded, leaving 411,219 probes included in the analysis; see Li et al. [33] for more details.

Epigenome-wide association analysis

We investigated the association using a linear mixed-effects model in which the methylation M value, a logit transformation of the percentage of methylation, as the outcome and smoking status (never, former and current smokers) as the predictor. The model was adjusted for age and estimated cell-type proportions [39] as fixed effects and for family and zygosity as random effects, fitted using the *lmer()* function from the R package *lme4* [40]. The likelihood ratio test was used to make inference, that is, a nested model without smoking status was fitted and a P value was calculated based on that, twice the difference in the log likelihoods between the full and nested models approximately follows the chi-squared distribution with two degrees of freedom. To account for multiple testing, associations with a false discovery rate (FDR) [41] < 0.05 were considered statistically significant and the corresponding CpGs were referred to as 'identified CpGs'.

For identified CpGs, we investigated their associations with cumulative smoke exposure indicated by pack-years for ever smokers and with years since quitting for former smokers. Pack-years were calculated as the average number of cigarettes smoked per day divided by 20 and multiplied by the number of years smoked, and were log-transformed to be approximately normal distributed. Years since quitting were calculated as age at interview minus age at stopping smoking. The covariates adjusted and statistical inference were the same as those for smoking status, except that the model for pack-years was additionally adjusted for smoking status (former and current smokers) to investigate associations independent of smoking status.

Replication of previously reported associations

After quality control, 18,671 CpGs reported from the largest meta-analysis performed by Joehanes et al. [11] were included in our study. For these CpGs, we investigated their associations with smoking status in our study. Given the sample size of our study and not to miss any potential replication, associations with a nominal $P < 0.05$ and the same direction as that reported by Joehanes et al. were considered to be replicated, and the corresponding CpGs were referred to as 'replicated CpGs'.

Familial confounding analysis

For the identified CpGs and replicated CpGs, we performed between- and within-sibship analyses [25, 42] to investigate if familial factors confound the associations. Given that never and former smokers had similar methylation levels for most of the CpGs, we combined them into one group. The new smoking status was thus analysed with current smokers as '1' and the rest as '0'.

In the analysis, the methylation M values, smoking exposures and covariates were orthogonally transformed within sibships to obtain sibship means and within-sibship differences for these variables; see Stone et al. [42] for more details about the transformation. The between-sibship analyses investigated associations between sibship means for methylation levels and those for smoking exposures, and the within-sibship analyses investigated associations between within-sibship differences for methylation levels and those for smoking exposures. Associations estimated from the within-sibship analyses are independent of familial confounding, as the confounding effects of familial factors shared by siblings, both known and unknown, were cancelled out when using within-sibship differences. Evidence for familial confounding can be obtained by comparing between-sibship coefficient (β_B) and within-sibship coefficient (β_W). When $\beta_B \neq \beta_W$ and $\beta_W \approx 0$,

i.e. the association disappears when familial factors are adjusted, the observation is consistent with the association being due to familial confounding. When $\beta_B \approx \beta_W \neq 0$, i.e. the association is similar regardless of whether familial factors are adjusted, the observation is consistent with absence of evidence for familial confounding; see Carlin et al. [43] for more details about the implications from comparing β_B and β_W.

Causal inference analysis

We performed causal inference between smoking status and methylation using Inference about Causation through Examination of FAmiliaL CONfounding (ICE FALCON), a regression-based methodology for analysing twin data [44–48]. By causal is meant, that if it were possible to vary a predictor measure experimentally, the expected value of the outcome measure would change.

As shown in Fig. 1, suppose there are two variables, X and Y, measured for pairs of twins, and for example, let X refer to smoking status and Y refer to methylation. Assume that X and Y are positively associated within an individual. Let S denote the unmeasured familial factors that affect both twins, S_X represents those factors that influence X values only, S_Y those that influence Y values only, and S_{XY} those that influence both X and Y values. For the purpose of explanation, let 'self' refer to an individual and 'co-twin' refer to the individual's twin, but recognise that these labels can be exchanged and both twins within a pair are used in the analysis.

If there is a correlation between Y_{self} and $X_{co-twin}$, it might be due to a familial confounder, S_{XY} (Fig. 1a). It could also be due to X having a causal effect on Y within an individual, provided X_{self} and $X_{co-twin}$ are correlated (Fig. 1b), or to Y having a casual effect on X, provided Y_{self} and $Y_{co-twin}$ are correlated (Fig. 1c). Note that the confounders specific to an individual, C_{self} and $C_{co-twin}$, do not of themselves result in a correlation between Y_{self} and $X_{co-twin}$.

Using the Generalised Estimating Equations (GEE), fitted using the *geeglm()* function from R package *geepack* [49], to take into account any correlation in Y between twins within the same pair, three models are fitted:

Model 1: $E(Y_{self}) = \alpha + \beta_{self}X_{self}$
Model 2: $E(Y_{self}) = \alpha + \beta_{co-twin}X_{co-twin}$
Model 3: $E(Y_{self}) = \alpha + \beta'_{self}X_{self} + \beta'_{co-twin}X_{co-twin}$

If the correlation between Y_{self} and $X_{co-twin}$ is solely due to familial confounders (Fig. 1a), the marginal association between Y_{self} and X_{self} (β_{self} in model 1) and the marginal association between Y_{self} and $X_{co-twin}$ ($\beta_{co-twin}$ in model 2) must both be non-zero. Adjusting for X_{self}, however, the conditional association between Y_{self} and $X_{co-twin}$ ($\beta'_{co-twin}$ in model 3) is expected to attenuate from $\beta_{co-twin}$ in model 2 towards the null. Similarly, adjusting for $X_{co-twin}$ (model 3), the conditional association

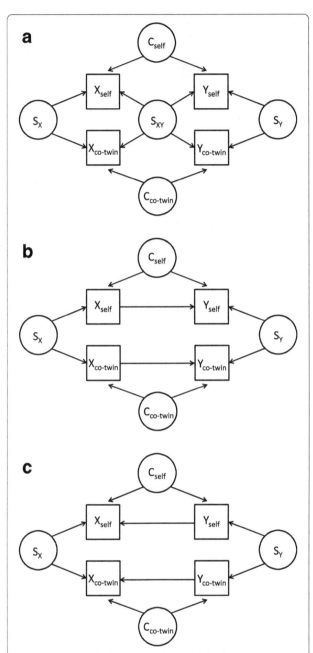

Fig. 1 Some possible directed acyclic graphs for the cross-twin cross-trait correlation. **a** The cross-twin cross-trait correlation is due to familial confounding. **b** The cross-twin cross-trait correlation is due to the causal effect of X on Y. **c** The cross-twin cross-trait correlation is due to the causal effect of Y on X

between $X_{\text{co-twin}}$ and Y_{self} (note that we assume X and Y are positively associated within an individual), so that $\beta_{\text{co-twin}}$ in model 2 depends on the within-pair correlations in X (ρ_X) and in Y (ρ_Y): if $\rho_X > \rho_Y$, $\beta_{\text{co-twin}}$ is expected to be positive; otherwise $\beta_{\text{co-twin}}$ to be negative. Conditioning on X_{self} (model 3), both pathways are blocked and the conditional association ($\beta'_{\text{co-twin}}$ in model 3) is expected to attenuate towards the null.

If the correlation between Y_{self} and $X_{\text{co-twin}}$ is solely due to a causal effect from Y to X (Fig. 1c), in model 2 the pathway through S_X is blocked due to X_{self} as a collider, and the pathway through S_Y is blocked due to that GEE analysis in effect conditions on $Y_{\text{co-twin}}$, so there is no marginal association between Y_{self} and $X_{\text{co-twin}}$, and $\beta_{\text{co-twin}}$ of model 2 is expected to be zero.

We studied methylation at the identified CpGs and replicated CpGs, respectively. For each group of CpGs, methylation was analysed as a weighted methylation score, calculated as the sum of the products of methylation level and weight of each CpG. For a locus containing multiple CpGs, only the CpG with the smallest P value was included in the methylation score. For the identified CpGs, the methylation level was the standardised M value and the weight was the log odds ratio for smoking status. For the replicated CpGs, the methylation level was the Beta value, the scale used in the meta-analysis, and the weight was the Z statistic reported by Joehanes et al. [11]. Smoking status was analysed as a binary variable with current smokers as '1' and the rest as '0'. We first used smoking status to be X and methylation score to be Y and regressed methylation score on smoking status. We then exchanged X and Y to regress smoking status on methylation score and undertook the same analyses. The data for 132 twin pairs were used. We made statistical inference about the change in regression coefficient using one-sided t test with a standard error computed using nonparametric bootstrap method. That is, twin pairs were randomly sampled with replacement to generate 1000 new datasets with the same sample size as the original dataset. ICE FALCON was then applied to each dataset to calculate the change in regression coefficient for that dataset and standard error was then estimated by computing the standard deviation.

Results

Characteristics of the sample

The mean (standard deviation [SD]) age for the 479 women was 56.4 (7.9) years. The women included 291 (60.8%) never smokers, 147 (30.7%) former smokers and 41 (8.5%) current smokers. Ever smokers had a median (interquartile range) of 7.0 (13.8) pack-years. Former smokers had an average (SD) of 21.5 (11.4) years since quitting.

between Y_{self} and X_{self} (β'_{self} in model 3) is expected to attenuate from β_{self} in model 1 towards the null.

If the correlation between Y_{self} and $X_{\text{co-twin}}$ is solely due to a causal effect from X to Y (Fig. 1b), Y_{self} and $X_{\text{co-twin}}$ in model 2 will be associated through two pathways: the confounder S_X, and conditioning on the collider $Y_{\text{co-twin}}$ (GEE analysis in effect conditions on $Y_{\text{co-twin}}$). Conditioning on $Y_{\text{co-twin}}$ induces a negative correlation

Epigenome-wide analysis results

Methylation at 39 CpGs located at 27 loci was found to be associated with smoking status (Table 1; Q-Q plot and Manhattan plot in Fig. 2). Associations for 37 of the 39 CpGs have been reported by at least two studies and associations for two CpGs, cg06226150 (*SLC2A4RG*) and cg21733098 (*12q24.32*), have only been reported from the meta-analysis performed by Joehanes et al. [11]. For

Table 1 39 CpGs at which methylation was found to be associated with smoking status with FDR < 0.05

CpG	CHR	Loci	Methylation level, mean (standard deviation)			P	FDR
			Never smokers	Former smokers	Current smokers		
cg05575921	5	AHRR	0.82 (0.04)	0.79 (0.05)	0.69 (0.08)	2.69E-41	1.11E-35
cg05951221	2	2q37.1	0.48 (0.05)	0.44 (0.06)	0.38 (0.06)	1.01E-28	2.08E-23
cg01940273	2	2q37.1	0.69 (0.04)	0.66 (0.05)	0.60 (0.05)	1.03E-25	1.41E-20
cg03636183	19	F2RL3	0.72 (0.04)	0.70 (0.05)	0.64 (0.06)	2.86E-22	2.94E-17
cg06126421	6	6p21.33	0.79 (0.05)	0.76 (0.06)	0.72 (0.06)	1.22E-17	1.00E-12
cg26703534	5	AHRR	0.68 (0.03)	0.69 (0.03)	0.64 (0.03)	4.44E-16	3.04E-11
cg21161138	5	AHRR	0.77 (0.03)	0.76 (0.04)	0.72 (0.05)	1.21E-14	7.11E-10
cg11660018	11	PRSS23	0.59 (0.04)	0.57 (0.04)	0.54 (0.04)	8.59E-12	4.42E-07
cg09935388	1	GFI1	0.82 (0.05)	0.81 (0.05)	0.75 (0.07)	5.90E-11	2.70E-06
cg25648203	5	AHRR	0.84 (0.02)	0.83 (0.02)	0.81 (0.03)	1.63E-10	6.71E-06
cg19859270	3	GPR15	0.93 (0.01)	0.93 (0.01)	0.92 (0.01)	2.77E-10	1.04E-05
cg03329539	2	2q37.1	0.47 (0.05)	0.46 (0.05)	0.42 (0.04)	5.04E-10	1.73E-05
cg24859433	6	6p21.33	0.88 (0.02)	0.88 (0.02)	0.86 (0.02)	6.02E-10	1.85E-05
cg14753356	6	6p21.33	0.47 (0.06)	0.45 (0.06)	0.43 (0.05)	6.28E-10	1.85E-05
cg07339236	20	ATP9A	0.17 (0.04)	0.16 (0.04)	0.13 (0.03)	3.68E-09	1.01E-04
cg04885881	1	1p36.22	0.48 (0.05)	0.47 (0.05)	0.44 (0.05)	4.46E-09	1.15E-04
cg23916896	5	AHRR	0.29 (0.07)	0.27 (0.06)	0.23 (0.06)	1.01E-08	2.43E-04
cg14817490	5	AHRR	0.30 (0.04)	0.03 (0.04)	0.26 (0.04)	1.37E-08	3.14E-04
cg11902777	5	AHRR	0.08 (0.02)	0.08 (0.02)	0.06 (0.02)	4.01E-08	8.55E-04
cg21611682	11	LRP5	0.61 (0.03)	0.60 (0.03)	0.58 (0.03)	4.16E-08	8.55E-04
cg01692968	9	9q31.1	0.41 (0.05)	0.39 (0.05)	0.38 (0.05)	5.57E-08	1.09E-03
cg08709672	1	AVPR1B	0.60 (0.03)	0.59 (0.03)	0.57 (0.03)	6.54E-08	1.22E-03
cg07826859	7	MYO1G	0.66 (0.04)	0.65 (0.04)	0.63 (0.03)	1.14E-07	2.04E-03
cg25189904	1	GNG12	0.53 (0.06)	0.51 (0.07)	0.47 (0.07)	1.36E-07	2.33E-03
cg17287155	5	AHRR	0.86 (0.03)	0.85 (0.03)	0.84 (0.03)	2.19E-07	3.61E-03
cg06226150	20	SLC2A4RG	0.28 (0.03)	0.28 (0.02)	0.26 (0.02)	2.85E-07	4.51E-03
cg23161492	15	ANPEP	0.30 (0.05)	0.29 (0.05)	0.26 (0.05)	6.19E-07	9.43E-03
cg09022230	7	TNRC18	0.76 (0.04)	0.75 (0.04)	0.73 (0.04)	6.57E-07	9.65E-03
cg19572487	17	RARA	0.63 (0.05)	0.61 (0.05)	0.60 (0.06)	7.54E-07	1.07E-02
cg03991871	5	AHRR	0.89 (0.03)	0.89 (0.03)	0.86 (0.04)	9.13E-07	1.25E-02
cg14580211	5	C5orf62	0.76 (0.04)	0.75 (0.04)	0.73 (0.04)	1.12E-06	1.48E-02
cg15187398	19	MOBKL2A	0.53 (0.05)	0.51 (0.05)	0.49 (0.04)	1.25E-06	1.60E-02
cg10750182	10	C10orf105	0.62 (0.03)	0.62 (0.03)	0.60 (0.03)	2.03E-06	2.53E-02
cg25949550	7	CNTNAP2	0.13 (0.02)	0.13 (0.02)	0.12 (0.02)	2.64E-06	3.19E-02
cg05284742	14	ITPK1	0.78 (0.03)	0.77 (0.03)	0.76 (0.04)	2.76E-06	3.24E-02
cg23931381	19	ARRDC2	0.89 (0.02)	0.88 (0.02)	0.87 (0.02)	2.98E-06	3.40E-02
cg26271591	2	NFE2L2	0.46 (0.06)	0.45 (0.06)	0.41 (0.06)	4.40E-06	4.72E-02
cg03646329	13	LPAR6	0.82 (0.04)	0.81 (0.05)	0.79 (0.05)	4.47E-06	4.72E-02
cg21733098	12	12q24.32	0.76 (0.06)	0.75 (0.07)	0.72 (0.06)	4.47E-06	4.72E-02

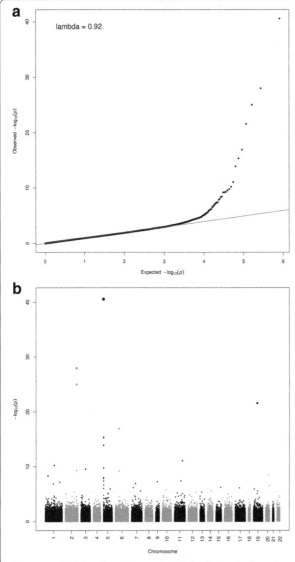

Fig. 2 Q-Q plot (**a**) and Manhattan plot (**b**) for the results from the epigenome-wide association analysis between DNA methylation and smoking status

all 39 CpGs, current smokers had the lowest methylation level (Table 1). The 27 loci included several consistently reported loci, such as *AHRR* (9 CpGs), *2q37.1* (3 CpGs), *6p21.33* (3 CpGs), and *F2RL3* (1 CpG).

Of the 39 CpGs and at a 5% FDR, methylation at 18 CpGs was negatively associated with pack-years and at 20 CpGs was positively associated with years since quitting. Methylation at 15 CpGs was associated with pack-years and years since quitting both (Table 2).

Replication for previously reported associations
For the associations for 18,671 CpGs reported by Joehanes et al. [11], 1882 were replicated with a nominal $P < 0.05$ and in the same direction, and the 133 most significant associations also had a FDR < 0.05.

Of the 1882 replications, 1154 were for the novel CpGs reported by Joehanes et al. (Additional file 1: Table S1).

Between- and within-sibship analyses results
For the 39 identified CpGs, no evidence for a difference between β_B and β_W was found for any CpG (Table 3; all P values > 0.05 from the β_B and β_W comparison). The same results were found from the analyses of pack-years and years since quitting (Table 3).

For the 1882 replicated CpGs, no evidence for a difference between β_B and β_W was found for any CpG (Additional file 2: Table S2; the smallest P value = 1.3×10^{-3} and the smallest FDR = 0.99 from the β_B and β_W comparison).

ICE FALCON analysis results
Within twin pairs, the correlation in smoking status was 0.11 (95% confidence interval (CI) – 0.06, 0.27), smaller than the correlations in methylation scores for the replicated CpGs and for the identified CpGs, which were 0.37 (95% CI 0.23, 0.50) and 0.22 (95% CI 0.05, 0.37), respectively.

The ICE FALCON results for methylation at the replicated CpGs are shown in Table 4. From the analysis in which smoking status was the predictor and methylation score the outcome, a women's methylation score was associated with her own smoking status (model 1; β_{self} = 74.6, 95% CI 55.3, 93.9), and negatively associated with her co-twin's smoking status (model 2; $\beta_{\text{co-twin}}$ = – 30.8, 95% CI – 57.7, – 4.0). Conditioning on her co-twin's smoking status (model 3), β'_{self} remained unchanged ($P = 0.41$) compared with β_{self} in model 1, while conditioning on her own smoking status (model 3), $\beta_{\text{co-twin}}$ in model 2 attenuated by 123.3% (95% CI 49.6%, 185.2%; $P = 0.002$) to be $\beta'_{\text{co-twin}}$ of 2.5 (95% CI – 16.3, 21.3). From the analysis in which methylation score was the predictor and smoking status the outcome, a woman's smoking status was associated with her own methylation score (model 1; β_{self} = 4.1, 95% CI 2.7, 5.4), but not with her co-twin's methylation score (model 2; $\beta_{\text{co-twin}}$ = 0.4, 95% CI – 1.0, 1.8). In model 3, β'_{self} and $\beta'_{\text{co-twin}}$ remained unchanged (both $P > 0.1$) compared with β_{self} in model 1 and $\beta_{\text{co-twin}}$ in model 2, respectively. These results were consistent with that smoking has a causal effect on the overall methylation level at these CpGs, but not in the opposite direction. Similar results were found and a similar causality was inferred for smoking status and the overall methylation level at the identified CpGs (Table 4).

Discussion
We performed an EWAS of smoking for a sample of middle-aged women and found 39 CpGs at which methylation was associated with smoking status. Our

Table 2 Associations of methylation at the 39 identified CpGs with pack-years and years since quitting

CpG	Pack-years			Years since quitting		
	Estimate (SE)	P value	FDR	Estimate (SE)	P value	FDR
cg05575921	− 10.68 (2.18)	1.11E-06	1.44E-05	1.63 (0.30)	1.11E-35	7.35E-08
cg05951221	− 10.32 (1.70)	3.04E-09	1.19E-07	1.66 (0.25)	2.08E-23	1.55E-10
cg01940273	− 8.99 (1.58)	1.93E-08	3.76E-07	1.47 (0.23)	1.41E-20	1.15E-09
cg03636183	− 4.65 (1.70)	5.35E-03	1.90E-02	0.98 (0.24)	2.94E-17	5.87E-05
cg06126421	− 5.92 (2.20)	6.15E-03	2.00E-02	1.31 (0.32)	1.00E-12	3.36E-05
cg26703534	0.28 (0.98)	8.02E-01	8.02E-01	− 0.3 (0.15)	3.04E-11	3.44E-02
cg21161138	− 4.30 (1.62)	6.89E-03	2.07E-02	0.65 (0.24)	7.11E-10	5.32E-03
cg11660018	− 3.74 (1.13)	7.82E-04	4.36E-03	0.83 (0.17)	4.42E-07	5.86E-07
cg09935388	− 6.87 (2.40)	3.06E-03	1.19E-02	1.10 (0.35)	2.70E-06	1.63E-03
cg25648203	− 2.25 (1.28)	7.19E-02	1.27E-01	0.37 (0.19)	6.71E-06	4.18E-02
cg19859270	− 2.69 (1.38)	4.65E-02	9.55E-02	0.58 (0.21)	1.04E-05	4.41E-03
cg03329539	− 6.27 (1.35)	3.59E-06	3.50E-05	0.83 (0.21)	1.73E-05	5.13E-05
cg24859433	− 1.45 (1.21)	2.04E-01	2.65E-01	0.41 (0.18)	1.85E-05	1.78E-02
cg14753356	− 2.64 (1.12)	1.63E-02	3.74E-02	0.38 (0.18)	1.85E-05	2.28E-02
cg07339236	− 2.71 (1.93)	1.51E-01	2.26E-01	0.89 (0.30)	1.01E-04	2.21E-03
cg04885881	− 2.16 (1.29)	7.80E-02	1.32E-01	0.28 (0.20)	1.15E-04	1.18E-01
cg23916896	− 4.32 (2.41)	6.61E-02	1.23E-01	0.47 (0.37)	2.43E-04	1.84E-01
cg14817490	− 3.31 (1.35)	1.26E-02	3.28E-02	0.35 (0.21)	3.14E-04	1.02E-01
cg11902777	− 7.19 (2.15)	6.99E-04	4.36E-03	0.65 (0.32)	8.55E-04	4.06E-02
cg21611682	− 2.04 (0.80)	9.75E-03	2.72E-02	0.27 (0.12)	8.55E-04	2.25E-02
cg01692968	− 2.37 (1.42)	8.83E-02	1.43E-01	0.72 (0.21)	1.09E-03	5.77E-04
cg08709672	− 1.05 (0.80)	1.75E-01	2.43E-01	0.29 (0.12)	1.22E-03	1.91E-02
cg07826859	− 1.75 (0.97)	6.23E-02	1.21E-01	0.17 (0.15)	2.04E-03	2.48E-01
cg25189904	− 7.02 (2.23)	1.38E-03	6.72E-03	0.74 (0.34)	2.33E-03	2.45E-02
cg17287155	− 1.67 (1.27)	1.74E-01	2.43E-01	0.31 (0.18)	3.61E-03	7.50E-02
cg06226150	− 1.37 (0.88)	1.12E-01	1.74E-01	0.21 (0.14)	4.51E-03	1.24E-01
cg23161492	− 5.41 (1.57)	5.18E-04	4.04E-03	0.68 (0.24)	9.43E-03	3.95E-03
cg09022230	0.73 (1.23)	5.42E-01	6.21E-01	0.15 (0.19)	9.65E-03	4.52E-01
cg19572487	− 4.19 (1.44)	2.96E-03	1.19E-02	0.60 (0.20)	1.07E-02	2.64E-03
cg03991871	− 5.59 (2.37)	1.63E-02	3.74E-02	0.40 (0.36)	1.25E-02	2.48E-01
cg14580211	− 0.43 (1.40)	7.46E-01	7.86E-01	0.43 (0.21)	1.48E-02	3.25E-02
cg15187398	− 1.28 (1.31)	3.10E-01	3.90E-01	0.15 (0.20)	1.60E-02	4.27E-01
cg10750182	− 0.64 (0.67)	3.20E-01	3.90E-01	0.15 (0.10)	2.53E-02	1.05E-01
cg25949550	− 2.80 (1.25)	2.24E-02	4.86E-02	0.59 (0.20)	3.19E-02	2.17E-03
cg05284742	− 0.49 (1.13)	6.58E-01	7.13E-01	0.21 (0.16)	3.24E-02	1.81E-01
cg23931381	0.71 (1.50)	6.03E-01	6.72E-01	0.20 (0.23)	3.40E-02	4.26E-01
cg26271591	− 0.44 (1.74)	7.87E-01	8.02E-01	0.34 (0.26)	4.72E-02	1.77E-01
cg03646329	− 2.47 (1.98)	1.99E-01	2.65E-01	0.63 (0.30)	4.72E-02	3.08E-02
cg21733098	− 1.78 (2.49)	4.55E-01	5.38E-01	− 0.03 (0.38)	4.72E-02	9.23E-01

Regression coefficients were reported as being multiplied by 100, as well as for standard errors

study confirmed the associations for several previously consistently reported loci including *AHRR*, *F2RL3*, *2q37.1*, and *6p21.33*, and for two novel CpGs, cg06226150 (*SLC2A4RG*) and cg21733098 (*12q24.32*), reported by the largest meta-analysis [11] so far. In addition, we replicated the associations for 1882 CpGs

Table 3 Associations of methylation at the 39 identified CpGs with smoking status, pack-years and years since quitting from the between- and within-sibship analyses

CpG	Smoking status			Pack-years			Years since quitting		
	Between-sibship coefficient (SE)	Within-sibship coefficient (SE)	P*	Between-sibship coefficient (SE)	Within-sibship coefficient (SE)	P*	Between-sibship coefficient (SE)	Within-sibship coefficient (SE)	P*
cg05575921	− 0.87 (0.12)	− 0.93 (0.08)	0.65	− 14.53 (4.65)	− 6.58 (3.76)	0.18	1.53 (0.57)	1.76 (0.52)	0.77
cg05951221	− 0.59 (0.10)	− 0.47 (0.07)	0.32	− 13.28 (4.08)	− 6.74 (2.87)	0.19	1.75 (0.43)	1.56 (0.41)	0.75
cg01940273	− 0.60 (0.10)	− 0.47 (0.06)	0.26	− 8.04 (3.30)	− 6.32 (2.88)	0.69	1.36 (0.35)	1.77 (0.41)	0.44
cg03636183	− 0.56 (0.09)	− 0.41 (0.06)	0.18	− 2.63 (4.55)	− 2.93 (2.90)	0.96	0.57 (0.50)	1.18 (0.41)	0.34
cg06126421	− 0.34 (0.14)	− 0.48 (0.08)	0.37	− 13.31 (6.51)	− 5.02 (3.69)	0.27	1.59 (0.68)	1.62 (0.52)	0.98
cg26703534	− 0.21 (0.06)	− 0.30 (0.04)	0.17	− 2.34 (2.07)	2.74 (1.62)	0.05	− 0.33 (0.28)	− 0.41 (0.24)	0.84
cg21161138	− 0.36 (0.09)	− 0.37 (0.06)	0.91	− 4.37 (4.08)	− 3.20 (2.41)	0.81	0.46 (0.47)	0.91 (0.36)	0.45
cg11660018	− 0.26 (0.08)	− 0.20 (0.04)	0.46	− 5.50 (2.48)	0.51 (2.11)	0.07	1.01 (0.28)	0.88 (0.29)	0.75
cg09935388	− 0.58 (0.14)	− 0.53 (0.10)	0.77	− 2.29 (5.39)	− 5.44 (4.01)	0.64	0.71 (0.68)	1.32 (0.59)	0.50
cg25648203	− 0.16 (0.08)	− 0.31 (0.05)	0.10	− 6.26 (2.96)	− 1.05 (2.12)	0.15	0.77 (0.36)	0.27 (0.30)	0.29
cg19859270	− 0.28 (0.08)	− 0.22 (0.05)	0.51	− 4.06 (3.48)	− 1.60 (2.63)	0.57	0.70 (0.34)	0.39 (0.38)	0.54
cg03329539	− 0.31 (0.08)	− 0.28 (0.06)	0.79	− 4.34 (3.26)	− 6.70 (2.45)	0.56	0.78 (0.35)	0.95 (0.35)	0.73
cg24859433	− 0.15 (0.07)	− 0.24 (0.05)	0.25	− 3.51 (2.81)	− 0.74 (2.12)	0.43	0.71 (0.30)	0.47 (0.32)	0.59
cg14753356	− 0.13 (0.07)	− 0.16 (0.04)	0.73	− 0.21 (3.36)	− 3.43 (1.78)	0.40	0.61 (0.34)	0.33 (0.27)	0.52
cg07339236	− 0.29 (0.11)	− 0.32 (0.07)	0.83	− 1.48 (4.44)	− 0.56 (3.24)	0.87	0.88 (0.48)	0.62 (0.48)	0.70
cg04885881	− 0.26 (0.08)	− 0.24 (0.05)	0.80	− 0.68 (2.88)	0.24 (2.12)	0.80	0.45 (0.32)	0.26 (0.32)	0.67
cg23916896	− 0.40 (0.15)	− 0.49 (0.10)	0.60	− 9.52 (5.34)	8.79 (4.31)	0.01	0.72 (0.62)	− 0.13 (0.67)	0.35
cg14817490	− 0.18 (0.09)	− 0.28 (0.05)	0.32	− 2.65 (3.72)	− 2.45 (2.15)	0.96	0.30 (0.42)	0.55 (0.30)	0.63
cg11902777	− 0.40 (0.14)	− 0.44 (0.09)	0.80	− 14.37 (4.30)	− 2.35 (3.70)	0.03	1.22 (0.54)	0.70 (0.56)	0.50
cg21611682	− 0.15 (0.05)	− 0.16 (0.03)	0.88	− 1.63 (1.75)	− 1.82 (1.48)	0.93	0.34 (0.22)	0.42 (0.22)	0.80
cg01692968	− 0.11 (0.09)	− 0.16 (0.06)	0.61	− 4.60 (3.02)	0.19 (2.32)	0.21	0.76 (0.39)	0.51 (0.34)	0.63
cg08709672	− 0.07 (0.06)	− 0.17 (0.03)	0.12	1.06 (2.28)	− 1.11 (1.30)	0.41	0.08 (0.26)	0.56 (0.19)	0.13
cg07826859	− 0.16 (0.06)	− 0.20 (0.04)	0.52	− 2.02 (2.59)	− 1.05 (1.69)	0.75	0.09 (0.26)	0.41 (0.24)	0.37
cg25189904	− 0.46 (0.11)	− 0.29 (0.09)	0.21	− 11.29 (4.40)	− 0.82 (4.21)	0.09	0.20 (0.57)	1.09 (0.61)	0.29
cg17287155	− 0.23 (0.08)	− 0.19 (0.05)	0.65	− 1.78 (2.87)	2.63 (2.03)	0.21	0.49 (0.3)	− 0.16 (0.32)	0.14
cg06226150	− 0.19 (0.05)	− 0.13 (0.04)	0.36	− 2.55 (2.48)	− 2.00 (1.44)	0.85	− 0.14 (0.26)	0.36 (0.22)	0.14
cg23161492	− 0.28 (0.11)	− 0.24 (0.06)	0.74	− 8.68 (4.48)	− 4.66 (2.44)	0.43	0.55 (0.51)	0.81 (0.37)	0.68
cg09022230	− 0.12 (0.08)	− 0.25 (0.05)	0.17	5.91 (2.41)	− 3.44 (1.92)	0.00	− 0.13 (0.34)	0.73 (0.29)	0.06
cg19572487	− 0.14 (0.07)	− 0.20 (0.06)	0.54	− 8.14 (3.10)	− 1.86 (2.16)	0.10	0.75 (0.37)	0.69 (0.33)	0.91
cg03991871	− 0.39 (0.16)	− 0.38 (0.08)	0.98	− 5.86 (5.57)	− 0.98 (4.13)	0.48	0.10 (0.63)	0.29 (0.63)	0.83
cg14580211	− 0.15 (0.09)	− 0.24 (0.05)	0.33	− 3.40 (3.54)	0.79 (2.41)	0.33	1.00 (0.37)	0.46 (0.33)	0.27
cg15187398	− 0.18 (0.08)	− 0.18 (0.05)	0.96	− 2.81 (3.07)	1.55 (2.24)	0.25	0.23 (0.39)	0.15 (0.35)	0.87
cg10750182	− 0.08 (0.04)	− 0.11 (0.03)	0.53	− 0.79 (1.75)	0.47 (1.18)	0.55	0.07 (0.19)	0.23 (0.17)	0.55
cg25949550	− 0.13 (0.07)	− 0.20 (0.05)	0.39	− 1.05 (2.47)	0.12 (2.36)	0.73	0.44 (0.30)	0.81 (0.34)	0.41
cg05284742	− 0.16 (0.06)	− 0.14 (0.05)	0.86	3.53 (2.18)	− 1.18 (1.94)	0.11	0.20 (0.30)	0.30 (0.28)	0.81
cg23931381	− 0.08 (0.08)	− 0.20 (0.06)	0.25	1.09 (3.58)	1.77 (2.65)	0.88	− 0.16 (0.43)	0.46 (0.37)	0.27
cg26271591	− 0.16 (0.10)	− 0.32 (0.07)	0.19	− 4.11 (4.56)	0.95 (3.10)	0.36	0.41 (0.46)	0.52 (0.45)	0.87
cg03646329	− 0.29 (0.13)	− 0.27 (0.08)	0.90	− 10.07 (5.29)	− 1.22 (3.22)	0.15	0.80 (0.64)	0.66 (0.46)	0.86
cg21733098	− 0.37 (0.16)	− 0.31 (0.09)	0.76	− 4.20 (6.47)	2.68 (4.37)	0.38	0.19 (0.70)	− 0.14 (0.67)	0.74

Regression coefficients from the analyses for pack-years and years since quitting were reported as being multiplied by 100, as well as for standard errors
*P-value from comparing the between-sibship coefficient with the within-sibship coefficient

Table 4 Results from the ICE FALCON analyses

CpGs	Coefficient	Model 1		Model 2		Model 3		Change	
		Estimate (SE)	P	Estimate (SE)	P	Estimate (SE)	P	Estimate (SE)	P
CpGs reported by Joehanes et al.									
Smoking as the predictor	β_{self}	74.61 (9.87)	4.0E-14	–	–	75.45 (9.29)	4.4E-16	0.84 (3.60)	4.1E-01
	$\beta_{co\text{-}twin}$	–	–	− 30.84 (13.69)	2.4E-02	2.50 (9.57)	7.9E-01	− 33.34 (11.60)	2.1E-03
Methylation score as the predictor	β_{self}	4.07 (0.70)	7.5E-09	–	–	4.45 (0.81)	3.6E-08	0.39 (0.47)	2.1E-01
	$\beta_{co\text{-}twin}$	–	–	0.41 (0.72)	5.7E-01	− 1.00 (0.82)	2.2E-01	− 1.42 (1.15)	1.1E-01
CpGs identified from our study									
Smoking as the predictor	β_{self}	27.70 (3.65)	3.4E-14	–	–	26.89 (3.79)	1.2E-12	− 0.81 (0.89)	1.8E-01
	$\beta_{co\text{-}twin}$	–	–	− 12.36 (3.86)	1.4E-03	− 3.45 (2.58)	1.8E-01	− 8.90 (5.52)	5.3E-02
Methylation score as the predictor	β_{self}	10.24 (2.19)	1.3E-08	–	–	11.14 (2.47)	6.7E-06	0.90 (1.27)	2.4E-01
	$\beta_{co\text{-}twin}$	–	–	− 4.48 (2.65)	9.2E-02	− 3.86 (2.66)	1.5E-01	0.61 (3.77)	4.4E-01

Regression coefficients from the analyses in which the methylation score as the predictor were reported as being multiplied by 100, as well as for standard errors

reported by the meta-analysis. The investigation of causation suggests that smoking has a causal effect on DNA methylation, not vice versa or being due to familial confounding.

To the best of our knowledge, our study is the first study to confirm the associations for cg06226150 and cg21733098. cg06226150 is located at the promoter of, and potentially regulates the expression of, SLC2A4RG (solute carrier family 2 member 4 regulator gene). SLC2A4RG is involved in the Gene Ontology pathway for regulation of transcription (GO:0006355). Protein encoded by SLC2A4RG regulates the activation of SLC2A4 (solute carrier family 2 member 4). SLC2A4 is involved in the glucose transportation across cell membranes stimulated by insulin. Genetic variants at SLC2A4RG have been found to be associated with inflammatory bowel disease [50] and prostate cancer [51]. cg21733098 is located at an intergenic region on 12q24.32. The region contains several long non-coding RNA genes. Little is known about the regulatory function of cg21733098. The biological relevance of smoking to blood methylation at these two CpGs is largely unknown, and more research are warranted.

We found evidence that 18 and 20 of the identified CpGs were also associated with pack-years and years since quitting, respectively. Given that smokers have lower methylation levels at the identified CpGs, the negative associations with pack-years imply that there appear to be dose-relationships between smoking and methylation at the 18 CpGs, and the positive associations with years quitting smoking imply that methylation changes at the 20 CpGs tend to reverse after cessation. The dose-relationship and reversion have also been reported by several studies [4, 9–12, 15, 16, 19, 20, 22].

Our study, as one of the first studies, provides insights into the causality underlying the cross-sectional association between smoking and blood DNA methylation. Our results are inconsistent with the proposition that the cross-sectional association is due to familial confounding, e.g. shared genes and/or environment. The roles of shared genes and/or environment are also in part unsupported by that certain smoking-related loci, such as AHRR and F2RL3, are observed across Europeans [3, 5, 8–11, 16, 19, 20, 22], South Asians [8], Arabian Asians [21], East Asians [12, 23], and African Americans [7, 11, 13, 18], who have different germline genetic backgrounds and environments. Our results support that smoking has a causal effect on the overall methylation at the identified CpGs and at the replicated CpGs, but not vice versa. Results from the two-step MR analysis performed by Jhun et al. [31] also suggest that differential methylation at cg03636183 (F2RL3) and cg19859270 (GPR15) between current and never smokers are consequential to smoking under the assumptions of MR.

That smoking causes changes in methylation is also supported to some extent by other evidence. The 'reversion' phenomenon is in line with the 'experimental evidence' criterion proposed by Bradford Hill, i.e. 'reducing or eliminating a putatively harmful exposure and seeing if the frequency of disease subsequently declines' [52]. The associations between cord blood methylation for newborns at some active-smoking-related loci, such as AHRR and GFI1, and maternal smoking in pregnancy [53] also imply that smoking is likely to cause methylation changes at these loci. Additionally, some smoking-related loci are involved in the metabolism of smoking-released chemicals. AHRR gene encodes a repressor of the aryl hydrocarbon receptor (AHR) gene, the protein encoded by which is involved in the regulation of biological response to planar aromatic hydrocarbons. Polycyclic aromatic hydrocarbons, one main smoking-related toxic and carcinogenic substance, trigger AHR signalling cascade [16, 22]. Protein coded by the *AHR* gene activates the expression of the *AHRR* gene, which in turn represses the function of AHR through a negative feedback mechanism [54]. That hypomethylation at *AHRR* gene caused by smoking is biologically plausible.

That smoking causes changes in blood methylation has great clinical and etiological implications: methylation might mediate the effects of smoking on smoking-related health outcomes. As introduced above, there have been a few studies [26–29] investigating the mediating role of methylation. A better understanding of the mechanisms of smoking affecting health is expected with more investigations on methylation.

Our study shows the value of ICE FALCON in causality assessment for observational associations. Associations from observational studies can be due to confounding and, although analyses of measured potential confounders can eliminate some confounding, there is always the possibility of unmeasured confounding, even with prospective studies. With recent discoveries of genetic markers that predict variation in risk factors, the MR concept has been explored by epidemiologists. MR uses measured genetic variants as the instrumental variable and the results of MR might be biased due to several factors such as strengthen of instrumental variable, directional pleiotropy, and unmeasured confounding [55]. ICE FALCON is a novel approach to making inference about causation. It in effect uses the familial causes of exposure and of outcome as instrumental variables. The familial causes are not measured but surrogated by co-twin's measured exposure and outcome. Thus, ICE FALCON resembles a bidirectional MR approach [56]. The instrumental variables consider all familial causes in exposure and in outcome, thus potentially less biased by their strengths than a finite number of genetic markers. More importantly, even should directional pleiotropy exist, the attenuation in the coefficient for co-twin's exposure after adjusting for an individual's own exposure also supports a causal effect.

Conclusions

We found evidence that in the peripheral blood from middle-aged women, DNA methylation at several loci is associated with smoking. By investigating causation underlying the association, our study found evidence consistent with smoking having a causal effect on methylation, but not vice versa.

Abbreviations

AHRR: Aryl hydrocarbon receptor repressor gene; CI: Confidence interval; CpG: Cytosine-guanine dinucleotide; EWAS: Epigenome-wide association study; F2RL3: F2R-like thrombin or trypsin receptor 3 gene; FDR: False discovery rate; GEE: Generalised estimating equations; GFI1: Growth factor independent 1 transcriptional repressor gene; GPR15: G protein-coupled receptor 15 gene; ICE FALCON: Inference about causation through examination of familial confounding; MR: Mendelian randomisation; SD: standard deviation; SLC2A4RG: Solute carrier family 2 member 4 regulator gene

Acknowledgements
We would like to thank all women participating in this study. The data analysis was facilitated by Spartan, the High Performance Computer and Cloud hybrid system of the University of Melbourne.

Funding
The Australian Mammographic Density Twins and Sisters Study was facilitated through the Australian Twin Registry, a national research resource in part supported by a Centre for Research Excellence Grant from the National Health and Medical Research Council (NHMRC) APP 1079102. The AMDTSS was supported by NHMRC (grant numbers 1050561 and 1079102) and Cancer Australia and National Breast Cancer Foundation (grant number 509307). SL is supported by the Australian Government Research Training Program Scholarship and the Richard Lovell Travelling Scholarship from the University of Melbourne. TLN is supported by a NHMRC Post-Graduate Scholarship and the Richard Lovell Travelling Scholarship from the University of Melbourne. MCS is a NHMRC Senior Research Fellow. JLH is a NHMRC Senior Principal Research Fellow.

Authors' contributions
SL and JLH conceived and designed the study. SL performed the statistical analyses. SL and JLH wrote the first draft of the manuscript. EMW, TLN, JEJ, JS, GSD, GGG, MCS, and JLH contributed to the data collection. MB contributed to the ICE FALCON analyses. RS contributed to the data interpretation. All authors participated in the manuscript revision and have read and approved the final manuscript.

Competing interests
The authors declare that they have no competing interests.

Author details
[1]Centre for Epidemiology and Biostatistics, Melbourne School of Population and Global Health, University of Melbourne, Parkville, Victoria, Australia. [2]Genetic Epidemiology Laboratory, Department of Pathology, University of Melbourne, Parkville, Victoria, Australia. [3]Precision Medicine, School of Clinical Sciences at Monash Health, Monash University, Clayton, Victoria, Australia. [4]Centre for Genetic Origins of Health and Disease, Curtin University and the University of Western Australia, Perth, Western Australia, Australia. [5]Cancer Epidemiology and Intelligence Division, Cancer Council Victoria, Melbourne, Victoria, Australia. [6]Murdoch Children's Research Institute, Royal Children's Hospital, Parkville, Victoria, Australia. [7]Department of Paediatrics, University of Melbourne, Parkville, Victoria, Australia.

References
1. Petronis A. Epigenetics as a unifying principle in the aetiology of complex traits and diseases. Nature. 2010;465:721–7.
2. Esteller M. Epigenetics in cancer. N Engl J Med. 2008;358:1148–59.
3. Allione A, Marcon F, Fiorito G, Guarrera S, Siniscalchi E, Zijno A, Crebelli R, Matullo G. Novel epigenetic changes unveiled by monozygotic twins discordant for smoking habits. PLoS One. 2015;10:e0128265.
4. Ambatipudi S, Cuenin C, Hernandez-Vargas H, Ghantous A, Le Calvez-Kelm F, Kaaks R, Barrdahl M, Boeing H, Aleksandrova K, Trichopoulou A, et al. Tobacco smoking-associated genome-wide DNA methylation changes in the EPIC study. Epigenomics. 2016;8:599–618.
5. Besingi W, Johansson A. Smoke-related DNA methylation changes in the etiology of human disease. Hum Mol Genet. 2014;23:2290–7.
6. Breitling LP, Yang R, Korn B, Burwinkel B, Brenner H. Tobacco-smoking-related differential DNA methylation: 27K discovery and replication. Am J Hum Genet. 2011;88:450–7.
7. Dogan MV, Shields B, Cutrona C, Gao L, Gibbons FX, Simons R, Monick M, Brody GH, Tan K, Beach SR, Philibert RA. The effect of smoking on DNA methylation of peripheral blood mononuclear cells from African American women. BMC Genomics. 2014;15:151.

8. Elliott HR, Tillin T, McArdle WL, Ho K, Duggirala A, Frayling TM, Davey Smith G, Hughes AD, Chaturvedi N, Relton CL. Differences in smoking associated DNA methylation patterns in South Asians and Europeans. Clin Epigenetics. 2014;6:4.

9. Guida F, Sandanger TM, Castagne R, Campanella G, Polidoro S, Palli D, Krogh V, Tumino R, Sacerdote C, Panico S, et al. Dynamics of smoking-induced genome-wide methylation changes with time since smoking cessation. Hum Mol Genet. 2015;24:2349–59.

10. Harlid S, Xu Z, Panduri V, Sandler DP, Taylor JA. CpG sites associated with cigarette smoking: analysis of epigenome-wide data from the Sister Study. Environ Health Perspect. 2014;122:673–8.

11. Joehanes R, Just AC, Marioni RE, Pilling LC, Reynolds LM, Mandaviya PR, Guan W, Xu T, Elks CE, Aslibekyan S, et al. Epigenetic signatures of cigarette smoking. Circ Cardiovasc Genet. 2016;9:436–47.

12. Lee MK, Hong Y, Kim SY, London SJ, Kim WJ. DNA methylation and smoking in Korean adults: epigenome-wide association study. Clin Epigenetics. 2016;8:103.

13. Philibert RA, Beach SR, Brody GH. Demethylation of the aryl hydrocarbon receptor repressor as a biomarker for nascent smokers. Epigenetics. 2012;7: 1331–8.

14. Philibert RA, Beach SR, Lei MK, Brody GH. Changes in DNA methylation at the aryl hydrocarbon receptor repressor may be a new biomarker for smoking. Clin Epigenetics. 2013;5:19.

15. Sayols-Baixeras S, Lluis-Ganella C, Subirana I, Salas LA, Vilahur N, Corella D, Munoz D, Segura A, Jimenez-Conde J, Moran S, et al. Identification of a new locus and validation of previously reported loci showing differential methylation associated with smoking. The REGICOR study. Epigenetics. 2015;10:1156–65.

16. Shenker NS, Polidoro S, van Veldhoven K, Sacerdote C, Ricceri F, Birrell MA, Belvisi MG, Brown R, Vineis P, Flanagan JM. Epigenome-wide association study in the European Prospective Investigation into Cancer and Nutrition (EPIC-Turin) identifies novel genetic loci associated with smoking. Hum Mol Genet. 2013;22:843–51.

17. Su D, Wang X, Campbell MR, Porter DK, Pittman GS, Bennett BD, Wan M, Englert NA, Crowl CL, Gimple RN, et al. Distinct epigenetic effects of tobacco smoking in whole blood and among leukocyte subtypes. PLoS One. 2016;11:e0166486.

18. Sun YV, Smith AK, Conneely KN, Chang Q, Li W, Lazarus A, Smith JA, Almli LM, Binder EB, Klengel T, et al. Epigenomic association analysis identifies smoking-related DNA methylation sites in African Americans. Hum Genet. 2013;132:1027–37.

19. Tsaprouni LG, Yang TP, Bell J, Dick KJ, Kanoni S, Nisbet J, Vinuela A, Grundberg E, Nelson CP, Meduri E, et al. Cigarette smoking reduces DNA methylation levels at multiple genomic loci but the effect is partially reversible upon cessation. Epigenetics. 2014;9:1382–96.

20. Wan ES, Qiu W, Baccarelli A, Carey VJ, Bacherman H, Rennard SI, Agusti A, Anderson W, Lomas DA, Demeo DL. Cigarette smoking behaviors and time since quitting are associated with differential DNA methylation across the human genome. Hum Mol Genet. 2012;21:3073–82.

21. Zaghlool SB, Al-Shafai M, Al Muftah WA, Kumar P, Falchi M, Suhre K. Association of DNA methylation with age, gender, and smoking in an Arab population. Clin Epigenetics. 2015;7:6.

22. Zeilinger S, Kuhnel B, Klopp N, Baurecht H, Kleinschmidt A, Gieger C, Weidinger S, Lattka E, Adamski J, Peters A, et al. Tobacco smoking leads to extensive genome-wide changes in DNA methylation. PLoS One. 2013;8:e63812.

23. Zhu X, Li J, Deng S, Yu K, Liu X, Deng Q, Sun H, Zhang X, He M, Guo H, et al. Genome-wide analysis of DNA methylation and cigarette smoking in a Chinese population. Environ Health Perspect. 2016;124:966–73.

24. Gao X, Jia M, Zhang Y, Breitling LP, Brenner H. DNA methylation changes of whole blood cells in response to active smoking exposure in adults: a systematic review of DNA methylation studies. Clin Epigenetics. 2015;7:113.

25. Li S, Wong EM, Southey MC, Hopper JL. Association between DNA methylation at SOCS3 gene and body mass index might be due to familial confounding. Int J Obes. 2017;41:995–6.

26. Fasanelli F, Baglietto L, Ponzi E, Guida F, Campanella G, Johansson M, Grankvist K, Johansson M, Assumma MB, Naccarati A, et al. Hypomethylation of smoking-related genes is associated with future lung cancer in four prospective cohorts. Nat Commun. 2015;6:10192.

27. Zhang Y, Elgizouli M, Schottker B, Holleczek B, Nieters A, Brenner H. Smoking-associated DNA methylation markers predict lung cancer incidence. Clin Epigenetics. 2016;8:127.

28. Zhang Y, Schottker B, Florath I, Stock C, Butterbach K, Holleczek B, Mons U, Brenner H. Smoking-associated DNA methylation biomarkers and their predictive value for all-cause and cardiovascular mortality. Environ Health Perspect. 2016;124:67–74.

29. Gao X, Mons U, Zhang Y, Breitling LP, Brenner H. DNA methylation changes in response to active smoking exposure are associated with leukocyte telomere length among older adults. Eur J Epidemiol. 2016;31:1231–41.

30. Reynolds LM, Wan M, Ding J, Taylor JR, Lohman K, Su D, Bennett BD, Porter DK, Gimple R, Pittman GS, et al. DNA methylation of the aryl hydrocarbon receptor repressor associations with cigarette smoking and subclinical atherosclerosis. Circ Cardiovasc Genet. 2015;8:707–16.

31. Jhun MA, Smith JA, Ware EB, Kardia SL, Mosley TH, Turner ST, Peyser PA, Kyun Park S. Modeling the causal role of DNA methylation in the association between cigarette smoking and inflammation in African Americans: a two-step epigenetic Mendelian randomization study. Am J Epidemiol. 2017;

32. Odefrey F, Stone J, Gurrin LC, Byrnes GB, Apicella C, Dite GS, Cawson JN, Giles GG, Treloar SA, English DR, et al. Common genetic variants associated with breast cancer and mammographic density measures that predict disease. Cancer Res. 2010;70:1449–58.

33. Li S, Wong EM, Joo JE, Jung CH, Chung J, Apicella C, Stone J, Dite GS, Giles GG, Southey MC, Hopper JL. Genetic and environmental causes of variation in the difference between biological age based on DNA methylation and chronological age for middle-aged women. Twin Res Hum Genet. 2015;18:720–6.

34. Joo JE, Wong EM, Baglietto L, Jung CH, Tsimiklis H, Park DJ, Wong NC, English DR, Hopper JL, Severi G, et al. The use of DNA from archival dried blood spots with the Infinium HumanMethylation450 array. BMC Biotechnol. 2013;13:23.

35. Aryee MJ, Jaffe AE, Corrada-Bravo H, Ladd-Acosta C, Feinberg AP, Hansen KD, Irizarry RA. Minfi: a flexible and comprehensive bioconductor package for the analysis of Infinium DNA methylation microarrays. Bioinformatics. 2014;30:1363–9.

36. Maksimovic J, Gordon L, Oshlack A. SWAN: subset-quantile within array normalization for illumina infinium HumanMethylation450 BeadChips. Genome Biol. 2012;13:R44.

37. Johnson WE, Li C, Rabinovic A. Adjusting batch effects in microarray expression data using empirical Bayes methods. Biostatistics. 2007;8:118–27.

38. Price ME, Cotton AM, Lam LL, Farre P, Emberly E, Brown CJ, Robinson WP, Kobor MS. Additional annotation enhances potential for biologically-relevant analysis of the Illumina Infinium HumanMethylation450 BeadChip array. Epigenetics Chromatin. 2013;6:4.

39. Houseman EA, Accomando WP, Koestler DC, Christensen BC, Marsit CJ, Nelson HH, Wiencke JK, Kelsey KT. DNA methylation arrays as surrogate measures of cell mixture distribution. BMC Bioinformatics. 2012;13:86.

40. Bates D, Mächler M, Bolker B, Walker S. Fitting linear mixed-effects models using lme4. J Stat Softw. 2015;67:48.

41. Benjamini Y, Hochberg Y. Controlling the false discovery rate—a practical and powerful approach to multiple testing. Journal of the Royal Statistical Society Series B-Methodological. 1995;57:289–300.

42. Stone J, Gurrin LC, Hayes VM, Southey MC, Hopper JL, Byrnes GB. Sibship analysis of associations between SNP haplotypes and a continuous trait with application to mammographic density. Genet Epidemiol. 2010;34:309–18.

43. Carlin JB, Gurrin LC, Sterne JA, Morley R, Dwyer T. Regression models for twin studies: a critical review. Int J Epidemiol. 2005;34:1089–99.

44. Hopper JL, Bui QM, Erbas B, Matheson MC, Gurrin LC, Burgess JA, Lowe AJ, Jenkins MA, Abramson MJ, Walters EH, et al. Does eczema in infancy cause hay fever, asthma, or both in childhood? Insights from a novel regression model of sibling data. J Allergy Clin Immunol. 2012;130:1117–22. e1111

45. Stone J, Dite GS, Giles GG, Cawson J, English DR, Hopper JL. Inference about causation from examination of familial confounding: application to longitudinal twin data on mammographic density measures that predict breast cancer risk. Cancer Epidemiol Biomark Prev. 2012;21:1149–55.

46. Bui M, Bjornerem A, Ghasem-Zadeh A, Dite GS, Hopper JL, Seeman E. Architecture of cortical bone determines in part its remodelling and structural decay. Bone. 2013;55:353–8.

47. Dite GS, Gurrin LC, Byrnes GB, Stone J, Gunasekara A, McCredie MR, English DR, Giles GG, Cawson J, Hegele RA, et al. Predictors of mammographic density: insights gained from a novel regression analysis of a twin study. Cancer Epidemiol Biomark Prev. 2008;17:3474–81.

48. Davey CG, Lopez-Sola C, Bui M, Hopper JL, Pantelis C, Fontenelle LF, Harrison BJ. The effects of stress-tension on depression and anxiety symptoms: evidence from a novel twin modelling analysis. Psychol Med. 2016;46:3213–8.

49. Højsgaard S, Halekoh U, Yan J. The R package geepack for generalized estimating equations. J Stat Softw. 2005;15:11.

50. de Lange KM, Moutsianas L, Lee JC, Lamb CA, Luo Y, Kennedy NA, Jostins L, Rice DL, Gutierrez-Achury J, Ji SG, et al. Genome-wide association study implicates immune activation of multiple integrin genes in inflammatory bowel disease. Nat Genet. 2017;49:256–61.

51. Eeles RA, Olama AA, Benlloch S, Saunders EJ, Leongamornlert DA, Tymrakiewicz M, Ghoussaini M, Luccarini C, Dennis J, Jugurnauth-Little S, et al. Identification of 23 new prostate cancer susceptibility loci using the iCOGS custom genotyping array. Nat Genet. 2013;45:385–91. 391e381-382

52. Rothman KJ, Greenland S, Lash TL. Modern epidemiology. 3rd. Philadephia: Lippincott Williams & Wilkins; 2008.

53. Joubert BR, Felix JF, Yousefi P, Bakulski KM, Just AC, Breton C, Reese SE, Markunas CA, Richmond RC, Xu CJ, et al. DNA methylation in newborns and maternal smoking in pregnancy: genome-wide consortium meta-analysis. Am J Hum Genet. 2016;98:680–96.

54. Mimura J, Ema M, Sogawa K, Fujii-Kuriyama Y. Identification of a novel mechanism of regulation of Ah (dioxin) receptor function. Genes Dev. 1999; 13:20–5.

55. VanderWeele TJ, Tchetgen Tchetgen EJ, Cornelis M, Kraft P. Methodological challenges in Mendelian randomization. Epidemiology. 2014;25:427–35.

56. Smith GD, Hemani G. Mendelian randomization: genetic anchors for causal inference in epidemiological studies. Hum Mol Genet. 2014;23:R89–98.

16

Interaction between prenatal pesticide exposure and a common polymorphism in the *PON1* gene on DNA methylation in genes associated with cardio-metabolic disease risk

Ken Declerck[1], Sylvie Remy[2,3], Christine Wohlfahrt-Veje[4], Katharina M. Main[4], Guy Van Camp[5], Greet Schoeters[3,6,7], Wim Vanden Berghe[1] and Helle R. Andersen[7*]

Abstract

Background: Prenatal environmental conditions may influence disease risk in later life. We previously found a gene-environment interaction between the paraoxonase 1 (*PON1*) Q192R genotype and prenatal pesticide exposure leading to an adverse cardio-metabolic risk profile at school age. However, the molecular mechanisms involved have not yet been resolved. It was hypothesized that epigenetics might be involved. The aim of the present study was therefore to investigate whether DNA methylation patterns in blood cells were related to prenatal pesticide exposure level, *PON1* Q192R genotype, and associated metabolic effects observed in the children.

Methods: Whole blood DNA methylation patterns in 48 children (6–11 years of age), whose mothers were occupationally unexposed or exposed to pesticides early in pregnancy, were determined by Illumina 450 K methylation arrays.

Results: A specific methylation profile was observed in prenatally pesticide exposed children carrying the *PON1* 192R-allele. Differentially methylated genes were enriched in several neuroendocrine signaling pathways including dopamine-DARPP32 feedback (appetite, reward pathways), corticotrophin releasing hormone signaling, nNOS, neuregulin signaling, mTOR signaling, and type II diabetes mellitus signaling. Furthermore, we were able to identify possible candidate genes which mediated the associations between pesticide exposure and increased leptin level, body fat percentage, and difference in BMI *Z* score between birth and school age.

Conclusions: DNA methylation may be an underlying mechanism explaining an adverse cardio-metabolic health profile in children carrying the *PON1* 192R-allele and prenatally exposed to pesticides.

Keywords: DNA methylation, Prenatal pesticide exposure, Paraoxonase 1, *PON1* Q192R genotype, Illumina 450 K methylation array, Cardio-metabolic health

* Correspondence: HRAndersen@health.sdu.dk
[7]Environmental Medicine, Institute of Public Health, University of Southern Denmark, Odense, Denmark
Full list of author information is available at the end of the article

Background

A considerable part of modern pesticides has neurotoxic and/or endocrine disrupting properties [1–3] and therefore the potential to disturb development of neurobehavioral, neuroendocrine, and reproductive functions [4–8] especially if exposure occurs during vulnerable time periods in fetal life or early childhood. To investigate potential health effects of prenatal pesticide exposure, we have followed a cohort of children, whose mothers were employed in greenhouse horticulture in pregnancy. Some of the mothers were occupationally exposed to mixtures of pesticides in the first trimester before the pregnancy was recognized, and preventive measures were taken. Findings from this cohort include associations between maternal pesticide exposure and lower birth weight followed by increased body fat accumulation during childhood [9], impaired reproductive development in boys [10, 11], and earlier breast development [12] and impaired neurobehavioral function in girls [13].

The HDL-associated enzyme paraoxonase 1 (PON1) catalyzes the hydrolysis of a wide range of substrates including some organophosphate insecticides [14, 15]. It also protects lipoproteins from oxidative modifications and hence against development of atherosclerosis [16, 17]. A common polymorphism in the coding sequence of the *PON1* gene substitutes glutamine (Q) to arginine (R) at position 192. This substitution seems to affect both properties of the enzyme, and several studies have indicated an increased risk of cardiovascular disease in R-allele carriers [17, 18]. To investigate if this polymorphism affected the sensitivity to prenatal pesticide exposure, the *PON1* Q192R genotype was determined in the children. We found a marked interaction between prenatal pesticide exposure and the *PON1* Q192R genotype. At school age, exposed children with the R-allele had significantly higher BMI, body fat percentage, abdominal circumference, and blood pressure compared to unexposed children with the same genotype. In the group of children with the QQ genotype, there was no effect of prenatal pesticide exposure on these parameters [19]. In addition, serum concentrations of leptin, glucagon, and plasminogen activator inhibitor type-1 (PAI-1) were enhanced in prenatally pesticide exposed children with the R-allele, also after adjusting for BMI [20] which also indicates disturbance of metabolic pathways related to development of metabolic syndrome [21–23]. In addition, leptin seemed to be a mediator of the increased fat accumulation during childhood related to prenatal pesticide exposure in children with the *PON1* 192R-allele [20]. Thus, the obtained results indicate a gene-environment interaction between pesticide exposure and *PON1* gene heterogeneities already in early prenatal life that might enhance the risk of cardio-metabolic diseases later in life.

The mechanism behind this interaction is not yet understood but might be mediated by epigenetic alterations depending on both genotype and prenatal exposure. Epigenetic marks, including DNA methylation and covalent histone modifications, are dynamic and can adapt to a variety of external stimuli [24]. Furthermore, during fetal development extensive de- and re-methylation events are taking place making this period highly vulnerable for epigenetic changes caused by environmental conditions [25]. Indeed, emerging evidence in experimental animals and in humans associate altered DNA methylation patterns with a variety of prenatal exposures including dietary factors, parental care, infections, smoking, and environmental pollutants [26–31]. In experimental animals, early life changes in DNA methylation have been associated with diet-induced obesity and insulin resistance [32]. Recently, also human studies have suggested that DNA methylation patterns at birth are related to birth weight and fat mass later in childhood [33, 34]. The aim of this exploratory study was to investigate whether methylation patterns in blood samples of school children were related to prenatal pesticide exposure, *PON1* Q192R genotype, and adverse health outcomes already observed in the children. We hypothesized that the health effects associated with early prenatal pesticide exposure were related to differential epigenetic modifications in children with the QQ-genotype and children carrying the R-allele.

Methods

Study population

This study is a part of an ongoing prospective study including 203 children born between 1996 and 2001 by female greenhouse workers. The children were examined for the first time at 3 months of age [11] and followed-up at school age when 44 new age-matched controls were included [9], and the *PON1* genotype was determined for 141 children [19]. For this exploratory study, 48 pre-pubertal (Tanner Stage 1) children, whose mothers reported not to have smoked during pregnancy, were selected equally distributed between the *PON1* 192QQ and QR/RR genotype. The QR/RR genotype group consisted of 3 children with the RR genotype and 21 with the QR genotype. After excluding children of mothers who smoked in pregnancy, the number of unexposed controls within each genotype was low, 20 with the QQ genotype and 16 with the QR/RR genotype. DNA qualified for methylation analysis was only available for 11 and 12 of these children, respectively. For each genotype, we then used individual matching to select one exposed child of same sex and age for each of the controls. For the QQ-genotype, two exposed children were selected for each of two controls to obtain 24 children. Thus, in total we used data from 13 exposed and 11 unexposed children with the QQ genotype, and 12 exposed and 12 unexposed children with the QR/RR genotype (Table 1).

Table 1 Population characteristics and anthropometric data for 48 pre-pubertal children examined at age 6–11 years stratified by *PON1* Q192R genotype and prenatal pesticide exposure

	PON1 192QQ		*PON1* QR/RR	
	Unexposed	Exposed	Unexposed	Exposed
N	11	13	12	12
Female sex	5 (45.5)	7 (53.8)	6 (50.0)	6 (50.0)
Maternal smoking in pregnancy	0 (0)	0 (0)	0 (0)	0 (0)
SES[a]	7/4 (63.6/36.4)	3/10 (23.1/76.9)*	5/7 (41.7/58.3)	2/10 (16.7/83.3)
Birth weight (g)	3640 (2600; 5412)	3382 (2750; 4573)	3789 (2984; 4345)	3500 (2900; 3914)*
Gestational age (days)	276 (257; 291)	283 (265; 295)	283 (261; 298)	281 (266; 291)
Age (years)	7.6 (6.2; 9.8)	8.4 (6.7; 10,0)	7.8 (6.6; 9.5)	7.7 (7.1; 9.4)
Height (cm)	133.3 (117.3; 145.2)	130.3 (109.7; 139.2)	130.9 (113.7; 149.1)	128.6 (119.3; 142,5)
Weight (kg)	30.9 (18.7; 38.0)	28.3 (18.0; 30.7)	26.3 (19.9; 36.5)	27.4 (19.5; 37.8)
BMI (kg/m^2)	16.2 (13.7; 20.5)	15.3 (14.9; 18.3)	15.5 (13.8; 16.9)	15.7 (13.8; 19.7)
BMI *Z* scores	0.66 (−1.03; 3.21)	−0.18 (−0.80; 1.49)	−0.04 (−1.31; 0.89)	−0.01 (−0.98; 3.14)
Delta BMI *Z* score since birth	−0.45 (−2.15; 2.97)	−0.71 (−2.57; 1.87)	−0.56 (−2.52; 1.03)	0.95 (−2.08; 2.97)*
Abdominal circumference (cm)	60.4 (52.0; 75.8)	58.7 (52.1; 66.8)	58.3 (52.0; 68.1)	60.8 (51.8; 70.6)*
Sum of four skin folds (mm)	38.4 (27.1; 85.4)	33.6 (25.4; 54.5)	34.0 (20.2; 45.2)	44.6 (28.8; 72.0)*
Systolic blood pressure (mmHg)	98.7 (93.7; 110.4)	97.2 (84.3; 105.3)	99.7 (84.7; 106.8)	101.7 (91.0; 108.6)
Diastolic blood pressure (mmHg)	54.7 (46.0; 69.9)	56.2 (46.0; 62.0)	56.3 (49.3; 69.1)	63.0 (57.3; 73.1)**
Leptin (ng/ml)	1.47 (0.70; 9.18)	4.40 (0.60; 15.29)	1.41 (0.67; 5.90)	4.69 (1.79; 12.25)**
Insulin (ng/ml)	0.36 (0.22; 1.15)	0.52 (0.23; 2.55)	0.34 (0.16; 1.62)	1.11 (0.24; 7.10)*
Paraoxonase activity (nmol/min/ml)	27.5 (9.9; 38.0)	30.9 (21.0; 38.9)	58.6 (41.9; 68.7)	59.6 (50.3; 71.5)

[a]*SES* socioeconomic status (social class 1–3/4–5). Differences between unexposed and exposed children for each *PON1* Q192R genotype were tested using Mann-Whitney *U* test for continuous variables and Fisher's exact test (dichotomous variables) or Likelihood ratio (categorical variables with > 2 categories). *P value ≤ 0.05, **P value ≤ 0.01. Values are presented as median (5–95%) for continuous variables and as *N* (%) for categorical variables

Recruitment, characteristics, exposure categorization, and clinical examinations of the children have previously been described in detail [9, 11, 19]. Briefly, we recruited pregnant women working in greenhouses and referred to the local Department of Occupational Health for risk assessment of their working conditions and guidance for safe work practices during pregnancy. Detailed information about working conditions inclusive pesticide use for the previous 3 months was obtained from maternal interview at enrollment (gestational weeks 4–10) and supplemented by telephone contact to the employers. For all women, re-entry activities (such as moving or packing potted plants or nipping cuttings) constituted their main work functions. Approximately 20% of the women reported having been directly involved in applying pesticides, mainly by irrigating fungicides or growth retardants. Only few (6%) of the women had applied insecticides. The women were categorized as occupationally exposed if pesticides were applied in the working area more than once a month, and the women handled treated plants within 1 week after treatment and/or the women were directly involved in applying pesticides. The women were categorized as occupationally unexposed if none of the above criteria was fulfilled. All

exposure assessments and categorization of the mothers as pesticide exposed or unexposed were performed independently by two toxicologists before the first examination of the children. Women categorized as pesticide exposed went on paid leave or were moved to work functions with less or no pesticide exposure shortly after enrollment. Hence, the exposure classification relates to the early weeks of the first trimester before study enrollment.

The exposure situation was complex since the use of specific pesticides varied with time and location, both within the same company and between companies, depending on the plant production and the type of pest to be controlled. Out of 124 different active pesticide ingredients used in the greenhouses were 59 insecticides (17 organophosphates, 12 pyrethroids, 9 carbamates, and 21 others), 40 fungicides, 11 growth regulators, and 14 herbicides. Some were used only in few greenhouses or in short periods, whereas others were used more often. Organophosphate insecticides were used to some extent in the working areas for 91% of the exposed mothers in the entire cohort, and for 24 out of the 25 exposed mothers whose children were included in this study. The most used organophosphates were dichlorvos, dimethoate, and chlorpyrifos. Other frequently used pesticides were the pyrethroid insecticides deltamethrin and fenpro-pathrin; the carbamate insecticides methiocarb, pirimicarb, and methomyl, and the fungicides fenarimol, prochloraz, tolclofos-methyl, vinclozolin, iprodion, and chlorothalonil. In general, the time interval between applying insecticides and working in the treated areas was longer (1–3 days) than for fungicides and growth regulators (often a few hours). Because of the complexity of the exposure situation and because most of the women at enrollment had been off work for some days while the risk assessment of their working conditions was performed, biomonitoring of the exposure was not feasible. A complete list of the pesticides used in the greenhouses can be obtained from the corresponding author.

At follow-up at age 6 to 11 years, 177 children underwent a standardized clinical examination in which systolic and diastolic blood pressure, pubertal staging, height, weight, thickness of skin folds, and other anthropometric parameters were measured [9]. The same pediatrician performed all clinical examinations blinded to information about maternal pesticide exposure during pregnancy.

Venous non-fasting blood samples were collected (between midmorning and late afternoon) in EDTA-coated and uncoated vials (Venoject). After centrifugation at $2000\,g$ for 10 min at 20 °C, buffy coat for genotyping and epigenetic analysis was separated from the EDTA-treated samples. Buffy coat and serum from the uncoated vials were stored at −80 °C until analysis.

As previously described [19], C-108T (rs705379) and Q192R (rs662) polymorphisms of the *PON1* gene was determined by the Taqman-based allele discrimination using the ABI Prism 7700 Sequence Detection System, serum activity of *PON1* was determined by spectrophotometry with paraoxon as substrate, and insulin (proinsulin and insulin) and leptin concentrations in serum were determined by commercial ELISA hormone kits from RayBio.

Genotyping and all serum analyses were performed blinded to both exposure information and examination outcomes.

Sample preparation
DNA from buffy coat samples was extracted using QIAamp DNA Blood Mini Kit (Qiagen, Hilden, Germany). The blood spin protocol was applied according to manufacturer's instructions. Samples were eluted in 100 μl elution buffer. DNA samples were bisulfite-converted using the EZ DNA methylation kit from Zymo according to manufacturer's instructions. Successful bisulfite conversion was checked using a bisulfite-specific PCR of an amplicon in the *SALL3* gene (see Additional file 1 for primer sequences). Only samples showing an intense band on agarose gel were

further analyzed by the 450 K methylation array. As a negative control non-converted gDNA was used.

DNA methylation and data preprocessing
The Infinium HumanMethylation450 BeadChip array (Illumina, San Diego, CA, USA) was used to measure DNA methylation genome-wide. 4 μL of bisulfite-converted DNA from each sample was amplified, fragmented, precipitated, resuspended, and subsequently hybridized onto the BeadChips. After overnight incubation of the BeadChips, unhybridized fragments were washed away, while hybridized fragments were extended using fluorescent nucleotide bases. Finally, the BeadChips were scanned using the Illumina iScan system to obtain raw methylation intensities for each probe.

We used the R package RnBeads to preprocess the Illumina 450 K methylation data [35]. Cg-probes were filtered before normalization based on following criteria: probes containing a SNP within 3 bp of the analyzed CpG site, bad quality probes based on an iterative Greedycut algorithm where a detection p value of 0.01 was set as a threshold for an unreliable measurement, and probes with missing values in at least one sample. After filtering these cg-probes, beta values (ratio of methylated probe intensity versus total probe intensity) were within-array normalized using the beta mixture quantile dilation (BMIQ) method [36]. Another filtering step was performed after normalization based on the following criteria: probes measuring methylation not at CpG sites and probes on sex chromosomes. The two filtering steps removed a total of 20,338 cg-probes and ended up with a data set of normalized methylation values for 465,239 cg-probes. Beta values were transformed to M values ($M = \log_2(\beta/(1-\beta))$) prior to further analyses. Principal component analysis (PCA) was conducted to detect possible batch effects. Associations between the first eight principal components and possible batch effect covariates were measured. The Kruskall-Wallis test was used to find associations with sentrix_ID (BeadChip), while the two-sided Wilcoxon sum rank test was used for associations with the processing date, exposure and *PON1* Q192R genotype. Significant associations between principal component 2 and sentrix_ID (BeadChip) and processing date were suggestive for batch effects and were therefore corrected using the ComBat function in the SVA R package [37] (Additional files 2 and 3). Raw and normalized array data were uploaded to the Gene Expression Omnibus (GEO) database and have accession number: GSE90177.

For each sample, the relative cell type contribution was measured using the approach described by Houseman et al. [38]. Reference methylomes of each blood cell type (granulocyte, CD4+ T-cell, CD8+ T-cell, B-cell, monocyte, NK-cell) were obtained from the study of

Reinius et al. using the FlowSorted.Blood.450 K R package [39]. The analysis was limited to the 100,000 most variable sites. The top 500 cg-probes associated with the cell types were used to estimate the relative cell type composition in each sample. One-way ANOVA was used to determine differences in relative cell type composition between the exposed and the unexposed children and between children with the QQ and QR/RR genotype. Associations between relative cell type composition and health outcomes (percentage body fat, delta BMI z-scores from birth to school age, and BMI Z scores), leptin levels and age were analyzed using simple linear regression.

Statistical analysis

Differential methylation was analyzed both at the single CpG site level and at the region level (Fig. 1). At the single CpG site level, multiple linear regression (Matlab version 2014b, The Mathworks®, Natick, MA, USA) was performed in which methylation was the dependent variable and *PON1* Q192R genotype and prenatal pesticide exposure (yes/no) were the independent variables. Our statistical approach was designed to explain—at the level of methylation—the previously reported gene-environment interaction between the paraoxonase 1 (*PON1*) Q192R genotype and prenatal pesticide exposure leading to an adverse cardio-metabolic risk profile at school age among children carrying the R-allele [19]. Thus, our primary interest was to identify methylation marks associated with exposure that were more altered in R-allele carriers than in QQ-homozygotes. Two statistical models were included in our statistical approach. In the first model, effect modification (interaction) of exposure by *PON1* Q192R genotype was allowed by including main effects (exposure and genotype) and cross-product terms (exposure*genotype) in the models. Statistical significant effects of exposure in the *PON1* 192QR/RR group were defined as follows: P value interaction term ≤ 0.1 and P value of exposure in the QR/RR group ≤ 0.001. This model allows studying synergistic effects where the combined effect of prenatal exposure and in the QR/RR group is greater than the sum of the effects of each factor alone. In the second model, effect modification of exposure by *PON1* Q192R genotype was not assumed (no cross product term included). Statistical significant effects of exposure were defined as follows: P value of exposure ≤ 0.001, P value of *PON1* genotype ≤ 0.1. In this model the combined effect of exposure and being R-allele carrier is equal to the sum of the effect of each factor separately. For both models, the associations were adjusted for child sex. To identify probes that were most aberrant in the exposed QR/RR group, we set an additional filter for both models in which we defined that the prenatally exposed QR/RR group should either be highest or lowest methylated (based on mean methylation level) as compared to the other three groups (exposed QQ, unexposed QR/RR and unexposed QQ). These sites are defined as significantly differentially methylated positions (sig-DMPs) in the remainder of this text. Sig-DMPs were annotated using the HumanMethylation450 v1.2 manifest file. The freely available EpiExplorer tool was used to add further annotation including chromatin state segmentation and histone modifications based on the UCSC hg19 browser [40]. Genomic locations of transcription factor binding sites (TFBS) were directly downloaded from the UCSC h19 genome browser. Enrichment or depletion of sig-DMPs in a particular genomic region was determined using the Fisher's exact test.

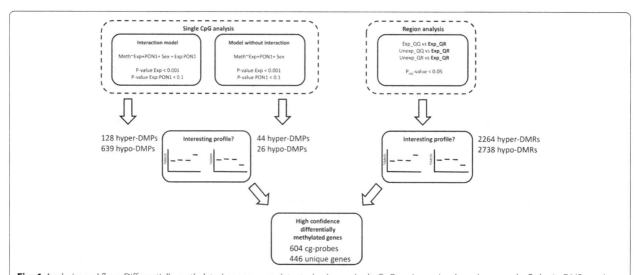

Fig. 1 Analysis workflow. Differentially methylated genes were detected using a single CpG and a region-based approach. Only sig-DMPs and sig-DMRs were selected in which the pesticide exposed QR carrier group was either hyper- or hypomethylated in comparison with the other groups (interesting profile). By overlapping the sig-DMPs with the sig-DMRs a high confidence list of differentially methylated genes could be generated

Differentially methylated regions (DMRs) were detected using the limma-based DMRcate R package [41]. We only looked for regions differentially methylated between the exposed QR/RR group and one of the other groups (exposed QQ, unexposed QR/RR and unexposed QQ). In line with identification of sig-DMPs, significant regions (P_{adj} value < 0.05) were selected in which the exposed R-allele carriers showed either the highest or the lowest methylation state which are called sig-DMRs in the remainder of this text. P values were corrected for multiple testing using the Benjamini-Hochberg method (P_{adj}).

Pyrosequencing

We used bisulfite pyrosequencing to further verify the methylation differences observed in the methylation array. We selected regions in four genes that are known to be involved in metabolism: *LEP*, *GPR39*, *PPARG*, and *OPCML* (Additional file 4). *LEP* DNA methylation has been associated with BMI, birth weight, and cholesterol levels [42–44]. Also, maternal conditions have an effect on the methylation status of the *LEP* promoter [45–48]. GPR39 belongs to the ghrelin receptor family and was shown to be associated with obesity [49]. PPARG is a nuclear receptor involved in regulation of lipid and metabolism as well as a target for some obesogenic endocrine disruptors [20, 50–53]. Furthermore, PPARγ is directly involved in the regulation of PON1 gene expression [54–56]. OPCML (Opioid Binding Protein/Cell Adhesion Molecule-like) is a member of the IgLON family. A SNP in the OPCML gene was associated with coronary artery calcified plaque in African Americans with type 2 diabetes [57]. A mouse and human GWAS analysis identified an OPCML SNP associated with obesity traits and visceral adipose/subcutaneous adipose ratio, respectively [58, 59]. 1 µg DNA from each sample was bisulfite-converted using the EpiTect Fast bisulfite Conversion Kit (Qiagen, Hilden, Germany) according to manufacturer's instructions. 15 ng of bisulfite-treated DNA was subsequently used in PCR amplification using the PyroMark PCR Kit (Qiagen, Hilden, Germany). Reverse primers were biotinylated to get biotin-labeled PCR products. Finally, DNA sequences were pyrosequenced using the PyroMark Q24 Advanced instrument (Qiagen, Hilden, Germany). First, streptavidin-coated Sepharose beads (High Performance, GE Healthcare, Uppsala, Sweden) were used to immobilize the biotin-labeled PCR products. Subsequently, PCR products were captured by the PyroMark vacuum Q24 workstation, washed and denaturated. The single stranded PCR products were mixed and were annealed with their corresponding sequencing primer. After the pyrosequencing run was finished, the results were analyzed using the Pyro-Mark Q24 Advanced software (Qiagen, Hilden, Germany). Biotinylated-reverse, forward, and sequencing primers

were designed using the PyroMark Assay Design 2.0 software (Qiagen, Hilden, Germany) (Additional file 1).

Mediation analysis

For a subset of sig-DMPs and sig-DMRs we analyzed (1) whether methylation is a mediator between exposure in *PON1* 192R-allele carriers and leptin levels; and (2) whether methylation is a mediator between exposure in *PON1* 192R-allele carriers and body fat accumulation (using delta BMI-score (from birth to school age), and percentage body fat as endpoints). Mediation analysis was restricted to the subset of the methylation data that overlap between the list of sig-DMPs (interaction model) and sig-DMRs. The analysis was performed by the procedure described by Baron and Kenny (1986) [60]. Leptin concentrations were logarithmically (ln) transformed prior to analysis. In mediation analysis considering body fat percentage and leptin, the models were adjusted for sex. As sex was already considered when calculating BMI Z score, associations considering mediation between pesticide exposure and BMI Z score were not adjusted for sex.

To demonstrate mediation, four requirements must be met: (model 1) the dependent outcome variable (leptin or a body fat measure) should be significantly associated with pesticide exposure (independent variable); (model 2) the DNA methylation mark (mediator) should be significantly associated with pesticide exposure; (model 3) the dependent variable should be significantly associated with the DNA methylation mark; and (model 4) the DNA methylation mark should be a significant predictor of the outcome variable, while controlling for pesticide exposure. The estimated exposure-related change in the outcome variables in model 4 should be less than in model 1 to demonstrate partial mediation, and drop to zero to demonstrate full mediation. A P value below 0.05 was used as a cut-off for statistical significance in each of the models.

Functionally relevant mediators, i.e., mediators that have been reported to be involved in development of weight gain/obesity, insulin resistance/diabetes, cardiovascular disease, and/or fetal growth retardation were subjected to further statistical analysis. R-package "mediation" was used to calculate the significance of the causal mediation effect using a bootstrapping approach [61]. It should be noted that the age of the children varied between 6 and 11 years at the follow-up examination where blood was collected. As child age might affect methylation levels, the exposed and unexposed children selected for this study were age-matched within each genotype.

Functional analysis

Ingenuity Pathway Analysis (IPA, Ingenuity Systems®) was used for biological interpretation. The overlap between sig-DMPs and sig-DMRs was determined and used as input for canonical pathway analysis. A Fisher's exact test was used to determine whether the gene lists include more genes associated with a given pathway as compared to random chance (P value ≤ 0.05).

The DisGeNet platform (http://www.disgenet.org/) was used to screen for gene disease associations [62]. The database (currently) contains 429111 gene disease associations for which the platform provides a reliability score (DisGeNET Score). This score ranges from 0 to 1 and takes into account the number and type of sources (level of curation, organisms), and the number of publications supporting the association (for further details we refer to the DisGeNet website). For this manuscript, we extracted the associations with a score above 0.1. By this criterion, 34180 gene disease associations remain in the database. Associated diseases were mapped to the overlapping list of genes between sig-DMPs and sig-DMRs.

Results

Descriptive statistics of the study population

Characteristics, inclusive anthropometric data, for the 48 children (6–11 years of age) are presented in Table 1. In accordance with the findings for the whole cohort [19], birth weights were significantly lower and measures of body composition (abdominal circumference, skin fold thickness), increase in BMI Z score from birth to school age (delta BMI Z score), diastolic blood pressure, and leptin and insulin concentrations at school age were significantly higher in the exposed $PON1$ 192QR/RR group compared with the unexposed QR/RR group. For children with the QQ genotype, none of the variables was significantly affected by prenatal pesticide exposure ($P > 0.05$).

Prenatal pesticide exposure-induced methylation changes at CpG sites enriched in promoter regions in *PON1* 192R-allele carriers

Genome-wide DNA methylation in whole blood samples from the children was determined by Illumina 450 K methylation arrays and differential methylation patterns related to prenatal pesticide exposure and *PON1* Q192R genotype were analyzed. First differential methylation was detected at the single CpG level using two multiple linear regression models (Fig. 1). Because relative cell type composition was not associated with pesticide exposure and *PON1* Q192R genotype (Additional file 5), differences in cellular composition were not further considered in the workflow of statistical analysis. Allowing effect modification by *PON1* Q192R genotype, 767 sig-DMPs were identified of which 128 were hypermethylated and 639 hypomethylated in prenatally exposed *PON1* 192R allele carriers. When effect modification was not assumed, and the interaction term between exposure and *PON1* genotype was removed from the models, 70 sig-DMPs of which 44 were hypermethylated and 26 hypomethylated in prenatally exposed *PON1* 192R-allele carriers were identified. Hierarchical clustering of the samples using all the sig-DMPs demonstrated a clear cluster of exposed *PON1* 192R-allele carriers (Fig. 2). Confidence in detection of differentially methylated genes was increased by further analysis showing that the changes in methylation were not restricted to single CpGs, but were often located in regions or so called differentially methylated regions (DMRs). 5002 sig-DMRs were identified, of which 2264 were hypermethylated and 2738 hypomethylated in the exposed *PON1* 192R carrier group compared to the other groups. Allowing interaction between exposure and *PON1* Q192R genotype to determine sig-DMPs, 547 out of 767 sites

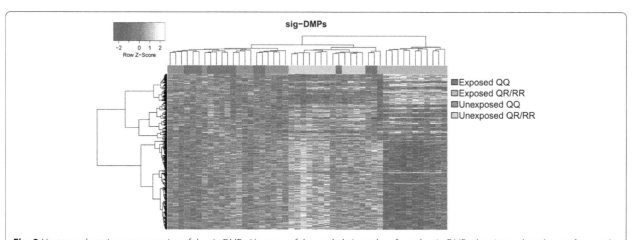

Fig. 2 Heatmap clustering representation of the sig-DMPs. Heatmap of the methylation values from the sig-DMPs showing a clear cluster of prenatal pesticide exposed *PON1*-192 R-carrier samples (*orange group*). Hierarchical clustering is based on the euclidean distance and average linkage metric. Higher methylation values are colored in *yellow*, while lower methylation values are colored in *blue*

(71.3%) were overlapping with the list of sig-DMRs (Additional file 6). When effect modification was not considered, 57 out of 70 sites (81.4%) were overlapping (Additional file 7).

The pyrosequencing methylation percentages confirmed the robustness of Illumina results. They showed significant positive correlations with the Illumina 450 K beta values for all measured CpG probes (Fig. 3), except for two probes in the *LEP* gene (cg00840332 and cg26814075) which were borderline significant (*P* value: 0.07 and 0.16, respectively). The reason for this less strong correlation between the Illumina and the pyrosequencing *LEP* methylation is probably the lower interindividual methylation variability in this region compared to *GPR39* and *PPARG*.

In accordance with the Illumina results, the pyrosequencing *LEP* methylation values were not associated with pesticide exposure and/or *PON1* Q192R genotype. Furthermore, the serum leptin concentrations were not correlated with *LEP* methylation status (data not shown). For *GPR39*, the region analyzed with pyrosequencing contained three Illumina cg-probes (cg17172683, cg11552903, and cg18444763), which showed a high correlation (*r* > 0.78) between the Illumina beta values and the pyrosequencing methylation percentages. For most CpGs in the pyrosequencing region, we could verify a significant exposure effect,

and in each CpG site, prenatally exposed children with the QR/RR genotype had the lowest mean methylation value (Additional files 8 and 9). In the *PPARG* promoter, a region was selected containing one Illumina cg-probe (cg01412654). Also here, the correlation between the 450 K Illumina beta values and the pyrosequencing methylation percentages was strong. However, DNA methylation in this region was not associated with pesticide exposure and/or *PON1* Q192R genotype and did not correlate with PON1 activity (data not shown). A region in the OPCML gene was found to be higher methylated in prenatal pesticide-exposed children carrying the *PON1* 192R-allele. The significant interaction effect between pesticide exposure and *PON1* Q192R genotype could be successfully verified by pyrosequencing. The pyrosequencing methylation values were significantly higher methylated in exposed children compared to unexposed children carrying the *PON1* 192R-allele for most of the CpG sites in the region (Additional file 9).

Next, we questioned whether the sig-DMPs were enriched or depleted in a specific genomic location (Fig. 4). Sig-DMPs for which interaction between exposure and *PON1* Q192R genotype was seen, were enriched in promoter regions (200 and 1500 bp upstream of transcription start sites) and depleted in gene bodies, 3′ UTRs and intergenic regions. This was also evident

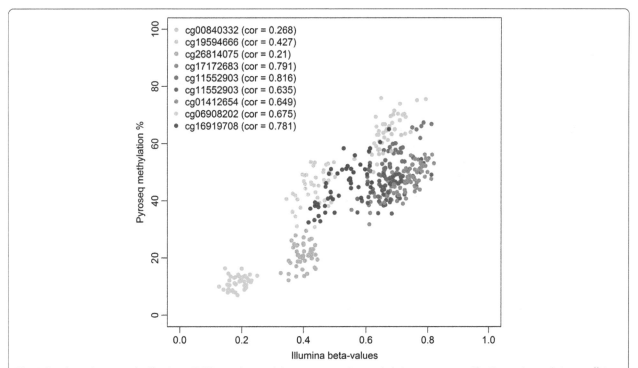

Fig. 3 Correlation between the Illumina 450 K beta values and the pyrosequencing methylation percentages. The Pearson's correlation coefficient for each CpG probe is indicated between brackets. CpG probes cg00840332, cg19594666, and cg26814075 are located in the *LEP* gene, cg17172683, cg11552903, and cg18444763 in the *GPR39* gene, cg01412654 in the *PPARG* gene, and cg06908202, and cg16919708 in the *OPCML* gene

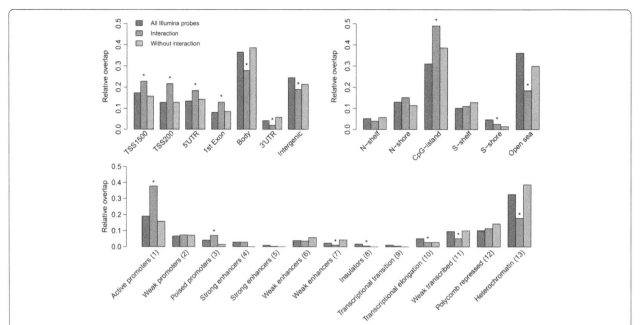

Fig. 4 Genomic location of sig-DMPs. DMPs were mapped to gene elements (*top left*), CpG islands (*top right*), and chromatin state segmentations (*bottom*). *Asterisks* indicate significant enrichment or depletion in comparison with all Illumina probes (*gray bars*) measured by the Fisher's exact test (*P* value < 0.05). Sig–DMPs where the interaction term was significant (*blue bars*) showed an enrichment in promoter regions and CpG islands. Sig-DMPs where no interaction was seen (*orange bars*) showed no significant enrichment or depletion in a particular genomic region

when we overlapped the sig-DMPs with different chromatin states, where we observed enrichment in active and poised promoters, while DMPs were depleted in regions like transcriptional elongation, weak transcribed, and heterochromatin regions. Furthermore, DMPs were significantly more located in CpG islands and less observed in CpG poor regions. Sig-DMPs found in the models without an interaction term were not enriched or depleted in a particular genomic region.

We also looked for enrichment in TFBS using available ChIP ENCODE data from the UCSC genome browser. Thirty-nine of the 161 TFBS were significantly enriched for the model with interaction (Bonferroni adjusted *P* value < 0.05) while no enrichment was found for the sig-DMPs found in the model without interaction (Additional file 10).

DNA methylation differences were enriched for genes involved in neuro-endocrine signaling pathways

Overlapping the list of sig-DMPs with the list of sig-DMRs we obtained a robust and a high confidence list of differentially methylated genes (*N* = 446). This list was used as an input for ingenuity pathway analysis. The top enriched canonical pathways (based on *P* value) were dopamine-DARPP32 feedback cAMP signaling, corticotrophin releasing hormone signaling, nNOS signaling in neurons, CDK5 signaling, and neuregulin signaling (Table 2). In the context of this manuscript, other significantly enriched pathways

such as mTOR signaling (rank 9, −log(*P* value) = 1.85) and type II diabetes mellitus signaling (rank 16, −log(*P* value) = 1.51) are also highly relevant.

DNA methylation (partially) mediates associations between pesticide exposure and higher leptin concentrations, body fat content, and delta BMI *Z* scores

The list of genes that overlaps between sig-DMPs (as identified by the interaction model) and sig-DMRs was also used as input for mediation analysis. We identified, respectively, 20, 31, and 45 candidate methylation marks that (partly) mediate the effect between pesticide exposure and serum leptin concentrations; delta BMI *Z* score; and body fat content. Based on applied cut-off criteria, we were not able to identify methylation marks that mediate the effect on BMI *Z* score. Currently known gene disease associations allowed to extract mediators that were reported to be involved in development of weight gain/obesity, insulin resistance/diabetes, cardiovascular disease, and/or fetal growth retardation. This subset of mediators is given in Table 3. Based on Baron and Kenny's steps to analyze mediation, the association between pesticide exposure and delta BMI *Z* score was partially mediated by hypomethylation of *UQCRC2*, *MTNR1B* and *GRIN2A*, and by hypermethylation of *FABP4* and *LRP8*. Methylation of *UQCRC2* and *LRP8* was also a partial mediator in the association between

Table 2 Significant enriched Ingenuity canonical pathways

Rank	Ingenuity canonical pathways	−log(P value)	Ratio	Hyper-genes	Hypo-genes
1	Dopamine-DARPP32 Feedback in cAMP signaling	3.98	0.07	CREB5, PPP2R2B, CACNA1A	KCNJ2, NOS1, GRIN2A, GUCY1B3, ADCY2, PRKCH, GNAI3, CACNA1D, PRKCG
2	Corticotropin releasing Hormone signaling	2.74	0.07	CREB5	JUND, NOS1, GUCY1B3, ADCY2, PRKCH, GNAI3, PRKCG
3	nNOS signaling in neurons	2.61	0.11	CAPN3	NOS1, GRIN2A, PRKCH, PRKCG
4	CDK5 signaling	2.41	0.07	PPP2R2B, CACNA1A	CDK5R1, NGFR, ITGA2, LAMB1, ADCY2
5	Neuregulin signaling	2.06	0.07	EGFR, ERBB3	CDK5R1, ITGA2, PRKCH, PRKCG
6	PCP pathway	2.06	0.08		JUND, FZD10, RSPO3, WNT7B, WNT9B
7	Maturity onset diabetes of young (MODY) signaling	2.03	0.14	CACNA1A	GAPDH, CACNA1D
8	Regulation of eIF4 and p70S6K signaling	2.02	0.05	PPP2R2B, FAU	RPS16, RPS13, RPS10, ITGA2, IRS1, RPS19
9	mTOR signaling	1.85	0.05	PPP2R2B, FAU	RPS16, RPS13, RPS10, IRS1, PRKCH, RPS19, PRKCG
10	Amyotrophic lateral sclerosis signaling	1.84	0.06	CAPN3, CACNA1A	NOS1, GRIN2A, NEFM, CACNA1D
11	NF-κB activation by viruses	1.8	0.07		ITGAV, CR2, ITGA2, PRKCH, PRKCG
12	Phosphatidylethanolamine biosynthesis III	1.7	1		PTDSS2
13	Role of CHK proteins in cell cycle checkpoint control	1.61	0.07	PPP2R2B, RFC4	E2F3, CHEK1
14	Synaptic long-term depression	1.6	0.05	IGF1R, PPP2R2B	NOS1, GUCY1B3, PRKCH, GNAI3, PRKCG
15	ErbB signaling	1.53	0.06	EGFR, ERBB3	NCK2, PRKCH, PRKCG
16	Type II diabetes mellitus signaling	1.51	0.05	PKM	NGFR, ADIPOR2, IRS1, PRKCH, PRKCG
17	G beta gamma signaling	1.49	0.06	EGFR	ADCY2, PRKCH, GNAI3, PRKCG
18	p70S6K signaling	1.48	0.05	EGFR, PPP2R2B	IRS1, PRKCH, GNAI3, PRKCG
19	Role of osteoblasts, osteoclasts, and chondrocytes in rheumatoid arthritis	1.47	0.04		FZD10, NGFR, SMAD5, WNT7B, ITGA2, IL1RAP, WNT9B, TCF7L2, NFATC1
20	Molecular mechanisms of cancer	1.46	0.04		RASGRF1, ITGA2, WNT7B, IRS1, E2F3, GNAI3, FZD10, SMAD5, ADCY2, WNT9B, PRKCH, CHEK1, PRKCG
21	nNOS signaling in skeletal muscle cells	1.45	0.13	CAPN3	NOS1
22	Factors promoting cardiogenesis in vertebrates	1.42	0.05		FZD10, SMAD5, PRKCH, TCF7L2, PRKCG
23	RAR activation	1.41	0.04		REL, ERCC2, SMAD5, NR2F1, ADCY2, PRKCH, RARB, PRKCG
24	Choline degradation I	1.4	0.5	CHDH	
25	Sulfate activation for sulfonation	1.4	0.5	PAPSS2	
26	Mismatch repair in eukaryotes	1.4	0.13	RFC4	MLH1
27	Glioma signaling	1.37	0.05	IGF1R, EGFR	PRKCH, E2F3, PRKCG
28	Netrin signaling	1.36	0.08		UNC5C, NCK2, NFATC1
29	Cellular effects of sildenafil (Viagra)	1.33	0.05	CACNG6, CACNA1A	KCNN1, GUCY1B3, ADCY2, CACNA1D
30	GNRH signaling	1.33	0.05	EGFR, CREB5	ADCY2, PRKCH, GNAI3, PRKCG
31	Protein kinase A signaling	1.31	0.03	HIST1H1A, CREB5	PTPN9, TIMM50, NFATC1, GNAI3, AKAP12, NGFR, PTP4A1, ADCY2, PRKCH, TCF7L2, PRKCG
32	Ovarian cancer signaling	1.31	0.05	EGFR	FZD10, WNT7B, MLH1, WNT9B, TCF7L2
33	Colorectal cancer metastasis signaling	1.3	0.04	EGFR	ADRBK1, APPL1, FZD10, WNT7B, MLH1, ADCY2, WNT9B, TCF7L2
34	Agrin interactions at neuromuscular junction	1.3	0.06	EGFR, ERBB3	ITGA2, LAMB1
35	Growth hormone signaling	1.3	0.06	IGF1R	IRS1, PRKCH, PRKCG

pesticide exposure and body fat percentage. *LRP8* was also found to mediate the association between pesticide exposure and serum leptin concentration. The *P* value for

significance of the causal mediation effect is included in Table 3 and was below 0.1 for all mediators except for *UQCRC2* and *GRIN2A*. Irrespective of disease association

Table 3 Methylation marks that partially mediate the association between pesticide exposure and leptin and body fat accumulation in *PON1*-192 R-allele carriers

Outcome	IlmnID	Nearest gene symbol	Gene name	Direction of methylation in exposed R carriers	Diseases	Significance of causal mediation effect (*P* value)
Leptin	cg03366858	*LRP8*	Low density lipoprotein receptor-related protein 8, apolipoprotein e receptor	Hyper	Myocardial infarction (0.22)|nerve degeneration (0.21)|Myocardial infarction, susceptibility to, 1 (finding) (0.2)	0.02
Leptin	cg18202502	*LRP8*	Low density lipoprotein receptor-related protein 8, apolipoprotein e receptor	Hyper	Myocardial infarction (0.22) | nerve degeneration (0.21)|myocardial infarction, susceptibility to, 1 (finding) (0.2)	0.024
Delta BMI Z score	cg00810945	*UQCRC2*	Ubiquinol-cytochrome c reductase core protein II	Hypo	Mitochondrial complex iii deficiency, nuclear type 5 (0.41) | obesity (0.21)	0.138
Delta BMI Z score	cg06337557	*MTNR1B*	Melatonin receptor 1B	Hypo	Diabetes mellitus, Type 2 (0.26)|polycystic ovary syndrome (0.21) | child development disorders, pervasive (0.21)|acute pancreatitis (0.1)	0.032
Delta BMI Z score	cg14152613	*FABP4*	Fatty acid-binding protein 4, adipocyte	Hyper	Carcinoma (0.21)|mammary neoplasms, experimental (0.21) | mammary neoplasms, animal (0.21)|insulin resistance (0.1)|erectile dysfunction (0.1)|diabetes mellitus, experimental (0.1)	0.068
Delta BMI Z score	cg15134033	*GRIN2A*	Glutamate receptor, ionotropic, N-methyl D-aspartate 2A	Hypo	Epilepsy (0.21)|colorectal neoplasms (0.21)|epilepsy, rolandic (0.21) |melanoma (0.21)|landau-kleffner syndrome (0.21)|autistic disorder (0.21)|morphine dependence (0.21)|language development disorders (0.21)|epilepsy, focal, with speech disorder and with or without mental retardation (0.21)|speech disorders (0.21)| substance withdrawal syndrome (0.21)|rolandic epilepsy, mental retardation, and speech dyspraxia, autosomal dominant (0.2)| reperfusion injury (0.1)|hypoxia-ischemia, brain (0.1)|sepsis (0.1)| fetal growth retardation (0.1)|central nervous system viral diseases (0.1)|placental insufficiency (0.1)	0.144
Delta BMI Z score	cg18202502	*LRP8*	Low density lipoprotein receptor-related protein 8, apolipoprotein e receptor	Hyper	Myocardial infarction (0.22)|nerve degeneration (0.21)|myocardial infarction, susceptibility to, 1 (finding) (0.2)	0.026
Bodyfat	cg00810945	*UQCRC2*	Ubiquinol-cytochrome c reductase core protein II	Hypo	Mitochondrial complex iii deficiency, nuclear type 5 (0.41) |obesity (0.21)	0.174
Bodyfat	cg03366858	*LRP8*	Low density lipoprotein receptor-related protein 8, apolipoprotein e receptor	Hyper	Myocardial infarction (0.22)|nerve degeneration (0.21)|myocardial infarction, susceptibility to, 1 (finding) (0.2)	<0.001
Bodyfat	cg18202502	*LRP8*	Low density lipoprotein receptor-related protein 8, apolipoprotein e receptor	Hyper	Myocardial infarction (0.22)|nerve degeneration (0.21)|myocardial infarction, susceptibility to, 1 (finding) (0.2)	<0.001

Only the subset of genes for which associations with metabolic disease have been reported is listed. DisGeNET Score—indicating reliability of the gene disease associations—is included between brackets

of interest, the full list of potential mediators is provided in Additional file 11 which also includes the outcome of the statistical analysis.

DNA methylation at the *PON1* promoter is affected by the *PON1*-108CT SNP (rs705379) and negatively correlated with paraoxonase 1 activity

Beside the genome-wide DNA methylation effects of the *PON1* Q192R genotype, we also observed a wide variation in DNA methylation in the *PON1* promoter itself for nine Illumina cg-probes. Prenatal pesticide exposure and/or *PON1* Q192R genotype did not affect *PON1* promoter methylation status. However, another polymorphism (rs705379, *PON1* -108CT) in the promoter region of PON1 could explain a large extent of this variation (Fig. 5). Individuals homozygous for the T-allele showed higher methylation values compared with the homozygous C-allele carriers. As expected, heterozygous individuals had an intermediate methylation value. Furthermore, the paraoxonase 1 activity was significantly associated with DNA methylation in the *PON1* promoter region, with higher methylation values resulting in lower paraoxonase 1 activity

(Fig. 6). *PON1* Q192R genotype had the strongest effect on PON1 activity, while variation in PON1 promoter methylation led to a smaller but significant effect on PON1 activity.

Discussion

We found that prenatal pesticide exposure was associated with a differential DNA methylation profile in children carrying the *PON1* 192R-allele compared to children with the *PON1* 192QQ genotype and unexposed children. 767 sig-DMPs were identified of which 128 were hypermethylated and 639 hypomethylated in prenatally exposed *PON1* 192R-allele carriers. The profiles of *PON1* 192R-allele carriers are clustered together. As far as we know, our study is the first one to demonstrate a link between epigenetics and genetic susceptibility towards pesticide exposure in fetal life. Our study supports a linkage of a differential methylation pattern and higher body fat content and serum leptin concentrations in school age children dependent on both *PON1* Q192R genotype and prenatal pesticide exposure.

The majority of the detected sig-DMPs were hypomethylated in exposed children with the *PON1* 192QR/

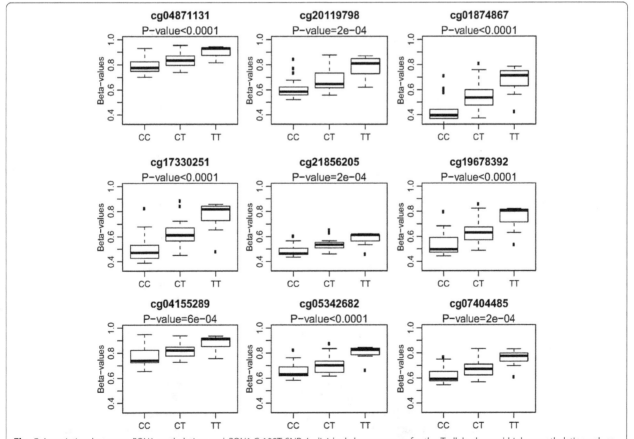

Fig. 5 Association between *PON1* methylation and *PON1* C-108T SNP. Individuals homozygous for the T allele showed higher methylation values (beta values) as compared with C-allele carriers. *P* values shown are those from the one-way ANOVA analysis

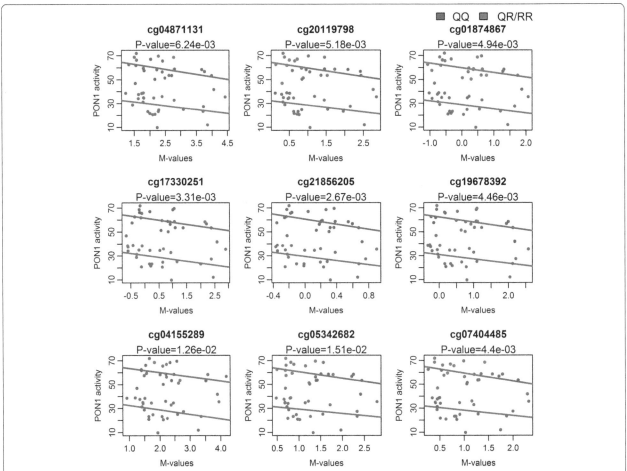

Fig. 6 Association between PON1 activity and *PON1* methylation. The *P* values of the main effect for methylation are displayed using the linear model PON1 activity ~ *M* value + *PON1*-192 genotype + sex. *Red* colored samples are *PON1* 192 R-allele carriers, and samples in *blue* are children with the *PON1* 192QQ genotype

RR genotype. Interestingly, these DMPs were mainly located in gene promoters, CpG islands and transcription factor-binding sites, suggesting a possible direct link with gene expression. To increase the confidence of our findings, we also screened for differentially methylated regions. Most of the single CpG sites were part of a DMR suggesting that these were independent of technical variation and could be considered as reliable.

Technical reliability of the outcomes from the 450 K Illumina methylation array was successfully confirmed by bisulfite pyrosequencing of corresponding CpG probe regions of four selected genes, i.e., *LEP*, *PPARG*, *GPR39*, and *OPCML* for which corresponding probes were available.

LEP was chosen because we previously found leptin to be a potential mediator of the association between prenatal pesticide exposure and body fat accumulation in children with the *PON1* 192R-allele [20]. In addition, multiple studies demonstrated associations between *LEP*

DNA methylation and BMI, birth weight, and cholesterol concentrations [42–44]. *LEP* was also found to be differentially methylated in the offspring of mothers suffering from the Dutch winter famine [45]. However, our pyrosequencing results did not demonstrate a correlation between leptin DNA methylation and leptin serum concentrations, and prenatal pesticide exposure was not associated with changes in leptin DNA methylation. This suggests that the higher leptin concentration observed in exposed children with the R-allele is not due to a direct effect on DNA methylation of the leptin gene itself. Another gene whose methylation was confirmed by pyrosequencing was *PPARG*, a nuclear receptor controlling the expression of genes involved in lipid storage and glucose metabolism and target for obesogenic compounds [50–53]. Furthermore, PPARγ is involved in the regulation of *PON1* expression [54–56]. However, we did not find a correlation between *PPARG* DNA methylation and PON1 activity (data not shown). In our dataset, prenatal pesticide exposure did not seem to change *PPARG* methylation levels irrespective of *PON1* Q192R genotype.

Reduced *GPR39* DNA methylation observed in prenatally pesticide exposed R-allele carriers was confirmed with pyrosequencing. GPR39 is receptor for obestatin (belonging to the ghrelin receptor family), involved in regulation of appetite and glucose homeostasis [63, 64] and associated with obesity [49]. Furthermore, GPR39 knockout mice showed an increased fat accumulation due to changes in lipolysis and energy expenditure [49]. So, misregulation of this gene due to methylation changes might lead to an obese phenotype. To our knowledge, no other study has yet reported methylation differences in this region associated with obesity or metabolic disorders or showed links with pesticide exposure.

The higher methylation values of the *OPCML* DMR in exposed children carrying the *PON1* 192R-allele could be confirmed by pyrosequencing. *OPCML* encodes for a protein belonging to the IgLON family. OPCML was shown to be a tumor suppressor and inactivated by DNA methylation in a variety of cancer types [65–68]. There is also a link with metabolic diseases, as SNPs in this gene were found to be associated with obesity traits, coronary artery calcified plaque, and visceral adipose/subcutaneous adipose ratio [57–59].

Further analysis revealed that the differences in DNA methylation were most pronounced in genes involved in neuro-endocrine signaling pathways, including "dopamine-DARPP32 feedback in cAMP signaling", "corticotropin releasing hormone signaling", "nNOS signaling in neurons", and "CDK5 signaling". These pathways are important in the control of food intake and energy balance. Dopamine signaling, for example, is one of the key players in the reward pathway, also controlling food intake and preferences. Reduced dopamine signaling is assumed to induce overeating [69, 70]. In mice, a high-fat diet during pregnancy resulted in altered gene expression and DNA methylation of the dopamine transporter gene in the offspring, leading to an increased preference for sucrose and fat [71]. Another study found similar results in prenatally stressed rats given a high fat-sucrose diet [72]. These studies suggest that prenatal and early life conditions may influence food intake and food preferences later in life through modulation of the dopamine pathway [73–77]. Organophosphate insecticides have been shown to modulate dopamine signaling [78]. Furthermore, low-dose exposure of neonatal rats caused metabolic dysfunction resembling prediabetes, and in adulthood, exposed animals gained excess weight when fed a high fat diet compared to unexposed rats on the same diet [79].

Corticotropin-releasing hormone (CRH) is a neuropeptide secreted in response to stress. However, a role for CRH in regulating energy balance and food intake has also been described [80–82] including a relation to the action of leptin [83].

Also NOS1 neurons are involved in energy balance and food intake [84–86]. Knock-out of *NOS1* in leptin receptor- and NOS1-expressing hypothalamic neurons results in hyperphagic obesity, decreased energy expenditure, and hyperglycemia in mice [85]. Interestingly, organophosphates have been shown to alter NOS1-expressing neurons during development in mice [87, 88].

Neureguline 1 treatment in rodents has been shown to increase serum leptin concentrations, prevent weight gain, and lower food intake. Hence, affecting this pathway may also change food intake and energy metabolism [89, 90].

A limitation of this study is that the methylation profile is measured at the same time as health outcomes and causality as such cannot be proven. Some of the genes that relate to the sig-DMPs are involved in neuro-endocrine pathways that regulate appetite and energy balance, but this study cannot rule out if these sig-DMPs are a consequence of alterations of food habits and physical activity among the exposed children with the *PON1* 192R-allele or an underlying mechanism. However, the mediation analysis suggested that some of the differentially methylated marks are on the mechanistic pathway between prenatal pesticide exposure and the measured outcomes. This result suggests that, at least in some, CpG sites a change in methylation might contribute to metabolic disturbances later in life. Furthermore, the association was not significant between pesticide exposure and BMI Z score as such, but between pesticide exposure and delta BMI Z score which integrates fat accumulation from birth and onwards to school age.

Interestingly, some of the mediator marks could be linked to specific genes that were reported earlier to play a role in the development of weight gain/obesity, insulin resistance/diabetes, cardiovascular disease, and/or fetal growth retardation: *UQCRC2*, *MTNR1B*, *GRIN2A*, *FABP4*, and *LRP8*. *FABP4* encodes for a member of the fatty acid-binding protein family regulating lipid trafficking, signaling, and metabolism. Different studies have demonstrated the role of this protein in obesity, type 2 diabetes and atherosclerosis development [91–93]. In ApoE deficient mice with hyperhomocysteine *FABP4* DNA methylation is reduced in the aorta compared to wild type mice, leading to a higher gene expression [94, 95]. *UQCRC2* encodes a protein which is part of the ubiquinol-cytochrome c reductase complex in the mitochondria. UQCRC2 was shown to be downregulated in individuals who were susceptible to weight gain and obesity development [96]. The melatonin receptor 1B (MTNR1B) has a main function in regulating circadian rhythm. Interestingly, several polymorphisms in the

MTNR1B gene are associated with type 2 diabetes, fasting glucose concentration, and insulin secretion [97–99]. *GRIN2A* encodes for a NMDA glutamate receptor subunit. Polymorphisms in the *GRIN2A* gene are associated with epilepsy and different neurological and mental disorders [100–104]. A decreased gene expression of GRIN2A in rats after intrauterine growth retardation suggests a possible role for this gene in fetal growth and development [105]. *LRP8* encodes for a member of the LDL receptor family. Common polymorphisms in the *LRP8* gene are associated with coronary artery disease, myocardial infarction, and high birth weight [106–110]. Thus, the mediation analysis suggests a mechanistic role of epigenetics in the development of an adverse metabolic risk profile among the prenatally exposed children with the *PON1* R-allele as previously reported for these children [19] and confirmed in the selected subset of children.

A few studies have investigated associations between PON1 genotype and metabolic disturbances in children. A recent study showed a higher risk of insulin resistance (HOMA-IR) in Mexican children with the RR-genotype as compared to children with the QQ or QR genotypes although BMI did not differ between the groups [111]. Among Mexican-American children from an agricultural community in California, a trend of increased BMI Z scores with increased number of PON1 192Q alleles was seen [112]. However, potential interactions between PON1 genotype and prenatal exposure to pesticides, or other environmental contaminants, were not investigated in these studies. In our cohort, unexposed QQ-homozygote children also tended to have higher body fat content than unexposed R-carriers, but prenatally pesticide exposed children with the R-allele accumulated more fat during childhood and had a more unhealthy metabolic risk profile at school age than unexposed children and exposed children with the QQ genotype [19].

We also demonstrated that methylation in the *PON1* promoter itself is affected by a SNP (*PON1* -108CT, rs705379). In addition, *PON1* methylation values were negatively associated with paraoxonase 1 activity. These results are in agreement with the outcome of a recent study from Huen and colleagues [113]. They found methylation in the same nine CpG sites to be associated with the *PON1* -108CT polymorphism and also reported an inverse association with AREase activity as a measure of *PON1* expression, both in newborns and 9-year-old children. Furthermore, they demonstrated that *PON1* methylation mediates the relationship between *PON1* expression and the promoter -108 genotype. However, the effect of prenatal pesticide exposure on the health outcomes shown in Table 1 was not modulated by PON1 -108CT genotype (data not shown).

Our findings indicate that the higher vulnerability among children with the R-allele towards prenatal pesticide exposure might be mediated by genotype-specific epigenetic alterations. However, a limitation of this study is that we cannot identify individual pesticides related to these findings, since the study design did not allow bio-monitoring of pesticide exposure in the mothers, and the exposure classification of the mothers encompassed more than 100 pesticides used in different mixtures [11].

However, the existence of mixed exposure is a real-world situation, and the longitudinal design, the blinded exposure classification, and the blinded clinical examinations, and genotyping minimized the possible impact of exposure misclassification and bias.

Since PON1 is known to detoxify some organophosphate insecticides (e.g., chlorpyrifos), and these substances were frequently applied in the mothers' working areas, organophosphate insecticides could be assumed to be responsible for the observed effects. However, the mechanism is unclear and does not seem to be related to the hydrolysis efficiency, since R-carriers have higher paraoxonase activity than QQ homozygotes. Besides, at relatively low exposure levels, as in this study, the capacity to detoxify organophosphates is considered to be independent of the *PON1* Q192R genotype [114], and furthermore, serum PON1 activity was reported to be low in newborns and may be even lower before birth, as indicated by lower activity in premature compared to term babies [115, 116]. Thus, differences in fetal detoxification of pesticides related to *PON1* genotype might not be a likely explanation of the exposure-related difference in methylation pattern between children with the QR/RR and QQ genotype.

Another limitation of the study is that DNA methylation analyses were performed in white blood cells as surrogates for the target tissues. We do not know whether the differences in DNA methylation patterns found in blood mirror a similar change in adipose tissue, for example. A recent study from Huang et al. demonstrated several potential limitations in using methylation profiles in blood to mirror the corresponding profile in target tissues by comparing paired blood and adipose tissue methylation profiles [117]. Furthermore, the composition of blood cell types may be variable and might affect the DNA methylation analyses. In our dataset, prenatal pesticide exposure and/or *PON1* Q192R genotype did not affect the relative blood cell counts determined by the reference-based method of Houseman. Cell counts were not included in the models due to the small sample size of the study. Since we found that some of the health effects (mainly leptin) were associated with cell type count (Additional file 12), we cannot exclude that the results of the mediation analysis were biased by differences in cell type composition. Based on the data of Reinius et al. [39], methylation of only two CpG probes (cg18202502 and cg15134033) in Table 3 were slightly associated with cell types (data not shown). Methylation in the other CpG probes in Table 3 was not significantly different between the blood cell types.

Finally, the small number of subjects included in this exploratory study is a clear weakness because of the limited statistical power. Despite these constraints, our findings suggest that DNA methylation might be a link between prenatal pesticide exposure and cardio-metabolic risk profile in children carrying the PON1 192R-allele. The findings deserve further investigation in a larger study with quantitative data on pesticide exposure. Whether this DNA methylation pattern is unique to pesticide exposure or is shared by other adverse prenatal environmental factors also needs further investigation.

Conclusions
In summary, our data indicate that DNA methylation may be an underlying mechanism explaining an adverse cardio-metabolic risk profile in prenatally pesticide-exposed children carrying the *PON1* 192R-allele.

Additional files

Additional file 1: Primer sequences.

Additional file 2: PCA before and after batch effect correction for Sentrix_ID and processing date using ComBat.

Additional file 3: Associations between the first eight principal components and covariates before and after ComBat batch correction. Associations between principal components and Sentrix_ID were measured using the Kruskal-Wallis test. Associations between principal components and processing date, exposure and *PON1* Q192R genotype were measured using the two-sided Wilcoxon sum rank test.

Additional file 4: Genomic location of the pyrosequencing assays represented as a UCSC genome browser track. The first track indicates the sequence analyzed by pyrosequencing (Seq_to_analyse). Other custom tracks include: CpG islands, Dnase I hypersensitivity clusters, H3K27ac histone marks, transcription factor-binding sites, and the Illumina 450 K methylation probes. A) *LEP* assay B) *GPR39* assay C) *PPARG* assay and D) *OPCML* assay.

Additional file 5: Relative cell type contribution estimated by the Houseman approach. Differences in cell type composition between the exposure groups were measured using one-way ANOVA.

Additional file 6: Sig-DMPs overlapping with DMRs (interaction model). For each DMP *P* values are given for the interaction between pesticide exposure and *PON1* genotype (P.Value.int_EXP:PON1), and for the exposure effect PON1 R-allele carrier group (P.Value.EXP_when PON1 QR/RR). The mean beta values in each exposure group are listed.

Additional file 7: Sig-DMPs overlapping with DMRs (model without interaction). For each DMP *P* values are given for the *PON1* effect (P.Value.PON1) and exposure effect (P.Value.EXP). The mean beta values in each exposure group are listed.

Additional file 8: Outcome of GPR39 DMR pyrosequencing. *Boxplots* showing methylation differences between the exposure groups in the *GPR39* pyrosequencing region. *P* values shown are those of the exposure effect.

Additional file 9: Outcome of GPR39 DMR pyrosequencing.

Additional file 10: Enrichment of TFBS for DMPs significant in the interaction model. *P* value from the Fisher's exact test were adjusted using the Bonferroni correction.

Additional file 11: Mediation analysis. Outcome statistics and gene disease associations of (partial) mediators between pesticide exposure and body fat measures in *PON1* R-allele carriers.

Additional file 12: Association between estimated blood cell counts and health outcomes. Simple linear regression was used to determine associations between the relative blood cell type composition and the health outcomes (body fat, BMI *Z* score, delta BMI *Z* score, leptin levels, and age).

Abbreviations
DMP: Differentially methylated posistion; DMR: Differentially methylated region; IPA: Ingenuity Pathway Analysis; PAI-1: Plasminogen activator inhibitor type-1; PON1: Paraoxonase 1; TFBS: Transcription factor binding site

Acknowledgements
We are grateful to the families for their participation in the greenhouse cohort study. We thank Mariann Bøllund and the greenhouse cohort study team for the skilled help with child examinations and the database. We thank Karen Hollanders (VITO, Belgium) for DNA extraction from blood samples.

Funding
The study was funded by The Danish Environmental Protection Agency (project number 667-00164). The funding organization had no role in study design, data collection and analysis, interpretation of the results, or preparation of the manuscript.

Authors' contributions
KD, SR, GS, WVB, and HRA conceived and designed the experiments. KD, SR, and HRA performed the experiments and analyzed the data. CWV, KMM, GVC, GS, WVB, and HRA contributed reagents/materials/analysis tools. KD, SR, GS, WVB, and HRA wrote the paper. KD, SR, CWV, KMM, GVC, GS, WVB, and HRA evaluated the manuscript text. All authors read and approved the final manuscript.

Competing interests
The authors declare that they have no competing interests.

Author details
[1]Laboratory of Protein Chemistry, Proteomics and Epigenetic Signalling (PPES), Department of Biomedical Sciences, University of Antwerp, Universiteitsplein 1, Antwerp, Belgium. [2]Department of Epidemiology and Social Medicine, Antwerp University, Universiteitsplein 1, Antwerp, Belgium. [3]Flemish Institute for Technological Research (VITO), Unit Environmental Risk and Health, Boeretang 200, Mol, Belgium. [4]Department of Growth and Reproduction, University Hospital of Copenhagen, Rigshospitalet, Copenhagen, Denmark. [5]Center of Medical Genetics, University of Antwerp and Antwerp University Hospital, Antwerp, Belgium. [6]Department of Biomedical Sciences, Antwerp University, Universiteitsplein 1, Antwerp, Belgium. [7]Environmental Medicine, Institute of Public Health, University of Southern Denmark, Odense, Denmark.

References
1. Bjorling-Poulsen M, Andersen HR, Grandjean P. Potential developmental neurotoxicity of pesticides used in Europe. Environ Health. 2008;7:50.
2. Andersen HR, Vinggaard AM, Rasmussen TH, Gjermandsen IM, Bonefeld-Jorgensen EC. Effects of currently used pesticides in assays for estrogenicity, androgenicity, and aromatase activity in vitro. Toxicol Appl Pharmacol. 2002; 179:1–12.

3. Orton F, Rosivatz E, Scholze M, Kortenkamp A. Widely used pesticides with previously unknown endocrine activity revealed as in vitro anti-androgens. Environ Health Perspect. 2011;119:794–800.

4. Grandjean P, Landrigan PJ. Neurobehavioural effects of developmental toxicity. Lancet Neurol. 2014;13:330–8.

5. London L, Beseler C, Bouchard MF, Bellinger DC, Colosio C, et al. Neurobehavioral and neurodevelopmental effects of pesticide exposures. Neurotoxicology. 2012;33:887–96.

6. Li AA, Baum MJ, McIntosh LJ, Day M, Liu F, et al. Building a scientific framework for studying hormonal effects on behavior and on the development of the sexually dimorphic nervous system. Neurotoxicology. 2008;29:504–19.

7. Gore AC. Neuroendocrine targets of endocrine disruptors. Hormones (Athens). 2010;9:16–27.

8. Jacobsen PR, Christiansen S, Boberg J, Nellemann C, Hass U. Combined exposure to endocrine disrupting pesticides impairs parturition, causes pup mortality and affects sexual differentiation in rats. Int J Androl. 2010;33:434–42.

9. Wohlfahrt-Veje C, Main KM, Schmidt IM, Boas M, Jensen TK, et al. Lower birth weight and increased body fat at school age in children prenatally exposed to modern pesticides: a prospective study. Environ Health. 2011;10:79.

10. Wohlfahrt-Veje C, Andersen HR, Jensen TK, Grandjean P, Skakkebaek NE, et al. Smaller genitals at school age in boys whose mothers were exposed to non-persistent pesticides in early pregnancy. Int J Androl. 2012;35:265–72.

11. Andersen HR, Schmidt IM, Grandjean P, Jensen TK, Budtz-Jorgensen E, et al. Impaired reproductive development in sons of women occupationally exposed to pesticides during pregnancy. Environ Health Perspect. 2008;116:566–72.

12. Wohlfahrt-Veje C, Andersen HR, Schmidt IM, Aksglaede L, Sorensen K, et al. Early breast development in girls after prenatal exposure to non-persistent pesticides. Int J Androl. 2012;35:273–82.

13. Andersen HR, Debes F, Wohlfahrt-Veje C, Murata K, Grandjean P. Occupational pesticide exposure in early pregnancy associated with sex-specific neurobehavioral deficits in the children at school age. Neurotoxicol Teratol. 2015;47:1–9.

14. Costa LG, Cole TB, Furlong CE. Polymorphisms of paraoxonase (PON1) and their significance in clinical toxicology of organophosphates. J Toxicol Clin Toxicol. 2003;41:37–45.

15. Mackness B, Durrington P, Povey A, Thomson S, Dippnall M, et al. Paraoxonase and susceptibility to organophosphorus poisoning in farmers dipping sheep. Pharmacogenetics. 2003;13:81–8.

16. Aviram M, Rosenblat M, Bisgaier CL, Newton RS, Primo-Parmo SL, et al. Paraoxonase inhibits high-density lipoprotein oxidation and preserves its functions. A possible peroxidative role for paraoxonase. J Clin Invest. 1998; 101:1581–90.

17. Durrington PN, Mackness B, Mackness MI. Paraoxonase and atherosclerosis. Arterioscler Thromb Vasc Biol. 2001;21:473–80.

18. Seo D, Goldschmidt-Clermont P. The paraoxonase gene family and atherosclerosis. Curr Atheroscler Rep. 2009;11:182–7.

19. Andersen HR, Wohlfahrt-Veje C, Dalgard C, Christiansen L, Main KM, et al. Paraoxonase 1 polymorphism and prenatal pesticide exposure associated with adverse cardiovascular risk profiles at school age. PLoS One. 2012;7:e36830.

20. Jorgensen A, Nellemann C, Wohlfahrt-Veje C, Jensen TK, Main KM, et al. Interaction between paraoxonase 1 polymorphism and prenatal pesticide exposure on metabolic markers in children using a multiplex approach. Reprod Toxicol. 2015;51:22–30.

21. Patel SB, Reams GP, Spear RM, Freeman RH, Villarreal D. Leptin: linking obesity, the metabolic syndrome, and cardiovascular disease. Curr Hypertens Rep. 2008;10:131–7.

22. Meas T, Deghmoun S, Chevenne D, Gaborit B, Alessi MC, et al. Plasminogen activator inhibitor type-1 is an independent marker of metabolic disorders in young adults born small for gestational age. J Thromb Haemost. 2010;8: 2608–13.

23. Huang KC, Lin RC, Kormas N, Lee LT, Chen CY, et al. Plasma leptin is associated with insulin resistance independent of age, body mass index, fat mass, lipids, and pubertal development in nondiabetic adolescents. Int J Obes Relat Metab Disord. 2004;28:470–5.

24. Ho SM, Johnson A, Tarapore P, Janakiram V, Zhang X, et al. Environmental epigenetics and its implication on disease risk and health outcomes. ILAR J. 2012;53:289–305.

25. Faulk C, Dolinoy DC. Timing is everything: the when and how of environmentally induced changes in the epigenome of animals. Epigenetics. 2011;6:791–7.

26. Saffery R, Novakovic B. Epigenetics as the mediator of fetal programming of adult onset disease: what is the evidence? Acta Obstet Gynecol Scand. 2014;93:1090–8.

27. Reynolds RM, Jacobsen GH, Drake AJ. What is the evidence in humans that DNA methylation changes link events in utero and later life disease? Clin Endocrinol (Oxf). 2013;78:814–22.

28. Perera F, Herbstman J. Prenatal environmental exposures, epigenetics, and disease. Reprod Toxicol. 2011;31:363–73.

29. Chmurzynska A. Fetal programming: link between early nutrition, DNA methylation, and complex diseases. Nutr Rev. 2010;68:87–98.

30. Knopik VS, Maccani MA, Francazio S, McGeary JE. The epigenetics of maternal cigarette smoking during pregnancy and effects on child development. Dev Psychopathol. 2012;24:1377–90.

31. Casati L, Sendra R, Sibilia V, Celotti F. Endocrine disrupters: the new players able to affect the epigenome. Front Cell Dev Biol. 2015;3:37.

32. Skinner MK. Environmental epigenomics and disease susceptibility. EMBO Rep. 2011;12:620–2.

33. Relton CL, Groom A, St Pourcain B, Sayers AE, Swan DC, et al. DNA methylation patterns in cord blood DNA and body size in childhood. PLoS One. 2012;7:e31821.

34. Godfrey KM, Sheppard A, Gluckman PD, Lillycrop KA, Burdge GC, et al. Epigenetic gene promoter methylation at birth is associated with child's later adiposity. Diabetes. 2011;60:1528–34.

35. Assenov Y, Muller F, Lutsik P, Walter J, Lengauer T, et al. Comprehensive analysis of DNA methylation data with RnBeads. Nat Methods. 2014;11:1138–40.

36. Teschendorff AE, Marabita F, Lechner M, Bartlett T, Tegner J, et al. A beta-mixture quantile normalization method for correcting probe design bias in Illumina Infinium 450 k DNA methylation data. Bioinformatics. 2013;29:189–96.

37. Leek JT, Johnson WE, Parker HS, Jaffe AE, Storey JD. The sva package for removing batch effects and other unwanted variation in high-throughput experiments. Bioinformatics. 2012;28:882–3.

38. Houseman EA, Accomando WP, Koestler DC, Christensen BC, Marsit CJ, et al. DNA methylation arrays as surrogate measures of cell mixture distribution. BMC Bioinformatics. 2012;13:86.

39. Reinius LE, Acevedo N, Joerink M, Pershagen G, Dahlen SE, et al. Differential DNA methylation in purified human blood cells: implications for cell lineage and studies on disease susceptibility. PLoS One. 2012;7:e41361.

40. Halachev K, Bast H, Albrecht F, Lengauer T, Bock C. EpiExplorer: live exploration and global analysis of large epigenomic datasets. Genome Biol. 2012;13:R96.

41. Peters TJ, Buckley MJ, Statham AL, Pidsley R, Samaras K, et al. De novo identification of differentially methylated regions in the human genome. Epigenetics Chromatin. 2015;8:6.

42. Obermann-Borst SA, Eilers PH, Tobi EW, de Jong FH, Slagboom PE, et al. Duration of breastfeeding and gender are associated with methylation of the LEPTIN gene in very young children. Pediatr Res. 2013;74:344–9.

43. Garcia-Cardona MC, Huang F, Garcia-Vivas JM, Lopez-Camarillo C, Del Rio Navarro BE, et al. DNA methylation of leptin and adiponectin promoters in children is reduced by the combined presence of obesity and insulin resistance. Int J Obes (Lond). 2014;38:1457–65.

44. Houde AA, Legare C, Biron S, Lescelleur O, Biertho L, et al. Leptin and adiponectin DNA methylation levels in adipose tissues and blood cells are associated with BMI, waist girth and LDL-cholesterol levels in severely obese men and women. BMC Med Genet. 2015;16:29.

45. Tobi EW, Lumey LH, Talens RP, Kremer D, Putter H, et al. DNA methylation differences after exposure to prenatal famine are common and timing- and sex-specific. Hum Mol Genet. 2009;18:4046–53.

46. Jousse C, Parry L, Lambert-Langlais S, Maurin AC, Averous J, et al. Perinatal undernutrition affects the methylation and expression of the leptin gene in adults: implication for the understanding of metabolic syndrome. FASEB J. 2011;25:3271–8.

47. Lesseur C, Armstrong DA, Paquette AG, Koestler DC, Padbury JF, et al. Tissue-specific Leptin promoter DNA methylation is associated with maternal and infant perinatal factors. Mol Cell Endocrinol. 2013;381:160–7.

48. Lesseur C, Armstrong DA, Paquette AG, Li Z, Padbury JF, et al. Maternal obesity and gestational diabetes are associated with placental leptin DNA methylation. Am J Obstet Gynecol. 2014;211(654):e651–9.

49. Petersen PS, Jin C, Madsen AN, Rasmussen M, Kuhre R, et al. Deficiency of the GPR39 receptor is associated with obesity and altered adipocyte metabolism. FASEB J. 2011;25:3803–14.

50. Janani C, Ranjitha Kumari BD. PPAR gamma gene–a review. Diabetes Metab Syndr. 2015;9:46–50.

51. Androutsopoulos VP, Hernandez AF, Liesivuori J, Tsatsakis AM. A mechanistic overview of health associated effects of low levels of organochlorine and organophosphorous pesticides. Toxicology. 2013;307:89–94.

52. Pillai HK, Fang M, Beglov D, Kozakov D, Vajda S, et al. Ligand binding and activation of PPARgamma by Firemaster(R) 550: effects on adipogenesis and osteogenesis in vitro. Environ Health Perspect. 2014;122:1225–32.

53. Grimaldi M, Boulahtouf A, Delfosse V, Thouennon E, Bourguet W, et al. Reporter cell lines for the characterization of the interactions between human nuclear receptors and endocrine disruptors. Front Endocrinol (Lausanne). 2015;6:62.

54. Khateeb J, Gantman A, Kreitenberg AJ, Aviram M, Fuhrman B. Paraoxonase 1 (PON1) expression in hepatocytes is upregulated by pomegranate polyphenols: a role for PPAR-gamma pathway. Atherosclerosis. 2010;208:119–25.

55. Camps J, Garcia-Heredia A, Rull A, Alonso-Villaverde C, Aragones G, et al. PPARs in regulation of paraoxonases: control of oxidative stress and inflammation pathways. PPAR Res. 2012;2012:616371.

56. Khateeb J, Kiyan Y, Aviram M, Tkachuk S, Dumler I, et al. Urokinase-type plasminogen activator downregulates paraoxonase 1 expression in hepatocytes by stimulating peroxisome proliferator-activated receptor-gamma nuclear export. Arterioscler Thromb Vasc Biol. 2012;32:449–58.

57. Divers J, Palmer ND, Lu L, Register TC, Carr JJ, et al. Admixture mapping of coronary artery calcified plaque in African Americans with type 2 diabetes mellitus. Circ Cardiovasc Genet. 2013;6:97–105.

58. Parks BW, Nam E, Org E, Kostem E, Norheim F, et al. Genetic control of obesity and gut microbiota composition in response to high-fat, high-sucrose diet in mice. Cell Metab. 2013;17:141–52.

59. Fox CS, Liu Y, White CC, Feitosa M, Smith AV, et al. Genome-wide association for abdominal subcutaneous and visceral adipose reveals a novel locus for visceral fat in women. PLoS Genet. 2012;8:e1002695.

60. Baron RM, Kenny DA. The moderator-mediator variable distinction in social psychological research: conceptual, strategic, and statistical considerations. J Pers Soc Psychol. 1986;51:1173–82.

61. Tingley D, Yamamoto T, Hirose K, Keele L, Imai K. mediation: R package for causal mediation analysis. J Stat Soft. 2014;59:1-38.

62. Pinero J, Queralt-Rosinach N, Bravo A, Deu-Pons J, Bauer-Mehren A, et al. DisGeNET: a discovery platform for the dynamical exploration of human diseases and their genes. Database (Oxford). 2015;2015:bav028.

63. Verhulst PJ, Lintermans A, Janssen S, Loeckx D, Himmelreich U, et al. GPR39, a receptor of the ghrelin receptor family, plays a role in the regulation of glucose homeostasis in a mouse model of early onset diet-induced obesity. J Neuroendocrinol. 2011;23:490–500.

64. Zhang JV, Ren PG, Avsian-Kretchmer O, Luo CW, Rauch R, et al. Obestatin, a peptide encoded by the ghrelin gene, opposes ghrelin's effects on food intake. Science. 2005;310:996–9.

65. Cui Y, Ying Y, van Hasselt A, Ng KM, Yu J, et al. OPCML is a broad tumor suppressor for multiple carcinomas and lymphomas with frequently epigenetic inactivation. PLoS One. 2008;3:e2990.

66. Sellar GC, Watt KP, Rabiasz GJ, Stronach EA, Li L, et al. OPCML at 11q25 is epigenetically inactivated and has tumor-suppressor function in epithelial ovarian cancer. Nat Genet. 2003;34:337–43.

67. Wu Y, Davison J, Qu X, Morrissey C, Storer B, et al. Methylation profiling identified novel differentially methylated markers including OPCML and FLRT2 in prostate cancer. Epigenetics. 2016;11:247–58.

68. Li C, Tang L, Zhao L, Li L, Xiao Q, et al. OPCML is frequently methylated in human colorectal cancer and its restored expression reverses EMT via downregulation of smad signaling. Am J Cancer Res. 2015;5:1635–48.

69. Volkow ND, Wang GJ, Baler RD. Reward, dopamine and the control of food intake: implications for obesity. Trends Cogn Sci. 2011;15:37–46.

70. Murray S, Tulloch A, Gold MS, Avena NM. Hormonal and neural mechanisms of food reward, eating behaviour and obesity. Nat Rev Endocrinol. 2014;10: 540–52.

71. Vucetic Z, Kimmel J, Totoki K, Hollenbeck E, Reyes TM. Maternal high-fat diet alters methylation and gene expression of dopamine and opioid-related genes. Endocrinology. 2010;151:4756–64.

72. Paternain L, Batlle MA, De la Garza AL, Milagro FI, Martinez JA, et al. Transcriptomic and epigenetic changes in the hypothalamus are involved in an increased susceptibility to a high-fat-sucrose diet in prenatally stressed female rats. Neuroendocrinology. 2012;96:249–60.

73. Ong ZY, Gugusheff JR, Muhlhausler BS. Perinatal overnutrition and the programming of food preferences: pathways and mechanisms. J Dev Orig Health Dis. 2012;3:299–308.

74. Palmer AA, Brown AS, Keegan D, Siska LD, Susser E, et al. Prenatal protein deprivation alters dopamine-mediated behaviors and dopaminergic and glutamatergic receptor binding. Brain Res. 2008;1237:62–74.

75. Vucetic Z, Totoki K, Schoch H, Whitaker KW, Hill-Smith T, et al. Early life protein restriction alters dopamine circuitry. Neuroscience. 2010;168:359–70.

76. Wright TM, Fone KC, Langley-Evans SC, Voigt JP. Exposure to maternal consumption of cafeteria diet during the lactation period programmes feeding behaviour in the rat. Int J Dev Neurosci. 2011;29:785–93.

77. Teegarden SL, Scott AN, Bale TL. Early life exposure to a high fat diet promotes long-term changes in dietary preferences and central reward signaling. Neuroscience. 2009;162:924–32.

78. Torres-Altoro MI, Mathur BN, Drerup JM, Thomas R, Lovinger DM, et al. Organophosphates dysregulate dopamine signaling, glutamatergic neurotransmission, and induce neuronal injury markers in striatum. J Neurochem. 2011;119:303–13.

79. Slotkin TA. Does early-life exposure to organophosphate insecticides lead to prediabetes and obesity? Reprod Toxicol. 2011;31:297–301.

80. Richard D, Huang Q, Timofeeva E. The corticotropin-releasing hormone system in the regulation of energy balance in obesity. Int J Obes Relat Metab Disord. 2000;24 Suppl 2:S36–9.

81. Richard D, Lin Q, Timofeeva E. The corticotropin-releasing factor family of peptides and CRF receptors: their roles in the regulation of energy balance. Eur J Pharmacol. 2002;440:189–97.

82. Sharma R, Banerji MA. Corticotropin releasing factor (CRF) and obesity. Maturitas. 2012;72:1–3.

83. Uehara Y, Shimizu H, Ohtani K, Sato N, Mori M. Hypothalamic corticotropin-releasing hormone is a mediator of the anorexigenic effect of leptin. Diabetes. 1998;47:890–3.

84. Sutton AK, Pei H, Burnett KH, Myers Jr MG, Rhodes CJ, et al. Control of food intake and energy expenditure by Nos1 neurons of the paraventricular hypothalamus. J Neurosci. 2014;34:15306–18.

85. Leshan RL, Greenwald-Yarnell M, Patterson CM, Gonzalez IE, Myers Jr MG. Leptin action through hypothalamic nitric oxide synthase-1-expressing neurons controls energy balance. Nat Med. 2012;18:820–3.

86. Sansbury BE, Hill BG. Regulation of obesity and insulin resistance by nitric oxide. Free Radic Biol Med. 2014;73:383–99.

87. Naseh M, Vatanparast J, Baniasadi M, Hamidi GA. Alterations in nitric oxide synthase-expressing neurons in the forebrain regions of rats after developmental exposure to organophosphates. Neurotoxicol Teratol. 2013;37:23–32.

88. Naseh M, Vatanparast J. Enhanced expression of hypothalamic nitric oxide synthase in rats developmentally exposed to organophosphates. Brain Res. 2014;1579:10–9.

89. Snodgrass-Belt P, Gilbert JL, Davis FC. Central administration of transforming growth factor-alpha and neuregulin-1 suppress active behaviors and cause weight loss in hamsters. Brain Res. 2005;1038:171–82.

90. Ennequin G, Boisseau N, Caillaud K, Chavanelle V, Etienne M, et al. Neuregulin 1 affects leptin levels, food intake and weight gain in normal-weight, but not obese, db/db mice. Diabetes Metab. 2015;41:168–72.

91. Wu G, Li H, Zhou M, Fang Q, Bao Y, et al. Mechanism and clinical evidence of lipocalin-2 and adipocyte fatty acid-binding protein linking obesity and atherosclerosis. Diabetes Metab Res Rev. 2014;30:447–56.

92. Hotamisligil GS, Bernlohr DA. Metabolic functions of FABPs—mechanisms and therapeutic implications. Nat Rev Endocrinol. 2015;11:592–605.

93. Furuhashi M, Saitoh S, Shimamoto K, Miura T. Fatty acid-binding protein 4 (FABP4): pathophysiological insights and potent clinical biomarker of metabolic and cardiovascular diseases. Clin Med Insights Cardiol. 2014;8:23–33.

94. Jiang Y, Ma S, Zhang H, Yang X, Lu GJ, et al. FABP4-mediated homocysteine-induced cholesterol accumulation in THP-1 monocyte-derived macrophages and the potential epigenetic mechanism. Mol Med Rep. 2016;14:969-76.

95. Yang AN, Zhang HP, Sun Y, Yang XL, Wang N, et al. High-methionine diets accelerate atherosclerosis by HHcy-mediated FABP4 gene demethylation pathway via DNMT1 in ApoE(-/-) mice. FEBS Lett. 2015;589:3998–4009.

96. Marrades MP, Gonzalez-Muniesa P, Arteta D, Martinez JA, Moreno-Aliaga MJ. Orchestrated downregulation of genes involved in oxidative metabolic pathways in obese vs. lean high-fat young male consumers. J Physiol Biochem. 2011;67:15–26.

97. Mussig K, Staiger H, Machicao F, Haring HU, Fritsche A. Genetic variants in MTNR1B affecting insulin secretion. Ann Med. 2010;42:387–93.

98. Nagorny C, Lyssenko V. Tired of diabetes genetics? Circadian rhythms and diabetes: the MTNR1B story? Curr Diab Rep. 2012;12:667–72.

99. Karamitri A, Renault N, Clement N, Guillaume JL, Jockers R. Minireview: Toward the establishment of a link between melatonin and glucose homeostasis: association of melatonin MT2 receptor variants with type 2 diabetes. Mol Endocrinol. 2013;27:1217–33.

100. Liu R, Dang W, Du Y, Zhou Q, Liu Z, et al. Correlation of functional GRIN2A gene promoter polymorphisms with schizophrenia and serum D-serine levels. Gene. 2015;568:25–30.

101. Endele S, Rosenberger G, Geider K, Popp B, Tamer C, et al. Mutations in GRIN2A and GRIN2B encoding regulatory subunits of NMDA receptors cause variable neurodevelopmental phenotypes. Nat Genet. 2010;42:1021–6.

102. Yoo HJ, Cho IH, Park M, Yang SY, Kim SA. Family based association of GRIN2A and GRIN2B with Korean autism spectrum disorders. Neurosci Lett. 2012;512:89–93.

103. Carvill GL, Regan BM, Yendle SC, O'Roak BJ, Lozovaya N, et al. GRIN2A mutations cause epilepsy-aphasia spectrum disorders. Nat Genet. 2013;45:1073–6.

104. Lal D, Steinbrucker S, Schubert J, Sander T, Becker F, et al. Investigation of GRIN2A in common epilepsy phenotypes. Epilepsy Res. 2015;115:95–9.

105. Ninomiya M, Numakawa T, Adachi N, Furuta M, Chiba S, et al. Cortical neurons from intrauterine growth retardation rats exhibit lower response to neurotrophin BDNF. Neurosci Lett. 2010;476:104–9.

106. Wang L, Wang X, Laird N, Zuckerman B, Stubblefield P, et al. Polymorphism in maternal LRP8 gene is associated with fetal growth. Am J Hum Genet. 2006;78:770–7.

107. Shen GQ, Li L, Wang QK. Genetic variant R952Q in LRP8 is associated with increased plasma triglyceride levels in patients with early-onset CAD and MI. Ann Hum Genet. 2012;76:193–9.

108. Shen GQ, Girelli D, Li L, Olivieri O, Martinelli N, et al. Multi-allelic haplotype association identifies novel information different from single-SNP analysis: a new protective haplotype in the LRP8 gene is against familial and early-onset CAD and MI. Gene. 2013;521:78–81.

109. Shen GQ, Girelli D, Li L, Rao S, Archacki S, et al. A novel molecular diagnostic marker for familial and early-onset coronary artery disease and myocardial infarction in the LRP8 gene. Circ Cardiovasc Genet. 2014;7:514–20.

110. Guo T, Yin RX, Yao LM, Huang F, Pan L, et al. Integrative mutation, haplotype and G x G interaction evidence connects ABGL4, LRP8 and PCSK9 genes to cardiometabolic risk. Sci Rep. 2016;6:37375.

111. Alegria-Torres JA, Garcia-Dominguez ML, Cruz M, Aradillas-Garcia C. Q192R polymorphism of paraoxonase 1 gene associated with insulin resistance in Mexican children. Arch Med Res. 2015;46:78–83.

112. Huen K, Harley K, Beckman K, Eskenazi B, Holland N. Associations of PON1 and genetic ancestry with obesity in early childhood. PLoS One. 2013;8: e62565.

113. Huen K, Yousefi P, Street K, Eskenazi B, Holland N. PON1 as a model for integration of genetic, epigenetic, and expression data on candidate susceptibility genes. Environ Epigenet. 2015;1:dvv003.

114. Coombes RH, Meek EC, Dail MB, Chambers HW, Chambers JE. Human paraoxonase 1 hydrolysis of nanomolar chlorpyrifos-oxon concentrations is unaffected by phenotype or Q192R genotype. Toxicol Lett. 2014;230:57–61.

115. Cole TB, Jampsa RL, Walter BJ, Arndt TL, Richter RJ, et al. Expression of human paraoxonase (PON1) during development. Pharmacogenetics. 2003; 13:357–64.

116. Costa LG, Giordano G, Cole TB, Marsillach J, Furlong CE. Paraoxonase 1 (PON1) as a genetic determinant of susceptibility to organophosphate toxicity. Toxicology. 2013;307:115–22.

117. Huang YT, Chu S, Loucks EB, Lin CL, Eaton CB, et al. Epigenome-wide profiling of DNA methylation in paired samples of adipose tissue and blood. Epigenetics. 2016;11:227–36.

17

Significantly altered peripheral blood cell DNA methylation profile as a result of immediate effect of metformin use in healthy individuals

Ilze Elbere[1†], Ivars Silamikelis[1†], Monta Ustinova[1], Ineta Kalnina[1], Linda Zaharenko[1], Raitis Peculis[1], Ilze Konrade[2], Diana Maria Ciuculete[3], Christina Zhukovsky[3], Dita Gudra[1], Ilze Radovica-Spalvina[1], Davids Fridmanis[1], Valdis Pirags[1], Helgi B. Schiöth[3] and Janis Klovins[1*]

Abstract

Background: Metformin is a widely prescribed antihyperglycemic agent that has been also associated with multiple therapeutic effects in various diseases, including several types of malignancies. There is growing evidence regarding the contribution of the epigenetic mechanisms in reaching metformin's therapeutic goals; however, the effect of metformin on human cells in vivo is not comprehensively studied. The aim of our study was to examine metformin-induced alterations of DNA methylation profiles in white blood cells of healthy volunteers, employing a longitudinal study design.

Results: Twelve healthy metformin-naïve individuals where enrolled in the study. Genome-wide DNA methylation pattern was estimated at baseline, 10 h and 7 days after the start of metformin administration. The whole-genome DNA methylation analysis in total revealed 125 differentially methylated CpGs, of which 11 CpGs and their associated genes with the most consistent changes in the DNA methylation profile were selected: *POFUT2, CAMKK1, EML3, KIAA1614, UPF1, MUC4, LOC727982, SIX3, ADAM8, SNORD12B, VPS8*, and several differentially methylated regions as novel potential epigenetic targets of metformin. The main functions of the majority of top-ranked differentially methylated loci and their representative cell signaling pathways were linked to the well-known metformin therapy targets: regulatory processes of energy homeostasis, inflammatory responses, tumorigenesis, and neurodegenerative diseases.

Conclusions: Here we demonstrate for the first time the immediate effect of short-term metformin administration at therapeutic doses on epigenetic regulation in human white blood cells. These findings suggest the DNA methylation process as one of the mechanisms involved in the action of metformin, thereby revealing novel targets and directions of the molecular mechanisms underlying the various beneficial effects of metformin.

Keywords: Metformin, Epigenetics, DNA methylation, White blood cells, Longitudinal study

* Correspondence: klovins@biomed.lu.lv
†Ilze Elbere and Ivars Silamikelis contributed equally to this work.
[1]Latvian Biomedical Research and Study Centre, Ratsupites Str. 1 k-1, Riga LV-1067, Latvia
Full list of author information is available at the end of the article

Background

Metformin is the first-line drug for type 2 diabetes (T2D) therapy, used since 1950s [1]. Although there are a great number of various studies on the metformin pharmacogenomics, pharmacokinetics, and lately its interaction with the gut microbiome, the details of the molecular mechanisms of metformin action have not been fully understood.

So far, there are only a few studies within the context of metformin action and changes in one of the most commonly studied epigenetic modifications—DNA methylation. One of the targeted studies has shown that metformin treatment of pregnant rats with gestational diabetes can reduce methylation level of peroxisome proliferator-activated receptor γ coactivator-1A (PPARGC1A), therefore preventing the abnormal glyco-lipid metabolism in their offspring [2]. In addition, a genome-wide study of metformin effects on lymphoblastoid cell lines has revealed potential biomarkers for metformin's anticancer response [3]. In the context of possible molecular mechanisms of how metformin induce changes in the methylation profile, a recent study has proved that, in cancer cells, metformin can exert its effects via regulation of the H19/SAHH axis [4]. This has been supported by data showing that metformin promotes global methylation by decreasing S-adenosyl-homocysteine (SAH) intracellular levels in various cell types, including non-cancerous [5]. One of the latest studies have specifically shown metformin's effect on lowering the methylation levels at the metformin transporter genes, resulting in higher expression levels in liver tissue [6]. Studies describing other epigenetic effects of metformin have shown its impact on various histone modifications via multiple mechanisms, mostly AMPK dependent, and effect on expression levels of numerous miRNAs through increase in DICER protein levels as well [7].

Nevertheless, there is a significant lack of information on how metformin affects global epigenetic regulation in non-cancerous cells or in cells obtained from metformin-treated humans. Therefore, our aim was to investigate the short-term effect of metformin on DNA methylation profiles in blood cells from healthy volunteers. Here we compared the changes in DNA methylation in the same subjects before and after the metformin intake.

Results

Characteristics of the study participants

We used Illumina Infinium 450k array to evaluate the effect of metformin on DNA methylation in 12 healthy volunteers. The characteristics of the study group are summarized in Table 1. Samples, for analysis of the methylation levels, from each participant were obtained

Table 1 Characteristics of the study group

Characteristic	Value
Female/male, n (%)	7 (58.3%)/5 (41.7%)
Age, years, mean ± SD	31.4 ± 6.7
BMI, mean ± SD	25.3 ± 3.5
ALAT*, U/l, mean ± SD	25 ± 13
Creatinine*, μmol/l, mean ± SD	68 ± 8.9
Fasting plasma glucose*, mmol/l, mean ± SD	5.1 ± 0.3

BMI body mass index, SD standard deviation, ALAT alanine aminotransferase
*Samples for hematological, biochemical tests were collected before metformin administration

at three time points, further marked as M0 (before starting a metformin therapy), M10h (10 h after the first metformin intake, before the second tablet), and M7d (time point after 7 days of metformin administration). M10h sample was chosen to evaluate effect of one metformin's dose; to ensure accuracy of this measurement, all study participants were strictly instructed to take the second metformin tablet only after the M10h blood sampling.

Differentially methylated CpGs

During the data preprocessing stage, 64,512 (13.29%) probes were filtered out, leaving 421,000 probes for downstream analysis. To detect differentially methylated CpG sites/probes (DMPs), we applied limma analysis between contrasts at all three time points, i.e., baseline, after 10 h and 7 days of metformin administration. The model included the methylation values at the contrasted time points, together with the cell-type estimations as covariates. Comparing methylation values at M10h and M0 samples, 72 differentially methylated CpG sites with a false discovery rate (FDR) of < 0.05 were identified after correction for multiple testing using the Benjamini-Hochberg method. In the same way, 52 DMPs were found applying contrast between methylation levels at M7d and M0 and only one (cg07026010—NUDCD3) in case of M7d with M10h comparison (full list of significant CpGs is available in Additional file 1). Of these, 43 (59.72%), 24 (46.15%), and 1 (100%) CpG sites were hypermethylated, and 29 (40.28%), 28 (53.85%), and 0 (0%) CpG sites were hypomethylated when contrast analyses were applied for M10h vs M0, M7d vs M0, and M7d vs M10h respectively (Fig. 1). The median absolute difference in beta values, comparing all contrasts, was 0.013 (interquartile range (IQR), 0.006–0.029) for statistically significant differentially methylated probes. The average estimated genomic inflation factor (λ) for all three contrasts before correction was 1.64 ± 0.28, and after including covariates, it was reduced to 1.30 ± 0.15. Additional evaluation of λ with qq-plots depicted the same improvement ensured by including covariates (data not shown).

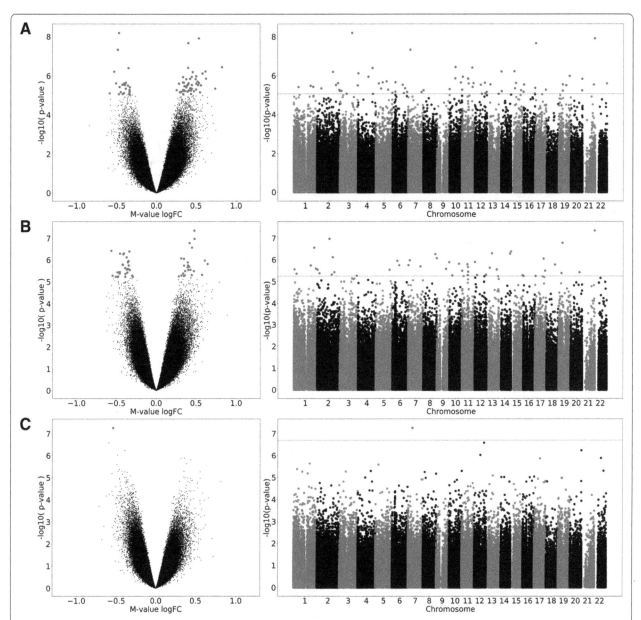

Fig. 1 Differentially methylated positions in all analyzed contrasts. Volcano plot showing raw − log10 (*p* value) versus log-fold change of *M* values and the Manhattan plot showing the position of probes with their corresponding unadjusted *p* values across the genome in **a** M10h vs M0, **b** M7d vs M0, and **c** M7d vs M10h sample comparisons. The significant CpG sites (after FDR correction) are highlighted in red. M0—before starting a metformin therapy; M10h—10 h after the first metformin intake, before the second tablet; M7d—time point after 7 days of metformin administration

Among the identified DMP, a total of 11 CpGs with the most consistent changes in the DNA methylation profile were emphasized (Fig. 2) based on two additional criteria. First, we included all overlapping DMP at both contrasts M10h vs M0 and M7d vs M0 (*n* = 5; cg03515060, cg18394557, cg16013966, cg05638165, cg18824330). Second, we selected probes if their median beta values at time points M10h and M7h overlapped IQRs of M7h and M10h, respectively. Also, IQRs of both time points could not overlap with IQR of time point M0 (*n* = 6; cg12740863, cg16843994, cg12162450, cg19176072, cg01644741,

cg02622542). Of these 11 CpGs, 8 (72.73%) CpG sites displayed hypermethylation, while 3 (27.27%) CpG sites showed hypomethylation when comparing methylation levels after the metformin use (at time points M10h and M7d) with methylation levels before the use of metformin (Fig. 2).

All identified 11 CpG sites corresponded to 11 genes according to the 450k annotation file published by Price [8] (Table 2). One of these CpG sites was located in high-density CpG island, 7 CpG sites—in intermediate-density CpG islands with 1 bordering

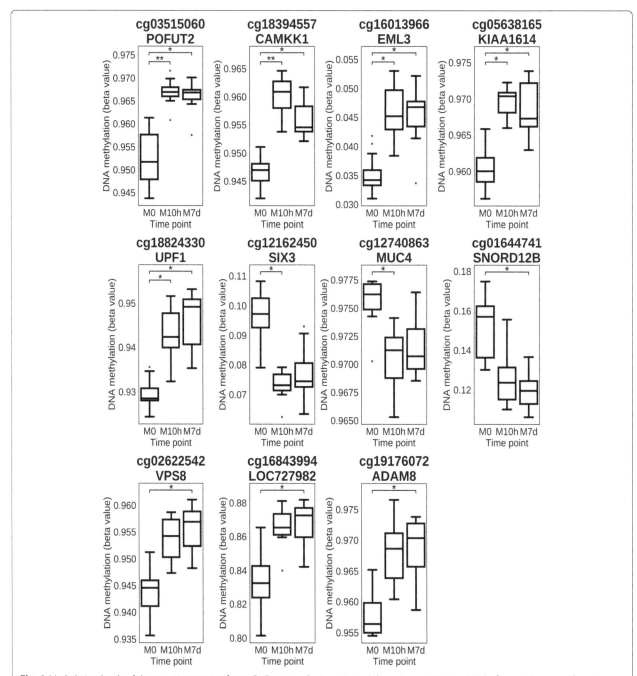

Fig. 2 Methylation levels of the top 11 most significant CpGs across the investigated three time points, i.e., M0 (before starting a metformin therapy), M10h (10 h after the first metformin intake, before the second tablet), and M7d (time point after 7 days of metformin administration), together with their associated genes. Box plots depict median, maximum, minimum, 25th percentile, and 75th percentile. Dots beyond the bounds of the whiskers represent outliers. * and ** denote significance levels 0.05 and 0.01 respectively

high-density CpG island, and 3—in non-islands according to HIL CpG classes.

To analyze the possible influence of circadian changes on the methylation profile, firstly, we searched our DMP list for the most common genes associated with regulation of circadian rhythm, such as *BMAL1, PER1, PER2, PER3, ARNTL, CRY1*, and

CRY2. Secondly, we evaluated the main known functional roles of genes associated with the 125 DMPs, and, thirdly, we used the results from pathway enrichment analysis to find any connections with the circadian regulation. In result of these steps, we did not find any significant associations between the DMPs and circadian rhythm.

Table 2 Characterization of the top 11 most significant CpG sites

Filter	CpG site	Chr	logFC M10h vs M0	logFC M7d vs M0	FDR M10h vs M0	FDR M7d vs M0	Gene	Distance to the closest TSS	Gene context[a]	Spearman's correlation between methylation and transcription
Significant in both of the following contrasts: M10h vs M0 and M7d vs M0	cg03515060	21	0.538	0.479	*0.003*	*0.018*	POFUT2	1984	Body	− 0.184
	cg18394557	17	0.406	0.286	*0.003*	*0.047*	CAMKK1	− 8799	Body	− 0.042
	cg16013966	11	0.428	0.395	*0.034*	*0.037*	EML3	− 308	1stExon;5'UTR;TSS1500	− 0.23
	cg05638165	1	0.358	0.351	*0.035*	*0.034*	KIAA1614	14,198	Body	− 0.337
	cg18824330	19	0.363	0.419	*0.043*	*0.022*	UPF1	− 9944	Body	− 0.382
Significant in one of the contrasts and medians for time points M10h or M7d in IQR	cg12740863	3	− 0.359	0.260	*0.034*	0.127	MUC4	− 26,158		0.37
	cg16843994	2	0.347	0.404	0.091	*0.038*	LOC727982	− 706		NA
	cg12162450	2	− 0.386	0.349	*0.040*	0.059	SIX3	7515		NA
	cg19176072	10	0.462	0.472	0.054	*0.040*	ADAM8	5756	Body	− 0.312
	cg01644741	20	− 0.287	0.366	0.137	*0.043*	SNORD12B	39	Body,TSS1500	0.036
	cg02622542	3	0.269	0.348	0.151	*0.047*	VPS8	− 2419		0.166

Statistically significant FDR values are marked in italics

5'UTR 5' untranslated region, *TSS* transcription starting site

[a]TSS1500: Region 200–1500 base pairs upstream of the transcription start site

The correlation between methylation and RNA expression level of the corresponding gene was verified using targeted data form RNA-seq. Out of 11 genes tested, only the expression of *UPF1* (p – 0.024), *MUC4* (p – 0.029), and *KIAA1614* (p – 0.048) showed significant correlation with the methylation of corresponding CpG sites (Table 2).

Differentially methylated regions (DMRs)

During the DMR analysis, we found 13 regions with significant differences in methylation levels (summarized in Table 3). Five of the identified regions overlapped with some of the significant DMPs but not with the 11 sites prioritized by us.

Enrichment analysis

To evaluate the potential biological significance of the impact of differentially methylated CpG sites, we performed a gene set pathway enrichment analysis by using the Ingenuity Pathway Analysis (IPA). All genes associated with significant differentially methylated probes (FDR < 0.05) from different contrasts were selected.

Table 3 Differentially methylated regions

Contrast	Gene	FDR	Number of probes	Chr	Start (bp)[a]	End (bp)[a]	Transcription factors[b]
M10h vs M0	EPHB1	1.60E−11	3	3	134,515,421	134,516,302	–
	CDCA7L	3.83E−07	5	7	21,985,276	21,985,628	Nr1h3
	CLVS2	8.21E−07	10	6	123,317,123	123,317,875	Nrsf
	BACE2, MIR3197	1.38E−06	3	21	42,539,960	42,540,409	CTCF
	EXPH5	5.76E−06	6	11	108,464,101	108,464,498	Cmyc; Egr1; FOXA1; MYC; Max; SP1;
	KCNE4	1.50E−05	3	2	223,916,686	223,916,861	USF1
	TTC38	1.50E−05	4	22	46,685,471	46,685,728	NA
	TTC39A	1.51E−05	5	1	51,810,626	51,811,022	–
	NA	2.17E−05	3	4	153,897,215	153,897,453	NA
	NA	2.33E−05	3	10	132,891,318	132,891,371	NA
M7d vs M0	SFRP2	1.18E−11	28	4	132,891,371	154,711,183	CTCF; Egr1
	GPR19	4.59E−10	11	12	12,848,515	12,849,588	E2F4; ZBTB33;
	TMEM216	3.46E−07	7	11	61,159,601	61,159,837	CTCF; Egr1; Gabp; Yy1

[a]Physical position (basepair, hg37)

[b]Data from Ensembl 91 regulation resources [98], hg38

Thus, 72 genes were selected from the M10h vs M0 contrast and included in the first pathway analysis, and 52 genes from the M7d vs M0 contrast and included in the second pathway analysis. We did not include the only significant result from the M7d vs M10h contrast. The top enriched canonical pathways are summarized in Table 4.

In addition to the canonical pathways, we identified nine enriched networks in the M10h vs M0 contrast, and four in the M7d vs M0 comparison. The top enriched networks with IPA score > 20 were as follows (score/focus molecules): M10h vs M0—hematological system development and function, cellular movement, cell-to-cell signaling and interaction (28/13); hereditary disorder, neurological disease, organismal injury and abnormalities (23/11). M7d vs M0—cell-to-cell signaling and interaction, cellular assembly and organization, cellular function and maintenance (48/19); cell morphology, cell-to-cell signaling and interaction, cellular assembly and organization (41/17). Two of the most relevant networks are visualized in Fig. 3.

Discussion

The aim of our study was to examine metformin-induced alterations in epigenetic regulation processes by performing genome-wide DNA methylation analysis in human white blood cells followed by estimation of RNA expression levels of identified genes. We conducted our study in order to understand the pathways affected by metformin at real life physiological conditions in humans. This is extremely important taking into account the pleiotropic effects of metformin, and such studies may pinpoint important novel targets not only for treatment of T2D but also for other diseases. Various studies have shown that the evaluated effects in the methylation profile of peripheral blood DNA, that is the only option to access repeated tissue sampling in humans, are highly representative to the changes in other organs [9–11]. It is known that the DNA methylation pattern is highly subject specific and is influenced by many factors making it very difficult to identify the metformin-specific effects in a case-control-based type of study. We therefore selected a longitudinal design for this study, using short response time in order to exclude the influence of other factors. We also involved healthy volunteers to avoid a background of any commonly studied diseases so far related with the beneficial effects of metformin. One of our goals was to detect the fastest practically measurable effect of metformin on DNA methylation. Taking into account the known high variability of metformin pharmacokinetics, the time point when to evaluate the immediate and at the same time most profound effect was chosen to be the impact of one dose, and sampling time was selected at 10 h, before the recommended administration time of the second dose. Furthermore, M10h vs M0 sample comparison revealed the highest number or DMPs, representing the significant effect of one metformin dose.

To our best knowledge, this is the first study showing the metformin-mediated change of DNA methylation in healthy individuals already 10 h after administration. From the pool of 125 significantly modified sites, we prioritized 11 differentially methylated CpG with the largest and most consistent changes in beta values at different contrasts.

We assumed that some methylation changes measured at 10 h (M10h) could be caused by the circadian rhythm, which has been well described before and proven to be a driver of dynamic gene expression [12]. To avoid any false conclusions about the epigenetic targets of metformin, we paid specific attention to the presence of genes involved in the circadian rhythm among regions covering DMPs. We also evaluated this possibility by focusing on two contrasts that represent the different methylation profiles of DNA purified from blood samples that were collected in two distinct time points of the day—M7d vs M0 and M7d vs M10h. We did not observe any overlapping DMPs between the particular contrasts, suggesting no significant influence of the circadian rhythm on the DNA methylation in our data. Surprisingly, we observed only one significant DMP comparing M7d and M10h time points, providing a strong support for the fact that

Table 4 Top enriched canonical pathways by IPA

Contrast	Pathway	p value
M10h vs M0	Unfolded protein response	8.82×10^{-3}
	Salvage Pathways of Pyrimidine Deoxyribonucleotides	0.021
	Glycogen Degradation II	0.031
	Glycogen Degradation III	0.036
	Granzyme B Signalling	0.041
	Gα12/13 Signalling	0.046
	Lipid Antigen Presentation by CD1	0.048
M7d vs M0	S-Methyl-5-thio-α-D-ribose 1-phosphate Degradation	6.82×10^{-3}
	Gustation Pathway	0.025

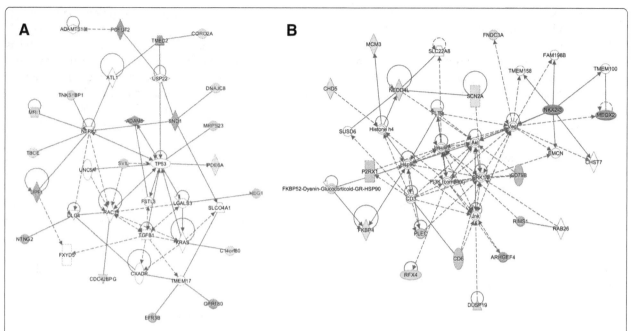

Fig. 3 Top enriched networks from IPA. Green nodes—hypermethylated; red nodes—hypomethylated. **a** Cell-to-cell signaling and interaction, cellular assembly and organization, and cellular function and maintenance (central predicted associated biological functions—tumorigenesis processes) had an IPA score of 48. **b** Cell morphology, cell-to-cell signaling and interaction, and cellular assembly and organization (central predicted associated biological functions—metabolism processes) had an IPA score of 41

observed methylation changes are indeed caused by metformin rather than other factors changing during the trial, such as diet or circadian cycle.

Genes corresponding to the top-ranked DMPs represent the main functional groups associated with previously described targets of metformin therapy: regulatory processes of energy homeostasis, inflammatory responses, tumorigenesis, and neurodegeneration. The criteria based on the comparison of median beta values and IQRs (see the "Results" section) were chosen to avoid bias in prioritization and would allow to include potentially important DMPs in addition to only those being significant at both M10h vs M0 and M7d vs M0 contrasts.

Interestingly, we found DMP within *CAMKK1* gene—one of two highly homologous genes coding for Ca2 +/calmodulin-dependent protein kinase kinases (CaMKK) [13]—with CaMKK2 being a known regulator of AMP-activated protein kinase (AMPK). Despite the fact that only CaMKK2 has been proven to form a stable complex with AMPK, both isoforms of the CaMKK are capable of phosphorylating the AMPKα subunit at Thr-172 in vitro [14, 15]. From our data, the differentially methylated CpG close to the *CAMKK1* TSS together with negatively correlated mRNA expression data as the result of metformin administration suggests a potential contribution of CaMKK1 in the AMPK-mediated mechanism of metformin anti-diabetic action.

Furthermore, it is known that metformin exerts its effects also via AMPK-independent mechanisms [16], as

shown by CaMKK1 ability to mediate glucose uptake in muscle cells independently from AMPK and Akt [17], in that way suggesting that methylation level changes in CaMKK1 could be a part from an alternative pathway responsible for the therapeutic effects of metformin.

Additionally, we identified a differentially methylated CpG site near the transcription factor coding gene *SIX3* [18]. Downregulation of *SIX3* due to the methylation of the *SIX3* promoter is observed in lung adenocarcinoma tissues and lung cancer cell lines, where mRNA expression of the gene is also associated with higher survival rate [19]. Some research suggest SIX3 linkage to diabetes from genetic studies [20] and show SIX3 as possible regulator of insulin production in β-cells in an age-dependent manner [21]. Lowered methylation level of CpG near the *SIX3* TSS shown in our data suggests the DNA methylation as another potential epigenetic mechanism involved in *SIX3* expression regulation. *SIX3* is not expressed in human white blood cells [22], explaining the absence of reads corresponding to *SIX3* in our RNA-seq data, but gene expression may manifest in other tissues. So far, normalized insulin production itself has not been considered as a therapeutic effect of metformin, although it might be affected along with metformin-induced improvements of insulin sensitivity [23].

Our data also show ADAM8 as a considerable potential contributor in the anti-inflammatory action of metformin, that is, one of the known beneficial effects of this medication [24]. ADAM8 is a cell surface protease, mainly expressed in granulocytes and monocytic cells, where it conducts the regulation of monocyte adhesion and migration [25–27]. Its contribution in the inflammatory responses regarding neurodegenerative disorders, allergy, asthma, and acute lung inflammation has been widely described before [28–31]. Our data justify the anti-inflammatory properties of metformin independently of diabetes status [24] and suggest the potential contribution of ADAM8 in the process. Due to its expression in human white blood cells, ADAM8 might be considered a promising biomarker for the detection of metformin-induced anti-inflammatory responses while reflecting inflammatory processes in adipocytes; however, further experimental evidence is required.

Many of the genes linked to the top-ranked DMPs are functionally associated with various malignancies. The most significant DMP in our study appeared to be situated in the body of *POFUT2*. O-Fucosyltransferase 2 encoded by *POFUT2* is proved to restrict epithelial-mesenchymal transition and affect cell motility in mouse embryos [32], and is considered as a useful prognostic biomarker in patients with glioblastoma and adenocarcinoma [33, 34]. To our knowledge, there are no reports yet describing *POFUT2*'s association with the beneficial effects of metformin. Our data also show several more DMPs located within or near the TSS of tumor-related genes, including SNORD12B—previously associated with colorectal and breast cancer pathogenesis [35–37], MUC4—promising prognostic marker and therapeutic target in the case of pancreatic cancer [38–40], KIAA1614 with promoter hypermethylation observed in colon tissues from patients with ulcerative collitis as well as in colon cancer cell lines [41], and UPF1 with indisputably crucial role in the maintenance of genome stability, significantly implicated in various malignancies [42–47].

The functions of two genes from the top DMPs' associated list are poorly defined. We identified increased DNA methylation level close to the TSS of VPS8 gene. VPS8 is an accessory subunit of CORVET complex, necessary for mediating multiple steps in the endocytic pathway and required for fusion of early endosomes [48]. Thus far, there is no conclusive data indicating the possible effects of VPS8 dysregulation on phenotype in humans [49–52]. Likewise, the function of long intergenic non-protein coding RNA 1249 (LINC01249/ LOC727982) is still not clear with only few reports on genetic association of SNPs within the gene with infectious disease and blood pressure [53, 54].

Overall, the DNA methylation has a repressive effect on transcription factor binding; therefore, we used EN-CODE data on transcription factor binding sites to identify such possible interactions [55, 56]. We detected transcription factors CTCF, CTCFL, and Egr1 binding to

the genomic region overlapping the differentially methylated CpG within *EML3* gene; out of these, CTCF is proved to mediate glucagon production [57] and Egr1 is responsible for insulin resistance [58]. Although there are no data available to date, supporting direct metformin impact on EML3 (nuclear microtubule-binding protein) [59] or describing EML3 contribution in metformin therapeutic effects, increased expression of *EML3* in cultured human cells as a result of metformin-1816 small molecule perturbation has been reported before [60]. Likewise, the genomic region within *UPF1* gene, covering the top-ranked CpG site is associated with CTCF, Egr1, and two more transcription factors: MYC—involved in the pathogenesis of diabetes [61], and PU1—initiating insulin resistance as well as regulating lipolysis [62].

The detected DMRs, as well, could essentially be grouped by connection to the processes currently known to be affected by metformin. For example, the most significant DMR was associated with *EPHB1*, which together with other Ephrin receptors forms the largest subgroup of the Eph receptor tyrosine kinase (RTK) family [63]. Underexpression of the EphB1 protein is significantly associated with tumor progression in gastric carcinomas and higher invasiveness of colorectal cancer cells, suggesting a tumor-suppressive role of the protein and possible implication in the beneficial effects of metformin [64, 65].

Another noteworthy DMR was associated with APP-cleaving enzyme 2 coding gene (*BACE2*) encoded protein that cleaves amyloid precursor protein into amyloid beta peptide, and is implicated in the pathogenesis of neurodegenerative diseases including Alzheimer's disease [66–68]. Interestingly, increased β-cell proliferation and glucose-stimulated insulin secretion resulting from reduced Bace2 levels have been previously reported [69]. In a mouse model of T2D, induced by the overexpression of human islet amyloid polypeptide, *BACE2* deficiency improved glucose tolerance, suggesting that *BACE2* inhibition might serve as a potential therapeutic strategy for T2D treatment [70].

Another DMR is associated with *SFRP2*, Secreted Frizzled Related Protein 2. Methylation changes in the promoter region of *SFRP2* have been proposed as a potential noninvasive biomarker for colorectal cancer [71, 72]. Its mRNA is also expressed in mouse and human adipose tissue, and elevated levels have positive correlation with BMI and with abnormal glucose tolerance [73].

The pathway enrichment analysis revealed metformin's association with various pathways some of which already has been described in connection with metformin action but not in the context of epigenetic regulation. The top enriched pathway after one dose of metformin—Unfolded Protein Response (UPR)—has been shown to be one of the main mechanisms of inducing apoptosis by metformin in acute lymphoblastic leukemia cells [74], and metformin-induced UPR inhibition in kidney cells can explain metformin's beneficial effects [75].

One of the products of the top enriched pathway describing changes after week long metformin administration (S-methyl-5-thio-α-D-ribose 1-phosphate Degradation) is L-methionine, an essential amino acid in human organism. Moreover, it is known that L-methionine is used for generation of S-adenosylmethionine (SAM) [76], which has been depicted to be an essential part of metformin-induced increase in global methylation levels as it accumulates in cells during metformin therapy [5]. Taking into account the results from enriched pathways and the fact that we mostly observe metformin-induced hypermethylation than hypomethylation, it is possible that activation of this particular canonical pathway may contribute to the previously described increase in SAM levels.

Although enriched networks (Fig. 3) are not directly related to known metformin effects, the downstream molecules of those associated with differential methylation levels in our study group are known to be involved in various pathways related with T2D (e.g., AKT, ERK1/2, JNK, P13K), insulin regulation processes [77], cancer development mechanisms [78], and other.

The correlation between DNA methylation and gene expression is complex and nonlinear [79]. The generally accepted consequence of DNA methylation is transcriptional repression; however, methylation in the transcribed region might also demonstrate positive correlation with mRNA expression [80]. In our study, we did not detect a convincing correlation between DNA methylation of top-ranked loci and transcription level of corresponding genes; however, the influence of methylation as well as gene expression itself are tissue-specific and might be missed by focusing on single type of cells only. Nevertheless, the significant correlation observed between the expression levels of *UPF1*, *MUC4*, *KIAA1614*, and the methylation level of the corresponding CpG sites provide evidence for a crucial contribution of epigenetic regulation in the mechanism of action of metformin, which results in specific alterations of gene expression profiles.

Currently, it is not fully known whether metformin has only an indirect effect on the epigenetic regulation processes in the human organism via the previously described H19/SAHH axis or through linking cellular metabolism to the mechanisms needed for DNA methylation [4, 5]. However, the methylation profile and concentration of metformin used in cell type specific in vitro experiments may significantly differ from the physiological levels and observations in the affected cells in human body. The large variation of SAH and SAM levels in various cell types has been described [5]. In addition, the previous studies evaluating the

metformin-induced methylation profile changes mostly have been targeted; thus, it is not surprising that we did not observe the DMPs at the same genes or pathways.

Major limitation of this study is the low sample size even though there are number of reports using the same number of individuals in their studies [81–84]. On the other hand, we believe that this weakness is compensated by the number of strengths in our design. First, we used a longitudinal study design and it has been recognized that, in similar time series studies, individuals can be treated as their own controls before and during treatment and sufficiently increase the power of the study [85] compared to case-control design especially accounting for the inter-individual variability among study participants. Secondly, the longitudinal design combined with observation of methylation changes in the shortest possible time allows us to minimize the effects of other factors that can induce changes in methylation unrelated to the metformin treatment. Thirdly, inclusion of healthy subjects should have minimized false associations and conclusions arising from unaccounted treatment status by metformin or other medications in T2D patients, including the unknown true duration of T2D before diagnosis. Finally, the use of genome-wide methylation analysis allows us to observe unbiased effects and find new metformin targets.

Another limitation in our study is the lack of clinical and biochemical measures at all time points. In the same time, it has been previously shown that metformin has small or no effect of such measures as plasma glucose level in healthy individuals [86, 87], and we decided not to include those in study protocol.

Unfortunately, due to the lack of similar studies, we were not able to support our findings from literature and replication in other cohorts is needed.

Conclusions

This is the first study showing the immediate effect of metformin on white blood cell DNA methylation in humans at therapeutic doses. The gained knowledge about the metformin-induced methylation profile changes in healthy individuals can be used as basis for further in vitro and in vivo studies, which are important due to the growing number of various metformin therapeutic application possibilities in non-diabetic patients.

Methods
Study design
Study group involved 12 healthy metformin-naïve voluntary individuals. The involvement and sample collection was organized in collaboration with the Genome Database of Latvian Population (LGDB) [88]. Exclusion/inclusion criteria (Additional file 2) were defined according to the requirements of concurrently ongoing

clinical trial (registration number: 2016-001092-74 (www.clinicaltrialsregister.eu)), which also involves gut microbiome analysis. Participants were included if they matched the following criteria: have not used antibiotics, immunosuppressive medicaments, corticosteroids, or pharmaceutical-grade probiotics during the time period of the past 2 months; have not been diagnosed with oncological, autoimmune, chronical gastrointestinal tract diseases, or T2D; have not had diarrhea in the past week; and are not taking any other medications incompatible with metformin. The research subjects received an 850-mg metformin tablet (Berlin-Chemie AG) twice a day for a week. Samples for hematological, biochemical tests were collected in certified clinical laboratory at fasting state 1–3 days before starting the metformin administration. Whole blood samples for methylation analysis were collected at three time points: (1) before starting metformin therapy (morning, fasting state)—M0, (2) 10 h after first metformin intake, before the second tablet (evening)—M10h, and (3) after 7 days of metformin administration (morning, fasting state)—M7d. Throughout the article, we have defined the measurement of 10-h sample as the immediate effect of metformin.

Sample analysis

DNA isolation from whole blood samples using the phenol-chloroform extraction method was performed by Genome Database of Latvian Population (briefly described before [89]). DNA samples were quantified with Qubit® 2.0 Fluorometer using Qubit dsDNA HS Assay Kit (TherfmoFisher Scientific, USA). For the bisulfite conversion, the EZ DNA Methylation-Gold TM kit (Zymo research, USA) was used according to the manufacturer's instructions. DNA methylation was determined by the Illumina Infinium HumanMethylation450 BeadChip Array (Illumina, USA), using 500 ng of each bisulfite-treated DNA sample.

Total RNA was isolated from whole blood samples using PerfectPure RNA Blood Kit (5Prime GmbH, Hamburg, Germany). Ribosomal RNS depletion was done with Low Input RiboMinus™ Eukaryote System v2 (Thermo Fisher Scientific, USA). For cDNA library preparation, we used Ion Total RNA-Seq Kit v2 (Thermo Fisher Scientific, USA), and sequencing was performed on the Ion Proton™ System and Ion PI™ Chip (Thermo Fisher Scientific, USA).

Data preprocessing and statistical analysis

IDAT files were imported using R package minfi [90]. Cell counts were estimated from methylation data using Houseman algorithm [91] implemented in minfi.

Data preprocessing and normalization was done using Enmix [92]. Briefly, probes with detection p value > 0.05 and probes with a multimodal distribution were filtered

out. Background correction was performed with the function preprocessENmix using unused color channels as a background parameter estimate. Probe intensities were normalized using a quantile normalization method, and probe type bias was adjusted using the Regression on Correlated Probes (RCP) method [93]. Probes having a SNP or single base extension annotation in CpG site were excluded. Due to interrupted use of metformin by one of the study subjects, the sample taken after 1 week of metformin administration for that particular subject was discarded.

Batch effect was removed from data using slide and subsequently subjects as covariates as they showed the strongest influence on the probe methylation variability. Batch effect was removed using ComBat [94] wrapped in the Enmix package. Differentially methylated probes between time points were identified using limma [95] on ComBat preprocessed data, adjusting for the following cell types estimated by minfi: CD8T, CD4T, NK, and Gran. Inflation factor of p-value distribution was estimated using R package GenABEL [96]. All analyses were performed using R (3.3.3).

Statistically significant DMRs were identified with DMRcate software [97], FDR < 0.05. Threshold for minimum number of probes within the region was set to three. DMRs were estimated from methylation M values using the individual CpG site significance threshold at FDR < 0.05. The interval between individual significant CpG sites had to be less than 1000 bp in the regions. The bandwidth scaling factor was set as suggested in the manual ($C = 2$). Regulatory information from Ensembl 91 regulation resources was added to identified DMPs and DMRs using Ensembl Regulation API [98].

Pathway enrichment analysis was performed with the IPA tool [99]. Information about enriched canonical pathways and networks was obtained performing the core analysis on all significant DMPs with FDR < 0.05.

RNA-seq data analysis

Reads were mapped against human reference genome GRCh38, and read quantification was performed using STAR (2.5.3a) [100]. Obtained per-gene read counts were normalized using trimmed mean normalization (TMM), and counts per million (CPM) values were calculated with edgeR [101]. ComBat [94] implemented in R package sva [102] was used to adjust CPM values for subject-specific effects, and the Spearman correlation was estimated for the adjusted CPM values and the beta values for 11 selected CpG sites with SciPy [103].

Abbreviations

CPM: Counts per million; DMP: Differentially methylated CpG site/probe; DMR: Differentially methylated region; IPA: Ingenuity Pathway Analysis; IQR: Interquartile range; RCP: Regression on Correlated Probes; T2D: Type 2 diabetes; TMM: Trimmed mean normalization; TSS: Transcription starting site

Acknowledgements
We acknowledge the Genome Database of Latvian Population (LGDB) and Latvian Biomedical Research and Study Centre for organization of participants' recruitment and bio-sample processing and data storage. The authors thank all the volunteers who participated in the clinical trial.

Funding
The work was supported by the European Regional Development Fund under the project "Investigation of interplay between multiple determinants influencing response to metformin: search for reliable predictors for efficacy of type 2 diabetes therapy" (Project Nr.: 1.1.1.1/16/A/091).

Authors' contributions
IKalnina, LZ, DF, HBS, and JK designed the research; IKonrade and VP oversaw patient recruitment; IE, IKalnina, DG, IRS, conducted the research; IE, IS, MU, and RP performed the analyses; IE, IS, MU wrote the manuscript; IK, DMC, CZ, HBS, and JK oversaw the research and reviewed the manuscript. All authors read and approved the final manuscript.

Competing interests
The authors declare that they have no competing interests.

Author details
[1]Latvian Biomedical Research and Study Centre, Ratsupites Str. 1 k-1, Riga LV-1067, Latvia. [2]Riga East Clinical University Hospital, 2 Hipokrata Street, Riga LV-1038, Latvia. [3]Department of Neuroscience, Functional Pharmacology, Uppsala University, BMC, Box 593, 751 24 Uppsala, Sweden.

References
1. Marshall SM. 60 years of metformin use: a glance at the past and a look to the future. Diabetologia. 2017;60:1561–5.
2. Song AQ, Sun LR, Zhao YX, Gao YH, Chen L. Effect of insulin and metformin on methylation and glycolipid metabolism of peroxisome proliferator-activated receptor gamma coactivator-1A of rat offspring with gestational diabetes mellitus. Asian Pac J Trop Med. 2016;9:91–5.
3. Niu N, Liu T, Cairns J, Ly RC, Tan X, Deng M, Fridley BL, Kalari KR, Abo RP, Jenkins G, et al. Metformin pharmacogenomics: a genome-wide association study to identify genetic and epigenetic biomarkers involved in metformin anticancer response using human lymphoblastoid cell lines. Hum Mol Genet. 2016;25:4819–34.
4. Zhong T, Men Y, Lu L, Geng T, Zhou J, Mitsuhashi A, Shozu M, Maihle NJ, Carmichael GG, Taylor HS, Huang Y. Metformin alters DNA methylation genome-wide via the H19/SAHH axis. Oncogene. 2017;36:2345–54.
5. Cuyas E, Fernandez-Arroyo S, Verdura S, Garcia RA, Stursa J, Werner L, Blanco-Gonzalez E, Montes-Bayon M, Joven J, Viollet B, et al. Metformin regulates global DNA methylation via mitochondrial one-carbon metabolism. Oncogene. 2017;37:963.
6. Garcia-Calzon S, Perfilyev A, Mannisto V, de Mello VD, Nilsson E, Pihlajamaki J, Ling C. Diabetes medication associates with DNA methylation of metformin transporter genes in the human liver. Clin Epigenetics. 2017;9:102.
7. Bridgeman SC, Ellison GC, Melton PE, Newsholme P, Mamotte CDS. Epigenetic effects of metformin: from molecular mechanisms to clinical implications. Diabetes Obes Metab. 2018;20:1553–62.
8. Price ME, Cotton AM, Lam LL, Farre P, Emberly E, Brown CJ, Robinson WP, Kobor MS. Additional annotation enhances potential for biologically-relevant analysis of the Illumina Infinium HumanMethylation450 BeadChip array. Epigenetics Chromatin. 2013;6:4.
9. Crujeiras AB, Diaz-Lagares A, Sandoval J, Milagro FI, Navas-Carretero S, Carreira MC, Gomez A, Hervas D, Monteiro MP, Casanueva FF, et al. DNA methylation map in circulating leukocytes mirrors subcutaneous adipose tissue methylation pattern: a genome-wide analysis from non-obese and obese patients. Sci Rep. 2017;7:41903.
10. Farre P, Jones MJ, Meaney MJ, Emberly E, Turecki G, Kobor MS. Concordant and discordant DNA methylation signatures of aging in human blood and brain. Epigenetics Chromatin. 2015;8:19.
11. Barault L, Ellsworth RE, Harris HR, Valente AL, Shriver CD, Michels KB. Leukocyte DNA as surrogate for the evaluation of imprinted Loci methylation in mammary tissue DNA. PLoS One. 2013;8:e55896.
12. Lim AS, Srivastava GP, Yu L, Chibnik LB, Xu J, Buchman AS, Schneider JA, Myers AJ, Bennett DA, De Jager PL. 24-hour rhythms of DNA methylation and their relation with rhythms of RNA expression in the human dorsolateral prefrontal cortex. PLoS Genet. 2014;10:e1004792.
13. Okuno S, Kitani T, Fujisawa H. Studies on the substrate specificity of Ca2+/calmodulin-dependent protein kinase kinase alpha. J Biochem. 1997;122:337–43.
14. Woods A, Dickerson K, Heath R, Hong SP, Momcilovic M, Johnstone SR, Carlson M, Carling D. Ca2+/calmodulin-dependent protein kinase kinase-beta acts upstream of AMP-activated protein kinase in mammalian cells. Cell Metab. 2005;2:21–33.
15. Green MF, Anderson KA, Means AR. Characterization of the CaMKKbeta-AMPK signaling complex. Cell Signal. 2011;23:2005–12.
16. Ben Sahra I, Laurent K, Loubat A, Giorgetti-Peraldi S, Colosetti P, Auberger P, Tanti JF, Le Marchand-Brustel Y, Bost F. The antidiabetic drug metformin exerts an antitumoral effect in vitro and in vivo through a decrease of cyclin D1 level. Oncogene. 2008;27:3576–86.
17. Witczak CA, Fujii N, Hirshman MF, Goodyear LJ. Ca2+/calmodulin-dependent protein kinase kinase-alpha regulates skeletal muscle glucose uptake independent of AMP-activated protein kinase and Akt activation. Diabetes. 2007;56:1403–9.
18. Granadino B, Gallardo ME, Lopez-Rios J, Sanz R, Ramos C, Ayuso C, Bovolenta P, Rodriguez de Cordoba S. Genomic cloning, structure, expression pattern, and chromosomal location of the human SIX3 gene. Genomics. 1999;55:100–5.
19. Mo ML, Okamoto J, Chen Z, Hirata T, Mikami I, Bosco-Clement G, Li H, Zhou HM, Jablons DM, He B. Down-regulation of SIX3 is associated with clinical outcome in lung adenocarcinoma. PLoS One. 2013;8:e71816.
20. Hwang JY, Sim X, Wu Y, Liang J, Tabara Y, Hu C, Hara K, Tam CH, Cai Q, Zhao Q, et al. Genome-wide association meta-analysis identifies novel variants associated with fasting plasma glucose in East Asians. Diabetes. 2015;64:291–8.
21. Arda HE, Li L, Tsai J, Torre EA, Rosli Y, Peiris H, Spitale RC, Dai C, Gu X, Qu K, et al. Age-dependent pancreatic gene regulation reveals mechanisms governing human beta cell function. Cell Metab. 2016;23:909–20.
22. Harrow J, Frankish A, Gonzalez JM, Tapanari E, Diekhans M, Kokocinski F, Aken BL, Barrell D, Zadissa A, Searle S, et al. GENCODE: the reference human genome annotation for the ENCODE project. Genome Res. 2012;22:1760–74.
23. Giannarelli R, Aragona M, Coppelli A, Del Prato S. Reducing insulin resistance with metformin: the evidence today. Diabetes Metab. 2003;29: 6S28–35.
24. Cameron AR, Morrison VL, Levin D, Mohan M, Forteath C, Beall C, McNeilly AD, Balfour DJ, Savinko T, Wong AK, et al. Anti-inflammatory effects of metformin irrespective of diabetes status. Circ Res. 2016;119:652–65.
25. Yoshiyama K, Higuchi Y, Kataoka M, Matsuura K, Yamamoto S. CD156 (human ADAM8): expression, primary amino acid sequence, and gene location. Genomics. 1997;41:56–62.
26. Yoshida S, Setoguchi M, Higuchi Y, Akizuki S, Yamamoto S. Molecular cloning of cDNA encoding MS2 antigen, a novel cell surface antigen strongly expressed in murine monocytic lineage. Int Immunol. 1990;2:585–91.
27. Hodgkinson CP, Ye S. Microarray analysis of peroxisome proliferator-activated receptor-gamma induced changes in gene expression in macrophages. Biochem Biophys Res Commun. 2003;308:505–10.
28. Dreymueller D, Pruessmeyer J, Schumacher J, Fellendorf S, Hess FM, Seifert A, Babendreyer A, Bartsch JW, Ludwig A. The metalloproteinase ADAM8 promotes leukocyte recruitment in vitro and in acute lung inflammation. Am J Physiol Lung Cell Mol Physiol. 2017;313:L602–14.
29. Schlomann U, Rathke-Hartlieb S, Yamamoto S, Jockusch H, Bartsch JW. Tumor necrosis factor alpha induces a metalloprotease-disintegrin, ADAM8 (CD 156): implications for neuron-glia interactions during neurodegeneration. J Neurosci. 2000;20:7964–71.
30. Fourie AM, Coles F, Moreno V, Karlsson L. Catalytic activity of ADAM8, ADAM15, and MDC-L (ADAM28) on synthetic peptide substrates and in ectodomain cleavage of CD23. J Biol Chem. 2003;278:30469–77.
31. Chen J, Jiang X, Duan Y, Long J, Bartsch JW, Deng L. ADAM8 in asthma. Friend or foe to airway inflammation? Am J Respir Cell Mol Biol. 2013;49:875–84.

32. Du J, Takeuchi H, Leonhard-Melief C, Shroyer KR, Dlugosz M, Haltiwanger RS, Holdener BC. O-fucosylation of thrombospondin type 1 repeats restricts epithelial to mesenchymal transition (EMT) and maintains epiblast pluripotency during mouse gastrulation. Dev Biol. 2010;346:25–38.

33. Dong S, Nutt CL, Betensky RA, Stemmer-Rachamimov AO, Denko NC, Ligon KL, Rowitch DH, Louis DN. Histology-based expression profiling yields novel prognostic markers in human glioblastoma. J Neuropathol Exp Neurol. 2005;64: 948–55.

34. Aramburu A, Zudaire I, Pajares MJ, Agorreta J, Orta A, Lozano MD, Gurpide A, Gomez-Roman J, Martinez-Climent JA, Jassem J, et al. Combined clinical and genomic signatures for the prognosis of early stage non-small cell lung cancer based on gene copy number alterations. BMC Genomics. 2015;16:752.

35. Gaedcke J, Grade M, Camps J, Sokilde R, Kaczkowski B, Schetter AJ, Difilippantonio MJ, Harris CC, Ghadimi BM, Moller S, et al. The rectal cancer microRNAome--microRNA expression in rectal cancer and matched normal mucosa. Clin Cancer Res. 2012;18:4919–30.

36. Xu L, Ziegelbauer J, Wang R, Wu WW, Shen RF, Juhl H, Zhang Y, Rosenberg A. Distinct profiles for mitochondrial t-RNAs and small nucleolar RNAs in locally invasive and metastatic colorectal cancer. Clin Cancer Res. 2016;22:773–84.

37. Askarian-Amiri ME, Crawford J, French JD, Smart CE, Smith MA, Clark MB, Ru K, Mercer TR, Thompson ER, Lakhani SR, et al. SNORD-host RNA Zfas1 is a regulator of mammary development and a potential marker for breast cancer. RNA. 2011;17:878–91.

38. Kaur S, Kumar S, Momi N, Sasson AR, Batra SK. Mucins in pancreatic cancer and its microenvironment. Nat Rev Gastroenterol Hepatol. 2013;10:607–20.

39. Andrianifahanana M, Moniaux N, Schmied BM, Ringel J, Friess H, Hollingsworth MA, Buchler MW, Aubert JP, Batra SK. Mucin (MUC) gene expression in human pancreatic adenocarcinoma and chronic pancreatitis: a potential role of MUC4 as a tumor marker of diagnostic significance. Clin Cancer Res. 2001;7:4033–40.

40. Gautam SK, Kumar S, Cannon A, Hall B, Bhatia R, Nasser MW, Mahapatra S, Batra SK, Jain M. MUC4 mucin- a therapeutic target for pancreatic ductal adenocarcinoma. Expert Opin Ther Targets. 2017;21:657–69.

41. Kang K, Bae JH, Han K, Kim ES, TO K, Yi JM. A genome-wide methylation approach identifies a new hypermethylated gene panel in ulcerative colitis. Int J Mol Sci. 2016;17:1291.

42. Azzalin CM, Lingner J. The human RNA surveillance factor UPF1 is required for S phase progression and genome stability. Curr Biol. 2006;16:433–9.

43. Chawla R, Redon S, Raftopoulou C, Wischnewski H, Gagos S, Azzalin CM. Human UPF1 interacts with TPP1 and telomerase and sustains telomere leading-strand replication. EMBO J. 2011;30:4047–58.

44. Chang L, Li C, Guo T, Wang H, Ma W, Yuan Y, Liu Q, Ye Q, Liu Z. The human RNA surveillance factor UPF1 regulates tumorigenesis by targeting Smad7 in hepatocellular carcinoma. J Exp Clin Cancer Res. 2016;35:8.

45. Li L, Geng Y, Feng R, Zhu Q, Miao B, Cao J, Fei S. The human RNA surveillance factor UPF1 modulates gastric cancer progression by targeting long non-coding RNA MALAT1. Cell Physiol Biochem. 2017;42:2194–206.

46. Liu C, Karam R, Zhou Y, Su F, Ji Y, Li G, Xu G, Lu L, Wang C, Song M, et al. The UPF1 RNA surveillance gene is commonly mutated in pancreatic adenosquamous carcinoma. Nat Med. 2014;20:596–8.

47. Wang D, Zavadil J, Martin L, Parisi F, Friedman E, Levy D, Harding H, Ron D, Gardner LB. Inhibition of nonsense-mediated RNA decay by the tumor microenvironment promotes tumorigenesis. Mol Cell Biol. 2011;31:3670 80.

48. Perini ED, Schaefer R, Stoter M, Kalaidzidis Y, Zerial M. Mammalian CORVET is required for fusion and conversion of distinct early endosome subpopulations. Traffic. 2014;15:1366–89.

49. Lunetta KL, D'Agostino RB Sr, Karasik D, Benjamin EJ, Guo CY, Govindaraju R, Kiel DP, Kelly-Hayes M, Massaro JM, Pencina MJ, et al. Genetic correlates of longevity and selected age-related phenotypes: a genome-wide association study in the Framingham Study. BMC Med Genet. 2007;8(Suppl 1):S13.

50. Pankratz N, Dumitriu A, Hetrick KN, Sun M, Latourelle JC, Wilk JB, Halter C, Doheny KF, Gusella JF, Nichols WC, et al. Copy number variation in familial Parkinson disease. PLoS One. 2011;6:e20988.

51. Antoni G, Oudot-Mellakh T, Dimitromanolakis A, Germain M, Cohen W, Wells P, Lathrop M, Gagnon F, Morange PE, Tregouet DA. Combined analysis of three genome-wide association studies on vWF and FVIII plasma levels. BMC Med Genet. 2011;12:102.

52. Cai DC, Fonteijn H, Guadalupe T, Zwiers M, Wittfeld K, Teumer A, Hoogman M, Arias-Vasquez A, Yang Y, Buitelaar J, et al. A genome-wide search for quantitative trait loci affecting the cortical surface area and thickness of Heschl's gyrus. Genes Brain Behav. 2014;13:675–85.

53. Manjurano A, Sepulveda N, Nadjm B, Mtove G, Wangai H, Maxwell C, Olomi R, Reyburn H, Drakeley CJ, Riley EM, et al. USP38, FREM3, SDC1, DDC, and LOC727982 gene polymorphisms and differential susceptibility to severe malaria in Tanzania. J Infect Dis. 2015;212:1129–39.

54. Simino J, Shi G, Bis JC, Chasman DI, Ehret GB, Gu X, Guo X, Hwang SJ, Sijbrands E, Smith AV, et al. Gene-age interactions in blood pressure regulation: a large-scale investigation with the CHARGE, global BPgen, and ICBP Consortia. Am J Hum Genet. 2014;95:24–38.

55. Maurano MT, Wang H, John S, Shafer A, Canfield T, Lee K, Stamatoyannopoulos JA. Role of DNA methylation in modulating transcription factor occupancy. Cell Rep. 2015;12:1184–95.

56. Rosenbloom KR, Sloan CA, Malladi VS, Dreszer TR, Learned K, Kirkup VM, Wong MC, Maddren M, Fang R, Heitner SG, et al. ENCODE data in the UCSC genome browser: year 5 update. Nucleic Acids Res. 2013;41:D56–63.

57. Saxena R, Saleheen D, Been LF, Garavito ML, Braun T, Bjonnes A, Young R, Ho WK, Rasheed A, Frossard P, et al. Genome-wide association study identifies a novel locus contributing to type 2 diabetes susceptibility in Sikhs of Punjabi origin from India. Diabetes. 2013;62:1746–55.

58. Shen N, Yu X, Pan FY, Gao X, Xue B, Li CJ. An early response transcription factor, Egr-1, enhances insulin resistance in type 2 diabetes with chronic hyperinsulinism. J Biol Chem. 2011;286:14508–15.

59. Tegha-Dunghu J, Neumann B, Reber S, Krause R, Erfle H, Walter T, Held M, Rogers P, Hupfeld K, Ruppert T, et al. EML3 is a nuclear microtubule-binding protein required for the correct alignment of chromosomes in metaphase. J Cell Sci. 2008;121:1718–26.

60. Lamb J, Crawford ED, Peck D, Modell JW, Blat IC, Wrobel MJ, Lerner J, Brunet JP, Subramanian A, Ross KN, et al. The connectivity map: using gene-expression signatures to connect small molecules, genes, and disease. Science. 2006;313:1929–35.

61. Cheung L, Zervou S, Mattsson G, Abouna S, Zhou L, Ifandi V, Pelengaris S, Khan M. c-Myc directly induces both impaired insulin secretion and loss of beta-cell mass, independently of hyperglycemia in vivo. Islets. 2010;2:37–45.

62. Lin L, Pang W, Chen K, Wang F, Gengler J, Sun Y, Tong Q. Adipocyte expression of PU.1 transcription factor causes insulin resistance through upregulation of inflammatory cytokine gene expression and ROS production. Am J Physiol Endocrinol Metab. 2012;302:E1550–9.

63. Himanen JP, Nikolov DB. Eph receptors and ephrins. Int J Biochem Cell Biol. 2003;35:130–4.

64. Wang JD, Dong YC, Sheng Z, Ma HH, Li GL, Wang XL, Lu GM, Sugimura H, Jin J, Zhou XJ. Loss of expression of EphB1 protein in gastric carcinoma associated with invasion and metastasis. Oncology. 2007;73:238–45.

65. Sheng Z, Wang J, Dong Y, Ma H, Zhou H, Sugimura H, Lu G, Zhou X. EphB1 is underexpressed in poorly differentiated colorectal cancers. Pathobiology. 2008;75:274–80.

66. Hussain I, Powell DJ, Howlett DR, Chapman GA, Gilmour L, Murdock PR, Tew DG, Meek TD, Chapman C, Schneider K, et al. ASP1 (BACE2) cleaves the amyloid precursor protein at the beta-secretase site. Mol Cell Neurosci. 2000;16:609–19.

67. Basi G, Frigon N, Barbour R, Doan T, Gordon G, McConlogue L, Sinha S, Zeller M. Antagonistic effects of beta-site amyloid precursor protein-cleaving enzymes 1 and 2 on beta-amyloid peptide production in cells. J Biol Chem. 2003;278:31512–20.

68. Murphy MP, LeVine H 3rd. Alzheimer's disease and the amyloid-beta peptide. J Alzheimers Dis. 2010;19:311–23.

69. Esterhazy D, Stutzer I, Wang H, Rechsteiner MP, Beauchamp J, Dobeli H, Hilpert H, Matile H, Prummer M, Schmidt A, et al. Bace2 is a beta cell-enriched protease that regulates pancreatic beta cell function and mass. Cell Metab. 2011;14:365–77.

70. Alcarraz-Vizan G, Castano C, Visa M, Montane J, Servitja JM, Novials A. BACE2 suppression promotes beta-cell survival and function in a model of type 2 diabetes induced by human islet amyloid polypeptide overexpression. Cell Mol Life Sci. 2017;74:2827–38.

71. Wang DR, Tang D. Hypermethylated SFRP2 gene in fecal DNA is a high potential biomarker for colorectal cancer noninvasive screening. World J Gastroenterol. 2008;14:524–31.

72. Pehlivan S, Artac M, Sever T, Bozcuk H, Kilincarslan C, Pehlivan M. Gene methylation of SFRP2, P16, DAPK1, HIC1, and MGMT and KRAS mutations in sporadic colorectal cancer. Cancer Genet Cytogenet. 2010;201:128–32.

73. Crowley RK, O'Reilly MW, Bujalska IJ, Hassan-Smith ZK, Hazlehurst JM, Foucault DR, Stewart PM, Tomlinson JW. SFRP2 is associated with increased adiposity and VEGF expression. PLoS One. 2016;11:e0163777.

74. Leclerc GM, Leclerc GJ, Kuznetsov JN, DeSalvo J, Barredo JC. Metformin induces apoptosis through AMPK-dependent inhibition of UPR signaling in ALL lymphoblasts. PLoS One. 2013;8:e74420.

75. Theriault JR, Palmer HJ, Pittman DD. Inhibition of the unfolded protein response by metformin in renal proximal tubular epithelial cells. Biochem Biophys Res Commun. 2011;409:500–5.

76. Mato JM, Lu SC. Role of S-adenosyl-L-methionine in liver health and injury. Hepatology. 2007;45:1306–12.

77. Frojdo S, Vidal H, Pirola L. Alterations of insulin signaling in type 2 diabetes: a review of the current evidence from humans. Biochim Biophys Acta. 2009;1792:83–92.

78. Kasznicki J, Sliwinska A, Drzewoski J. Metformin in cancer prevention and therapy. Ann Transl Med. 2014;2:57.

79. Lim YC, Li J, Ni Y, Liang Q, Zhang J, Yeo GSH, Lyu J, Jin S, Ding C. A complex association between DNA methylation and gene expression in human placenta at first and third trimesters. PLoS One. 2017;12:e0181155.

80. Jones PA. The DNA methylation paradox. Trends Genet. 1999;15:34–7.

81. Benton MC, Johnstone A, Eccles D, Harmon B, Hayes MT, Lea RA, Griffiths L, Hoffman EP, Stubbs RS, Macartney-Coxson D. An analysis of DNA methylation in human adipose tissue reveals differential modification of obesity genes before and after gastric bypass and weight loss. Genome Biol. 2015;16:8.

82. Alkhaled Y, Laqqan M, Tierling S, Lo Porto C, Amor H, Hammadeh ME. Impact of cigarette-smoking on sperm DNA methylation and its effect on sperm parameters. Andrologia. 2018;50:e12950.

83. Cheng Q, Zhao B, Huang Z, Su Y, Chen B, Yang S, Peng X, Ma Q, Yu X, Zhao B, Ke X. Epigenome-wide study for the offspring exposed to maternal HBV infection during pregnancy, a pilot study. Gene. 2018;658:76–85.

84. Urdinguio RG, Torro MI, Bayon GF, Alvarez-Pitti J, Fernandez AF, Redon P, Fraga MF, Lurbe E. Longitudinal study of DNA methylation during the first 5 years of life. J Transl Med. 2016;14:160.

85. Goodrich JK, Di Rienzi SC, Poole AC, Koren O, Walters WA, Caporaso JG, Knight R, Ley RE. Conducting a microbiome study. Cell. 2014;158:250–62.

86. Sambol NC, Chiang J, O'Conner M, Liu CY, Lin ET, Goodman AM, Benet LZ, Karam JH. Pharmacokinetics and pharmacodynamics of metformin in healthy subjects and patients with noninsulin-dependent diabetes mellitus. J Clin Pharmacol. 1996;36:1012–21.

87. Tokubuchi I, Tajiri Y, Iwata S, Hara K, Wada N, Hashinaga T, Nakayama H, Mifune H, Yamada K. Beneficial effects of metformin on energy metabolism and visceral fat volume through a possible mechanism of fatty acid oxidation in human subjects and rats. PLoS One. 2017;12:e0171293.

88. Rovite V, Wolff-Sagi Y, Zaharenko L, Nikitina-Zake L, Grens E, Klovins J. Genome Database of the Latvian Population (LGDB): design, goals, and primary results. J Epidemiol. 2018;28:353–60.

89. Ignatovica V, Latkovskis G, Peculis R, Megnis K, Schioth HB, Vaivade I, Fridmanis D, Pirags V, Erglis A, Klovins J. Single nucleotide polymorphisms of the purinergic 1 receptor are not associated with myocardial infarction in a Latvian population. Mol Biol Rep. 2012;39:1917–25.

90. Aryee MJ, Jaffe AE, Corrada-Bravo H, Ladd-Acosta C, Feinberg AP, Hansen KD, Irizarry RA. Minfi: a flexible and comprehensive Bioconductor package for the analysis of Infinium DNA methylation microarrays. Bioinformatics. 2014;30:1363–9.

91. Houseman EA, Accomando WP, Koestler DC, Christensen BC, Marsit CJ, Nelson HH, Wiencke JK, Kelsey KT. DNA methylation arrays as surrogate measures of cell mixture distribution. BMC Bioinformatics. 2012;13:86.

92. Xu Z, Niu L, Li L, Taylor JA. ENmix: a novel background correction method for Illumina HumanMethylation450 BeadChip. Nucleic Acids Res. 2016;44:e20.

93. Niu L, Xu Z, Taylor JA. RCP: a novel probe design bias correction method for Illumina Methylation BeadChip. Bioinformatics. 2016;32:2659–63.

94. Johnson WE, Li C, Rabinovic A. Adjusting batch effects in microarray expression data using empirical Bayes methods. Biostatistics. 2007;8:118–27.

95. Ritchie ME, Phipson B, Wu D, Hu Y, Law CW, Shi W, Smyth GK. limma powers differential expression analyses for RNA-sequencing and microarray studies. Nucleic Acids Res. 2015;43:e47.

96. Aulchenko YS, Ripke S, Isaacs A, van Duijn CM. GenABEL: an R library for genome-wide association analysis. Bioinformatics. 2007;23:1294–6.

97. Peters TJ, Buckley MJ, Statham AL, Pidsley R, Samaras K, R VL, Clark SJ, Molloy PL. De novo identification of differentially methylated regions in the human genome. Epigenetics Chromatin. 2015;8:6.

98. Zerbino DR, Johnson N, Juetteman T, Sheppard D, Wilder SP, Lavidas I, Nuhn M, Perry E, Raffaillac-Desfosses Q, Sobral D, et al. Ensembl regulation resources. Database (Oxford). 2016;2016:bav119.

99. Kramer A, Green J, Pollard J Jr, Tugendreich S. Causal analysis approaches in ingenuity pathway analysis. Bioinformatics. 2014;30:523–30.

100. Dobin A, Davis CA, Schlesinger F, Drenkow J, Zaleski C, Jha S, Batut P, Chaisson M, Gingeras TR. STAR: ultrafast universal RNA-seq aligner. Bioinformatics. 2013;29:15–21.

101. Robinson MD, McCarthy DJ, Smyth GK. edgeR: a Bioconductor package for differential expression analysis of digital gene expression data. Bioinformatics. 2010;26:139–40.

102. Leek JT, Johnson WE, Parker HS, Jaffe AE, Storey JD. The sva package for removing batch effects and other unwanted variation in high-throughput experiments. Bioinformatics. 2012;28:882–3.

103. Olivier BG, Rohwer JM, Hofmeyr JH. Modelling cellular processes with Python and Scipy. Mol Biol Rep. 2002;29:249–54.

Permissions

List of Contributors

Milena Magalhães, Mélodie Thomasset and Albertina De Sario
Laboratoire de Génétique de Maladies Rares, EA7402 Montpellier University, Montpellier, France

Isabelle Rivals
Equipe de Statistique Appliquée—ESPCI ParisTech, PSL Research University—UMRS1158, Paris, France

Mireille Claustres, Jessica Varilh, Anne Bergougnoux and Emmanuelle Beyne
Laboratoire de Génétique de Maladies Rares, EA7402 Montpellier University, Montpellier, France
Laboratoire de Génétique Moléculaire—CHU Montpellier, Montpellier, France

Laurent Mely
CRCM, Renée Sabran Hospital—CHU Lyon, Hyères, France

Sylvie Leroy
CRCM, Pasteur Hospital—CHU Nice, Nice, France

Loïc Guillot
Sorbonne Universités, UPMC Univ Paris 06, Paris, France
INSERM U938—CRSA, Paris, France

Harriet Corvol
Sorbonne Universités, UPMC Univ Paris 06, Paris, France
INSERM U938—CRSA, Paris, France
APHP, Trousseau Hospital, Paris, France

Marlène Murris
CRCM, Larrey Hospital—CHU Toulouse, Toulouse, France

Davide Caimmi, Isabelle Vachier and Raphaël Chiron
CRCM, Arnaud de Villeneuve Hospital—CHU Montpellier, Montpellier, France

Bradley J. Toghill, Athanasios Saratzis and Matthew J. Bown
Department of Cardiovascular Sciences and the NIHR Leicester Biomedical Research Centre, University of Leicester, Leicester LE2 7LX, UK

Peter J. Freeman and Nicolas Sylvius
Department of Genetics and Genome Biology, University of Leicester, Leicester LE1 7RH, UK

Olivia M. de Goede and Wendy P. Robinson
BC Children's Hospital Research Institute, Room 2082, 950W 28th Avenue, Vancouver, BC V5Z 4H4, Canada
Department of Medical Genetics, University of British Columbia, Vancouver, BC V6T 1Z3, Canada

Pascal M. Lavoie
BC Children's Hospital Research Institute, Room 2082, 950W 28th Avenue, Vancouver, BC V5Z 4H4, Canada
Department of Pediatrics, University of British Columbia, Vancouver, BC V6T 1Z3, Canada

A. Spanò
UOC of Clinical Biochemistry, Sandro Pertini Hospital, Rome, Italy

T. Guastafierro
UOC of Clinical Biochemistry, Sandro Pertini Hospital, Rome, Italy
CRIIS (Interdisciplinary, Interdepartmental and Specialistic Reference Center for Early Diagnosis of Scleroderma, Treatment of Sclerodermic Ulcers and Videocapillaroscopy), Sandro Pertini Hospital, Rome, Italy

M. G. Bacalini
IRCCS Institute of Neurological Sciences, Bologna, Italy

A. Marcoccia
CRIIS (Interdisciplinary, Interdepartmental and Specialistic Reference Center for Early Diagnosis of Scleroderma, Treatment of Sclerodermic Ulcers and Videocapillaroscopy), Sandro Pertini Hospital, Rome, Italy
UOSD Ischemic Microangiopathy and Sclerodermic Ulcers, Sandro Pertini Hospital, Rome, Italy

D. Gentilini, S. Pisoni and A. M. Di Blasio
Centre for Biomedical Research and Technologies, Italian Auxologic Institute, IRCCS, Milan, Italy

A. Corsi and D. Raimondo
Department of Molecular Medicine, Sapienza University of Rome, Rome, Italy

C. Franceschi
IRCCS Institute of Neurological Sciences, Bologna, Italy
Department of Experimental, Diagnostic and Specialty Medicine, University of Bologna, Bologna, Italy
Interdepartmental Center "L. Galvani", University of Bologna, Bologna, Italy

P. Garagnani
Department of Experimental, Diagnostic and Specialty Medicine, University of Bologna, Bologna, Italy
Interdepartmental Center "L. Galvani", University of Bologna, Bologna, Italy
Center for Applied Biomedical Research (CRBA), St. Orsola-Malpighi University Hospital, Bologna, Italy
Clinical Chemistry, Department of Laboratory Medicine, Karolinska Institute at Huddinge University Hospital, S-141 86 Stockholm, Sweden
CNR Institute for Molecular Genetics, Unit of Bologna, Bologna, Italy
Laboratory of Musculoskeletal Cell Biology, Rizzoli Orthopedic Institute, Bologna, Italy

F. Bondanini
CRIIS (Interdisciplinary, Interdepartmental and Specialistic Reference Center for Early Diagnosis of Scleroderma, Treatment of Sclerodermic Ulcers and Videocapillaroscopy), Sandro Pertini Hospital, Rome, Italy
UOC of Clinical Pathology, Saint' Eugenio Hospital, Rome, Italy

Alexander M. Morin and Julia L. MacIsaac
Department of Medical Genetics, Centre for Molecular Medicine and Therapeutics, British Columbia Children's Hospital Research Institute, University of British Columbia, Vancouver, British Columbia, Canada

Alexandre A. Lussier
Department of Medical Genetics, Centre for Molecular Medicine and Therapeutics, British Columbia Children's Hospital Research Institute, University of British Columbia, Vancouver, British Columbia, Canada

Department of Cellular and Physiological Sciences, Life Sciences Institute, University of British Columbia, Vancouver, British Columbia, Canada

Michael S. Kobor
Department of Medical Genetics, Centre for Molecular Medicine and Therapeutics, British Columbia Children's Hospital Research Institute, University of British Columbia, Vancouver, British Columbia, Canada
Human Early Learning Partnership, University of British Columbia, Vancouver, British Columbia, Canada

Joanne Weinberg
Department of Cellular and Physiological Sciences, Life Sciences Institute, University of British Columbia, Vancouver, British Columbia, Canada

Jenny Salmon and Albert E. Chudley
Department of Pediatrics and Child Health, Faculty of Medicine, University of Manitoba, Winnipeg, Manitoba, Canada
Department of Biochemistry and Medical Genetics, Faculty of Medicine, University of Manitoba, Winnipeg, Manitoba, Canada

James N. Reynolds
Department of Biomedical and Molecular Sciences, Centre for Neuroscience Studies, Queen's University, Kingston, Ontario, Canada

Paul Pavlidis
Michael Smith Laboratories, University of British Columbia, Vancouver, British Columnbia, Canada
Department of Psychiatry, University of British Columbia, Vancouver, British Columbia, Canada

Manosij Ghosh and Radu Corneliu Duca
Department of Public Health and Primary Care, Environment and Health, KU
Leuven - University of Leuven, Kapucijnenvoer 35 blok D box 7001, 3000 Leuven, Belgium

Sara Pauwels
Department of Public Health and Primary Care, Environment and Health, KU
Leuven - University of Leuven, Kapucijnenvoer 35 blok D box 7001, 3000 Leuven, Belgium
Flemish Institute of Technological Research (VITO), Unit Environmental Risk and Health, Boeretang 200, 2400 Mol, Belgium

Gudrun Koppen
Flemish Institute of Technological Research (VITO), Unit Environmental Risk and Health, Boeretang 200, 2400 Mol, Belgium

Bram Bekaert
Department of Imaging & Pathology, KU Leuven - University of Leuven, 3000 Leuven, Belgium
University Hospitals Leuven; Department of Forensic Medicine; Laboratory of Forensic Genetics and Molecular Archeology, KU Leuven - University of Leuven, 3000 Leuven, Belgium

Kathleen Freson
Center for Molecular and Vascular Biology, KU Leuven - University of Leuven, UZ Herestraat 49 - box 911, 3000 Leuven, Belgium

Inge Huybrechts
International Agency for Research on Cancer, 150 Cours Albert Thomas, 69372 LyonCEDEX 08, France

Sabine A. S. Langie
Flemish Institute of Technological Research (VITO), Unit Environmental Risk and Health, Boeretang 200, 2400 Mol, Belgium
Faculty of Sciences, Hasselt University, 3590 Diepenbeek, Belgium

Roland Devlieger
Department of Development and Regeneration, KU Leuven - University of Leuven, 3000 Leuven, Belgium
Department of Obstetrics and Gynecology, University Hospitals of Leuven, 3000 Leuven, Belgium

Lode Godderis
Department of Public Health and Primary Care, Environment and Health, KU
Leuven - University of Leuven, Kapucijnenvoer 35 blok D box 7001, 3000 Leuven, Belgium
IDEWE, External Service for Prevention and Protection at Work, Interleuvenlaan 58, 3001 Heverlee, Belgium

Julius C. Pape, Tania Carrillo-Roa and Darina Czamara
Department of Translational Research in Psychiatry, Max Planck Institute of Psychiatry, Munich, Germany

Elisabeth B. Binder
Department of Translational Research in Psychiatry, Max Planck Institute of Psychiatry, Munich, Germany

Department of Psychiatry and Behavioral Sciences, Emory University School of Medicine, Atlanta, GA, USA

Barbara O. Rothbaum, Helen S. Mayberg and Boadie W. Dunlop
Department of Psychiatry and Behavioral Sciences, Emory University School of Medicine, Atlanta, GA, USA

Charles B. Nemeroff
Department of Psychiatry and Behavioral Sciences, University of Miami Miller School of Medicine, Miami, FL, USA

Anthony S. Zannas
Department of Translational Research in Psychiatry, Max Planck Institute of Psychiatry, Munich, Germany
Department of Psychiatry and Behavioral Sciences, Duke University Medical Center, Durham, NC, USA

Dan Iosifescu
Department of Psychiatry, Icahn School of Medicine at Mount Sinai, New York, NY, USA
New York University School of Medicine, New York, NY, USA
Nathan Kline Institute for Psychiatric Research, Orangeburg, NY, USA

Sanjay J. Mathew
Menninger Department of Psychiatry & Behavioral Sciences, Baylor College of Medicine & Michael E. Debakey VA Medical Center, Houston, TX, USA

Thomas C. Neylan
Department of Psychiatry, University of California, San Francisco, San Francisco, CA, USA
The San Francisco Veterans Affairs Medical Center, San Francisco, CA, USA

Yonghong Zhang, Jinhua Liu, Kang Li and Ning Li
Beijing Youan Hospital, Capital Medical School, Beijing, China

Rudy Zhou, Sergiy Dymov and Moshe Szyf
Department of Pharmacology and Therapeutics, McGill University, 3655 Sir William Osler Promenade, Montreal, Quebec H3G 1Y6, Canada

Sophie Petropoulos
Department of Pharmacology and Therapeutics, McGill University, 3655 Sir William Osler Promenade, Montreal, Quebec H3G 1Y6, Canada
Deparment of Clinical Science, Karolinska Institutet, Alfred Nobels Allé 8, 141 52 Huddinge, Sweden

David Cheishvili
Department of Pharmacology and Therapeutics, McGill University, 3655 Sir William Osler Promenade, Montreal, Quebec H3G 1Y6, Canada
Montreal EpiTerapia Inc., 4567 Cecile, H9K1N2, Montreal, QC, Canada

Renu Jeyapala and Julia Garcia
Lunenfeld-Tanenbaum Research Institute, Sinai Health System, Toronto, Canada

Fang Zhao, Ekaterina Olkhov-Mitsel and Shivani Kamdar
Lunenfeld-Tanenbaum Research Institute, Sinai Health System, Toronto, Canada
Department of Laboratory Medicine & Pathobiology, University of Toronto, Toronto, Canada

Rachel Hurst, Robert Mills, Jeremy Clark and Colin Cooper
Schools of Medicine and Biological Sciences, University of East Anglia, Norwich, Norfolk, UK

Marcelino Yazbek Hanna
Norfolk and Norwich University Hospital, Norwich, Norfolk, UK

Alexandra V. Tuzova, Eve O'Reilly, Sarah Kelly and Antoinette S. Perry
Cancer Biology and Therapeutics Laboratory, School of Biomolecular and Biomedical Science, Conway Institute, University College Dublin, Dublin 4, Ireland

Daniel Brewer
Schools of Medicine and Biological Sciences, University of East Anglia, Norwich, Norfolk, UK
The Earlham Institute, Norwich, Norfolk, UK

Neil Fleshner
Division of Urology, University Health Network, University of Toronto, Toronto, Canada

Bharati Bapat
Lunenfeld-Tanenbaum Research Institute, Sinai Health System, Toronto, Canada

Department of Laboratory Medicine & Pathobiology, University of Toronto, Toronto, Canada
Division of Urology, University Health Network, University of Toronto, Toronto, Canada

Eun-Young Choi and Youn-Jae Kim
Translational Research Branch, Research Institute, National Cancer Center, Goyang, Gyeonggi 10408, Republic of Korea

Seong-Min Park
Translational Research Branch, Research Institute, National Cancer Center, Goyang, Gyeonggi 10408, Republic of Korea
Personalized Genomic Medicine Research Center, KRIBB, Daejeon 34141, Republic of Korea

Mingyun Bae and Jung Kyoon Choi
Department of Bio and Brain Engineering, KAIST, Daejeon 34141, Republic of Korea

Mihoko Shimada-Sugimoto and Katsushi Tokunaga
Department of Human Genetics, Graduate School of Medicine, The University of Tokyo, 7-3-1 Hongo, Bunkyo Ward, Tokyo 113-0033, Japan

Taku Miyagawa
Department of Human Genetics, Graduate School of Medicine, The University of Tokyo, 7-3-1 Hongo, Bunkyo Ward, Tokyo 113-0033, Japan
Department of Psychiatry and Behavioral Sciences, Tokyo Metropolitan Institute of Medical Science, 2-1-6 Kamikitazawa, Setagaya Ward, Tokyo 156-8506, Japan

Takeshi Otowa
Graduate School of Clinical Psychology, Teikyo Heisei University Major of Professional Clinical Psychology, 2-51-4 Higashiikebukuro, Toshima Ward, Tokyo 171-0014, Japan

Tadashi Umekage
Division for Environment, Health and Safety, The University of Tokyo, 7-3-1 Hongo, Bunkyo Ward, Tokyo 113-0033, Japan

Yoshiya Kawamura
Department of Psychiatry, Shonan Kamakura General Hospital, 1370-1 Okamoto, Kamakura City, Kanagawa 247-8533, Japan

Miki Bundo and Kazuya Iwamoto
Department of Molecular Brain Science, Graduate School of Medical Sciences, Kumamoto University, 1-1-1 Honjo, Chuo Ward, Kumamoto City, Kumamoto 860-8556, Japan

Mamoru Tochigi
Department of Neuropsychiatry, Teikyo University School of Medicine, 2-11-1 Kaga, Itabashi Ward, Tokyo 173-0003, Japan

Kiyoto Kasai
Department of Neuropsychiatry, Graduate School of Medicine, The University of Tokyo, 7-3-1 Hongo, Bunkyo Ward, Tokyo 113-0033, Japan

Hisanobu Kaiya
Panic Disorder Research Center, Warakukai Med Corp, 3-9-18 Akasaka, Minato Ward, Tokyo 107-0052, Japan

Hisashi Tanii
Department of Psychiatry, Institute of Medical Life Science, Graduate School of Medicine, Mie University, 2-174 Edobashi, Tsu City, Mie 514-8502, Japan

Yuji Okazaki
Department of Psychiatry, Koseikai Michinoo Hospital, 1-1 Nijigaokamachi, Nagasaki City, Nagasaki 852-8055, Japan

Tsukasa Sasaki
Department of Physical and Health Education, Graduate School of Education, The University of Tokyo, 7-3-1 Hongo, Bunkyo Ward, Tokyo 113-0033, Japan

Youdinghuan Chen and Lucas A. Salas
Department of Epidemiology, Geisel School of Medicine at Dartmouth, Lebanon, NH, USA
Department of Molecular and Systems Biology, Geisel School of Medicine at Dartmouth, Lebanon, NH, USA

Brock C. Christensen
Department of Epidemiology, Geisel School of Medicine at Dartmouth, Lebanon, NH, USA
Department of Molecular and Systems Biology, Geisel School of Medicine at Dartmouth, Lebanon, NH, USA
Department of Community and Family Medicine, Geisel School of Medicine at Dartmouth, Lebanon, NH, USA

David A. Armstrong
Department of Medicine, Dartmouth-Hitchcock Medical Center, Lebanon, NH, USA

Alix Ashare
Department of Medicine, Dartmouth-Hitchcock Medical Center, Lebanon, NH, USA
Program in Experimental and Molecular Medicine, Geisel School of Medicine at Dartmouth, Hanover, NH, USA
Department of Microbiology and Immunology, Geisel School of Medicine at Dartmouth, Dartmouth-Hitchcock Medical Center, Lebanon, NH, USA

Haley F. Hazlett
Program in Experimental and Molecular Medicine, Geisel School of Medicine at Dartmouth, Hanover, NH, USA

Amanda B. Nymon
Department of Microbiology and Immunology, Geisel School of Medicine at Dartmouth, Dartmouth-Hitchcock Medical Center, Lebanon, NH, USA

John A. Dessaint, Daniel S. Aridgides and Diane L. Mellinger
Department of Medicine, Dartmouth-Hitchcock Medical Center, Lebanon, NH, USA

Xiaoying Liu
Department of Pathology and Laboratory Medicine, Dartmouth-Hitchcock Medical Center, Lebanon, NH, USA

Xu Gao, Yan Zhang and Lutz Philipp Breitling
Division of Clinical Epidemiology and Aging Research, German Cancer Research Center (DKFZ), Im Neuenheimer Feld 581, 69120 Heidelberg, Germany

Hauke Thomsen
Division of Molecular Genetic Epidemiology, German Cancer Research Center (DKFZ), Im Neuenheimer Feld 580, 69120 Heidelberg, Germany

Ee Ming Wong, Ji-Hoon Eric Joo and Melissa C. Southey
Genetic Epidemiology Laboratory, Department of Pathology, University of Melbourne, Parkville, Victoria, Australia
Precision Medicine, School of Clinical Sciences at Monash Health, Monash University, Clayton, Victoria, Australia

Hermann Brenner
Division of Clinical Epidemiology and Aging Research, German Cancer Research Center (DKFZ), Im Neuenheimer Feld 581, 69120 Heidelberg, Germany
Division of Preventive Oncology, German Cancer Research Center (DKFZ) and National Center for Tumor Diseases (NCT), Im Neuenheimer Feld 460, 69120 Heidelberg, Germany
German Cancer Consortium (DKTK), German Cancer Research Center (DKFZ), Im Neuenheimer Feld 280, 69120 Heidelberg, Germany

Yashna Paul, Baisakhi Mondal, Vikas Patil and Kumaravel Somasundaram
Department of Microbiology and Cell Biology, Indian Institute of Science, Bangalore 560012, India

Shuai Li, Minh Bui, Tuong L. Nguyen, Gillian S. Dite and John L. Hopper
Centre for Epidemiology and Biostatistics, Melbourne School of Population and Global Health, University of Melbourne, Parkville, Victoria, Australia

Jennifer Stone
Centre for Genetic Origins of Health and Disease, Curtin University and the University of Western Australia, Perth, Western Australia, Australia

Graham G. Giles
Centre for Epidemiology and Biostatistics, Melbourne School of Population and Global Health, University of Melbourne, Parkville, Victoria, Australia
Cancer Epidemiology and Intelligence Division, Cancer Council Victoria, Melbourne, Victoria, Australia

Richard Saffery
Murdoch Children's Research Institute, Royal Children's Hospital, Parkville, Victoria, Australia
Department of Paediatrics, University of Melbourne, Parkville, Victoria, Australia

Ken Declerck and Wim Vanden Berghe
Laboratory of Protein Chemistry, Proteomics and Epigenetic Signalling (PPES), Department of Biomedical Sciences, University of Antwerp, Universiteitsplein 1, Antwerp, Belgium

Sylvie Remy
Department of Epidemiology and Social Medicine, Antwerp University, Universiteitsplein 1, Antwerp, Belgium
Flemish Institute for Technological Research (VITO), Unit Environmental Risk and Health, Boeretang 200, Mol, Belgium

Greet Schoeters
Flemish Institute for Technological Research (VITO), Unit Environmental Risk and Health, Boeretang 200, Mol, Belgium
Department of Declerck et al. Clinical Epigenetics Biomedical Sciences, Antwerp University, Universiteitsplein 1, Antwerp, Belgium Environmental Medicine, Institute of Public Health, University of Southern Denmark, Odense, Denmark

Christine Wohlfahrt-Veje and Katharina M. Main
Department of Growth and Reproduction, University Hospital of Copenhagen, Rigshospitalet, Copenhagen, Denmark

Guy Van Camp
Center of Medical Genetics, University of Antwerp and Antwerp University Hospital, Antwerp, Belgium

Helle R. Andersen
Environmental Medicine, Institute of Public Health, University of Southern Denmark, Odense, Denmark

Ilze Elbere, Ivars Silamikelis, Monta Ustinova, Ineta Kalnina, Linda Zaharenko, Raitis Peculis, Dita Gudra, Ilze Radovica-Spalvina, Davids Fridmanis, Valdis Pirags and Janis Klovins
Latvian Biomedical Research and Study Centre, Ratsupites Str. 1 k-1, Riga LV-1067, Latvia

Ilze Konrade
Riga East Clinical University Hospital, 2 Hipokrata Street, Riga LV-1038, Latvia

Diana Maria Ciuculete, Christina Zhukovsky and Helgi B. Schiöth
Department of Neuroscience, Functional Pharmacology, Uppsala University, BMC, Box 593, 751 24 Uppsala, Sweden

Index

CPSIA information can be obtained
at www.ICGtesting.com
Printed in the USA
BVHW011017190622
640137BV00003B/30